MECHANICS

by

WILLIAM FOGG OSGOOD

LATE PERKINS PROFESSOR OF MATHEMATICS
IN HARVARD UNIVERSITY

NEW YORK

DOVER PUBLICATIONS, INC.

Published in Canada by General Publishing Company, Ltd., 30 Lesmill Road, Don Mills, Toronto, Ontario.

Published in the United Kingdom by Constable and Company, Ltd., 10 Orange Street, London W. C. 2.

This Dover edition, first published in 1965, is an unabridged and unaltered republication of the work first published by The Macmillan Company in 1937.

Library of Congress Catalog Card Number: 65-17671

Manufactured in the United States of America

Dover Publications, Inc.
180 Varick Street
New York 14, N.Y.

PREFACE

Mechanics is a natural science, and like any natural science requires for its comprehension the observation and knowledge of a vast fund of individual cases. And so the solution of problems is of prime importance throughout all the study of this subject.

But Mechanics is not an empirical subject in the sense in which physics and chemistry, when dealing with the border region of the human knowledge of the day are empirical. The latter take cognizance of a great number of isolated facts, which it is not as yet possible to arrange under a few laws, or postulates. The laws of Mechanics, like the laws of Geometry, so far as first approximations go — the laws that explain the motion of the golf ball or the gyroscope or the skidding automobile, and which make possible the calculation of lunar tables and the prediction of eclipses — these laws are known, and will be as new and important two thousand years hence, as in the recent past of science when first they emerged into the light of day.

Here, then, is the problem of training the student in Mechanics — to provide him with a vast fund of case material and to develop in him the habits of thought which refer a new problem back to the few fundamental laws of the subject. The physicist is keenly alive to the first requirement and tries to meet it both by simple laboratory experiments and by problems in the part of a general course on physics which is especially devoted to "Mechanics." The interest of the mathematician too often begins with virtual velocities and d'Alembert's Principle, and the variational principles, of which Hamilton's Principle is the most important. Both are right, in the sense that they are doing nothing that is wrong; but each takes such a fragmentary view of the whole subject, that his work is ineffectual.

The world in which the boy and girl have lived is the true laboratory of elementary mechanics. The tennis ball, the golf ball, the shell on the river; the automobile — good old Model T, in its day, and the home-made autos and motor boats which youngsters construct and will continue to construct — the amateur printing press; the games in which the mechanics of the body is a part; all these things go to provide the student with rich laboratory experience before he begins a systematic study of mechanics. It is this experience on which the teacher of Mechanics can draw, and draw, and draw again.

The Cambridge Tripos of fifty years and more ago has been discredited in recent years, and the criticism was not without foundation. It was a method which turned out problem solvers — so said its opponents. But it turned out a Clerk Maxwell and it vitally influenced the training of the whole group of English physicists, whose work became so illustrious. In his interesting autobiography, *From Emigrant to Inventor*, Pupin acknowledges in no uncertain terms the debt he owes to just this training, and to Arthur Gordon Webster, through whom he first came to know this method — a method which Benjamin Osgood Peirce also prized highly in his work as a physicist. And so we make no apologies for availing ourselves to the fullest extent of that which the old Tripos Papers contributed to training in Mechanics. But we do not stop there. After all, it is the laws of Mechanics, their comprehension, their passing over into the flesh and blood of our scientific thought, and the mathematical technique and theory, that is our ultimate goal. To attain to this goal the mathematical theory, absurdly simple as it is at the start, must be systematically inculcated into the student from the beginning. In this respect the physicists fail us. Because the mathematics is simple, they do not think it important to insist on it. Any way to get an answer is good enough for them. But a day of reckoning comes. The physicist of to-day is in desperate need of mathematics, and at best all he can do is to grope, trying one mathematical expedient after another and holding to no one of these long enough to test it mathematically. Nor is he to be blamed. It is the old (and most useful) method of trial and error he is employing, and must continue to employ for the present.

Is the writer on Mechanics, per contra, to accept the challenge of preparing the physicist to solve these problems? That is too large a task. Rather, it is the wisdom of Pasteur who said: "Fortune favors the prepared mind" that may well be a guide for us now and in the future. What *can* be done, and what we have attempted in the present work, is to unite a broad and deep knowledge of the most elementary physical phenomena in the field of Mechanics with the best mathematical methods of the present day, treating with completeness, clarity, and rigor the beginnings of the subject; in scope not restricted, in detail not involved, in spirit scientific.

The book is adapted to the needs of a first course in Mechanics, given for sophomores, and culminating in a thorough study of the dynamics of a rigid body in two dimensions. This course may be followed by a half-course or a full course which begins with the kinematics and kinetics of a rigid body in three dimensions and proceeds to Lagrange's Equations and the variational principles. So important are Hamilton's Equations and their solution by means of Jacobi's Equation, that this subject has also been included. It appears that there is a special need for treating this theory, for although it is exceedingly simple, the current text-books are unsatisfactory. They assume an undefined knowledge of the theory of partial differential equations of the first order, but they do not show how the theory is applied. As a matter of fact, no theory of these equations at all is required for understanding the solution just mentioned. What is needed is the fact that Hamilton's Equations are invariant of a contact transformation. A simple proof is given in Chapter XIV, in which the method most important for the physicist, namely, the method of separating the variables, is set forth with no involved preliminaries. But even this proof may be omitted or postponed, and the student may strike in at once with Chapter XV.

The concept of the vector is essential throughout Mechanics, but intricate vector analysis is wholly unnecessary. A certain minute amount of the latter is however helpful, and has been set forth in Appendix A.

Appendix D contains a definitive formulation of a class of problems which is most important in physics, and shows how d'Alembert's Principle and Lagrange's Equations apply. It ties together the various detailed studies of the text and gives the reader a comprehensive view of the subject as a whole.

The book is designed as a careful and thorough introduction to Mechanics, but not of course, in this brief compass, as a treatise. With the principles of Mechanics once firmly established and clearly illustrated by numerous examples the student is well equipped for further study in the current text-books, of which may be mentioned: Routh: *An Elementary Treatise on Rigid Dynamics* and also *Advanced Dynamics*, by the same author; particularly valuable for its many problems. Webster, *Dynamics* — good material, and excellent for the student who is well trained in the rudiments, but hard reading for the beginner through poor presentation and lacunae in the theory; Appell, *Mécanique rationelle*, vols. i and ii — a charming book, which the student may open at any chapter for supplementary reading and examples. Jeans, *Mechanics*, may also be mentioned for supplementary exercises; as a text it is unnecessarily hard mathematically for the Sophomore, and it does not go far enough physically for the upperclassman. It is unnecessary to emphasize the importance of further study by the problem method of more advanced and difficult exercises, such as are found in these books. But to go further in incorporating these problems into the present work would increase its size unduly.

It is not merely a formal tribute, but one of deep appreciation, which I wish to pay to The Macmillan Company and to The Norwood Press for their hearty cooperation in all the many difficult details of the typography. Good composition is a distinct aid in setting forth the thought which the formulas are designed to express. Its beauty is its own reward.

To his teacher, Benjamin Osgood Peirce, who first blazed the trail in his course, Mathematics 4, given at Harvard in the middle of the eighties the Author wishes to acknowledge his profound gratitude. Out of these beginnings the book has grown, developed through the Author's courses at Harvard, extending over more than forty years, and out of courses given later at The National University of Peking. May it prove a help to the beginner in his first approach to the subject of Mechanics.

William Fogg Osgood

May 1937

This Dover edition is dedicated
to the memory of
WILLIAM FOGG OSGOOD
(1864-1943)

CONTENTS

CHAPTER I

STATICS OF A PARTICLE

CHAPTER II

STATICS OF A RIGID BODY

CHAPTER III

MOTION OF A PARTICLE

CHAPTER IV

DYNAMICS OF A RIGID BODY

CHAPTER V

KINEMATICS IN TWO DIMENSIONS

CHAPTER VI

ROTATION

CHAPTER VII

WORK AND ENERGY

CHAPTER VIII

IMPACT

CHAPTER IX

RELATIVE MOTION AND MOVING AXES

CHAPTER XIII

HAMILTON'S PRINCIPLE AND THE PRINCIPLE OF LEAST ACTION

CHAPTER XIV

CONTACT TRANSFORMATIONS

CHAPTER XV

SOLUTION OF HAMILTON'S EQUATIONS

APPENDIX

MECHANICS

CHAPTER I

STATICS OF A PARTICLE

1. Parallelogram of Forces. By a *force* is meant a push or a pull. A stretched elastic band exerts a force. A spiral spring, like those used in the upholstered seats of automobiles, when compressed by a load, exerts a force. The earth exerts a force of attraction on a falling rain drop.

The effect of a force acting at a given point, O, depends not merely on the magnitude, or intensity, of the force, but also on the direction in which it acts. Lay off a right line from O in the direction of the force, and make the length of the line proportional to the intensity of the force; for example, if F is 10 lbs., the length may be taken as 10 in., or 10 cm., or more generally, ten times the length which represents the unit force. Then this directed right line, or *vector*, gives a complete geometric picture of the force. Thus if a barrel of flour is suspended by a rope (and is at rest), the attraction of gravity — the pull of the earth — will be represented by a vector pointing downward and of length W, the weight of the barrel. On the other hand, the force which the rope exerts on the barrel will be represented by an equal and opposite vector, pointing upward. For, action and reaction are equal and opposite.

Fig. 1

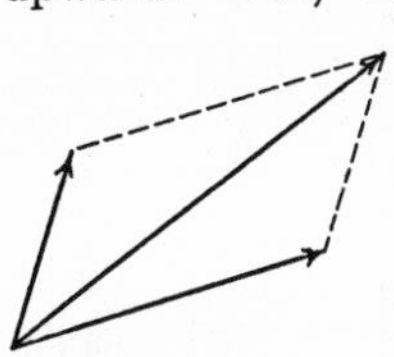

When two forces act at a point, they are equivalent to a single force, which is found as follows. Lay off from the point the two vectors, P and Q, which represent the given forces, and construct the parallelogram, of which the right line segments determined by P and Q are two adjacent sides. The diagonal of the parallelogram drawn from O determines a vector, R, which represents

Fig. 2

the combined effect of P and Q. This force, R, is called the *resultant* of P and Q, and the figure just described is known as the *parallelogram of forces.*

Example 1. Two forces of 20 pounds each make an angle of 60° with each other. To find their resultant.

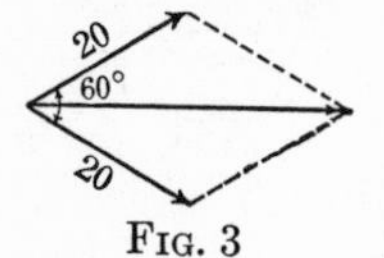

Fig. 3

Here, it is obvious from the geometry of the figure that the parallelogram is a rhombus, and that the length of the diagonal in question is $20\sqrt{3} = 34.64$. Hence the resultant is a force of 34.64 pounds, its line of action bisecting the angle between the given forces.

Example 2. Two forces of 7 pounds and 9 pounds act at a point and make an angle of 70° with each other. To find their resultant.

Graphical Solution. Draw the forces to scale, constructing the angle by means of a protractor. Then complete the parallelogram and measure the diagonal. Find its direction with the protractor.

Example 3. A picture weighing 15 lbs. hangs from a nail in the wall by a wire, the two segments of which make angles of 30° with the horizon. Find the tension in the wire.

Here, the resultant, 15, of the two unknown tensions, T and T, is given, and the angles are known. It is evident from the figure that T also has the value 15. So the answer is: 15 lbs.

Decomposition of Forces. Conversely, a given force can be decomposed along any two directions whatever. All that is needed is, to construct the parallelogram, of which the given force is the diagonal and whose sides lie along the given lines.

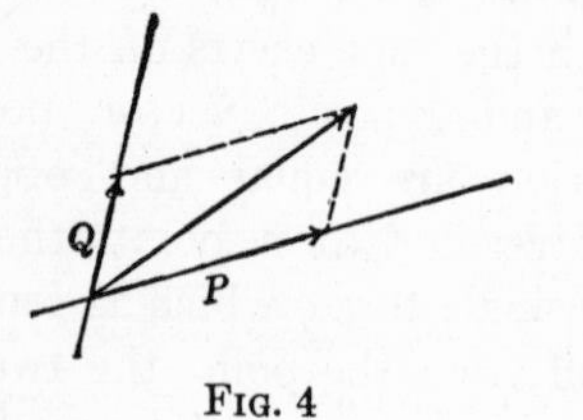

Fig. 4

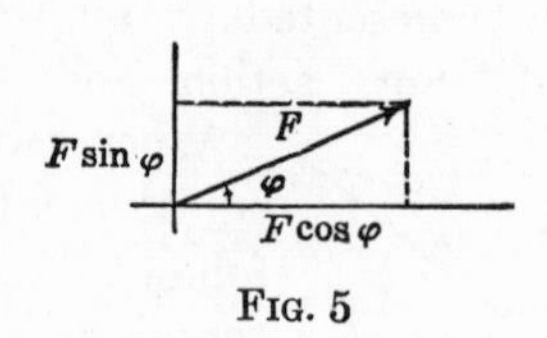

Fig. 5

If, in particular, the lines are perpendicular to each other, the components will evidently be:

$$F\cos\varphi, \qquad F\sin\varphi.$$

EXERCISES

1. Two forces of 5 lbs. and 12 lbs. make a right angle with each other. Show that the resultant force is 13 lbs. and that it makes an angle of $22° 37'$ with the larger force.

2. Forces of 5 lbs. and 7 lbs. make an angle of 100° with each other. Determine the resultant force graphically.

3. If the forces in Question 1 make an angle of 60° with each other, find the resultant.

Give first a graphical solution. Then obtain an analytical solution, using however no trigonometry beyond a table of natural sines, cosines, and tangents.

4. If two forces of 12 lbs. and 16 lbs. have a resultant of 20 lbs., what angle must they make with each other and with the resultant?

5. A force of 100 lbs. acts north. Resolve it into an easterly and a north-westerly component.

6. A force of 50 lbs. acts east north-east. Resolve it into an easterly and a northerly component. *Ans.* 46.20 lbs.; 19.14 lbs.

7. A force of 12 lbs. acts in a given direction. Resolve it into two forces that make angles of 30° and 40° with its line of action. Only a graphical solution is required.

2. Analytic Treatment by Trigonometry. The problem of finding the resultant calls for the determination of one side of a triangle when the other two sides and the included angle are known; and also of finding the remaining angles. The first problem is solved by the *Law of Cosines* in Trigonometry:

(1) $$c^2 = a^2 + b^2 - 2ab \cos C.$$

Here, $a = P, \quad b = Q, \quad c = R,$

$$C = 180° - \omega$$

and hence

(2) $$R^2 = P^2 + Q^2 + 2PQ \cos \omega.$$

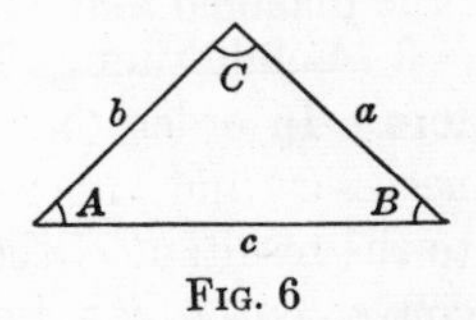

FIG. 6

Example. Forces of 5 lbs. and 8 lbs. make an angle of 120° with each other. Find their resultant. Here,

$$R^2 = 25 + 64 - 2 \times 5 \times 8 \times \tfrac{1}{2} = 49;$$

$$R = 7 \text{ lbs.}$$

To complete the solution and find the remaining angles we can use the *Law of Sines:*

(3)
$$\frac{a}{\sin A} = \frac{b}{\sin B} = \frac{c}{\sin C}.$$

Thus

(4)
$$\frac{Q}{\sin \varphi} = \frac{R}{\sin \omega}.$$

There is no difficulty here about the sign when the adjacent angle is used, since $\sin (180 - \omega) = \sin \omega$.

In the numerical example above, Equation (4) becomes:

$$\frac{8}{\sin \varphi} = \frac{7}{\frac{1}{2}\sqrt{3}}.$$

Thus

$$\sin \varphi = \tfrac{4}{7}\sqrt{3}, \qquad \cos \varphi = \tfrac{1}{7}, \qquad \varphi = 81^\circ \, 47'.$$

The third angle is found from the fact that the sum of the angles of a triangle is two right angles:

$$A + B + C = 180^\circ.$$

To sum up, then: Compute the resultant by the Law of Cosines and complete the solution by the Law of Sines.

EXERCISES

Give both a graphical and an analytical solution each time.

1. Forces of 2 lbs. and 3 lbs. act at right angles to each other. Find their resultant in magnitude and direction.*

2. Forces of 4 lbs. and 5 lbs. make an angle of 70° with each other. Find their resultant.

3. Equilibrium. The Triangle of Forces. Addition of Vectors. In order that three forces be in equilibrium, it is clearly necessary and sufficient that any one of them be equal and opposite to the resultant of the other two. The condition can be expressed conveniently by aid of the idea of the *addition of vectors.*

First of all, two vectors are *defined* as equal if they have the same magnitude, direction, and sense, no matter where in the plane (or in space) they may lie.

* Observe that in this case it is easier to determine the angle from its *tangent.* Square roots should be computed from a Table of Square Roots. Huntington's *Four-Place Tables* are convenient, and are adequate for the ordinary cases that arise in practice. But cases not infrequently arise in which more elaborate tables are needed, and Barlow's will be found useful.

Vector Addition. Let **A** and **B** be any two vectors. Construct **A** with any point, O, as its initial point. Then, with the terminal point of **A** as its initial point, construct **B**. The vector, **C**, whose initial point is the initial point of **A** and whose terminal point is the terminal point of **B**, is defined as the *vector sum*, or, simply, the *sum* of **A** and **B**:

$$\mathbf{C} = \mathbf{A} + \mathbf{B}.$$

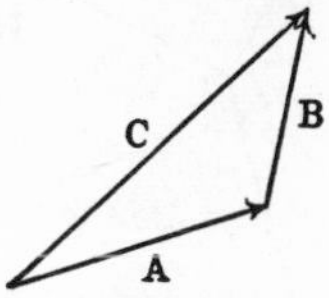

FIG. 7

It is obvious that

$$\mathbf{B} + \mathbf{A} = \mathbf{A} + \mathbf{B}.$$

Any number of vectors can be added by applying the definition successively. It is easily seen that

$$(\mathbf{A} + \mathbf{B}) + \mathbf{C} = \mathbf{A} + (\mathbf{B} + \mathbf{C}).$$

Consequently the sum

$$\mathbf{A}_1 + \mathbf{A}_2 + \cdots + \mathbf{A}_n$$

is independent of the order in which the terms are added.

For accuracy and completeness it is necessary to introduce the *nil vector*. Suppose, for example, that **A** and **B** are equal and opposite. Then their sum is not a vector in any sense as yet considered, for the terminal point coincides with the initial point. When this situation occurs, we say that we have a *nil vector*, and denote it by 0:

$$\mathbf{A} + \mathbf{B} = 0.$$

We write, furthermore,*

$$\mathbf{B} = -\mathbf{A}.$$

Equilibrium. The condition, necessary and sufficient, that three forces be in equilibrium is that their vector sum be 0. Geometrically this is equivalent to saying that the vectors which represent the forces can be drawn so that the figure will close and form a triangle. From the Law of Sines we have:

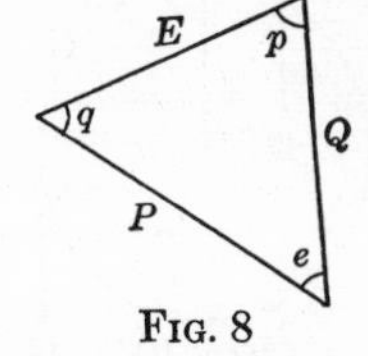

FIG. 8

$$(1) \qquad \frac{P}{\sin p} = \frac{Q}{\sin q} = \frac{E}{\sin e},$$

$$p + q + e = 180°.$$

* It is not necessary for the present to go further into vector analysis than the above definitions imply. Later, the two forms of *product* will be needed, and the student may be interested even at this stage in reading Chapter XIII of the author's *Advanced Calculus*, or Appendix A.

Since $\sin(180° - A) = \sin A$, we can state the result in the following form. Let three forces, P, Q, and E, acting on a particle, be in equilibrium. Denote the angles between the forces, as indicated, by p, q, e. Then Equation (1) represents a necessary condition for equilibrium. Conversely, this condition is sufficient.

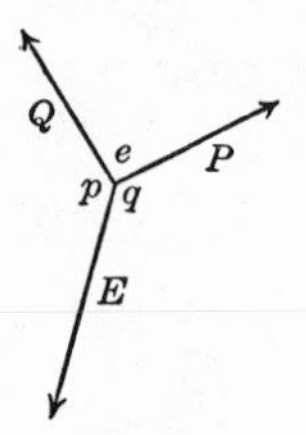

FIG. 9

We thus obtain a convenient solution in all cases except the one in which the magnitudes of the forces, but no angles, are given. Here, the Law of Cosines * gives one angle, and then a second angle can be computed by the Law of Sines.

Example 1. Forces of 4, 5, and 6 are in equilibrium. Find the angles between them.

First, solve the problem graphically, measuring the angles. Next, apply the Law of Cosines:

$$4^2 = 5^2 + 6^2 - 2 \times 5 \times 6 \times \cos\varphi,$$

$$\cos\varphi = \tfrac{3}{4}, \qquad \varphi = 41° \, 23'.$$

A second angle is now computed by the Law of Sines:

$$\frac{5}{\sin\psi} = \frac{4}{\sin\varphi},$$

$$\sin\psi = \frac{5\sqrt{7}}{16}, \qquad \psi = 55° \, 46'.$$

The third angle is 82° 51′.

Example 2. A 40 lb. weight rests on a smooth horizontal cylinder and is kept from slipping by a cord that passes over the cylinder and carries a 10 lb. weight at its other end. Find the position of equilibrium.

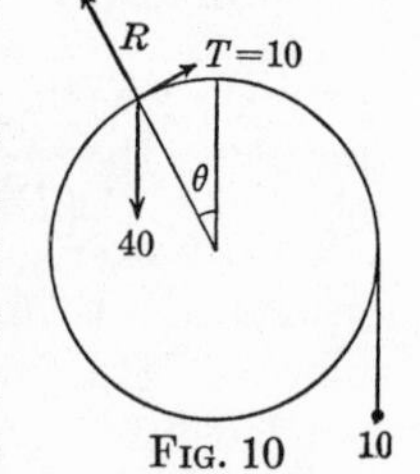

FIG. 10

The cord is assumed weightless, and since it passes over a smooth surface, the tension in it is the same at all points. The surface of the cylinder is smooth, hence its reaction is normal to its surface. Let θ be the unknown angle that the radius drawn to the weight

* If we had a large number of numerical problems to solve, it would pay to use the more elaborate theorems of Trigonometry (*e.g.* Law of Tangents). But for ordinary household purposes the more familiar law is enough.

makes with the vertical. Then from inspection of the figure we see that

$$\frac{10}{\sin\theta} = \frac{40}{\sin 90^\circ},$$

or

$$\sin\theta = \tfrac{1}{4}, \qquad \theta = 14^\circ\, 29'.$$

EXERCISES

1. Forces of 7, 8, and 9 pounds keep a particle at rest. Find the angles they make with one another.

2. Forces of 51.42, 63.81, and 71.93 grs. keep a particle at rest. What angle do the first two forces make with each other? Find the other angles.

3. A weightless string passes over two smooth pegs at the same level and carries weights of P and P at its ends. In the middle, there is knotted a weight W. What angle do the segments of the string make with the vertical, when the system is at rest?

Ans. $\sin\theta = \dfrac{W}{2P}$.

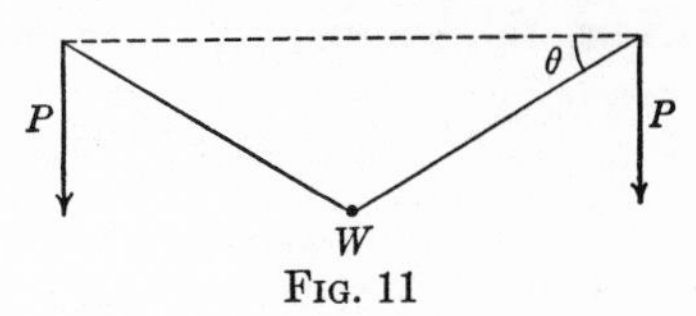

FIG. 11

4. A boat is prevented from drifting down stream by two ropes tied to the bow of the boat, and to stakes at opposite points on the banks. One rope is 125 ft. long; the other, 150 ft.; and the stream is 200 ft. broad. If the tension in the shorter rope is 20 lbs., what is the tension in the other rope?

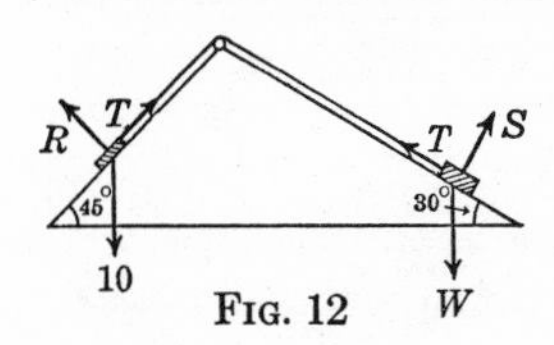

FIG. 12

5. Two smooth inclined planes, back to back, meet along a horizontal straight line, and make angles of 30° and 45° with the horizon. A weight of 10 lbs. placed on the first plane is held by a cord that passes over the top of the planes and carries a weight W, resting on the other plane and attached to the end of the cord. Find what value W must have.

4. The Polygon of Forces. From the case of three forces the generalization to the case of n forces acting at a point presents no difficulty. Add the forces geometrically, *i.e.* by the vector law. The vector sum represents the resultant of all n forces. Thus, in the figure, the resultant is given by

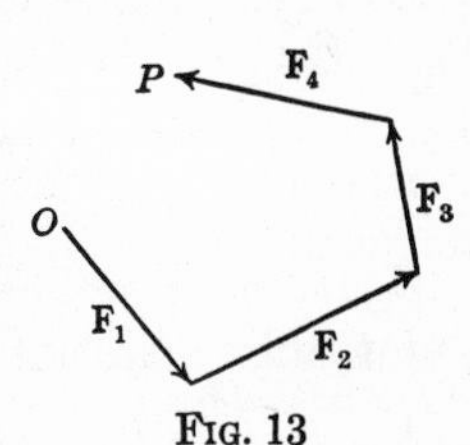

FIG. 13

the vector whose initial point is the point O and whose terminal point is P.

The condition for equilibrium is clearly that the resultant be a nil vector, or that the broken line close and form a polygon — but not necessarily a polygon in the sense of elementary geometry, since its sides may intersect, as in Figure 14.

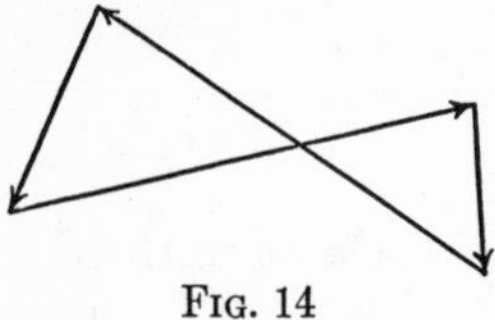

FIG. 14

This condition will obviously be fulfilled if and only if the sum of the projections of the forces along each of two lines that intersect is zero.*

Analytically, the resultant can be represented as follows. Let a Cartesian system of coordinates be assumed, and let the components of the force $\mathbf{F}_k$ along the axes be X_k and Y_k. If, now, the components of the resultant are denoted by X and Y, we have:

$$X = X_1 + X_2 + \cdots + X_n,$$
$$Y = Y_1 + Y_2 + \cdots + Y_n.$$

Such sums are written as

$$(1) \qquad \sum_{k=1}^{n} X_k, \quad \text{or} \quad \sum_{k} X_k, \quad \text{or} \quad \sum X_k,$$

depending on how elaborate the notation should be to insure clearness.

The forces will be in equilibrium if, and only if, the resultant force is nil, and this will be the case if

$$(2) \qquad \sum_{k=1}^{n} X_k = 0, \qquad \sum_{k=1}^{n} Y_k = 0.$$

EXERCISE

A 4 lb. weight is acted on by three forces, all of which lie in the same vertical plane: a force of 10 lbs. making an angle of 30° with the vertical, and forces of 8 lbs. and 12 lbs. on the other side of the vertical and making angles of 20° with the upward vertical and 15° with the downward vertical respectively. Find the force that will keep the system at rest.

Space of Three Dimensions. If more than two forces act at a point, they need not lie in a plane. But they can be added two

* Cf. Osgood and Graustein, *Analytic Geometry*, pp. 1–6.

at a time by the parallelogram law, the first two thus being replaced by a single force — their resultant — and this force in turn compounded with the third force; etc. The broken line that represents the addition of the vectors no longer lies in a plane, but becomes a skew broken line in space, and the polygon of forces becomes a skew polygon. The components of the resultant force along the three axes are:

$$(3) \qquad X = \sum_{k=1}^{n} X_k, \qquad Y = \sum_{k=1}^{n} Y_k, \qquad Z = \sum_{k=1}^{n} Z_k.$$

The condition for equilibrium is:

$$(4) \qquad \sum_{k=1}^{n} X_k = 0, \qquad \sum_{k=1}^{n} Y_k = 0, \qquad \sum_{k=1}^{n} Z_k = 0.$$

In all of these formulas, X_k, Y_k, Z_k are algebraic quantities, being positive when the component has the sense of the positive axis of coordinates, and negative when the sense is the opposite. In solving problems in equilibrium it is frequently simpler to single out the components that have one sense along the line in question and equate their sum, each being taken as positive, to the sum of the components in the opposite direction, each of these being taken as positive, also. The method will be illustrated by the examples in friction of the next paragraph.

5. Friction. Let a brick be placed on a table; let a string be fastened to the brick, and let the string be pulled horizontally with a force F just sufficient to move the brick. Then the law of physics is that

$$F = \mu R,$$

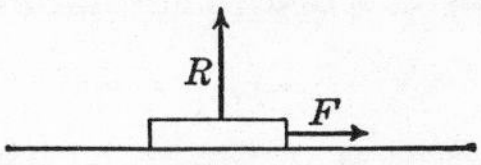

R
F

FIG. 15

where R (here, the weight of the brick) is the normal* pressure of the table on the brick, and μ (the coefficient of friction) is a constant for the two surfaces in contact. Thus, if a second brick were placed on top of the first, R would be doubled, and so would F.

We can state the law of friction generally by saying: When two surfaces are in contact and one is just on the point of slipping

* *Normal* means, at right angles to the surface in question. The normal to a surface at a point is the line perpendicular to the tangent plane of the surface at the point in question.

over the other, the tangential force F due to friction is proportional to the normal pressure R between the surfaces, or

$$F = \mu R,$$

where μ, the *coefficient of friction*, is independent of F and R, and depends only on the substances in contact, but not on the area of the surfaces which touch each other. For metals on metals μ usually lies between 0.15 and 0.25 in the case of statical friction. For sliding friction μ is about 0.15; cf. Rankine, *Applied Mechanics*.

A simple experiment often performed in the laboratory for determining μ is the following. Let one of the surfaces be represented by an inclined plane, the angle of which can be varied. Let the other surface be represented by a rider, or small block of the substance in question, placed on the plane. If the plane is gradually tilted from a horizontal position, the rider will not slip for a time. Finally, a position will be reached for which the rider just slips. This angle of the plane is known as the *angle of friction* and is usually denoted by λ. Let us show that

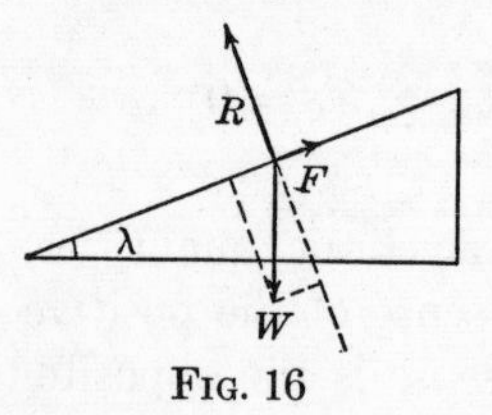

FIG. 16

$$\mu = \tan \lambda.$$

Resolve the force of gravity, W, into its two components along the plane and normal to the plane. These are:

$$W \sin \lambda, \qquad W \cos \lambda.$$

And now the forces acting up the plane (*i.e.* the *components* directed up the plane) must equal the forces down the plane, or

$$F = W \sin \lambda;$$

and the forces normal to the plane and upward must equal the forces normal to the plane and downward, or

$$R = W \cos \lambda.$$

Hence

$$\frac{F}{R} = \frac{\sin \lambda}{\cos \lambda} = \tan \lambda.$$

But

$$F = \mu R.$$

Consequently

$$\mu = \tan \lambda.$$

Example. A 50 lb. weight is placed on a rough inclined plane, angle of elevation, 30°. A cord attached to the weight passes over a smooth pulley at the top of the plane and carries a weight W at its lower end. For what values of W will the system be in equilibrium if $\mu = \frac{1}{6}$?

Here, $\lambda < 30°$, and so, if W is very small, the 50 lb. weight will slip down the plane. Suppose W is just large enough to prevent slipping. Then friction acts up the plane, and the forces which produce equilibrium are those indicated. Hence

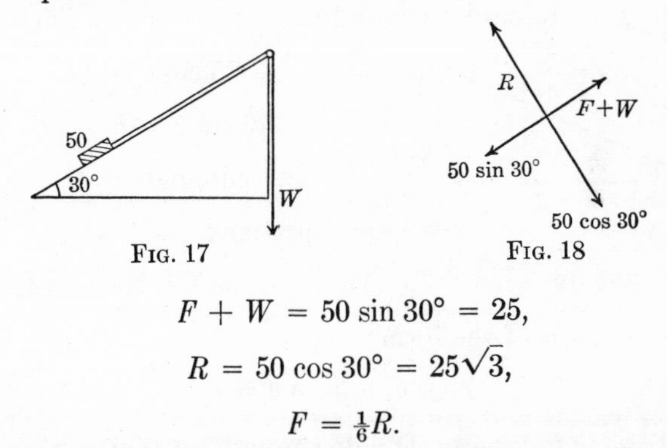

FIG. 17 FIG. 18

$$F + W = 50 \sin 30° = 25,$$

$$R = 50 \cos 30° = 25\sqrt{3},$$

and

$$F = \tfrac{1}{6}R.$$

It follows, then, that

$$W = \frac{6 - \sqrt{3}}{6} 25 = 17.8 \text{ lbs.}$$

If, now, W is slightly increased, the 50 lb. weight will obviously still be in equilibrium, and this will continue to be the case until the 50 lb. weight is just on the point of slipping up the plane. This will occur when $W = 32.2$ lbs. as the student can now prove for himself. Consequently, the values of W for which there is equilibrium are those for which

$$17.8 \leqq W \leqq 32.2.$$

EXERCISES

1. If the cylinder of Example 2, § 3, is rough, $\mu = \frac{1}{6}$, find the total range of equilibrium. *Ans.* $4° 49' \leqq \theta \leqq 23° 44'$.

2. Consider the inclined planes of Exercise 5, § 3. If the one on which the 10 lb. weight rests is rough, $\mu = \frac{1}{10}$, find the range of values for W that will yield equilibrium.

3. Prove the formula

$$\mu = \tan \lambda$$

by means of the triangle of forces, *i.e.* the Law of Sines, § 3, (1).

6. Solution of a Trigonometric Equation. Problem. A 50 lb. weight rests on a rough horizontal plane, $\mu = \frac{1}{6}$. A cord is fastened to the weight, passes over a smooth pulley 2 ft. above the plane, and carries a weight of 25 lbs. which hangs freely at its other end. To find all the positions of equilibrium.

Resolving the forces horizontally and vertically, we find:

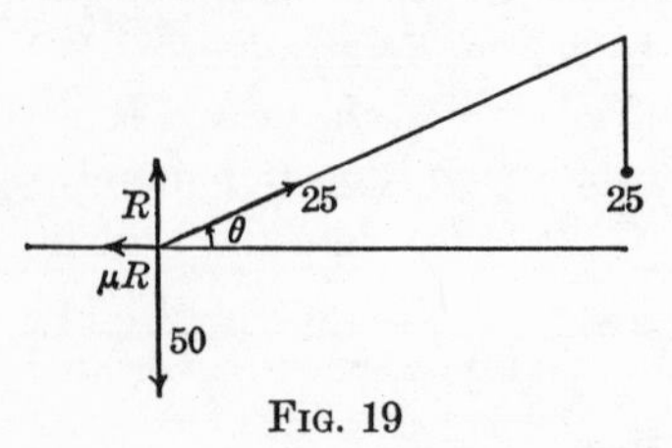

FIG. 19

$$25 \cos \theta = \tfrac{1}{6}R,$$

$$25 \sin \theta + R = 50.$$

Hence, eliminating R, we obtain the equation:

$$(1) \qquad 6 \cos \theta + \sin \theta = 2.$$

This equation is of the form:

$$(2) \qquad a \cos \theta + b \sin \theta = c,$$

and is solved as follows.* Divide through by $\sqrt{a^2 + b^2}$:

$$(3) \qquad \frac{a}{\sqrt{a^2 + b^2}} \cos \theta + \frac{b}{\sqrt{a^2 + b^2}} \sin \theta = \frac{c}{\sqrt{a^2 + b^2}},$$

and then set

$$(4) \qquad \frac{a}{\sqrt{a^2 + b^2}} = \cos \alpha, \qquad \frac{b}{\sqrt{a^2 + b^2}} = \sin \alpha.$$

Thus

$$\cos \alpha \cos \theta + \sin \alpha \sin \theta = \frac{c}{\sqrt{a^2 + b^2}},$$

or

$$(5) \qquad \cos (\theta - \alpha) = \frac{c}{\sqrt{a^2 + b^2}}.$$

* The student should observe carefully the trigonometric technique set forth in this paragraph, not merely because equations of this type are important in themselves, but because the practical value of a working knowledge of trigonometry is not confined to solving numerical triangles. Of far greater scope and importance in practice are the purely analytical reductions to other trigonometric forms, and the solution of trigonometric equations. That is one of the reasons why the harder examples at the end of the chapter are valuable. They not only give needed practice in formulating mathematically physical data; they require also the ability to handle analytical trigonometry according to the demands of practice.

The angle α is most easily determined from the equation:

$$\tan\alpha = \frac{b}{a}. \tag{6}$$

Thus α is seen to be one of two angles — which one is rendered clear by plotting the point on the unit circle:

$$x^2 + y^2 = 1,$$

whose coordinates are

$$x = \frac{a}{\sqrt{a^2+b^2}}, \qquad y = \frac{b}{\sqrt{a^2+b^2}}.$$

The angle from the positive axis of x to the radius drawn to this point is α. Thus we have a graphical determination of α. It is not necessary to compute the coordinates accurately, but merely to observe in which quadrant the point lies, so as to know which root of the equation for $\tan\alpha$ to take. Thus if $b > 0$, α must be an angle of the first or second quadrant. Finally, if $c/\sqrt{a^2+b^2}$ is numerically greater than 1, the equation has no solution.

In defining α, it would, of course, have answered just as well if $\sin\alpha$ and $\cos\alpha$ had been interchanged, and if either or both the ratios in (4) had been replaced by their negative values.*

Returning now to the numerical equation above, we see that

$$\tan\alpha = \tfrac{1}{6}, \qquad \sin\alpha > 0, \qquad \alpha = 9^\circ\,28';$$

$$\cos(\theta - 9^\circ\,28') = \frac{2}{\sqrt{37}},$$

$$\theta - 9^\circ\,28' = \pm\,70^\circ\,48'.$$

Since θ in the problem before us must be an angle of the first quadrant, the lower sign is impossible, and

$$\theta = 80^\circ\,16'.$$

We have determined the point of the plane at which all the friction is called into play and the 50 lb. weight is just on the point of slipping. For other positions, F will not equal μR. Such a position will be one of equilibrium if the amount of friction actually called into play, or F, is *less* than the amount that could

* Equation (2) might also have been solved by transposing one term from the left- to the right-hand side and squaring. On using the Pythagorean Identity:

$$\sin^2\theta + \cos^2\theta = 1,$$

we should be led to a quadratic equation in the sine or cosine. This equation will, in general, have *four* roots between 0° and 360°, and three of them must be excluded. Moreover, the actual computation by this method is more laborious.

be called into play, or μR. It seems plausible that such points lie to the right of the critical point; but this conclusion is not immediately justified, for, although the amount of friction required, or

$$F = 25 \cos \theta,$$

is less for a larger θ, — still, the amount available, or

$$\mu R = \frac{50 - 25 \sin \theta}{6},$$

is also less. We must prove, therefore, that

$$F < \mu R$$

or

$$25 \cos \theta < \frac{50 - 25 \sin \theta}{6}.$$

This will obviously be so if

$$6 \cos \theta < 2 - \sin \theta, \qquad 80° \, 16' < \theta < 90°,$$

or if

$$6 \cos \theta + \sin \theta < 2,$$

or if

$$\frac{6}{\sqrt{37}} \cos \theta + \frac{1}{\sqrt{37}} \sin \theta < \frac{2}{\sqrt{37}},$$

or if

$$\cos (\theta - 9° \, 28') < \frac{2}{\sqrt{37}}.$$

As θ, starting with the value $80° \, 16'$, increases, $\theta - 9° \, 28'$ also increases, and consequently $\cos (\theta - 9° \, 28')$ decreases. Consequently our guess is borne out by the facts, and the 50 lb. weight will be in equilibrium at all points on the table within a circle of radius .343 ft., or a little over 4 in., whose centre is directly under the pulley.

EXERCISES

1. Solve the same problem if the plane is inclined at an angle of 15° with the horizon, and the vertical plane through the weight and the pulley is at right angles to the rough plane.

Ans. $43° \, 33' \leqq \theta \leqq 69° \, 24'$.

2. At what angle should the plane be tilted, in order that the region of equilibrium may just extend indefinitely down the plane?

3. Find the angle of the third quadrant determined by the equation:

$$3 \sin \theta - 2 \cos \theta = 1.$$

4. Show that there are in all eight ways of solving Equation (2), given by setting the right-hand sides of Equations (4) equal to $\pm \cos\alpha$, $\pm \sin\alpha$ and $\pm \sin\alpha$, $\pm \cos\alpha$, where the $\pm$ signs are independent of each other.

5. Evaluate the integral:

$$\int \frac{dx}{a\cos x + b\sin x}.$$

6. Solve the equation:

$$2\cos^2\varphi - 4\cos\varphi\sin\varphi - 3\sin^2\varphi = -5.$$

Suggestion. Introduce the double angle, 2φ.

EXERCISES ON CHAPTER I*

1. A rope runs through a block, to which another rope is attached. The tension in the first rope is 120 lbs., and the angle it includes is 70°. What is the tension in the second rope?

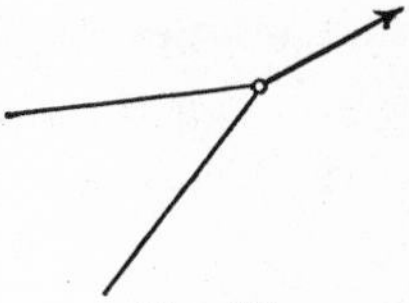

FIG. 20

2. A man weighing 160 lbs. is lying in a hammock. The rope at his head makes an angle of 30° with the horizon, and the rope at his feet, an angle of 15°. Find the tensions in the two ropes.

3. A load of furniture is being moved. The rope that binds it passes over the round of a chair. The tension on one side of the round is 40 lbs. and on the other side, 50 lbs.; and the angle is 100°. What force does the round have to withstand?

4. A canal boat is being towed by a hawser pulled by horses on the bank. The tension in the hawser is 400 lbs. and it makes an angle of 15° with the bank. What is the effective pull on the boat in the direction of the canal?

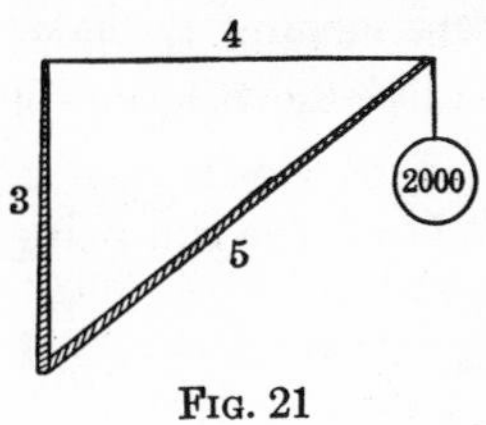

FIG. 21

5. A crane supports a weight of a ton as shown in the figure. What are the forces in the horizontal and in the oblique member?

6. Three smooth pulleys can be set at pleasure on a horizontal circular wire. Three strings, knotted together, pass over the

* The student should begin each time by drawing an adequate figure, illustrating the physical objects involved, and he should put in the forces with colored ink or pencil. A bottle of red ink, used sparingly, contributes tremendously to clear thinking.

pulleys and carry weights of 7, 8, and 9 lbs. at their free ends. How must the pulleys be set, in order that the knot may be at rest at the centre of the circle?

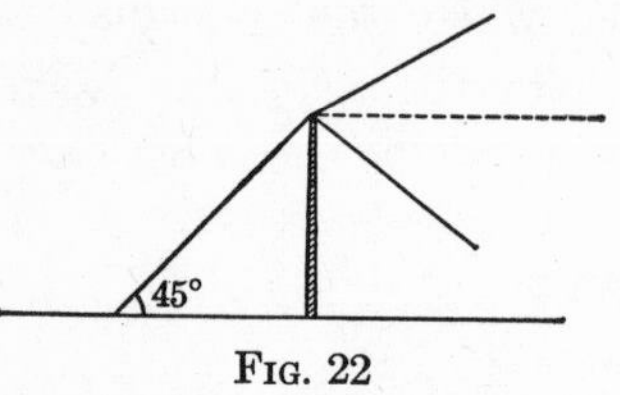

Fig. 22

7. A telegraph pole at two cross-roads supports a cable, the tension in which is a ton. The cable lies in a horizontal plane and is turned through a right angle at the pole. The pole is kept from tipping by a stay from its top to the ground, the stay making an angle of 45° with the vertical. What is the tension in the stay?

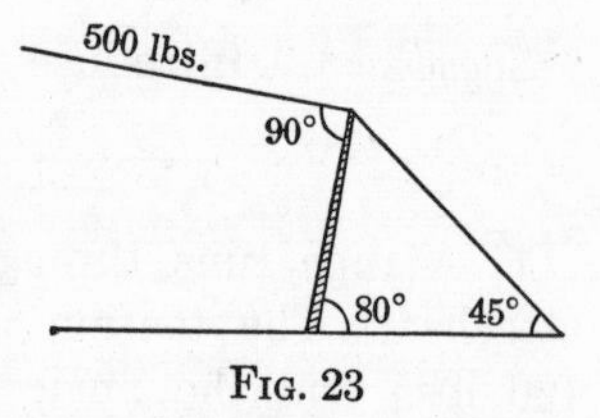

Fig. 23

8. The figure suggests a stake of a circus tent, with a tension of 500 lbs. to be held. What is the tension in the stay, if the stake could turn freely?

9. Two men are raising a weight of 150 lbs. by a rope that passes over two smooth pulleys and is knotted at A. How hard are they pulling?

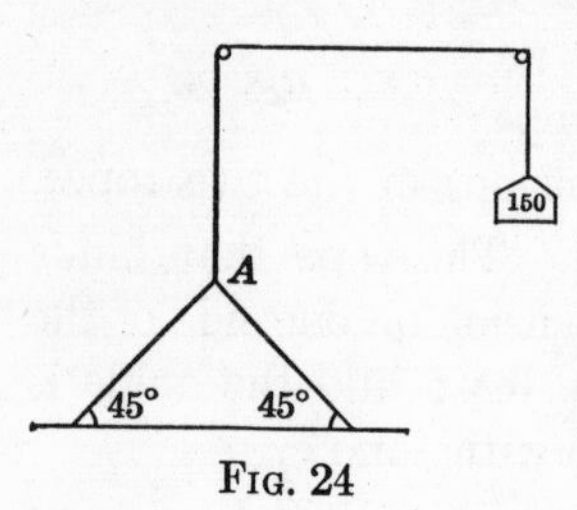

Fig. 24

10. If, in the preceding question, instead of being knotted at A, the two ropes the men have hold of passed over pulleys at A and were vertical above A, how hard would the men then have to pull?

11. A weight W is placed in a smooth hemispherical bowl; a string, attached to the weight, passes over the edge of the bowl and carries a weight P at its other end. Find the position of equilibrium.

12. Solve the same problem for a parabolic bowl, the rim being at the level of the focus.

13. One end of a string is made fast to a peg at A. The string passes over a smooth peg at B, at the same level as A, and carries a weight P at its free end. A smooth heavy bead, of weight W, can slide on the string. Find the position of equilibrium and the pressure on the peg at B.

14. A bead weighing W lbs. can slide on a smooth vertical circle of radius a. To the bead is attached a string that passes over a smooth peg situated at a distance $\frac{1}{2}a$ above the centre of the circle, and has attached to its other end a weight P. Find all the positions of equilibrium.

15. A 50 lb. weight rests on a smooth inclined plane (angle with the horizontal, 20°) and is kept from slipping by a cord which passes over a smooth peg 1 ft. above the top of the plane, and which carries a weight of 25 lbs. at its other end. Find the position of equilibrium.

16. A heavy bead can slide on a smooth wire in the form of a parabola with vertical axis and vertex at the highest point. A string attached to the bead passes over a smooth peg at the focus of the parabola and carries a weight at its other end. Show that in general there is only one position of equilibrium; but sometimes all positions are positions of equilibrium.

17. A weightless bead * can slide on a smooth wire in the form of an ellipse whose plane is vertical. A string is knotted to the bead and passes over two smooth pegs at the foci, which are at the same horizontal height. Weights of P and Q are attached to the two ends of the string. Find the positions of equilibrium.

18. An inextensible flexible string has its ends made fast at two points and carries a weightless smooth bead. Another string is fastened to the bead and drawn taut. Show that every position of the bead is one of equilibrium, if the second string is properly directed.

19. Give a mechanical proof, based on the preceding question, that the focal radii of an ellipse make equal angles with the tangent.

20. A bead of weight P can slide on a smooth, vertical rod. To the bead is attached an inextensible string of length $2a$, carrying at its middle point a weight W and having its other end made fast to a peg at a horizontal distance a from the rod. Show that the position of equilibrium is given by the equations:

$$P \tan \varphi = (P + W) \tan \theta, \qquad \sin \theta + \sin \varphi = 1,$$

where θ, φ are the angles the segments of the string make with the vertical.

* Questions of this type may be objected to on the ground that a force must act on mass, and so there is no sense in speaking of forces which act on a massless ring. But if the ring has minute mass, the difficulty is removed. The problem may be thought of, then, as referring to a heavy bead, whose weight is just supported by a vertical string. Since the weight of the bead now has no influence on the position of equilibrium, the mass of the bead may be taken as very small, and so, physically negligible.

21. If, in the preceding question, $P = W$, show that $\theta = 21° 55'$, $\varphi = 38° 49'$. Determine these angles when $W = 2P$.

22. A flexible inextensible string in the form of a loop 60 in. long is laid over two smooth pegs 20 in. apart and carries two smooth beads of weight P and W. Find the position of equilibrium, if the beads cannot come together.

Ans.
$$W \sin\theta = P \sin\varphi,$$
$$\cos\theta + \cos\varphi = 3\cos\theta\cos\varphi;$$
$$\cos^4\theta - \frac{2}{3}\cos^3\theta - \frac{8}{9}\left(1 - \frac{P^2}{W^2}\right)\cos^2\theta + \frac{2}{3}\left(1 - \frac{P^2}{W^2}\right)\cos\theta - \frac{1}{9}\left(1 - \frac{P^2}{W^2}\right) = 0;$$
hence
$$9\cos^4\theta - 6\cos^3\theta - 8\lambda\cos^2\theta + 6\lambda\cos\theta - \lambda = 0,$$
where
$$\lambda = \frac{(W^2 - P^2)}{W^2}.$$

23. Show that if, in the preceding question, $P = 5$ and $W = 10$, $\theta = 25° 8'$, and find the reaction on the peg.

24. One end of an inextensible string a in. long is made fast to a peg A and at the other end is knotted a weight W. A second string, attached to W, passes over a smooth peg at B, distant b in. from A and at the same level, and carries a weight P at its other end. Find the position of equilibrium.

If $P = W$, how far below the level of the pegs will the first weight rest?

Fig. 25

25.* Observe the braces that stiffen the frame of a railroad car. Formulate a reasonable problem suggested by what you see, and solve it.

26. A bridge of simple type is suggested by the figure. In designing such a structure, the stiffness of the members at a point

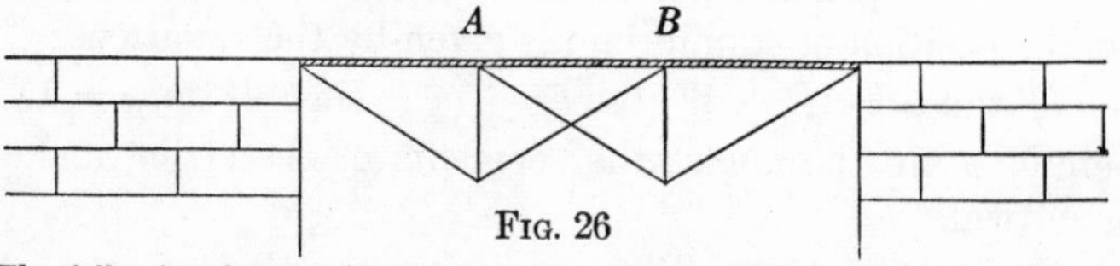

Fig. 26

* The following four problems are given only in outline, and the student thus has the opportunity of filling in reasonable numerical data and formulating a clean-cut question. It is not necessary that he respond to all the problems; but he should demand of himself that he develop a number of them and supplement these by others of like kind which he finds of his own initiative in everyday life. For, *imagination* is one of the highest of the intellectual gifts, and too much effort cannot be spent in developing it.

where these come together is not to be utilized, but the frame is planned as if the members were all pivoted there. Draw such a bridge to scale and find what the tensions and thrusts will be if it is to support a weight of 20 tons at each of the points A, B. Make a reasonable assumption about the weight of the road bed, but neglect the weight of the tie rods, etc.

27. The tension in each of the traces attached to a whiffle-tree 3 ft. long is 100 lbs. The distance from the ring to the whiffle-tree is 10 in. What is the tension in the chains?

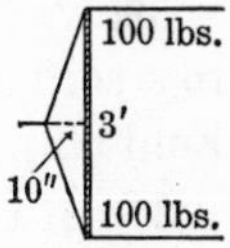

100 lbs.
3′
10″
100 lbs.

FIG. 27

28. Have you ever seen a *funicular* — a small passenger car, hauled up a steep mountain by a cable? How is the tension in the cable related to the weight of the car? When the direction of the cable is changed by a friction pin, or roller, over which the cable passes, what is the pressure on the pin?

FRICTION

29. Consider the inclined planes of Question 5, § 3. If both are rough and $\mu = \frac{1}{10}$ for each, what is the range of values for W consistent with equilibrium?

30. A weightless bead can slide on a rough horizontal wire, $\mu = 0.1$. A cord is attached to the bead and carries a weight at its other end, thus forming a simple pendulum. Through what angle can the pendulum swing without causing the bead to slip?

31. A water main 5 ft. in diameter is filled with water to a depth of 1 ft. A mouse tumbles in and swims to the nearest point on the wall. If the coefficient of friction between her feet and the pipe is $\frac{1}{3}$, can she clamber up, or will she be drowned?

32. A heavy bead is placed on a rough vertical circle, the coefficient of friction being $\frac{2}{7}$. If the angle between the radius drawn to the bead and the vertical is 16°, find whether the bead will slip when released.

33. A rope is fastened to a weight that rests on a rough horizontal plane, and pulled until the weight just moves. Find the tension in the rope, and show that it will be least when the rope makes, with the horizontal, the angle of friction.

34. The same question for an inclined plane.

35. A 50 lb. weight is placed on a rough inclined plane, $\mu = \frac{1}{4}$, angle of inclination, 10°. A string tied to the weight passes over a smooth peg at the same level as the weight and carries a weight of 7 lbs. at its lower end. When the system is released from rest, will it slip?

36. A weight is placed on a rough inclined plane and is attached to a cord, the other end of which is made fast to a peg in the plane. Find all positions of equilibrium.

37. If the parabolic wire described in Question 16 is rough, and the weights are P and W, find all positions of equilibrium.

Ans. When $P \neq W$, the limiting position is given by one or the other of the equations:

$$\tan\frac{\theta}{2} = \frac{P - W}{P + W} \cdot \frac{1}{\mu}, \qquad \tan\frac{\theta}{2} = \frac{W - P}{W + P} \cdot \frac{1}{\mu}.$$

Find the other positions of equilibrium, and discuss the case $P = W$.

38. Cast iron rings weighing 1 lb. each can slide on a rough horizontal rod, $\mu = \frac{1}{6}$. A string 6 ft. long is attached to each of these beads and carries a smooth bead weighing 5 lbs. How far apart can the two beads on the rod be placed, if the system is to remain at rest when released?

39. An elastic string 6 ft. long, obeying Hooke's Law, is stretched to a length of 6 ft. 6 in. by a force of 20 lbs. The ends of the string are made fast at two points 6 ft. apart and on the same level. A weight of 4 lbs. is attached to the mid-point of the string and carefully lowered. Find the position of equilibrium, neglecting the weight of the string. *Ans.* θ is given by the equation:

$$\cot\theta = 120\,(1 - \cos\theta).$$

40. Solve the preceding equation for θ, to one-tenth of a degree.

Ans. $\theta = 14\frac{1}{2}°$.

41. The mast of a derrick is 40 ft. high, and a stay is fastened to a block of stone weighing 4 tons and resting on a pavement, $\mu = \frac{3}{4}$. The boom is 35 ft. long, and its end is distant 20 ft. from the top of the mast. Is it possible to raise a 5 ton weight, without the derrick's being pulled over, the distance from the stone to the derrick being 120 ft.?

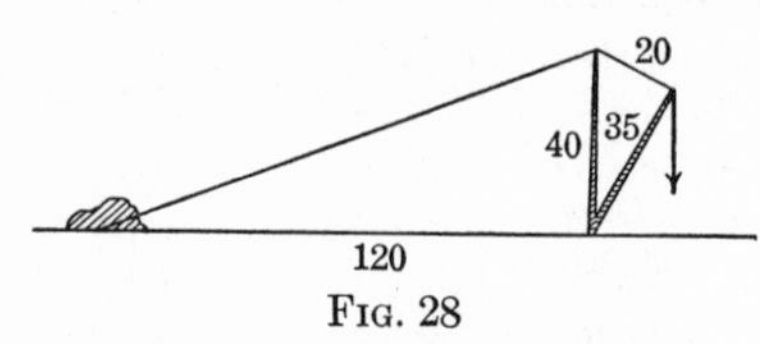

Fig. 28

CHAPTER II

STATICS OF A RIGID BODY

1. Parallel Forces in a Plane. Let two parallel forces, P and Q, act on a body at A and B, and let them have the same sense. Introduce two equal and opposite forces, S and S', at A and B as shown in the figure, and, compounding them with P and Q respectively, carry the resulting forces back to the point D in which their lines of action meet. These latter forces are now seen to have a resultant,

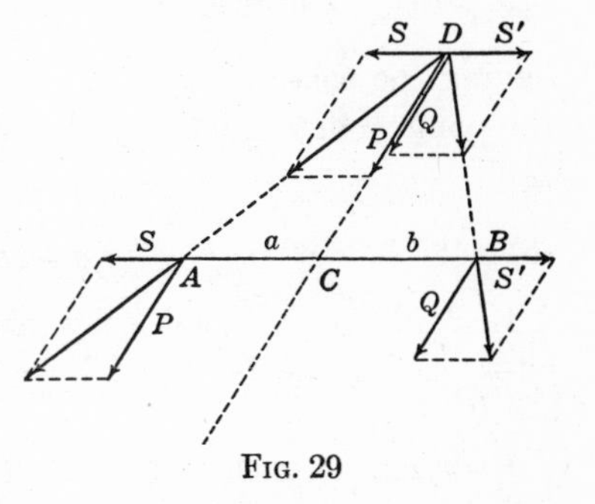

Fig. 29

$$(1) \qquad R = P + Q,$$

parallel to the given forces and having the same sense, its line of action dividing the line AB into two segments, AC and CB. Let the lengths of the segments be denoted as follows: $AC = a$, $CB = b$, $AB = c$, $DC = h$. From similar triangles it is seen that

$$\frac{P}{S} = \frac{h}{a}, \qquad \frac{Q}{S'} = \frac{h}{b}.$$

Hence

$$(2) \qquad aP = bQ.$$

Moreover,

$$(3) \qquad a + b = c.$$

To sum up, then: *The original forces, P and Q, have a resultant determined by the equations* (1), (2), *and* (3).

Example. The familiar gravity balance, in which one arm, a, from which the weight P to be determined is suspended, is short, and the other arm, b, from which the rider Q hangs is long, is a case in point.

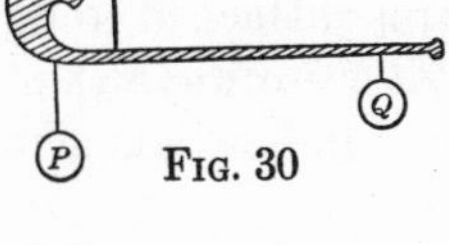

Fig. 30

Opposite Forces. If P and Q are opposite in direction, and unequal ($Q > P$, say), they also have a resultant. Introduce a force E (Equilibriant) parallel to P and Q and having the sense of P, determining it so that Q will be equal and opposite to the resultant of P and E. Then

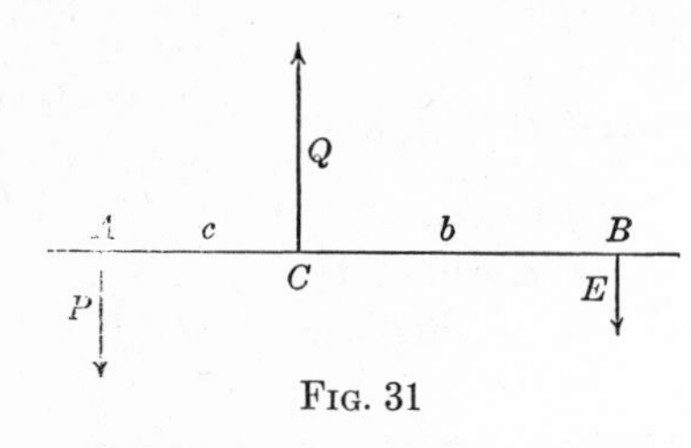

FIG. 31

$$Q = P + E,$$

$$cP = bE,$$

$$a = b + c.$$

Thus P and Q are seen to have a resultant,

$$R = Q - P, \tag{4}$$

having the sense of Q, its line of action cutting AC produced in the point B determined by the equations:

$$aP = bQ, \tag{5}$$

$$a = b + c. \tag{6}$$

If P and Q are equal, they form a *couple* and, as we shall show later, cannot be balanced by a single force; *i.e.* they have no *resultant* (force).

Example. Consider a pair of nut crackers. The forces that act on one of the members are *i*) P, the pull of the hinge; *ii*) Q, the pressure of the nut; and *iii*) the force E the hand exerts, balancing the resultant, R, of P and Q.

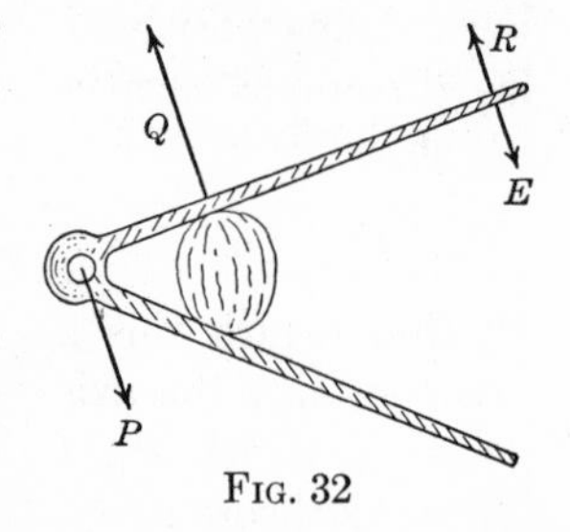

FIG. 32

We have here made use of the so-called *Principle of the Transmissibility of Force*, which says that the effect of a force on a body is the same, no matter at what point in its line it acts. Thus a service truck will tow a mired car as effectively (but no more effectively) when the tow-rope is long, as when it is short, provided that in each case the rope is parallel to the road bed.

Moreover, it is not necessary to think of the point of application as lying in the material body. It might be the centre of a ring. For we can always imagine a rigid weightless truss attached to the body and extending to the desired point. But we always think of a *body*, i.e. *mass*, on which the system of forces in question acts.

EXERCISES

1. A 10 ton truck passes over a bridge that is 450 ft. long. When the truck is one-third of the way over, how much of the load goes to one end of the bridge, and how much to the other end? *Ans.* $6\frac{2}{3}$ tons to the nearer end.

2. A weight of 200 lbs. is to be raised by a lever 6 ft. long, the fulcrum being at one end of the lever, and the weight distant 9 in. from the fulcrum. What force at the other end is needed, if the weight of the lever is negligible?

3. A coolie carries two baskets of pottery by a pole 6 ft. long. If one basket weighs 50 pounds and the other, 70 pounds, how far are the ends of the pole from his shoulder?

2. Analytic Formulation; *n* Forces. Suppose that n parallel forces act. Then two, which are not equal and opposite, can be replaced by their resultant, and this in turn combined with a third one of the given forces, until the number has been reduced to two. These will in general have a resultant, but, in particular, may form a couple or be in equilibrium. Thus the problem could be solved piecemeal in any given case.

An explicit analytic solution can be obtained as follows. Begin with $n = 2$ and denote the forces by P_1 and P_2. Moreover, let P_1 and P_2 be taken as *algebraic* quantities, being positive if they act in one direction; negative, if they act in the opposite direction.

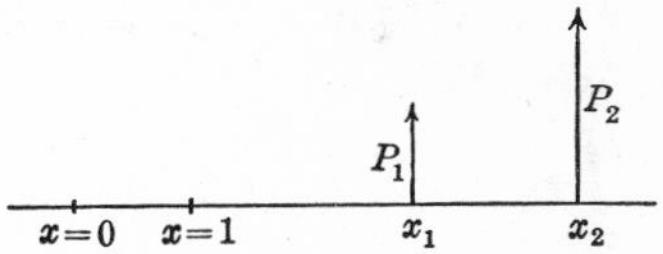

FIG. 33

Next draw a line perpendicular* to the lines of action of P_1 and P_2, and regard this line as the scale of (positive and negative) numbers, like the axis of x. Let x_1, x_2 be the coordinates of the points in which P_1, P_2 cut the line. We proceed to prove the following theorem.

The forces P_1 and P_2 have a resultant,

$$R = P_1 + P_2, \tag{1}$$

provided $P_1 + P_2 \neq 0$. Its line of action has the coordinate:

$$\bar{x} = \frac{P_1 x_1 + P_2 x_2}{P_1 + P_2}. \tag{2}$$

* An oblique direction could be used, but in the absence of any need for such a generalization, the orthogonal direction is more concrete.

Suppose, first, that P_1 and P_2 are both positive. Then, by § 1,

$$R = P_1 + P_2$$

and

$$aP_1 = bP_2,$$

where

$$a = \bar{x} - x_1, \qquad b = x_2 - \bar{x},$$

provided $x_1 < x_2$ (algebraically). Hence

$$(\bar{x} - x_1) P_1 = (x_2 - \bar{x}) P_2,$$

and from this equation, the relation (2) follows at once.

The case that $x_2 < x_1$ is dealt with in a similar manner, as is also the case that P_1 and P_2 are both negative.

Next, suppose P_1 and P_2 have opposite senses, but

$$P_1 + P_2 \neq 0.$$

Let

$$P_1 < 0, \quad P_2 > 0, \quad |P_1| < P_2,$$

where $|x|$ means the *numerical* or *absolute value* of x. Thus $|-3| = 3$, $|3| = 3$. Moreover, let $x_1 < x_2$. Then, by § 1, P_1 and P_2 have a resultant,

$$R = P_1 + P_2,$$

and the coordinate, $\bar{x}$, corresponding to it is obtained as follows:

$$a = \bar{x} - x_1, \qquad b = \bar{x} - x_2,$$

and hence, from § 1, (5):

$$(\bar{x} - x_1)(-P_1) = (\bar{x} - x_2) P_2.$$

or

$$\bar{x} = \frac{P_1 x_1 + P_2 x_2}{P_1 + P_2}.$$

So again we arrive at the same formulas, (1) and (2), as the solution of the problem.

It remains merely to treat the remaining cases in like manner. The final result will always be expressed by formulas (1) and (2). We are now ready to proceed to the general case.

THEOREM 1. *Let n parallel forces, $P_1, \cdots, P_n$, act. They will have a resultant,*

$$R = P_1 + \cdots + P_n,$$

provided this sum $\neq 0$, and its line of action will correspond to $\bar{x}$, where

$$\bar{x} = \frac{P_1 x_1 + \cdots + P_n x_n}{P_1 + \cdots + P_n}. \tag{3}$$

The proof can be given by the method of mathematical induction. The theorem is known to be true for $n = 2$. Suppose it were not true for all values of n. Let m be the smallest value of n for which it is false. We now proceed to deduce a contradiction.

Suppose, then, that $P_1, \cdots, P_m$ is a system of parallel forces, for which the theorem is false, although it is true for $n = 2, 3, \cdots, m - 1$. By hypothesis,

$$P_1 + \cdots + P_m \neq 0.$$

Now, it is possible to find $m - 1$ of the P_r's whose sum is not 0:

$$P_1 + \cdots + P_{m-1} \neq 0,$$

let us say. These $m - 1$ forces have, by hypothesis, a resultant:

$$R' = P_1 + \cdots + P_{m-1},$$

and its $\bar{x}$ has the value:

$$\bar{x}' = \frac{P_1 x_1 + \cdots + P_{m-1} x_{m-1}}{P_1 + \cdots + P_{m-1}},$$

since the theorem holds by hypothesis for all values of $n < m$.

Next, combine this force with P_m. Since

$$R' + P_m = P_1 + \cdots + P_m \neq 0,$$

the two forces have a resultant,

$$R = R' + P_m = P_1 + \cdots + P_m,$$

and its line of action is given by the equation

$$\bar{x} = \frac{R'\bar{x}' + P_m x_m}{R' + P_m} = \frac{P_1 x_1 + \cdots + P_m x_m}{P_1 + \cdots + P_m}.$$

But this result contradicts the assumption that the theorem is false for $n = m$. Hence the theorem is true for all values of n.

Couples. Let

$$(4) \qquad P_1 + \cdots + P_n = 0, \qquad P_n \neq 0.$$

Then

$$P_1 + \cdots + P_{n-1} \neq 0,$$

and the forces $P_1, \cdots, P_{n-1}$ have a resultant,

$$R' = P_1 + \cdots + P_{n-1},$$

whose line of action is given by the equation:

$$\bar{x}' = \frac{P_1 x_1 + \cdots + P_{n-1} x_{n-1}}{P_1 + \cdots + P_{n-1}}.$$

If, in particular, $\bar{x}' = x_n$, this resultant, R', will have the same line of action as P_n; and since

$$R' + P_n = 0, \qquad \text{or} \qquad R' = -P_n,$$

the n forces will be in equilibrium. We then have:

$$x_n = \frac{P_1 x_1 + \cdots + P_{n-1} x_{n-1}}{-P_n},$$

(5)
$$P_1 x_1 + \cdots + P_n x_n = 0.$$

And conversely, if this condition holds, we can retrace our steps and infer equilibrium. But, in general, $x' \neq \bar{x}_n$. Hence

$$x_n \neq \frac{P_1 x_1 + \cdots + P_{n-1} x_{n-1}}{-P_n}.$$

(6)
$$P_1 x_1 + \cdots + P_n x_n \neq 0.$$

We then have a couple. And conversely, if (4) and (6) hold, we can retrace our steps and infer that we have a couple. We have thus proved the following theorem.

THEOREM 2. *The n parallel forces $P_1, \cdots, P_n$ form a couple if, and only if*

$$P_1 + \cdots + P_n = 0.$$

$$P_1 x_1 + \cdots + P_n x_n \neq 0.$$

Equilibrium. The case of equilibrium includes not only the case above considered ($P_n \neq 0$), but also the case in which all n forces vanish. We thus have the following theorem.

THEOREM 3. *The n parallel forces $P_1, \cdots, P_n$ are in equilibrium if, and only if*

$$P_1 + \cdots + P_n = 0,$$

$$P_1 x_1 + \cdots + P_n x_n = 0.$$

3. Centre of Gravity. Let n particles, of masses $m_1, \cdots, m_n$, be fastened to a rigid rod, the weight of which may be neglected, and let them be acted on by the force of gravity. If the rod is supported at a suitable point, G, and is at rest, there will be no tendency to turn in any direction. This point is called the *centre of gravity* of the n particles, and its position is determined by the equation:

(1)
$$\bar{x} = \frac{m_1 x_1 + \cdots + m_n x_n}{m_1 + \cdots + m_n}.$$

If the particles lie anywhere in a plane, being rigidly connected by a truss work of weightless rods, and if we denote the coordinates of m_k by (x_k, y_k), the centre of gravity is defined in a similar manner (see below) and its coordinates, $(\bar{x}, \bar{y})$, are given by Equations (1) and

$$(2) \qquad \bar{y} = \frac{m_1 y_1 + \cdots + m_n y_n}{m_1 + \cdots + m_n}.$$

For, let the plane of the particles be vertical, the axis of x being horizontal. Then the system is acted on by n parallel forces, whose lines of action cut the axis of x at right angles in the points $x_1, \cdots, x_n$, and their resultant is determined in position by Equation (1). On rotating the plane through a right angle and repeating the reasoning, Equation (2) is obtained.

The centre of gravity of any material system, made up of particles and line, surface, and volume distributions, is defined as a point, G, such that, if the parts of the system be rigidly connected by weightless rods, and if G be supported, there will be no tendency of the system to rotate, no matter how it be oriented. We have proved the existence of such a point in the case of n particles lying on a line. For n particles in a plane we have assumed that a centre of gravity exists and lies in the plane, and then we have computed its coordinates. We shall prove later that n particles always have a centre of gravity, and that its coordinates are given by Equations (1), (2), and

$$(3) \qquad \bar{z} = \frac{m_1 z_1 + \cdots + m_n z_n}{m_1 + \cdots + m_n}.$$

In the case of a continuous distribution of matter, like a triangular lamina or a solid hemisphere, the methods of the Calculus lead to the solution. It is the definite integral, defined as the limit of a sum, that is here employed, and Duhamel's Principle is essential in the formulation. In the simpler cases, simple integrals suffice. But even in some of these cases, surface and volume integrals simplify the computation.

The following centres of gravity are given for reference.

a) Solid hemisphere: $\bar{x} = \frac{3}{8}a$.

b) Hemispherical surface: $\bar{x} = \frac{1}{2}a$.

c) Solid cone: $\bar{x} = \frac{3}{4}h$.

d) Conical surface: $\bar{x} = \frac{2}{3}h$.

e) Triangle: Intersection of the medians.

4. Moment of a Force. Let $\mathbf{F}$ be a force lying in a given plane, and let O be a point of the plane. By the *moment* of $\mathbf{F}$ about O is meant the product of the force by the distance from O of its line of action, or hF. A moment may furthermore be defined as an algebraic quantity, being taken as positive when it tends to turn the body in one direction (chosen arbitrarily as the positive direction), and negative in the other case. Finally, if O lies on the line of action of the force, the moment is defined as 0.

Let a force $\mathbf{F}$ act at a point (x, y), and let the components of $\mathbf{F}$ along the axes be denoted by X, Y. Then the moment (taken algebraically) of $\mathbf{F}$ about the origin, O, is:

$$xY - yX. \tag{1}$$

Proof. Let the equation of the line of action of $\mathbf{F}$ be written in Hesse's Normal Form:

$$x \cos \alpha + y \sin \alpha = h.$$

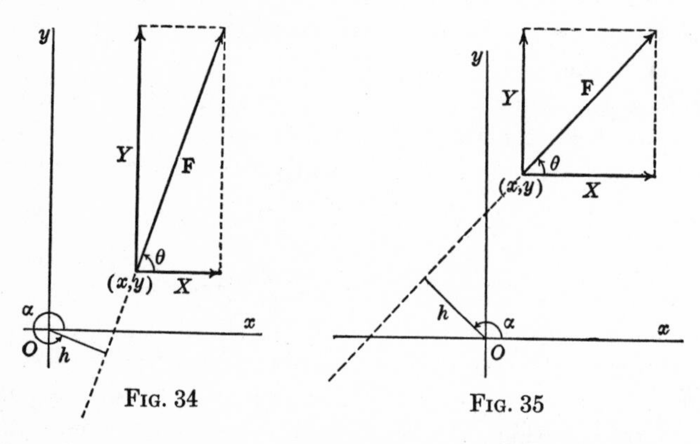

Fig. 34

Fig. 35

Suppose, first, that the moment is positive. Then it will be

$$hF = x(F \cos \alpha) + y(F \sin \alpha).$$

Here, $2\pi - \alpha$ is the complement of θ, Fig. 34:

$$2\pi - \alpha = \frac{\pi}{2} - \theta, \qquad \alpha = \theta - \frac{\pi}{2} + 2\pi.$$

Hence

$$\cos \alpha = \sin \theta, \qquad \sin \alpha = -\cos \theta,$$

and since

$$X = F \cos \theta, \qquad Y = F \sin \theta,$$

the proof is complete.

If, however, the moment is negative, α and θ will be connected by the relation, Fig. 35:

$$\pi - \alpha = \frac{\pi}{2} - \theta, \qquad \alpha = \theta + \frac{\pi}{2}.$$

Hence

$$\cos\alpha = -\sin\theta, \qquad \sin\alpha = \cos\theta.$$

The moment will now be repressed as

$$-hF = x(-F\cos\alpha) + y(-F\sin\alpha),$$

and thus we arrive at the same expression, (1), as before. The same is true for a nil moment. Hence the formula (1) holds in all cases.

From (1) we prove at once that the moment of the resultant of two forces acting at a point is the sum of the moments of the two given forces. Let the latter be $\mathbf{F}_1$, $\mathbf{F}_2$, with the moments

$$xY_1 - yX_1 \qquad \text{and} \qquad xY_2 - yX_2.$$

The components of the resultant force are seen to take the form: $X_1 + X_2$ and $Y_1 + Y_2$, and its moment is

$$x(Y_1 + Y_2) - y(X_1 + X_2).$$

From this expression the truth of the theorem is at once obvious.

Finally, the moment of a force about an arbitrary point, (x_0, y_0), is seen to be:

$$(x - x_0)Y - (y - y_0)X. \tag{2}$$

The physical meaning of the moment of a force about a point is a measure of the *turning effect of the force*. Suppose the body were pivoted at O. Then the tendency to turn about O, due to the force $\mathbf{F}$, is expressed quantitatively by the moment. And a set of forces augment or reduce one another in their combined turning effect according to the magnitude and sense of the sum of the moments of the individual forces. From this point of view a moment is often described in physics and engineering as a *torque*.

5. Couples in a Plane. A couple has already been defined as a system of two equal and opposite parallel forces. A couple cannot be balanced by a single force, but is an independent mechanical entity; the proof is given below.

By the *moment* of a couple, taken numerically, is meant the product of either force by the distance between the lines of action of the forces.

THEOREM. *Two couples having the same moment and sense are equivalent.*

Suppose first that the forces of the one couple are parallel to the forces of the other couple. Then, by proper choice of the axis of x, we can represent the couples as indicated, where

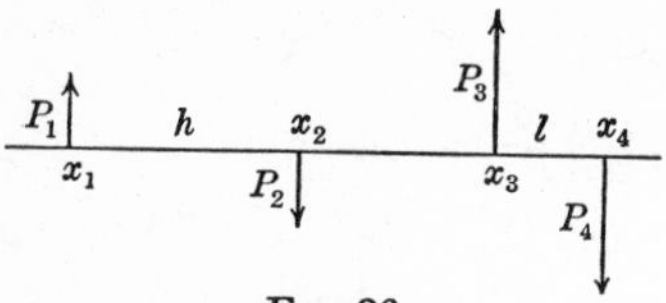

$$P_1 + P_2 = 0, \qquad 0 < P_1;$$
$$P_3 + P_4 = 0, \qquad 0 < P_3;$$
$$x_2 = x_1 + h, \qquad 0 < h;$$
$$x_4 = x_3 + l, \qquad 0 < l;$$

FIG. 36

$$P_1 h = P_3 l.$$

Now consider the system of four forces, P_1, P_2, $-P_3$, $-P_4$. These are in equilibrium. For,

$$P_1 + P_2 - P_3 - P_4 = 0$$

and

$$P_1 x_1 + P_2 x_2 - P_3 x_3 - P_4 x_4 =$$
$$(P_1 + P_2)\, x_2 - P_1 h - (P_3 + P_4)\, x_4 + P_3 l = 0.$$

Hence the first couple is balanced by the negative of the second couple, and thus the theorem is proved for the case that all the forces are parallel.

If the forces of the two couples are oblique to each other, let OA and OB be two lines at right angles to the forces of the first couple and to those of the second couple respectively. Lay off two equal distances, $OA = h$ and $OB = h$, on these lines. Then by the theorem just proved the first couple can be represented as indicated by the forces P and P.* Furthermore, the second couple, reversed in sense, can be represented by the forces Q and Q. Let the lines of action of P at A and Q at B meet in C, and carry these forces forward so that each acts at C. Then the four

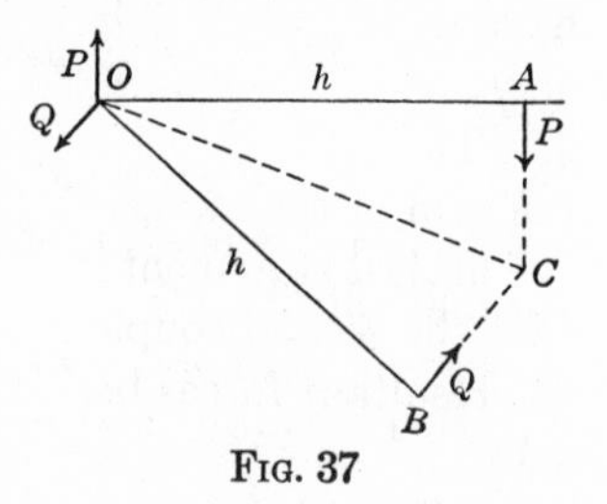

FIG. 37

* It might seem that there are two cases to be considered, for cannot the vectors that represent the forces of the first couple be opposite in sense? True. But then we can begin with the second couple. Its forces will be represented by the Q and Q of the diagram; and the forces of the first couple, reversed in sense, will now appear as P and P.

forces obviously are in equilibrium, for all four are equal in magnitude, since by hypothesis the moments of the two given couples are equal; and the forces make equal angles with the indefinite line OC, in such a manner that the resultant of one pair is equal and opposite to that of the other pair, the lines of action of these resultants coinciding.

To sum up, then, the effect of a couple in a given plane is the same, no matter where its forces act, and no matter how large or small the forces may be, provided only that the *moment* of the couple is preserved both in *magnitude* and in *sense.*

Composition of Couples. From the foregoing it appears that two couples in a plane can be compounded into a single couple, whose moment is the sum of the moments of the constituent couples; all moments being taken algebraically. For, assume a system of Cartesian axes in the plane, and mark the point A : (1, 0). The first couple can be realized by a force P_1 at A parallel to the axis of y (and either positive or negative) and an equal and opposite force $-P_1$ at the origin, acting along the axis of y. The moment of this couple, taken algebraically, is obviously P_1.

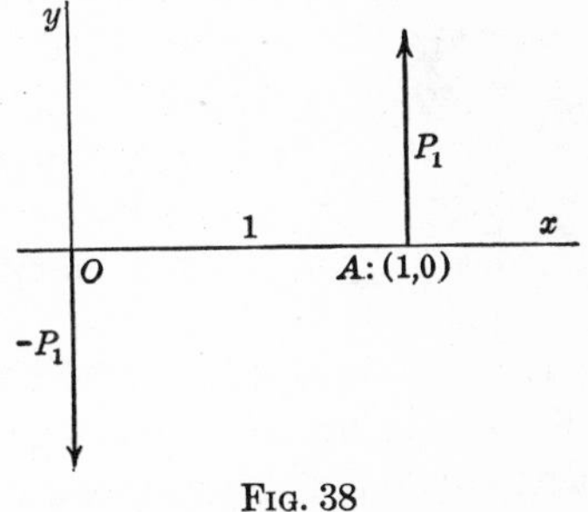

FIG. 38

Dealing with the second couple in a similar manner, we now have as the result two forces, P_1 and P_2, at A parallel to the axis of y; and two equal and opposite forces at O along the axis of y. These forces constitute a resultant couple, whose moment is the sum of the moments of the given couple.

This last statement is at fault in one particular. It may happen that the second couple is equal and opposite to the first, and then the resultant forces both vanish. In order that this case may not cause an exception, we extend the notion of couple to include a *nil couple*: *i.e.* a couple whose forces are both zero, or whose forces lie in the same straight line; and we define its moment to be 0. We are thus led to the following theorem.

THEOREM. *If n couples act in a plane, their combined effect is equivalent to a single couple, whose moment is the sum of the moments of the given couples.*

Remark. The moment of a couple is equal to the sum of the moments of its forces about an arbitrary point of the plane. This is seen directly geometrically from the definition of a moment. In particular, let a point O be chosen at pleasure. The couple can be realized by two forces, one of which passes through O. The moment of the couple is then equal to the moment of the other force about O.

6. Resultant of Forces in a Plane. Equilibrium. Let any forces act in a plane. Then they are equivalent *i*) to a single force, or *ii*) to a single couple; or, finally, *iii*) they are in equilibrium. Let O be an arbitrary point of the plane. Beginning with the force $\mathbf{F}_1$, let us introduce at O two forces equal and opposite to $\mathbf{F}_1$. The two forces checked form a couple, and the remaining force is the original force $\mathbf{F}_1$, transferred to the point O.

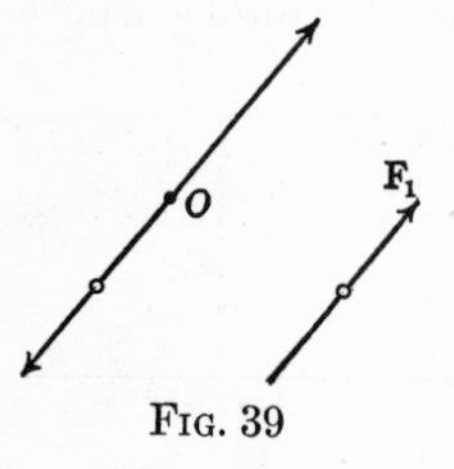

Fig. 39

Proceeding in this manner with each of the remaining forces, $\mathbf{F}_2, \cdots, \mathbf{F}_n$, we arrive at a new system of forces and couples equivalent to the original system of forces and consisting of those n forces, all acting at O, plus n couples. These n forces are equivalent to a single force, $\mathbf{R}$, at O; or are in equilibrium. And the n couples are equivalent to a single couple, or are in equilibrium. In general, the resultant force, $\mathbf{R}$, will not vanish, nor will the resultant couple disappear. The latter can, in particular, be realized as a force equal and opposite to $\mathbf{R}$ and acting at O, and a second force equal to $\mathbf{R}$, but having a different line of action. Thus the resultant of all n forces is here a single force. Incidentally we have shown that a non-vanishing couple can not be balanced by a non-vanishing force; for, the effect of such a force and such a couple is a force equal to the given force, but transferred to a new line of action, parallel to the old line.

It may happen that the resultant force vanishes, but the resultant couple does not. For equilibrium, it is necessary and sufficient that both the resultant force and the resultant couple vanish. This condition can, with the help of the Remark at the close of § 5, be expressed in the following form.

Equilibrium. *A system of n forces in a plane will be in equilibrium if, and only if*

i) they are such as would keep a particle at rest if they all acted at a point; and

ii) the sum of the moments of the forces about a point (one point is enough, and it may be chosen anywhere) of the plane is zero.

Analytically, the condition can be formulated as follows. Let the point about which moments are to be taken, be chosen as the origin, and let the force $\mathbf{F}_r$ act at the point (x_r, y_r). Then

$$i) \qquad \sum_{r=1}^{n} X_r = 0, \qquad \sum_{r=1}^{n} Y_r = 0;$$

$$ii) \qquad \sum_{r=1}^{n} (x_r Y_r - y_r X_r) = 0.$$

Example. A ladder rests against a wall, the coefficient of friction for both ladder and wall being the same, μ. If the ladder is just on the point of slipping when inclined at an angle of 60° with the horizontal, what is the value of μ?

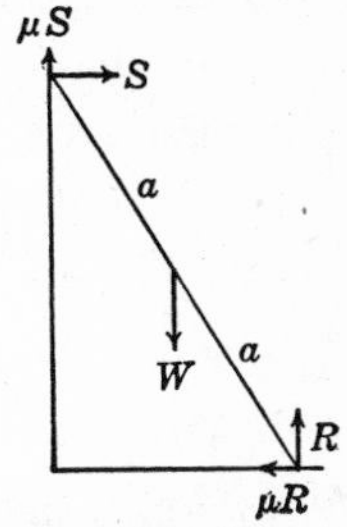

Since all the friction is called into play, the forces are as indicated in the figure, R and S being unknown, and μ also unknown.

Condition *i*) tells us that the sum of the vertical components upward must equal the sum of the vertical components downward, or

FIG. 40

$$R + \mu S = W.$$

Furthermore, the sum of the horizontal components to the right must equal the sum of the horizontal components to the left, or

$$S = \mu R.$$

Finally, the moments about a point, O, of the plane must balance. It is convenient to choose as O a point through which a number of unknown forces pass; for example, one end of the ladder, say the upper end. Thus

$$2a \cos 60^\circ\, R = 2a \sin 60^\circ\, \mu R + a \cos 60^\circ\, W,$$

or

$$R = \sqrt{3}\, \mu R + \tfrac{1}{2} W.$$

Hence

$$R = \frac{W}{2\,(1 - \mu\sqrt{3})}, \qquad S = \frac{\mu W}{2\,(1 - \mu\sqrt{3})}.$$

We can now eliminate R and S. The resulting equation is

$$\mu^2 + 2\sqrt{3}\,\mu = 1.$$

Thus

$$\mu = 2 - \sqrt{3} = 0.27.$$

The other root, being negative, has no physical meaning.

EXERCISES

1. If in the example just discussed the wall is smooth, but the floor is rough, and if $\mu = \frac{1}{6}$, find all positions of equilibrium.

2. If in the example of the text μ could be as great as 1, show that all positions would be positions of equilibrium.

3. A ladder 12 ft. long and weighing 30 lbs. rests at an angle of 60° with the horizontal against a smooth wall, the floor being rough, $\mu = \frac{1}{3}$. A man weighing 160 lbs. goes up the ladder. How far will he get before the ladder slips?

4. In the last question, how rough must the floor be to enable the man to reach the top?

5. Show that a necessary and sufficient condition for equilibrium is that the sum of the moments about each of three points, A, B, and C, not lying in a line, shall vanish for each point separately.

7. Couples in Space. THEOREM I. *A couple may be transferred to a parallel plane without altering its effect, provided merely that its moment and sense are preserved.*

It is sufficient to consider two couples in parallel planes, whose moments are equal and opposite, and to show that their forces are in equilibrium. Construct a cube with two of its faces in the planes of the couples. Then one couple can be represented by the forces marked P and P, and the reversed couple, by the forces Q and Q $(P = Q)$.

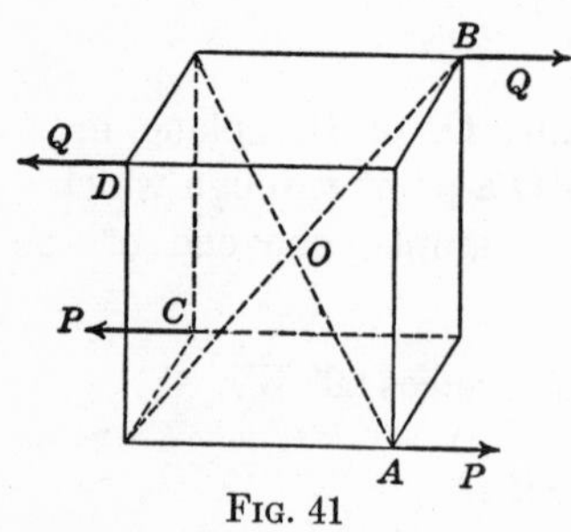

FIG. 41

Consider the resultant of P at A and Q at B. It is a force of $P + Q$ $(= 2P)$ parallel to P and having the same sense, and passing through the centre, O, of the cube. Turn next to P at C and Q at D. The resultant of these forces is obviously equal and opposite to the resultant just considered and having the same line of action. The four forces are, then, in equilibrium. This completes the proof.

Example. In a certain type of auto (Buick 45–6–23) the last bolt in the engine head was so near the cowl that a flat wrench could not be used. The garage man immediately bent a flat wrench through a right angle, applied one end of the wrench to the nut and, passing a screw driver through the opening in the other end, turned the nut. Thus the applied couple was transferred from the horizontal plane through the screw driver to the plane of the nut.

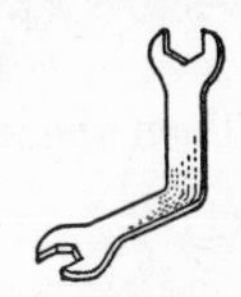

FIG. 42

Vector Representation of Couples. A couple can be represented by a vector as follows. Construct a vector perpendicular to the plane of the couple and of length equal to the moment of the couple. As regards the sense of the vector, either convention is permissible. Let us think of ourselves as standing upright on the plane of the couple and looking down on the plane. If we are on the proper side of the plane, we shall see the couple tending to produce rotation in the clock-wise sense. And now the direction from our feet to our head may be taken as the positive sense of the vector — or, equally well, the opposite direction.

THEOREM II. *The combined effect of two couples is the same as that of a single couple represented by the vector obtained by adding geometrically the two vectors which represent respectively the given couples.*

The theorem has already been proved for the case that the planes of the given couples are parallel or coincident. If they intersect, lay off a line segment of unit length, AB, on their line of intersection, and take the forces of the couples so that they act at A and B perpendicularly to the line AB. Then it is easily seen that the resultant of the two forces at A and the resultant of the two forces at B form a new couple.

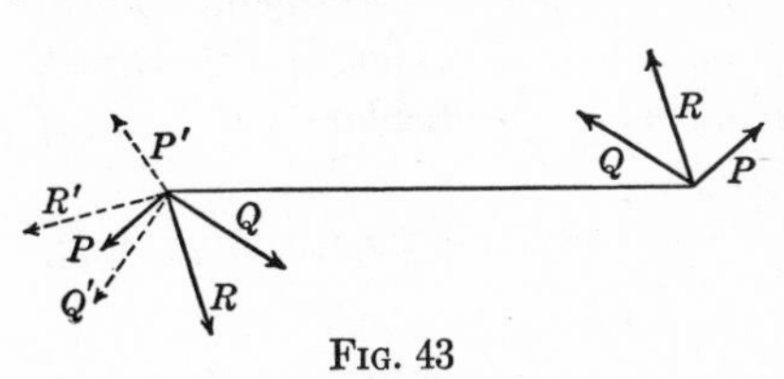

FIG. 43

Finally, the vector representations of these three couples are three vectors perpendicular respectively to the three planes of the couples, equal in length to the forces of the couples, and so oriented as to give the same figure yielded by three of the forces, properly chosen, only turned through 90°.

8. Resultant of Forces in Space. Equilibrium. Let any n forces act on a body in space. Let them be represented by the vectors $\mathbf{F}_1, \cdots, \mathbf{F}_n$. Let O be an arbitrary point of space. Introduce at O two forces that are equal and opposite to the force $\mathbf{F}_k$. Then the n forces $\mathbf{F}_1, \cdots, \mathbf{F}_n$ at O have a resultant:

$$\text{(1)} \qquad \mathbf{R} = \mathbf{F}_1 + \cdots + \mathbf{F}_n,$$

acting at O, or are in equilibrium. And the remaining forces, combined suitably in pairs, yield n couples, $\mathbf{C}_1, \cdots, \mathbf{C}_n$, whose resultant couple, $\mathbf{C}$, is:

$$\text{(2)} \qquad \mathbf{C} = \mathbf{C}_1 + \cdots + \mathbf{C}_n,$$

or, in particular, vanishes; the couples being then in equilibrium.

In general, neither $\mathbf{R}$ nor $\mathbf{C}$ will vanish. Thus the n given forces reduce to a force and a couple. The plane of the resultant couple, $\mathbf{C}$, will in general be oblique to the line of action of the resultant force, and hence the vector $\mathbf{C}$ oblique to the vector $\mathbf{R}$. Let

$$\mathbf{C} = \mathbf{C}_1 + \mathbf{C}_2,$$

where $\mathbf{C}_1$ is collinear with $\mathbf{R}$, and $\mathbf{C}_2$ is perpendicular to $\mathbf{R}$. The couple represented by $\mathbf{C}_2$ can be realized by two forces in a plane containing the resultant force, $\mathbf{R}$; and its forces can be combined with $\mathbf{R}$, thus yielding a single force $\mathbf{R}$, whose line of action, however, has been displaced. This leaves only the couple $\mathbf{C}_1$. We have, therefore, obtained the following theorem.

THEOREM. *Any system of forces in space is in general equivalent to a single force whose line of action is uniquely determined, and to a single couple, whose plane is perpendicular to the line of action of the resultant force.*

In particular, the resultant force may vanish, or the resultant couple may vanish, or both may vanish.

Equilibrium. The given forces are said to be in *equilibrium* if and only if both the resultant force and the resultant couple vanish.

For completeness it is necessary to show that the resultant force, $\mathbf{R}$, together with its line of action, and the resultant couple, $\mathbf{C}_1$, are uniquely determined. For it is conceivable that a different choice, O', of the point O might have led to a different result. Now, the vector **R** is uniquely determined by (1), and so is the same in each case; but **C** depends on the choice of O', and so $\mathbf{C}_1$ might conceivably be different from $\mathbf{C}_1'$, though each would be collinear with **R**. This is, however, not the case. For, reverse **R** in the second case, and also the couple $\mathbf{C}_1'$. Then the reversed force and couple must balance the first force and couple. But this situation leads to a contradiction, as the reader will at once perceive.

9. Moment of a Vector. Couples. Given a force, **F**, acting along a line L, and any point O in space. By the *vector moment of* **F** *with respect to* (or *about*) O is meant the vector product *

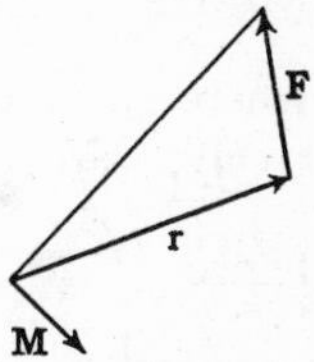

$$\mathbf{M} = \mathbf{r} \times \mathbf{F}, \tag{1}$$

where **r** is a vector drawn from O to a point of L. It is a vector at right angles to the plane of O and L, and its length is numerically equal to the moment of **F** about O in that plane. Its sense depends on whether we are using a right-handed or a left-handed system. Referred to Cartesian axes

FIG. 44

$$\mathbf{M} = \begin{vmatrix} \mathbf{i} & \mathbf{j} & \mathbf{k} \\ x-a & y-b & z-c \\ X & Y & Z \end{vmatrix} = L\,\mathbf{i} + M\,\mathbf{j} + N\,\mathbf{k}, \tag{2}$$

$$\begin{cases} L = (y-b)Z - (z-c)Y \\ M = (z-c)X - (x-a)Z \\ N = (x-a)Y - (y-b)X \end{cases} \tag{3}$$

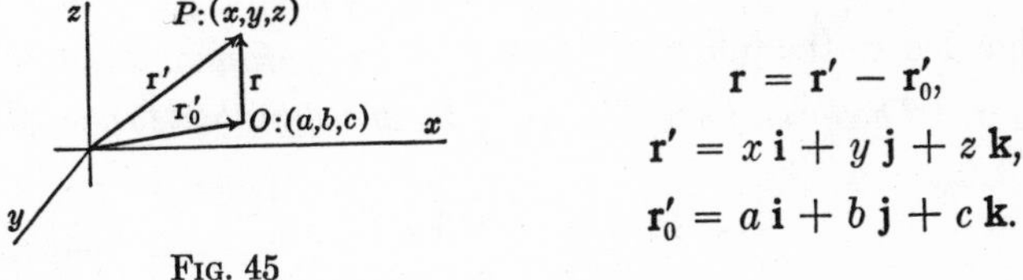

FIG. 45

$$\mathbf{r} = \mathbf{r}' - \mathbf{r}_0',$$
$$\mathbf{r}' = x\,\mathbf{i} + y\,\mathbf{j} + z\,\mathbf{k},$$
$$\mathbf{r}_0' = a\,\mathbf{i} + b\,\mathbf{j} + c\,\mathbf{k}.$$

* The student should read § 3 of Appendix A. This, together with the mere definitions that have gone before, is all of Vector Analysis which he will need for the present.

Vector Representation of a Couple. Let a couple consist of two forces, $\mathbf{F}_1$ and $\mathbf{F}_2$:

$$\mathbf{F}_1 + \mathbf{F}_2 = 0,$$

acting respectively along two lines L_1 and L_2. The vector $\mathbf{C}$ which represents the couple is seen from the definition of the vector product to be:

$$\mathbf{C} = \mathbf{r} \times \mathbf{F}_1, \tag{4}$$

where $\mathbf{r}$ represents any vector drawn from a point of $\mathbf{L}_2$ to a point of $\mathbf{L}_1$. Let O be any point of space. Then the sum of the vector moments of $\mathbf{F}_1$ and $\mathbf{F}_2$ with respect to O yields the vector couple:

$$\mathbf{C} = \mathbf{r}_1 \times \mathbf{F}_1 + \mathbf{r}_2 \times \mathbf{F}_2, \tag{5}$$

where $\mathbf{r}_1$, $\mathbf{r}_2$ are any vectors drawn from O to L_1 and L_2 respectively. For

$$\mathbf{r} = \mathbf{r}_1 - \mathbf{r}_2;$$

hence

$$\mathbf{r} \times \mathbf{F}_1 = \mathbf{r}_1 \times \mathbf{F}_1 - \mathbf{r}_2 \times \mathbf{F}_1 = \mathbf{r}_1 \times \mathbf{F}_1 + \mathbf{r}_2 \times \mathbf{F}_2.$$

10. Vector Representation of Resultant Force and Couple. Resultant Axis. Wrench. Let $\mathbf{F}_1, \cdots, \mathbf{F}_n$ be any system of forces in space. Let P be any point of space, and let equal and opposite forces, $\mathbf{F}_k$ and $-\mathbf{F}_k$, $k = 1, \cdots, n$, be applied at P. Consider the n forces $\mathbf{F}_1, \cdots, \mathbf{F}_n$ which act at P. Their resultant is

$$\mathbf{R} = \mathbf{F}_1 + \cdots + \mathbf{F}_n.$$

The remaining forces yield n couples, consisting each of $\mathbf{F}_k$ at $P_k: (x_k, y_k, z_k)$ and $-\mathbf{F}_k$ at $P: (x, y, z)$. Let $\mathbf{r}_k$, $\mathbf{r}$ be the vectors drawn from the origin of coordinates (chosen arbitrarily) to P_k and P respectively. Then the k-th couple, $\mathbf{C}_k$, is represented by the equation:

$$\mathbf{C}_k = \mathbf{r}_k \times \mathbf{F}_k - \mathbf{r} \times \mathbf{F}_k.$$

We are thus led to the following theorem.

THEOREM. *The given forces* $\mathbf{F}_1, \cdots, \mathbf{F}_n$ *are equivalent to a single force,*

$$\mathbf{R} = \mathbf{F}_1 + \cdots + \mathbf{F}_n, \tag{6}$$

acting at P; *and to a couple,*

$$\mathbf{C} = \sum_{k=1}^{n} \mathbf{r}_k \times \mathbf{F}_k - \mathbf{r} \times \mathbf{R}. \tag{7}$$

Resultant Axis. The resultant axis is the locus of points P, for which $\mathbf{C}$ lies along $\mathbf{R}$; *i.e.* is collinear with $\mathbf{R}$. The condition for this is obviously the vanishing of the vector product:

$$\text{(8)} \qquad \mathbf{R} \times \mathbf{C} = 0, \qquad \mathbf{R} \neq 0.$$

Let

$$\text{(9)} \qquad \sum_{k=1}^{n} \mathbf{r}_k \times \mathbf{F}_k = L\,\mathbf{i} + M\,\mathbf{j} + N\,\mathbf{k},$$

$$\text{(10)} \qquad \mathbf{R} = X\,\mathbf{i} + Y\,\mathbf{j} + Z\,\mathbf{k}.$$

Thus

$$\text{(11)} \quad L = \sum_{k=1}^{n} (y_k Z_k - z_k Y_k), \qquad M = \sum_{k=1}^{n} (z_k X_k - x_k Z_k),$$

$$N = \sum_{k=1}^{n} (x_k Y_k - y_k X_k).$$

Then the condition (8) becomes, by virtue of (7):

$$\mathbf{R} \times (L\,\mathbf{i} + M\,\mathbf{j} + N\,\mathbf{k}) = \mathbf{R} \times (\mathbf{r} \times \mathbf{R}),$$

or:

$$\text{(12)} \quad \begin{vmatrix} \mathbf{i} & \mathbf{j} & \mathbf{k} \\ X & Y & Z \\ yZ - zY & zX - xZ & xY - yX \end{vmatrix} = \begin{vmatrix} \mathbf{i} & \mathbf{j} & \mathbf{k} \\ X & Y & Z \\ L & M & N \end{vmatrix}$$

This is the equation of the resultant axis in vector form. To reduce to ordinary Cartesian form, equate the coefficients of $\mathbf{i}$, $\mathbf{j}$, $\mathbf{k}$ respectively. Thus we find:

$$\text{(13)} \quad \begin{cases} R^2 x - X(xX + yY + zZ) = YN - ZM \\ R^2 y - Y(xX + yY + zZ) = ZL - XN \\ R^2 z - Z(xX + yY + zZ) = XM - YL \end{cases}$$

One of these equations may become illusory through the vanishing of all the coefficients; but *some* two always define intersecting planes, for the rank of the determinant is 2, since $R > 0$; and between the three equations there exists an identical relation.

Let (ξ, η, ζ) be the coordinates of the nearest point of the line to the origin. Then

$$\text{(14)} \qquad \xi X + \eta Y + \zeta Z = 0.$$

Hence

$$\text{(15)} \quad \xi = \frac{YN - ZM}{R^2}, \qquad \eta = \frac{ZL - XN}{R^2}, \qquad \zeta = \frac{XM - YL}{R^2}.$$

Thus we have found one point of the resultant axis, and the direction of the axis is that of **R**. The resultant couple is given by (7), where

$$\mathbf{r} = \xi\,\mathbf{i} + \eta\,\mathbf{j} + \zeta\,\mathbf{k}. \tag{16}$$

Wrench. A *wrench* is defined as two forces, acting at arbitrary points; moreover, neither force shall vanish, and their lines of action shall be skew.

Let the forces be $\mathbf{F}_k$, acting at (x_k, y_k, z_k), $k = 1, 2$. The reader will do well to compute the resultant force, axis, and couple. Suppose, in particular, that $\mathbf{F}_1$ is a unit force along the positive axis of Z, and $\mathbf{F}_2$ is a force of 2, parallel to the axis of y and acting at the point (1, 0, 0).

EXERCISE

Let $\mathbf{F}_1$ and $\mathbf{F}_2$ be two forces, the sum of whose moments about a point O is 0. Show that $\mathbf{F}_1$, $\mathbf{F}_2$, and O lie in a plane.

11. Moment of a Vector about a Line. Let a line, L, and a vector, **F**, be given. Let L' be the line of **F**, and let O, O' be the points of L and L' nearest together. Let **r** be the vector from O to O', and let $|\mathbf{r}| = h$. Let **a** be a unit vector along L. Assume coördinate axes as shown. Then

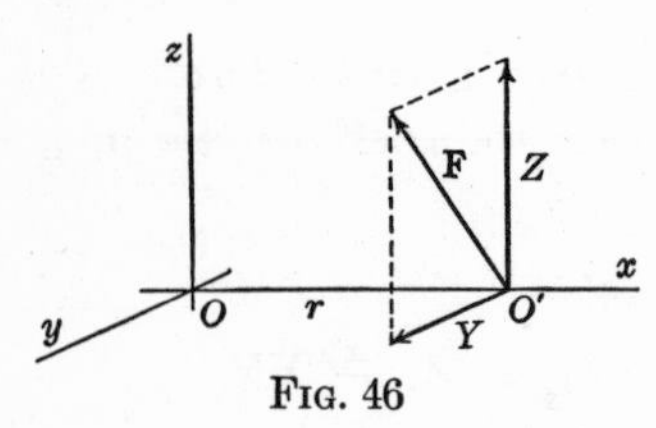

FIG. 46

$$\mathbf{F} = Y\,\mathbf{j} + Z\,\mathbf{k}.$$

By the *moment of* **F** *about* L is meant:

$$\mathbf{M} = hY\,\mathbf{a},$$

where $\mathbf{a} = \mathbf{k}$. The moment **M** can be expressed in invariant form as follows. Since

$$\mathbf{r} = h\,\mathbf{i},$$

we have:

$$\mathbf{r} \times \mathbf{F} = -\,hZ\,\mathbf{j} + hY\,\mathbf{k};$$

$$\mathbf{k} \cdot (\mathbf{r} \times \mathbf{F}) = hY.$$

Hence

$$\mathbf{M} = \{\mathbf{a} \cdot (\mathbf{r} \times \mathbf{F})\}\,\mathbf{a}. \tag{1}$$

More generally, **r** may be any vector drawn from a point of L to a point of L'. We have thus arrived at the following result.

The moment of a vector **F** *about a line L is given by the formula:*

$$\mathbf{M} = \{\mathbf{a} \cdot (\mathbf{r} \times \mathbf{F})\}\, \mathbf{a},$$

where **r** *is any vector drawn from a point of L to a point of the line of F, and* **a** *is a unit vector collinear with L and having the sine attributed to L.*

In particular, the moments of **F** about the three axes are respectively:

$$yZ - zY, \quad zX - xZ, \quad xY - yX.$$

EXERCISE

A force of 12 kgs. acts at the point $(-1, 3, -2)$, and its direction cosines are $(-3, 4, -12)$. Find its moment about the principal diagonal of the unit cube; *i.e.* the line through the origin, making equal angles with the positive axes.

12. Equilibrium. In § 8 we have obtained a necessary and sufficient condition for the equilibrium of n forces, $\mathbf{F}_1, \cdots, \mathbf{F}_n$, in terms of the vanishing of the resultant force and the resultant couple. By means of Equations (6) and (7) of § 10 we can formulate these conditions analytically. The first, namely, $\mathbf{R} = 0$, gives:

$$(1) \qquad \sum_k X_k = 0, \qquad \sum_k Y_k = 0, \qquad \sum_k Z_k = 0;$$

and now the second, namely, $\mathbf{C} = 0$, reduces Equation (7) to the vanishing of the first term on the right, or

$$(2) \qquad \sum_k (y_k Z_k - z_k Y_k) = 0, \qquad \sum_k (z_k X_k - x_k Z_k) = 0,$$

$$\sum_k (x_k Y_k - y_k X_k) = 0.$$

This last condition, which was obtained from the vanishing of a couple, admits two further interpretations in terms of the vanishing of vector moments, namely:

i) The sum of the vector moments of the given forces with respect to an arbitrary point O of space is 0.

ii) The sum of the vector moments of the given forces about an arbitrary line of space is 0.

The condition *ii*) is equivalent to the following:

ii') The sum of the vector moments of the given forces about each of three particular non-complanar lines is 0.

Necessary and Sufficient Conditions. It is important for clearness to analyse these conditions further, as to whether they are necessary or sufficient or both.

Condition *i*), regarded as a *necessary* condition, is broadest when O is taken as *any* point of space. But Condition *ii*) is *sufficient* if it holds for the lines through just one particular point O, the condition (1) being fulfilled.

Condition *ii*), regarded as a *necessary* condition, is broadest when the line is taken as *any* line in space. But as a *sufficient* condition, though true as formulated, it is less general than (1) and Condition *ii'*), which may, therefore, be taken as the broadest formulation of the sufficient condition.

EXERCISES

1. Show that Condition *i*) is sufficient for equilibrium.

2. Show that Condition *ii*) is sufficient for equilibrium.

13. Centre of Gravity of n Particles. Let the n particles $m_1, \cdots, m_n$ be acted on by gravity. Thus n parallel forces arise, and since they have the same sense, they have a resultant not 0.

Let the axis of z be vertical and directed downward. Then the resultant is a force directed downward and of magnitude

$$(1) \qquad R = m_1 g + \cdots + m_n g = g \sum_{k=1}^{n} m_k,$$

the resultant axis being vertical. Furthermore, $X_k = 0$, $Y_k = 0$, $Z_k = m_k g$. Thus, § 10, (11):

$$(2) \quad L = g \sum_{k=1}^{n} m_k y_k, \qquad M = -g \sum_{k=1}^{n} m_k x_k, \qquad N = 0.$$

The nearest point of the resultant axis to the origin has the coordinates given by (15), § 10:

$$(3) \qquad x = \xi = \frac{\Sigma\, m_k x_k}{\Sigma\, m_k}, \qquad y = \eta = \frac{\Sigma\, m_k y_k}{\Sigma\, m_k}, \qquad z = \zeta = 0.$$

If any point of this line is sustained, the system of particles (thought of as rigidly connected) will be supported, and the system will remain at rest. In particular, one point on this line has the coordinates:

$$(4) \qquad \bar{x} = \frac{\Sigma\, m_k x_k}{\Sigma\, m_k}, \qquad \bar{y} = \frac{\Sigma\, m_k y_k}{\Sigma\, m_k}, \qquad \bar{z} = \frac{\Sigma\, m_k z_k}{\Sigma\, m_k}.$$

If, secondly, we allow gravity to act parallel to the axis of x, the resultant axis now becomes parallel to that axis, and the nearest point to the origin is found by advancing the letters cyclically in Equations (3). Again the point whose coordinates are given by (4) lies on this axis. And, similarly, when gravity acts parallel to the axis of y. It seems plausible, then, that if this point be supported, the system will be at rest, no matter in what direction gravity acts. This is, in fact, the case. To prove the statement, let the point P of § 10 be taken as $(\bar{x}, \bar{y}, \bar{z})$. Then $\mathbf{C} = 0$. For

$$0 = \sum_k \begin{vmatrix} \mathbf{i} & \mathbf{j} & \mathbf{k} \\ x_k & y_k & z_k \\ m_k\,\alpha & m_k\,\beta & m_k\,\gamma \end{vmatrix} - \begin{vmatrix} \mathbf{i} & \mathbf{j} & \mathbf{k} \\ \bar{x} & \bar{y} & \bar{z} \\ \Sigma\, m_k\,\alpha & \Sigma\, m_k\,\beta & \Sigma\, m_k\,\gamma \end{vmatrix},$$

since the coefficient of each of the unit vectors **i**, **j**, **k** is seen at once to vanish, no matter what values α, β, γ may have.

Thus the existence of a centre of gravity for n particles is established. It is a point such that, no matter how the system be oriented, the resultant couple due to gravity is nil.

14. Three Forces. If three non-vanishing forces, acting on a rigid body, are in equilibrium, they lie in a plane and either pass through a point or are parallel.

Proof. If two forces in space are in equilibrium, they must be equal and opposite, and have the same line of action; or else each must vanish. Exclude the latter case as trivial. Take vector moments about an arbitrary point, O, in the line of action of one of the forces. Then the vector moment of the other force must vanish by § 12. Thus the second force either vanishes or passes through O; *i.e.* through every point of the line of action of the first force. Finally, they must be equal and opposite.

In the case of three forces, no one of which vanishes, and no two of which have the same line of action, take vector moments about a point O in the line of action of the first force, but of no other force. The sum of the second and third vector moments about O must be zero. Hence the second and third forces lie in a plane through O. They are, therefore, equivalent to a single force — a couple is impossible, since it could not be balanced by the first force. Thus the first force reduces to the resultant, reversed in sense, of the second and third forces, and the theorem is proved.

A Trigonometric Theorem. The following trigonometric theorem is useful in many problems of the equilibrium of a body acted on by three forces. Let a line be drawn from the vertex of a triangle, dividing the opposite side into two segments of lengths m and n, and making angles θ and φ with these sides. Then

$$(m + n) \cot \psi = m \cot \theta - n \cot \varphi,$$

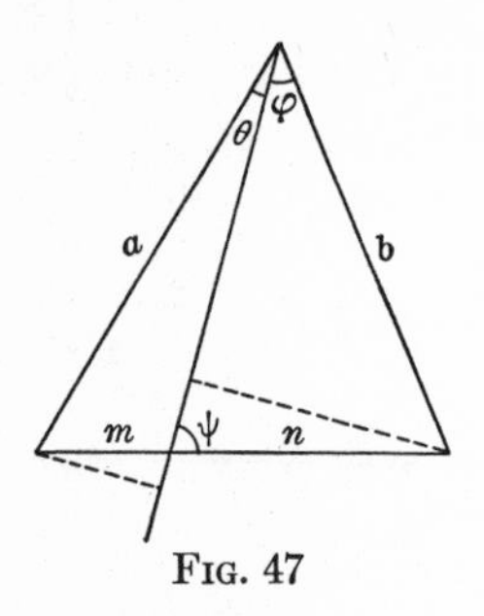

FIG. 47

where ψ is the angle this line makes with the segment n.

The proof is immediate. Project the sides of the triangle on this line, produced:

$$(m + n) \cos \psi = a \cos \theta - b \cos \varphi,$$

and then apply the law of sines:

$$\frac{a}{\sin \psi} = \frac{m}{\sin \theta}, \qquad \frac{b}{\sin \psi} = \frac{n}{\sin \varphi}.$$

Example 1. A uniform rod of length $2a$ is held by a string of length $2l$ attached to one end of the rod and to a peg in a smooth vertical wall, the other end of the rod resting against the wall. Find all the positions of equilibrium.

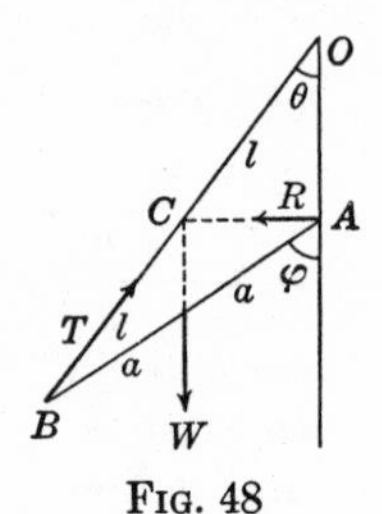

FIG. 48

The three forces of W, T, and R must pass through a point, and this must be the mid-point of the string. Hence, applying the above trigonometric theorem to either of the triangles ABO or ABC, we have:

$$(1) \qquad 2 \tan \theta = \tan \varphi.$$

A second relation is obtained from purely geometrical considerations, namely: *

$$(2) \qquad l \cos \theta = 2a \cos \varphi.$$

It remains to solve these equations. Squaring (1) and reducing, we have:

$$4 \sec^2 \theta = 3 + \sec^2 \varphi,$$

or:

$$\cos^2 \theta = \frac{4 \cos^2 \varphi}{1 + 3 \cos^2 \varphi}.$$

* It would be possible to use the geometric relation

$$l \sin \theta = a \sin \varphi.$$

But the further computation of the solution would be less simple.

Combining with (2), we get

$$\frac{4l^2 \cos^2 \varphi}{1 + 3 \cos^2 \varphi} = 4a^2 \cos^2 \varphi.$$

Since $\cos \varphi$ cannot vanish, it follows that

$$\cos \varphi = \frac{\sqrt{l^2 - a^2}}{a\sqrt{3}}.$$

But a and l are not unrestricted, for $0 < \cos \varphi < 1$. Hence

$$0 < \frac{l^2 - a^2}{3a^2} < 1,$$

and so $$a < l < 2a,$$

or, the string must be longer than the rod, but not twice as long. Furthermore, there are always two positions of equilibrium, in which the rod is vertical, regardless of l and a.

Remark. What the trigonometric theorem has done for us is to eliminate the forces. Without it, we should have been obliged to write down two or three equations involving T and R, and then eliminate these unknowns, with which we have no concern so far as the position of equilibrium goes.

Example 2. Suppose that, in the last example, the wall is rough. Then there is, in addition, an upward force of friction, $F = \mu R$, making four forces in all, — when the rod is just on the point of slipping down the wall. But the forces R and F can be compounded into a single force S making an angle λ with the normal to the wall, and so the problem is reduced to a three-force problem. Applying the trigonometric theorem to the triangle ABC we find:

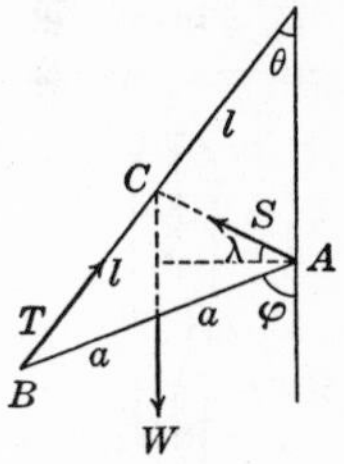

$$2a \cot \varphi = a \cot \theta - a \tan \lambda,$$

$$(3) \qquad 2 \cot \varphi = \cot \theta - \mu.$$

FIG. 49

It is better here to take the geometric relation in the form:

$$(4) \qquad l \sin \theta = a \sin \varphi.$$

From (3) we now have:

$$\csc^2 \theta = 4 \cot^2 \varphi + 4\mu \cot \varphi + \mu^2 + 1.$$

Hence

$$l^2 \sin^2 \theta = \frac{l^2}{4 \cot^2 \varphi + 4\mu \cot \varphi + \mu^2 + 1} = a^2 \sin^2 \varphi.$$

This last equation can be given the form:

$$4\cos^2\varphi + 4\mu\cos\varphi\sin\varphi + (1+\mu^2)\sin^2\varphi = \frac{l^2}{a^2}.$$

This equation, in turn, could be reduced to a quartic in $\sin\varphi$ or $\cos\varphi$; but such procedure would be bad technique. Rather, let

$$2\cos^2\varphi = 1 + \cos 2\varphi, \qquad 2\cos\varphi\sin\varphi = \sin 2\varphi,$$
$$2\sin^2\varphi = 1 - \cos 2\varphi.$$

The equation is thus reduced to an equation of the form:

$$A\cos 2\varphi + B\sin 2\varphi = C,$$

and now can be solved by the method of Chapter I, § 6.

EXERCISE

1. Complete the study of Example 2, *i*) computing A, B, C, and *ii*) finding when the rod is just on the point of slipping up.

Ans. *i*) $(3-\mu^2)\cos 2\varphi + 4\mu\sin 2\varphi = 2\left(\frac{l}{a}\right)^2 - 5 - \mu^2$;

ii) the same equation with the sign of μ reversed.

2. If $a = 1$, $l = 1\frac{3}{4}$, $\mu = 0.1$, find all positions of equilibrium.

EXERCISES ON CHAPTER II *

1. Show that, in a tackle and fall which has n pulleys in each block, the power, P, exerted is $2n+1$ times the tension in the rope.

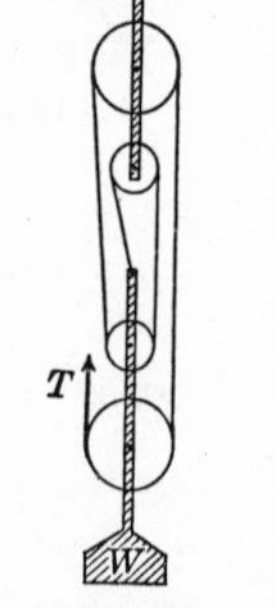

Fig. 50

2. What force applied horizontally to the hub of a wheel (at rest) will just cause the wheel to surmount an obstacle of height h?

3. Two heavy beads of the same weight can slide on a rough horizontal rod. To the bead is attached a string that carries a smooth heavy bead. How far apart can the beads on the rod be placed if they are to remain at rest when released?

4. A gate is raised on its hinges and does not fall back. How rough are the hinges?

* Begin each problem by drawing a figure showing the forces, and the lengths and angles which enter.

5. There has been a light fall of snow on the gate. A cat weighing 5 lbs. walks along the top of the gate, and the gate drops. The disconcerted cat springs off. It is observed from her tracks in the snow that she reached a point 2 ft. from the end of the gate. The distance between the hinges is $2\frac{1}{2}$ ft., and the centre of gravity of the gate is 5 ft. from the vertical line through the hinges. If the gate weighs 100 lbs., what is the value of μ?

6. A rod rests in a smooth hemispherical bowl, one end inside the bowl and the rim of the bowl in contact with the rod. Find the position of equilibrium.

Ans. $\cos\theta = \dfrac{a + \sqrt{32r^2 + a^2}}{8r}$,

where the radius of the bowl is r, the distance of the centre of gravity of the rod from its lower end is a, and the inclination of the rod to the horizon, θ, provided $a < 2r$.

7. A uniform rod rests with one end on a rough floor and the other end on a smooth plane inclined to the horizon at an angle α. Find all positions of equilibrium.

8. The same problem where both floor and plane are rough.

9. A picture hangs on a wall. Formulate the problem of equilibrium when the wall is smooth, and solve it.

10. The same question where the wall is rough.

11. A smooth rod rests with one end against a vertical wall, a peg distant h from the wall supporting the rod. Find the position of equilibrium.

Ans. $\cos\theta = \sqrt[3]{\dfrac{h}{a}}$.

12. The same problem where the wall is rough, the peg being smooth. Find *all* positions of equilibrium.

13. A barrel is lying on its side. A board is laid on the barrel, with its lower end resting on the floor. Find all positions of equilibrium. (Barrel, floor, and board are all rough.)

14. A plank 8 ft. long is stood up against a carpenter's work-bench, which is 2 ft. 8 in. high. The coefficient of friction between either the floor or the bench and the plank is $\frac{1}{4}$. If the plank makes an angle of 15° with the vertical, will it slip down when let go?

15. A smooth uniform rod rests in a test-tube. Find the position of equilibrium.

Ans. The solution is given by the equations:

$$2 \tan \theta = \cot \varphi, \qquad r \sin \varphi + r = 2a \cos \theta.$$

16. A uniform rod 2 ft. long rests with one end on a rough table. To the other end of the rod is attached a string 1 ft. long, made fast to a peg 2 ft. above the table. Find all positions of equilibrium.

Ans. One system of limiting positions is given for $\mu = \frac{1}{2}$ by the equations:

$$\cot \varphi = 2 + 2 \cot \theta, \qquad 2 \cos \theta + \cos \varphi = 2.$$

Solve these equations by means of the Method of Successive Approximations.

17. A water tower is 100 ft. high and 100 ft. in diameter. Find approximately the tension in the plates near the base.

18. Water is gradually poured into a tumbler. Show that the centre of gravity of the glass and the water is lowest when it is in the surface of the water.

19. If one attempts to pull out a two-handled drawer by one handle, what is the condition that the drawer will stick fast?

CHAPTER III

MOTION OF A PARTICLE

1. Rectilinear Motion.* The simplest case of motion of matter under the action of force is that in which a rigid body moves without rotation, each point of the body describing a right line, and the forces that act being resolved along that line. Consider, for example, a train of cars, and neglect the rotation of the wheels and axles. The train is moved by the draw-bar pull of the locomotive, and the motion is resisted by the friction of the tracks and the wind pressure. Obviously, it is only the components of the forces parallel to the tracks that count, and the problem of Dynamics, or Kinetics, as it is more specifically called, is to determine the relation between the forces and the motion; or, if one will: — Given the forces, to find the distance traversed as a function of the time.

A more conventional example, coming nearer to possible experimentation in the laboratory, would be that of a block of iron

* The student must not feel obliged to finish this chapter before going on. What is needed is a thorough drill in the treatment of the early problems *by the present methods*, for these are the general methods of Mechanics, to inculcate which is a prime object of this book. Elementary text-books in Physics sometimes write down three equations:

$$s = \tfrac{1}{2}at^2, \qquad v = at, \qquad v^2 = 2as,$$

and give an unconscionable number of problems to be solved by this device. The pedagogy of this procedure is totally wrong, since it replaces ideas by a rule of thumb, and even this rule is badly chosen, since it disguises, instead of revealing, the mechanical intuition. Now, a *feeling* for Mechanics is the great object to be obtained, and the habits of thought which promote such intuition are, fortunately, cultivated by just the same mathematical treatment which applies in the more advanced parts of Mechanics. It is a happy circumstance that here there is no conflict, but the closest union, between the physics of the subject and the mathematical analysis.

A thorough study of §§ 1–12 through working each problem by the present general methods is most important. Moreover, § 22 should here be included with, of course, the definition of vector acceleration given in § 16 and the statement of Newton's Second Law in § 17. The student should then turn to Chapter IV, the most revealing chapter in the whole elementary part of the book, and study it in all detail. The remaining sections of the present chapter should be read casually at an early stage, so as not to impede progress. Ultimately, they are important; but they are most useful when the student comes to recognize their importance through his experience gathered from the later work above referred to.

placed on a table and drawn along by cords, so applied that the block does not rotate and that each point of it describes a right line — with varying velocity.

It is clear that a block of platinum having the same mass, *i.e.* containing the same amount of matter, if acted on by the same forces, would move just like the block of iron, if the two were started side by side from rest or with the same initial velocities. We can conceive physical substances of still greater density, and the same would be true. On compressing the given amount of matter into smaller and ever smaller volume, we are led to the idea of a *particle*, or *material point*, *i.e.* a geometrical *point*, to which the property of *mass* is attached. This conception has the advantage that such a particle would move exactly as the actual body does if acted on by the same forces; but we need say nothing about rotation, since this idea does not enter when we consider only particles. Moreover, there is no doubt about where the forces are applied — they must be applied at the one point, the particle.

2. Newton's Laws of Motion. Sir Isaac Newton (1642–1727), who was one of the chief founders of the Calculus, stated three laws governing the motion of a body.

First Law. *A body at rest remains at rest and a body in motion moves in a straight line with unchanging velocity, unless some external force acts on it.*

Second Law. *The rate of change of the momentum of a body is proportional to the resultant external force that acts on the body.*

Third Law. *Action and reaction are equal and opposite.*

The meaning of the First Law is clear enough, if we restrict ourselves for the present to bodies and particles as described and moving in § 1.* The Third Law, too, is self-explanatory. Consider, for example, two particles of unequal mass, connected by a spring, the mass of which is negligible. Then the pull (or push) of the spring on the one particle is equal and opposite to its pull (or push) on the other particle.

The Second Law is expressed in terms of *momentum*, and the momentum of a particle is defined as the product of its mass by

* We might consider, furthermore, such material distributions as *laminae, i.e.* material surfaces; and also *wires*, or material curves. Finally, rigid combinations of all these bodies.

its velocity, or mv. Here, v is not an essentially positive quantity — the mere speed. We must think of the position of the body as described by a suitable coordinate, s. The latter may be the distance actually traversed by the particle; or we may think of the path of the particle as the axis of x, and s as the coordinate of the particle. The velocity, v, will then be defined as ds/dt:

$$v = \frac{ds}{dt}, \tag{1}$$

and is positive when s is increasing; negative, when s is decreasing.

The Second Law can now be stated in the form:

$$\frac{d(mv)}{dt} \propto f. \tag{2}$$

Here, f denotes the resultant force, and is positive when it tends to increase s; negative, when it tends to decrease s.

Ordinarily, m is constant — always, in the case of the bodies cited above — and so

$$\frac{d(mv)}{dt} = m\frac{dv}{dt}.$$

The quantity dv/dt is defined as the *acceleration*, and is often represented by α:

$$\alpha = \frac{dv}{dt}. \tag{3}$$

It is positive when v is increasing, negative when v is decreasing.

Newton's Second Law can now be stated in the form: *The mass times the acceleration is proportional to the force:*

$$m\alpha \propto f. \tag{4}$$

From the proportion we now pass to an equation:

$$m\alpha = \lambda f, \tag{5}$$

where λ is a physical constant. The value of λ depends on the units used. If these are the English units, the *pound* being the unit of mass, the *foot* the unit of length, the *second* the unit of time, and the *pound* the gravitational unit of force, then λ has the value 32 (or, more precisely, 32.2), and Newton's Second Law of Motion becomes here:

$$m\frac{dv}{dt} = 32f. \tag{A_1}$$

In the decimal system, the *gramme* being the unit of mass, the *centimetre* the unit of length, the *second* the unit of time, and the *gramme* the gravitational unit of force, $\lambda = 981$, and Newton's Second Law of Motion becomes here:

$$\text{A}_2) \qquad m\frac{dv}{dt} = 981 f.$$

In § 3 we shall discuss the absolute units. In particular, the units of mass, length, and time having been chosen arbitrarily, as in Physics, the so-called "absolute unit of force" is that unit which makes $\lambda = 1$ in Newton's Equation, so that here:

$$\text{A}_3) \qquad m\frac{dv}{dt} = f.$$

Three Forms for the Acceleration. The acceleration is defined as dv/dt, and since $v = ds/dt$, we have:

$$(6) \qquad \alpha = \frac{d^2 s}{dt^2}.$$

A third form is obtained by starting with the equation:

$$(7) \qquad \frac{dv}{dt} = \frac{ds}{dt}\frac{dv}{ds}$$

and then replacing ds/dt by its value, v. Thus

$$(8) \qquad \alpha = v\frac{dv}{ds}.$$

These three forms for the acceleration:

$$(9) \qquad \alpha = \frac{dv}{dt} = \frac{d^2 s}{dt^2} = v\frac{dv}{ds},$$

connect the three letters s, t, v in pairs in all possible ways. Which form it is better to use in a given case, will become clear from practice in solving problems.

Example 1. A freight train weighing 200 tons is drawn by a locomotive that exerts a draw-bar pull of 9 tons. 5 tons of this force are expended in overcoming frictional resistances. How much speed will the train have acquired at the end of a minute, if it starts from rest?

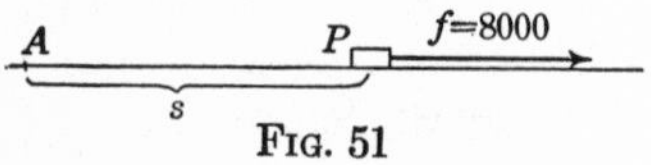

A P f=8000 s

Fig. 51

Here we have

$$m = 200 \times 2000 = 400{,}000 \text{ lbs.},$$

$$f = 9 \times 2000 - 5 \times 2000 = 8000 \text{ lbs.*}$$

and hence Equation A_1) becomes:

$$400{,}000 \frac{dv}{dt} = 32 \times 8000,$$

or

$$\frac{dv}{dt} = \frac{16}{25}.$$

Integrating with respect to t, we find:

$$v = \tfrac{16}{25}t + C.$$

Since $v = 0$ when $t = 0$, we must have $C = 0$, and hence

$$v = \tfrac{16}{25}t.$$

At the end of a minute, $t = 60$, and so

$$v = \tfrac{16}{25} \times 60 = 38.4 \text{ ft. per sec.}$$

To reduce feet per second to miles per hour it is convenient to notice that 30 miles an hour is equivalent to 44 ft. a second, as the student can readily verify; or roughly, 2 miles an hour corresponds to 3 ft. a second. Hence the speed in the present case is about two-thirds of 38.4, or 26 miles an hour.

Example 2. A stone is sent gliding over the ice with an initial velocity of 30 ft. a sec. If the coefficient of friction between the stone and the ice is $\frac{1}{10}$, how far will the stone go?

Here, the only force that we take account of is the retarding force of friction, and this amounts to one-tenth of a pound of force for every pound of mass there is in the stone. Hence, if there are m pounds of mass in the stone the force will be $\frac{1}{10}m$ lbs.,† and since it tends to decrease s, it is to be taken as negative:

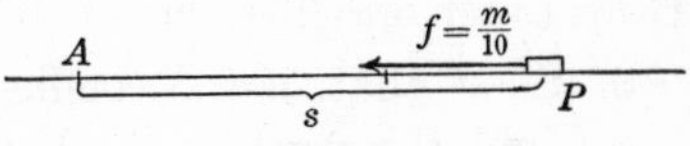

FIG. 52

$$m\alpha = 32\left(-\frac{m}{10}\right),$$

$$\alpha = -\tfrac{16}{5}.$$

* The student must distinguish carefully between the two meanings of the word *pound*, namely (*a*) a *mass*, and (*b*) a *force* — two totally different physical objects. Thus a pound of lead is a certain quantity of *matter*. If it is hung up by a string, the tension in the string is a pound of *force*.

† The student should notice that m is neither a mass nor a force, but a *number*, like all the other letters of Algebra, the Calculus, and Physics.

Now what we want is a relation between v and s, for the question is: How *far* ($s = ?$), when the stone *stops* ($v = 0$)? So we use the value (8) of α and thus obtain the equation:

$$v\frac{dv}{ds} = -\frac{16}{5},$$

or

$$v\,dv = -\tfrac{16}{5}ds.$$

Hence

$$\frac{v^2}{2} = -\frac{16}{5}s + C.$$

To determine C we have the data that, when $s = 0$, $v = 30$. Since in particular the equation must hold for these values,

$$\frac{30^2}{2} = 0 + C, \qquad C = 450,$$

and so

$$v^2 = 900 - \tfrac{32}{5}s.$$

When the stone stops, $v = 0$, and we have

$$0 = 900 - \tfrac{32}{5}s, \qquad s = 141 \text{ ft.}$$

EXERCISES *

1. If the train of Example 1 was moving at the rate of 4 m. an hour when we began to take notice, how fast would it be moving half a minute later? Give a complete solution, beginning with drawing the figure. *Ans.* About 17 m. an h.

2. A small boy sees a slide on the ice ahead, and runs for it. He reaches it with a speed of 8 miles an hour and slides 15 feet. How rough are his shoes? *Ans.* $\mu = .15$.

3. Show that, if the coefficient of friction between a sprinter's shoes and the track is $\frac{1}{12}$, his best possible record in a hundred-yard dash cannot be less than 15 seconds.

4. An electric car weighing 12 tons gets up a speed of 15 miles an hour in 10 seconds. Find the average force that acts on it,

* It is important that the student should work these exercises by the method set forth in the text, beginning each time by drawing a figure and marking (*i*) the *force*, by means of a directed right line, or vector, drawn preferably in red ink; and (*ii*) the *coordinate* used, as s or x, etc. He should not try to adapt such formulas of Elementary Physics as

$$v = at, \qquad s = \tfrac{1}{2}at^2, \qquad v^2 = 2as$$

to present purposes. For, the object of these simple exercises is to prepare the way for applications in which the force is not constant, and here the formulas just cited do not hold.

i.e. the constant force which would produce the same velocity in the same time.

5. In the preceding problem, assume that the given speed is acquired after running 200 feet. Find the time required and the average force.

6. A train weighing 500 tons and running at the rate of 30 miles an hour is brought to rest by the brakes after running 600 feet. While it is being stopped it passes over a bridge. Find the force with which the bridge pulls on its anchorage. *Ans.* 25.2 tons.

7. An electric car is starting on an icy track. The wheels skid and it takes the car 15 seconds to get up a speed of two miles an hour. Compute the coefficient of friction between the wheels and the track.

3. Absolute Units of Force. The units in terms of which we measure mass, space, time, and force are arbitrary, as was pointed out in § 2. If we change one of them, we thereby change the value of λ in Newton's Second Law. Consequently, by changing the unit of force properly, the units of mass, space, and time being held fast, we can make $\lambda = 1$. Hence the definition above given:

DEFINITION. The absolute unit of force is that unit which makes $\lambda = 1$ in Newton's Second Law of Motion: *

$$m\alpha = f. \tag{1}$$

In order to determine experimentally the absolute unit of force, we may allow a body to fall freely and observe how far it goes in a known time. It is a physical law that the force with which gravity attracts any body is proportional to the mass of that body. Let the number g be the number of absolute units of force with

* We have already met a precisely similar question twice in the Calculus. In differentiating the function $\sin x$ we obtain the formula

$$D_x \sin x = \cos x$$

only when we measure angles in radians. Otherwise the formula reads:

$$D_x \sin x = \lambda \cos x.$$

In particular, if the unit is a degree, $\lambda = \pi/180$. We may, therefore, define a radian as follows: The absolute unit of angle (the radian) is that unit which makes $\lambda = 1$ in the above equation.

Again, in differentiating the logarithm, we found;

$$D_x \log_a x = (\log_a e)\,\frac{1}{x}.$$

This multiplier reduces to unity when we take $a = e$. Hence the definition: The absolute (natural) base of logarithms is that base which makes the multiplier $\log_a e$ in the above equation equal to unity.

which gravity attracts the unit of mass. Then the force, measured in absolute units, with which gravity attracts a body of m units of mass will be mg. Newton's Second Law A_3) gives for this case:

$$m\frac{dv}{dt} = mg, \qquad \text{hence} \qquad \frac{dv}{dt} = g;$$

$$v = gt + C, \qquad C = 0;$$

$$v = \frac{ds}{dt} = gt,$$

$$s = \tfrac{1}{2}gt^2 + K, \qquad K = 0,$$

and we have the law for freely falling bodies deduced directly from Newton's Second Law of Motion, the hypothesis being merely that the force of gravity is constant. Substituting in the last equation the observed values $s = S$, $t = T$, we get:

$$g = \frac{2S}{T^2}.$$

If we use English units for mass, space, and time, g has, to two significant figures, the value 32, *i.e.* the absolute unit of force in this system, a *poundal,* is equal nearly to *half an ounce.* If we use c.g.s. units, g ranges from 978 to 983 at different parts of the earth, and has in Cambridge the value 980. The absolute unit of force in this system is called the *dyne.*

Since g is equal to the acceleration with which a body falls freely under the attraction of gravity, g is called the *acceleration of gravity.* But this is not our definition of g; it is a theorem about g that follows from Newton's Second Law of Motion.

The student can now readily prove the following theorem, which is often taken as the definition of the absolute unit of force in elementary physics: The absolute unit of force is that force which, acting on the unit of mass for the unit of time, generates the unit of velocity.

Incidentally we have obtained two of the equations for a freely falling body:

$$v = gt, \qquad s = \tfrac{1}{2}gt^2.$$

The third is found by setting $\alpha = v\,dv/ds$ and integrating:

$$v\frac{dv}{ds} = g,$$

$$\tfrac{1}{2}v^2 = gs + C; \qquad 0 = 0 + C,$$

$$v^2 = 2gs.$$

Example. A body is projected down a rough inclined plane with an initial velocity of v_0 feet per second. Determine the motion completely.

The forces which act are: the component of gravity, $mg \sin \gamma$ absolute units, down the plane, and the force of friction, $\mu R = \mu mg \cos \gamma$ up the plane. Hence

$$m\alpha = mg \sin \gamma - \mu mg \cos \gamma,$$

$$\frac{dv}{dt} = g \sin \gamma - \mu g \cos \gamma.$$

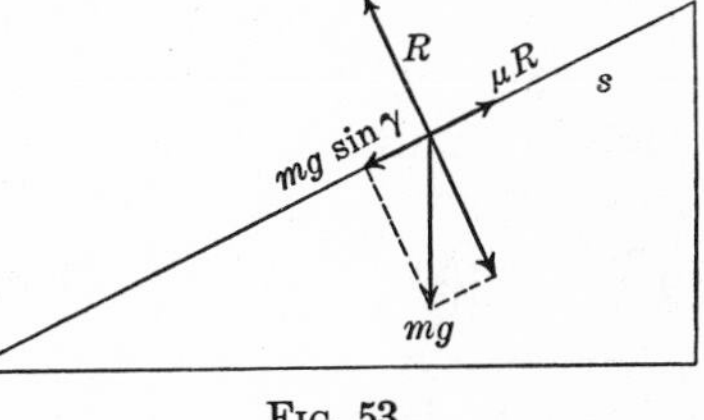

FIG. 53

Integrating this equation, we get

$$v = g (\sin \gamma - \mu \cos \gamma) t + C,$$

$$v_0 = 0 + C,$$

A) $$v = \frac{ds}{dt} = g (\sin \gamma - \mu \cos \gamma) t + v_0.$$

A second integration gives

B) $$s = \tfrac{1}{2} g (\sin \gamma - \mu \cos \gamma) t^2 + v_0 t,$$

the constant of integration here being 0.

To find v in terms of s we may eliminate t between A) and B). Or we can begin by using formula (8), § 2, for the acceleration:

$$v \frac{dv}{ds} = g (\sin \gamma - \mu \cos \gamma),$$

$$\tfrac{1}{2} v^2 = g (\sin \gamma - \mu \cos \gamma) s + K,$$

$$\tfrac{1}{2} v_0^2 = 0 + K,$$

$$v^2 = 2g (\sin \gamma - \mu \cos \gamma) s + v_0^2.$$

EXERCISES

1. If, in the example discussed in the text, the body is projected up the plane, find how far it will go up.

2. Determine the time it takes the body in Question 1 to reach the highest point.

3. Obtain the usual formulas for the motion of a body projected vertically:

$$v^2 = 2gs + v_0^2 \quad \text{or} \quad = - 2gs + v_0^2;$$

$$v = gt + v_0 \quad \text{or} \quad = - gt + v_0;$$

$$s = \tfrac{1}{2} gt^2 + v_0 t \quad \text{or} \quad = - \tfrac{1}{2} gt^2 + v_0 t.$$

4. On the surface of the moon a pound weighs only one-sixth as much as on the surface of the earth. If a mouse can jump up 1 foot on the surface of the earth, how high could she jump on the surface of the moon? Compare the time she is in the air in the two cases.

5. A block of iron weighing 100 pounds rests on a smooth table. A cord, attached to the iron, runs over a smooth pulley at the edge of the table and carries a weight of 15 pounds, which hangs vertically. The system is released with the iron 10 feet from the pulley. How long will it be before the iron reaches the pulley, and how fast will it be moving?

Ans. 2.19 sec.; 9.1 ft. a sec.

6. Solve the same problem on the assumption that the table is rough, $\mu = \frac{1}{20}$, and that the pulley exerts a constant retarding force of 4 ounces.

7. Regarding the big locomotive exhibited at the World's Fair in 1905 by the Baltimore and Ohio Railroad the *Scientific American* said: "Previous to sending the engine to St. Louis, the engine was tested at Schenectady, where she took a 63-car train weighing 3150 tons up a one-per-cent. grade."

Find how long it would take the engine to develop a speed of 15 m. per h. in the same train on the level, starting from rest, the draw-bar pull being assumed to be the same as on the grade.

8. If Sir Isaac Newton registered 170 pounds on a spring balance in an elevator at rest, and if, when the elevator was moving, he weighed only 169 pounds, what inference would he draw about the motion of the elevator?

9. What does a man whose weight is 180 pounds weigh in an elevator that is descending with an acceleration of 2 feet per second per second?

4. Elastic Strings. When an elastic string is stretched by a moderate amount, the tension, T, in the string is proportional to the stretching, *i.e.* to the difference, s, between the stretched and the unstretched length of the string:

$$(1) \qquad T \propto s, \qquad \text{or} \qquad T = ks,$$

where k is a physical constant, whose value depends both on the particular string and on the units used.

Suppose, for example, that a string is stretched 6 in. by a force of 12 lbs.; to determine k. If we measure the force in gravitational units, *i.e.* pounds, then

$$T = 12 \qquad \text{when} \qquad s = \tfrac{1}{2}.$$

Hence, substituting these values in equation (1), we have:

$$12 = k\tfrac{1}{2}, \qquad \text{or} \qquad k = 24,$$

$$(2) \qquad T = 24s.$$

If we had chosen to measure the force in absolute units, *i.e.* poundals, then, since it takes (nearly) 32 of these units to make a pound, the given force of 12 pounds would be expressed as (nearly) 12×32, or precisely $12g$, poundals. Hence, substituting the present value of the force in (1), which, to avoid confusion, we will now write in the form:

$$T' = k's,$$

we have:

$$12g = k'\tfrac{1}{2} \qquad \text{or} \qquad k' = 24g,$$

$$(3) \qquad T' = 24gs.$$

When the string is stretched 1 in., $s = \frac{1}{12}$, and the tension as given by (2) is $T = 2$, *i.e.* 2 pounds. Formula (3), on the other hand, gives $2g$, or 64 (nearly) as the value of the tension, expressed in terms of poundals, and this is right; for it takes 64 half-ounces to make 2 pounds, and so we should have $T' = 2g$.*

The law of strings stated above is familiar to the student in the form of *Hooke's Law:*

$$T = E\frac{l' - l}{l},$$

where l is the natural, or unstretched, length of the string, and l', the stretched length; the coefficient E being Young's Modulus. For a given string, $E/l = k$ is constant, and $l' - l = s$ is variable.

* It is easy to check an answer in any numerical case. The student has only to ask himself the question: "Have I expressed my force in pounds, or have I expressed it in terms of half-ounces?" Just as five dollars is expressed by the number 5 when we use the dollar as the unit, but by the number 500 when we use the cent, so, generally, the *smaller* the unit, the *larger* the number which expresses a given quantity.

EXERCISES

1. An elastic string is stretched 2 in. by a force of 3 lbs. Find the tension (*a*) in pounds; (*b*) in poundals, when it is stretched s ft.

Ans. (*a*) $T = 18s$; (*b*) $T = 18gs$.

2. When the string of Question 1 is stretched 4 in., what is the tension (*a*) in terms of gravitational units; (*b*) in terms of absolute units?

Ans. (*a*) 6 pounds; (*b*) 192 poundals.

3. An elastic string is stretched 1 cm. by a force of 100 grs. Find the tension (*a*) in grs.; (*b*) in dynes, when it is stretched s cm.

Ans. (*a*) $100s$; (*b*) $98{,}000s$.

4. One end of an elastic string 3 ft. long is fastened to a peg at A, and a 2-pound weight is attached to the other end. The weight is gradually lowered till it is just supported by the string, and it is found that the length of the string has thus been doubled. Find the tension in the string when it is stretched s ft.

Ans. $\frac{2}{3}s$ lbs.; $\frac{64}{3}s$ poundals.

5. A Problem of Motion. One end of the string considered in the text of § 4 is fastened to a peg at a point O of a smooth horizontal table; a weight of 3 lbs. is attached to the other end of the string and released from rest on the table with the string stretched one foot. How fast will the weight be moving when the string becomes slack?

The weight evidently describes a straight line from the starting point, A, toward the peg O, and we wish to know its velocity when it has reached a point B, one foot from A.

The solution is based on Newton's Second Law of Motion. It is convenient here to take as the coordinate, not the distance AP that the particle has travelled at any instant, but its distance s from B. The force which acts is the tension of the string; measured in absolute units it is $24gs$. Since it tends to decrease s, it is negative. Hence Newton's Law becomes:

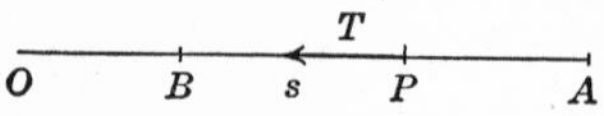

FIG. 54

$$3\frac{d^2 s}{dt^2} = -\ 24gs. \tag{1}$$

To integrate this equation, replace $\frac{d^2 s}{dt^2}$ by its value $v\frac{dv}{ds}$:

$$v\frac{dv}{ds} = -\ 8gs. \tag{2}$$

Hence $$v\,dv = -\,8gs\,ds,$$

$$\int v\,dv = -\,8g\int s\,ds,$$

(3) $$\frac{v^2}{2} = -\,4gs^2 + C.$$

To determine C, observe that initially, *i.e.* when the particle was released at A, $v = 0$ and $s = 1$. Hence

$$0 = -\,4g + C, \qquad C = 4g,$$

and (3) becomes

(4) $$v^2 = 8g\,(1 - s^2).$$

We have now determined the velocity of the particle at an arbitrary point of its path, and thus are in a position to find its velocity at the one point specified in the question proposed, namely, at B. Here, $s = 0$, and

$$v^2|_{s=0} = 8g = 8 \times 32, \qquad v|_{s=0} = 16 \text{ (ft. per sec.)}$$

EXERCISES *

1. The weight in the problem just discussed is projected from B along the table in the direction of OB produced with a velocity of 8 ft. per sec. Find how far it will go before it begins to return.

Ans. Newton's equation is the same as before, and the integral, (3), is the same; but initially $s = 0$ and $v = 8$. Hence $C = 32$, and the answer is 6 inches.

2. If, in the example worked in the text, the table is rough and the coefficient of friction, μ, has the value $\frac{1}{4}$, how fast will the body be moving when it reaches B?

Ans. Newton's equation now becomes:

$$3\,\frac{d^2 s}{dt^2} = -\,24gs + \tfrac{1}{4}\cdot 3g,$$

and the answer is: $4\sqrt{15} = 15.49$ ft. per sec.

3. Solve the problem of Question 1, for a rough table, $\mu = \frac{1}{4}$.

Ans. The required distance is the positive root of the equation $16s^2 + s - 4 = 0$, or $s = .4698$ ft., or about $5\frac{5}{8}$ in.

* In the following exercises and examples, it will be convenient to take the value of g as exactly 32 when English units are used. Begin each exercise by drawing a figure showing the coordinate used, and mark the forces in red ink.

4. Find where the weight in Question 2 will come to rest if the string, after becoming slack, does not get in the way.

5. The 2 lb. weight of Question 4, § 4, is released from rest at a point B directly under the peg A and at a distance of 3 ft. from A; the string thus being taut, but not stretched. Find how far it will fall before it begins to rise. *Ans.* 6 ft.

6. If, in the last question, the weight is dropped from the peg at A, find how far it descends before it begins to rise.

Ans. To a distance of $6 + 3\sqrt{3} = 11.196$ ft. below A.

7. If the weight in the last two questions is carried to a point 7 ft. below A and released, show that it will rise to a distance of 5 ft. below A before beginning to fall.

8. If, in the last question, the weight is released from a point 10 ft. below A, show that it will rise to a height of 1 ft. and 10 in. below A.

9. The string of the example studied in the text of § 4 is placed on a smooth inclined plane making an angle of 30° with the horizon, and one end is made fast to a peg at A in the plane. If a weight of $1\frac{1}{3}$ lbs. be attached to the other end of the string and released from rest at A, find how far down the plane it will slide. Assume the unstretched length of the string to be 4 ft.

10. The same question if the plane is rough, $\mu = \frac{1}{6}\sqrt{3}$.

11. A cylindrical spar buoy (specific gravity $\frac{1}{2}$) is anchored so that it is just submerged at high water. If the cable should break at high tide, show that the spar would jump entirely out of the water.

Assume that the buoyancy of the water is always just equal to the weight of water displaced.

12. A particle of mass 2 lbs. lies on a rough horizontal table, and is fastened to a post by an elastic band whose unstretched length is 10 inches. The coefficient of friction is $\frac{1}{3}$, and the band is doubled in length by hanging it vertically with the weight at its lower end. If the particle be drawn out to a distance of 15 inches from the post and then projected directly away from the post with an initial velocity of 5 ft. a sec., find where it will stop for good.

6. Continuation; the Time. The time required by the body whose motion was studied in § 5 to reach the point B can be found as follows. From equation (4) we have:

$$v = \frac{ds}{dt} = \pm\sqrt{8g}\,\sqrt{1 - s^2}. \tag{5}$$

Since s decreases as t increases, ds/dt is negative, and the lower sign holds. Replacing $\sqrt{8g}$ by its value, 16, we see that

$$\frac{ds}{dt} = -\,16\sqrt{1 - s^2}. \tag{6}$$

This differential equation is readily solved by *separating the variables, i.e.* by transforming the equation so that only the variable s occurs on one side of the new equation, and only t on the other; thus

$$16\,dt = -\,\frac{ds}{\sqrt{1 - s^2}}. \tag{7}$$

Hence $$16t = -\int\frac{ds}{\sqrt{1 - s^2}} = -\,\sin^{-1} s + C.$$

If we measure the time from the instant when the body was released at A, then $t = 0$ and $s = 1$ are the initial values which determine C:

$$0 = -\,\sin^{-1} 1 + C, \qquad C = \frac{\pi}{2}.$$

Thus $$16t = \frac{\pi}{2} - \sin^{-1} s.$$

The right-hand side of this equation has the value $\cos^{-1} s$. Hence we have, as the final result,*

$$16t = \cos^{-1} s, \qquad \text{or} \qquad s = \cos 16t. \tag{8}$$

* In evaluating the above integral we might equally well have used the formula

$$\int\frac{ds}{\sqrt{1 - s^2}} = -\,\cos^{-1} s + C'.$$

We should then have had:

$$16t = \cos^{-1} s - C'.$$

Substituting the initial values $t = 0$, $s = 1$ in this equation, we find:

$$0 = \cos^{-1} 1 - C', \qquad \text{or} \qquad C' = 0,$$

and the final result is the same as before.

This equation gives the time it takes the body to reach an arbitrary point of its path. In particular, the time from A to B is found by putting $s = 0$:

$$(9) \qquad 16t = \cos^{-1} 0 = \frac{\pi}{2}, \qquad t = \frac{\pi}{32} = .09818 \text{ sec.}$$

EXERCISES

1. Show that if the body, in the case just discussed, had been released from rest at any other distance from the peg, the string being stretched, the time to the point at which the string becomes slack would have been the same.

2. Show that it takes the body twice as long to cover the first half of its total path as it does to cover the remainder.

Find the time required to cover the entire path in the case of the following exercises at the close of § 5.

3. Exercise 1. *Ans.* $\frac{\pi}{32} = .09818.$

4. Exercise 5.

Ans. $t = \sqrt{\frac{3}{32}} \int \frac{ds}{\sqrt{6s - s^2}}$; total time, $\pi\sqrt{\frac{3}{32}} = .9618$ sec.

5. Exercise 6. *Ans.* $t = \sqrt{\frac{3}{g}} \left[\frac{\pi}{2} + \sin^{-1}\frac{1}{\sqrt{3}}\right] = ?$

6. Exercise 7. *Ans.* .9618 sec.

7. Exercise 9. **8.** Exercise 10. **9.** Exercise 8.

7. Simple Harmonic Motion. The simplest and most important case of oscillatory motion which occurs in nature is that known as *Simple Harmonic Motion.* It is illustrated with the least amount of technical detail by the following example, or by the first Exercise below.

Example. A hole is bored through the centre of the earth, a stone is inserted, the air is exhausted, and the stone is released from rest at the surface of the earth. To determine the motion.

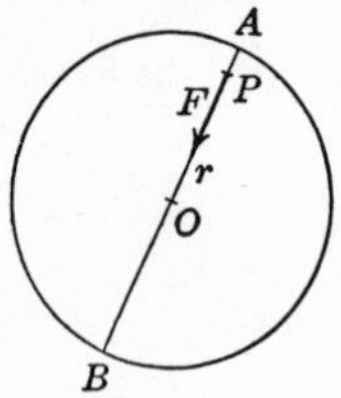

Fig. 55

The earth is here considered as a homogeneous sphere, at rest in space. Its attraction, F, on the stone diminishes as the stone nears the centre, and it can be shown to be proportional, at any point of the hole, to the distance of the stone from the centre:

$$F \propto r, \qquad \text{or} \qquad F = kr.$$

To determine the constant k, observe that, at the surface, $r = R$ (the radius of the earth), and, if we measure F in absolute units, $F = mg$, where m denotes the mass of the stone. Hence

$$mg = kR \qquad \text{or} \qquad k = \frac{mg}{R},$$

and

$$F = \frac{mg}{R} r.$$

As the coordinate of the stone we will take its distance, r, from the centre of the earth. Then Newton's Second Law gives us:

$$(1) \qquad m \frac{d^2 r}{dt^2} = - \frac{mg}{R} r.$$

For, when r is positive, the force tends to decrease r, and so is negative. When r is negative, the force tends to increase r algebraically, and so is positive. Hence (1) is right in all cases.

In order to integrate Equation (1), which can be written in the form:

$$(2) \qquad \frac{d^2 r}{dt^2} = - \frac{g}{R} r,$$

we employ the device of multiplying through by $2\,dr/dt$:

$$2 \frac{dr}{dt} \frac{d^2 r}{dt^2} = - \frac{2g}{R} r \frac{dr}{dt}.$$

The left-hand side thus becomes $\dfrac{d}{dt}\left(\dfrac{dr}{dt}\right)^2$. Hence each side can be integrated with respect to t: *

* This method can be applied to any differential equation of the form:

$$\frac{d^2 y}{dx^2} = f(y).$$

Multiply through by $2\,dy/dx$:

$$2 \frac{dy}{dx} \frac{d^2 y}{dx^2} = 2 f(y) \frac{dy}{dx}.$$

The left-hand side thus becomes $\dfrac{d}{dx}\left(\dfrac{dy}{dx}\right)^2$. Hence

$$\frac{d}{dx}\left(\frac{dy}{dx}\right)^2 = 2 f(y) \frac{dy}{dx}, \qquad \text{or} \qquad d\left(\frac{dy}{dx}\right)^2 = 2 f(y)\, dy.$$

Integrating, we have

$$\left(\frac{dy}{dx}\right)^2 = 2 \int f(y)\, dy.$$

$$\int \frac{d}{dt}\left(\frac{dr}{dt}\right)^2 dt = -\frac{2g}{R}\int r\frac{dr}{dt}dt,$$

or

$$\left(\frac{dr}{dt}\right)^2 = -\frac{2g}{R}\int r\,dr = -\frac{g}{R}\cdot r^2 + C.$$

To determine C, observe that initially, *i.e.* when the stone was at A, $r = R$ and the velocity, dr/dt, $= 0$. Hence

$$0 = -\frac{g}{R}R^2 + C, \qquad \text{or} \qquad C = \frac{g}{R}R^2.$$

Thus finally:

$$\left(\frac{dr}{dt}\right)^2 = \frac{g}{R}(R^2 - r^2). \tag{3}$$

At the centre of the earth, $r = 0$, and $(dr/dt)^2 = gR$. If we take the radius of the earth as 4000 miles, then $R = 4000 \times 5280$, $g = 32$, and the velocity is about 26,000 ft. a sec., or approximately 5 miles a second.

The stone keeps on with diminishing speed and comes to rest for an instant when $r = -R$, *i.e.* it just reaches the other side of the earth, and then falls back. Thus it oscillates throughout the whole length of the hole, reaching the surface at the end of each excursion, and continuing this motion forever. The result is not unreasonable, for there is no damping of any sort, — no friction or air resistance.

The Time. To find the time we proceed as in § 6. From Equation (3) it follows that

$$\frac{dr}{dt} = -\sqrt{\frac{g}{R}}\sqrt{R^2 - r^2}.$$

Hence, separating the variables, we have:

$$dt = -\sqrt{\frac{R}{g}}\frac{dr}{\sqrt{R^2 - r^2}}, \qquad t = -\sqrt{\frac{R}{g}}\int\frac{dr}{\sqrt{R^2 - r^2}},$$

or

$$t = \sqrt{\frac{R}{g}}\cos^{-1}\frac{r}{R} + C.$$

Initially, $t = 0$ and $r = R$; thus $C = 0$, and

$$t = \sqrt{\frac{R}{g}}\cos^{-1}\frac{r}{R}, \qquad \text{or} \qquad r = R\cos\left(t\sqrt{\frac{g}{R}}\right). \tag{4}$$

The time from A to O is found by putting $r = 0$:

$$t|_{r=0} = \frac{\pi}{2}\sqrt{\frac{R}{g}}.$$

On computing the value of this expression it is seen to be 21 min. and 16 sec. The time from A to B is twice the above. Hence the time of a complete excursion, from A to B and back to A is

$$2\pi\sqrt{\frac{R}{g}}.$$

This time is known as the *period* of the oscillation.*

The General Case. Simple Harmonic Motion is always dominated by the differential equation

$$\text{A)} \qquad \frac{d^2x}{dt^2} = -n^2x,$$

where the coordinate x characterizes the displacement from the position of no force. This equation can be integrated as in the special case above, and it is found that

$$\text{B)} \qquad \left(\frac{dx}{dt}\right)^2 = n^2(h^2 - x^2),$$

where h denotes the value of x which corresponds to the extreme displacement. The velocity when $x = 0$ is numerically nh, and thus is proportional both to n and to h. A second integration gives

$$\text{C)} \qquad x = h \cos nt,$$

provided the time is measured from an instant when $x = h$. The period, T, is inversely proportional to n:

$$\text{D)} \qquad T = \frac{2\pi}{n},$$

and the amplitude is $2h$. Thus the period is independent of the amplitude.

The motion represented by Equation C) is known as *Simple Harmonic Motion.* The graph of the function is obtained from

* In the first equation (4) the principal value of the anti-cosine holds during the first passage of the stone from A to B. The second equation (4) holds without restriction.

the graph of the cosine curve by plotting the latter to one scale on the axis of t, and to another scale on the axis of x.

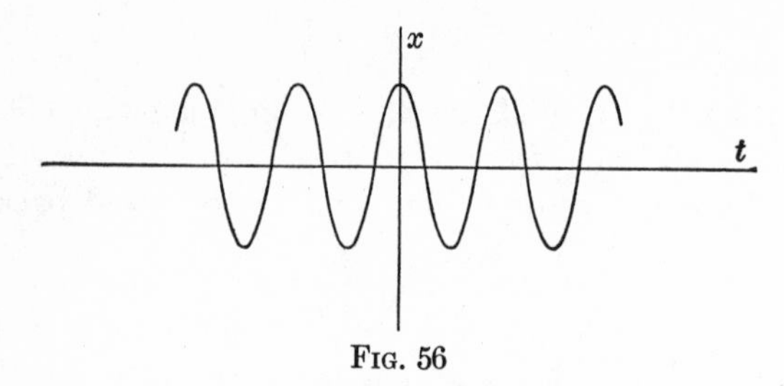

FIG. 56

EXERCISES

1. Two strings like the one described in the text of § 4 are fastened, one end of each, to two pegs, A and B, on a smooth horizontal table, the distance AB being double the length of either string, and the other end of each string is made fast to a 3 lb. weight, which is placed at O, the mid-point of AB. Thus each string is taut, but not stretched. The weight being moved to a point C between O and A and then released from rest, show that it oscillates with simple harmonic motion. Find the velocity with which it passes O and the period of the oscillation. It is assumed that the string which is slack in no wise interferes with or influences the motion.

Ans. The differential equation which dominates the motion is $\frac{d^2x}{dt^2} = -256x$, where x denotes the displacement of the 3 lb. weight; hence the motion is simple harmonic motion. The required velocity is numerically $16h$, where h denotes the maximum displacement. The period is $2\pi/16 = .3927$ sec.

2. Work the same problem for two strings like the one of Question 4, § 4, and a 2 lb. weight.

3. Show that the motion of Example 7, § 5, is simple harmonic motion, and find the period.

4. If a straight hole were bored through the earth from Boston to London, a smooth tube containing a letter inserted, the air exhausted from the tube, and the letter released at Boston, how long would it take the letter to reach London?

5. If in the problem of Question 9, § 5, the weight were released with the string taut, but not stretched, and directed straight down the plane, show that the weight would execute simple harmonic motion. Determine the amplitude and the period.

6. Work the problem of the text for the moon; cf. the data in § 8.

7. A steel wire of one square millimeter cross-section is hung up in Bunker Hill Monument, and a weight of 25 kilogrammes is fastened to the lower end of the wire and carefully brought to rest. The weight is then given a slight vertical displacement. Determine the period of the oscillation.

Given that the force required to double the length of the wire is 21,000 kilogrammes, and that the length of the wire is 210 feet.

Ans. A little over half a second.

8. A number of iron weights are attached to one end of a long round wooden spar, so that, when left to itself, the spar floats vertically in water. A ten-kilogramme weight having become accidentally detached, the spar is seen to oscillate with a period of 4 seconds. The radius of the spar is 10 centimetres. Find the sum of the weights of the spar and attached iron. Through what distance does the spar oscillate?

Ans. (*a*) About 125 kilogrammes; (*b*) 0.64 metre.

8. Motion under the Attraction of Gravitation. *Problem.* To find the velocity which a stone acquires in falling to the earth from interstellar space.

Assume the earth to be at rest and consider only the force which the earth exerts. Let the stone be released from rest at A, and let r be its distance from the centre O of the earth at any subsequent instant. Then the force, F, acting on it is, by the law of gravitation, inversely proportional to r:

$$F = \frac{\lambda}{r^2}.$$

Since $F = mg$ when $r = R$, the radius of the earth,

$$mg = \frac{\lambda}{R^2} \qquad \text{and} \qquad F = \frac{mg\,R^2}{r^2}.$$

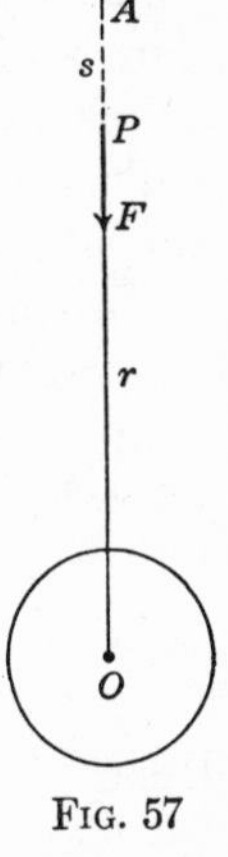

FIG. 57

Newton's Second Law of Motion here takes on the form:

$$m\frac{d^2r}{dt^2} = -\frac{mg\,R^2}{r^2}.$$

Hence

$$\frac{d^2r}{dt^2} = -\frac{g\,R^2}{r^2}. \tag{1}$$

To integrate this equation, we employ the method of § 7 and multiply by $2\,dr/dt$:

$$2\frac{dr}{dt}\frac{d^2r}{dt^2} = -\frac{2\,g\,R^2}{r^2}\frac{dr}{dt}, \qquad \text{or} \qquad \frac{d}{dt}\left(\frac{dr}{dt}\right)^2 = -\frac{2\,g\,R^2}{r^2}\frac{dr}{dt}.$$

Integrating with respect to t we find:

$$\frac{dr^2}{dt^2} = -2\,g\,R^2\int\frac{dr}{r^2} = \frac{2\,g\,R^2}{r} + C.$$

Initially $dr/dt = 0$ and $r = l$; hence

$$0 = \frac{2\,g\,R^2}{l} + C, \qquad C = -\frac{2\,g\,R^2}{l}.$$

$$\frac{dr^2}{dt^2} = 2\,g\,R^2\left(\frac{1}{r} - \frac{1}{l}\right). \tag{2}$$

Since dr/dt is numerically equal to the velocity, the velocity V at the surface of the earth is given by the equation:

$$V^2 = 2\,g\,R^2\left(\frac{1}{R} - \frac{1}{l}\right).$$

If l is very great, the last term in the parenthesis is small, and so, no matter how great l is, V can never quite equal $\sqrt{2\,g\,R}$. Here $g = 32$, $R = 4000 \times 5280$, and hence the velocity in question is about 36,000 feet, or 7 miles, a second.

This solution neglects the retarding effect of the atmosphere; but as the atmosphere is very rare at a height of 50 miles from the earth's surface, the result is reliable down to a point comparatively near the earth.

In order to find the time it would take the stone to fall, consider the equation derived from (2):

$$\frac{dr}{dt} = -\sqrt{2\,g\,R^2}\sqrt{\frac{l-r}{lr}}.$$

Hence

$$dt = -\frac{\sqrt{l}}{8\,R}\frac{r\,dr}{\sqrt{lr - r^2}}$$

and
$$t = -\frac{\sqrt{l}}{8R}\int \frac{r\,dr}{\sqrt{lr - r^2}}.$$

Turning to Peirce's *Tables*, No. 169, we find:

$$\int \frac{r\,dr}{\sqrt{lr - r^2}} = -\sqrt{lr - r^2} + \frac{l}{2}\int \frac{dr}{\sqrt{lr - r^2}}$$
$$= -\sqrt{lr - r^2} + \frac{l}{2}\sin^{-1}\frac{2r - l}{l}.$$

Thus
$$t = \frac{\sqrt{l}}{8R}\left\{\sqrt{lr - r^2} - \frac{l}{2}\sin^{-1}\frac{2r - l}{l}\right\} + K.$$

Initially $t = 0$ and $r = l$:

$$0 = \frac{\sqrt{l}}{8R}\left\{0 - \frac{l}{2}\frac{\pi}{2}\right\} + K.$$

Finally, then,

$$t = \frac{\sqrt{l}}{8R}\left\{\sqrt{lr - r^2} + \frac{l}{2}\left[\frac{\pi}{2} - \sin^{-1}\frac{2r - l}{l}\right]\right\}.$$

For purposes of computation, a better form of this equation is the following:

$$(3) \qquad t = \frac{l^{\frac{3}{2}}}{8R}\left\{\sqrt{\frac{r}{l} - \left(\frac{r}{l}\right)^2} + \frac{1}{2}\cos^{-1}\left(2\frac{r}{l} - 1\right)\right\}.$$

EXERCISES*

1. If the earth had no atmosphere, with what velocity would a stone have to be projected from the earth's surface, in order not to come back?

2. If the moon were stopped in its course, how long would it take it to fall to the earth? Regard the earth as stationary.

Ans. 4 days, 18 hrs., 10 min.

* In working these exercises, the following data may be used:
Radius of the moon, $\frac{3}{11}$ that of the earth.
Mass of moon, $\frac{1}{81}$ that of earth.
Mean distance of moon from earth, 237,000 miles.
Acceleration of gravity on the surface of the moon, $\frac{1}{6}$ that on the surface of the earth.
Diameter of sun, 860,000 miles.
Mass of sun, 333,000 that of the earth.
Mean distance of earth from sun, 93,000,000 miles.
Acceleration of gravity on the surface of the sun, 905 ft. per sec. per sec.

3. Solve the preceding problem accurately, assuming that the earth and the moon are released from rest in interstellar space at their present mean distance apart. Their common centre of gravity will then remain stationary.

4. The same problem for the earth and the sun.

5. If the earth and the moon were held at rest at their present mean distance apart, with what velocity would a projectile have to be shot from the surface of the moon, in order to reach the earth?

6. If the earth and the moon were held at rest at their present mean distance apart, and a stone were placed between them at the point of no force and then slightly displaced toward the earth, with what velocity would it reach the earth?

7. If a hole were bored through the centre of the moon, assumed spherical, homogeneous, and at rest in interstellar space, and a stone dropped in, how long would it take the stone to reach the other side?

8. Show that if two spheres, each one foot in diameter and of density equal to the earth's mean density (specific gravity, 5.6) were placed with their surfaces $\frac{1}{4}$ of an inch apart and were acted on by no other forces than their mutual attractions, they would come together in about five minutes and a half. Given that the spheres attract as if all their mass were concentrated at their centres.

9. Work Done by a Variable Force. If a force, F, constant in magnitude and always acting along a fixed line AB in the same sense, be applied to a particle,* and if the particle be displaced along the line in the direction of the force, the work done by the force on the particle is defined in elementary physics as

$$W = Fl,$$

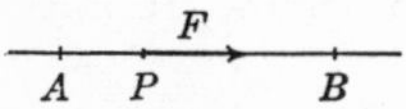

F

$A \quad P \quad B$

FIG. 58

where l denotes the distance through which the particle has been displaced.

Suppose, however, that the force is variable, but varying continuously and always acting along the same fixed line. How shall the work now be defined?

* Or, more generally, to one and the same point P of a rigid or deformable material body.

Let a coordinate be assumed on the line; *i.e.* think of the line as the axis of x. Let the particle be displaced from A: $x = a$ to B: $x = b$, and let $a < b$. Let F, to begin with, always act in the direction of the positive sense along the axis. Then

$$F = f(x),$$

where $f(x)$ denotes a positive continuous function of x.

Divide the interval (a, b) up into n parts by the points x_1, $x_2, \cdots, x_{n-1}$, and let $x_0 = a$, $x_n = b$. Then, if

$$x_{k+1} - x_k = \Delta x_k,$$

the work, ΔW_k, done by the force in displacing the particle through the k-th interval ought, in order to correspond to the general physical conception of work, to lie between the quantities

$$F'_k \Delta x \qquad \text{and} \qquad F''_k \Delta x,$$

where F'_k and F''_k denote respectively the smallest and the largest values of $f(x)$ in this interval.* We have, then:

$$F'_k \Delta x \leqq \Delta W_k \leqq F''_k \Delta x. \tag{1}$$

On writing out the double inequality (1) for $k = 0, 1, \cdots, n - 1$ and adding the n relations thus resulting together, we find that $W = \Sigma \Delta W_k$ lies between the two sums:

$$F'_0 \Delta x + F'_1 \Delta x + \cdots + F'_{n-1} \Delta x, \tag{2}$$

$$F''_0 \Delta x + F''_1 \Delta x + \cdots + F''_{n-1} \Delta x. \tag{3}$$

Each of these sums suggests the sum

$$f(x_0) \Delta x_0 + f(x_1) \Delta x_1 + \cdots + f(x_{n-1}) \Delta x_n, \tag{4}$$

whose limit is the definite integral,

$$\lim_{n=\infty} [f(x_0) \Delta x + f(x_1) \Delta x + \cdots + f(x_{n-1}) \Delta x] = \int_a^b f(x)\, dx. \tag{5}$$

That W is in fact equal to this integral:

$$W = \int_a^b f(x)\, dx, \tag{6}$$

follows from Duhamel's Theorem.

* This statement is pure physics. It is the physical axiom on which the generalization of the definition of *work* is based. More precisely, it is one of two physical axioms, the other being that the total work, W, for the complete interval is the sum of the partial works, ΔW_k, for the subintervals.

If the force F acts in the direction opposite to that in which the point of application is moved, we extend the definition and say that negative work is done. For the case that F is constant, the work is now defined as follows:

$$W = F(b - a). \tag{7}$$

Here, F is to be taken as a negative number equal numerically to the intensity of the force.

Thus (7) is seen to hold in whichever direction the force acts, provided that $a < b$. Will (7) still hold if $b < a$? It will. There are in all four possible cases:

i) $++$ *ii*) $--$ *iii*) $+-$ *iv*) $-+$

In cases *i*) and *ii*) the force overcomes resistance, and positive work is done. In cases *iii*) and *iv*) the force is overcome, and negative work is done. Hence (7) holds in all cases.

It is now easy to see how the definition of work should be laid down when F varies in any continuous manner. The considerations are precisely similar to those which led to Equation (6), and that same equation is the final result in this, the most general, case:

$$W = \int_a^b f(x)\,dx.$$

Example. To find the work done in stretching a wire. Let the natural (or unstretched) length of the wire be l, the stretched length, l'. Then the tension, T, is given by *Hooke's Law:*

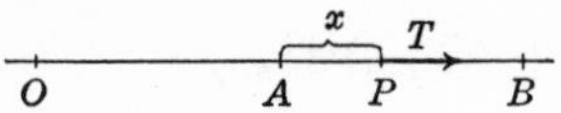

FIG. 59

$$T = \lambda \frac{l' - l}{l},$$

where λ is independent of l and l', and is known as *Young's Modulus.*

Let the wire, in its natural state, lie along the line OA, and let it, when stretched, lie along OB, OP being an arbitrary intermediate position. Let x be measured from A, and let $x = h$ at B. Then

$$T = \lambda \frac{x}{l}$$

and

$$W = \int_0^h \lambda \frac{x}{l}\,dx = \frac{\lambda}{l}\int_0^h x\,dx = \frac{\lambda h^2}{2l}.$$

This is the work done *on the wire* by the force that stretches it. If the wire contracts, the work done by the wire *on the body* to which its end P is attached will be

$$\int_h^0 \left(-\lambda \frac{x}{l}\right) dx = -\left.\frac{\lambda x^2}{2l}\right|_h^0 = \frac{\lambda h^2}{2l}.$$

EXERCISES

1. In the problem of § 7 compute the work done by the earth on the particle when the latter reaches the centre.

2. A particle of mass m moves down an inclined plane. Show that the work done on it by the component of gravity down the plane is the same as the work done by gravity on the particle when it descends vertically a distance equal to the change in level which the particle undergoes.

3. A particle is attracted toward a point O by a force which is inversely proportional to the square of the distance from O. How much work is done on the particle when it moves from a distance a to a distance b along a right line through O?

4. If the earth and the moon were stopped in their courses and allowed to come together by their mutual attraction, how much work would the earth have done on the moon when they meet?

5. Find the work done by the sun on a meteor when the latter moves along a straight line passing through the centre of the sun, from an initial distance R to a final distance r.

10. Kinetic Energy and Work. Let a particle of mass m describe a right line with velocity $v = ds/dt$. Its *kinetic energy* is defined as the quantity:

$$\frac{mv^2}{2}.$$

Let the particle move under the action of any force F which varies continuously: $F = f(s)$. Then Newton's Second Law can be written in the form:

$$mv\frac{dv}{ds} = f(s).$$

Hence

$$mv\,dv = f(s)\,ds.$$

Integrate this equation between the limits a and b, denoting the corresponding values of v by v_1 and v_2:

$$\int_{v_1}^{v_2} mv\, dv = \int_a^b f(s)\, ds.$$

The left-hand side has the value:

$$\left.\frac{mv^2}{2}\right|_{v_1}^{v_2}$$

The right-hand side is, by definition, the work W done on the particle by the force F. Hence

$$(1) \qquad \frac{mv_2^2}{2} - \frac{mv_1^2}{2} = W,$$

and we infer the result:

THEOREM. *The change in the kinetic energy of a particle is equal to the work done on it by the force which acts on it.*

This theorem expresses, in this the simplest case imaginable, the Principle of Work and Energy in Mechanics. By means of it a first integral of the equation arising from Newton's Second Law can be found in the case of a particle, when the force is known as a function of the position, and the student will do well to go back over the foregoing problems and exercises, and examine their solution from this new point of view; *e.g.* Equation (4) in § 5, Equation (3) in § 7, and Equation (2) in § 8 are, save for the factor $m/2$, the Equation of Energy, as (1) is often called.

EXERCISES

Work the Exercises of §§ 5, 7, 8, so far as possible, by the Method of Work and Energy.

11. Change of Units in Physics.* To *measure* a quantity is to determine how many times a certain amount of that substance, chosen arbitrarily as the *unit*, is contained in a given

* The introduction of this paragraph and the next at this stage seems to require justification. If these two purely physical subjects are sufficiently important to be taken up here, then why not, at the beginning of this chapter, the first time they are needed? But if they are merely for reference, why break the unity, coherence, of the presentation by placing them here rather than at the end of the chapter? The Author feels that this is about the time when the beginner in Mechanics should turn his attention systematically to these subjects, for until he has some knowledge of the problems studied in this chapter, he can hardly be expected to recognize the importance of Change of Units and of the Check of Dimensions.

amount of the substance.* Thus to measure the length of right lines is to find how many times a right line chosen arbitrarily as the *unit of length* — a foot or a centimetre or a cubit — is contained in a given right-line segment. The number, s, thus resulting is called the *length* of the line. It depends on *two* things — the particular line and the unit chosen. If a different unit of length be chosen, the same line will have a different number, s', assigned to it, and its length then becomes s'.† Now, for all lines, s' *will be proportional to* s:

$$(1) \qquad s' \propto s \qquad \text{or} \qquad s' = c\,s,$$

where c is a constant depending on the units. It is determined in any given case by substituting particular values for s and s', known to correspond. Thus if we wish to transform from feet to yards, consider in particular a line which is a yard long. Here, s' will equal 1 and s will equal 3, so

$$1 = 3c, \qquad c = \tfrac{1}{3},$$

and,‡

$$(2) \qquad s' = \tfrac{1}{3}s.$$

Example. If a yard is the unit of length, a minute the unit of time, a ton the unit of mass, and a kilogramme the unit of force, find λ in Newton's Second Law.

We will start with Newton's Equation in the English units:

$$(3) \qquad m\frac{d^2 s}{dt^2} = 32 f,$$

* The word *substance* here may be too narrow in its connotations, for we want a word that will include every measurable quantity, from the length of a light-wave to the wheat crop of the world. Such a word obviously does not exist, and so we agree to use *substance* in this sense as a *terminus technicus*.

† It would seem paradoxical to say that the *same* line has a length of 6 when the foot is the unit, and a length of 2 when the yard is the unit. But it must be remembered that the length is a function of *two* variables, the unit being one of them. The attempt is sometimes made to meet the apparent difficulty by saying "3 ft. = 1 yd." But this makes confusion worse confounded; for 3 = 1 is not true, while on the other hand to try to introduce "concrete numbers," like 3 ft., 10 lbs., 5 secs., into mathematics, is not feasible. To try to change units in this way leads to blunders and wrong numerical results. There is only one kind of number in elementary mathematics. To attempt to qualify it as abstract, is to qualify that which is unique. The denominate attribute (3 ft., 10 lbs., etc.) is part of the *physical thing* conceived; it does not pertain to the *mathematical counterpart*, which is purely arithmetical.

‡ Compare this equation with the attempted form of statement mentioned in the last footnote: "1 yd. = 3 ft." It would seem to follow from that statement that s' yds. = $3s$ ft. But $s' = \frac{1}{3}s$. What a cheerful prospect for getting the right answer by that method!

and write the transformed equation in the form:

$$m'\frac{d^2s'}{dt'^2} = \lambda' f'. \tag{4}$$

Then the problem is to determine λ'. Here, from (2):

$$s' = \tfrac{1}{3}s.$$

Next,

$$m' = km,$$

$$1 = k \times 2000, \qquad k = \frac{1}{2000}, \qquad m' = \frac{m}{2000}.$$

Similarly,

$$t' = \frac{t}{60}, \qquad f' = \frac{f}{2.20}.$$

Thus

$$m'\frac{d^2s'}{dt'^2} = \frac{60^2}{2000 \times 3} m \frac{d^2s}{dt^2},$$

$$\lambda' f' = \frac{1}{2.20}\lambda' f.$$

The left-hand sides of these equations are equal by (4). On equating the right-hand sides and dividing by (3) we find:

$$\frac{60^2}{2000 \times 3} = \frac{\lambda'}{2.20 \times 32}, \qquad \lambda' = 422.4.*$$

On dropping the primes, Newton's Second Law, written in the new units, appears in the form:

$$m\frac{d^2s}{dt^2} = 420 f.$$

EXERCISES

1. If the units of length, time, and mass are respectively a mile, a day, and a ton, compute the absolute unit of force in pounds.

2. If the acceleration of gravity is 981 in the c.g.s. system, compute g in the English system.

3. If the acceleration of gravity is 32.2 in the English system, compute g in the c.g.s. system.

* More precisely, the result should be tabulated as:

$$\lambda' = 4.2 \times 10^2,$$

since the data, namely, $\lambda = 32$, are correct only to two significant figures.

4. If the unit of force be a pound, the unit of time a second, and the unit of length a foot, explain what is meant by the *absolute unit of mass*, and show that it is equal (nearly) to 32 lbs.

5. Formulate and solve the same problem in the decimal system.

6. If the unit of mass is a pound, the unit of length, a foot, and the unit of force, a pound, find the absolute unit of time.

Ans. .176 secs.

12. The Check of Dimensions. The physical quantities that enter in Mechanics can be expressed in terms of the units of Mass $[M]$, Length $[L]$, and Time $[T]$. Thus velocity is of the dimension length/time, or $L/T = LT^{-1}$. Acceleration has the dimension LT^{-2}, and force, the dimension ML/T^{-2}.

When an equation is written in literal form, as

$$m\frac{d^2 s}{dt^2} = f,$$

each term must have the same dimension. For, such an equation remains true, no matter what the units of mass, length, and time may be; and if two terms had different dimensions in any one of the fundamental quantities (mass, length, time), a change of units would lead to a new equation not in general equivalent to the old one.

This principle affords a useful check on computation. Thus, if an ellipse is given by the equation:

$$\frac{x^2}{a^2} + \frac{y^2}{b^2} = 1,$$

all the quantities x, y, a, b are of dimension one in length, or L. The dimension of its *area* must be L^2; and it is, for $A = \pi ab$. The volume of the ellipsoid of revolution corresponding to rotation about the axis of x should be of dimension L^3, and it is:

$$V = \tfrac{4}{3}\pi ab^2.$$

This principle affords a useful check on putting in or leaving out g, when problems are formulated literally. Thus in the Example of § 3, if we had forgotten our g in writing down the right-hand side, the check of dimensions would immediately have shown up the oversight. For, the left-hand member is of dimension ML/T^{-2}; hence every term on the right must have this same dimension. It does, in the correct equation of the text. It is, of course, only when all the quantities which enter are in *literal* form, that the check can be used. If some are replaced by numbers, the check does not apply.

Observe that in computing the dimension of a derivative, like d^2s/dt^2, we may think of the latter as a quotient, the numerator being a *difference*, and hence of the dimension of the dependent variable, while the denominator is thought of as a *power*.

EXERCISES

Determine the dimension of each of the following quantities:

1. Kinetic energy. *Ans.* ML^2T^{-2}.

2. Work. *Ans.* ML^2T^{-2}.

3. Moment of inertia. *Ans.* ML^2.

4. Momentum. *Ans.* MLT^{-1}.

5. Couples. *Ans.* ML^2T^{-2}.

6. Volume density. *Ans.* ML^{-3}.

7. Surface density. *Ans.* ML^{-2}.

8. Line density. *Ans.* ML^{-1}.

9. The acceleration of gravity. *Ans.* LT^{-2}.

10. The wind resistance can often be assumed proportional to the square of the velocity. If it is written as cv^2, what is the dimension of c? *Ans.* ML^{-1}.

11. In Question 10, what is the answer when the wind resistance is taken per square foot of surface exposed?

12. Check the dimensions in each equation occurring in § 3.

13. In § 4, Equation (3), the check fails. Explain why.

14. What are the dimensions of Young's Modulus?

15. In the Example treated in § 8, we wished to find the velocity of the stone at the centre of the earth in miles per second. But if we substituted for R, in the formula $(dr/dt)^2 = gR$, the value of R in miles (*i.e.* 4000), we obtained a wrong answer, even though the dimensions of both sides of this equation are the same, namely, L^2/T^{-2}. Explain why, and show how Formula (2), which is one hundred per cent literal, can be used to yield a correct result, when $R = 4000$.

16. Examine each equation in § 8 as to whether the Check of Dimensions is applicable.

13. Motion in a Resisting Medium. When a body moves through the air or through the water, these media oppose resistance, the magnitude of which depends on the velocity, but does not follow any simple mathematical law. For low velocities up to 5 or 10 miles per hour, the resistance R can be expressed approximately by the formula:

$$R = av, \tag{1}$$

where a is a constant depending both on the medium and on the size and shape of the body, but not on its mass. For higher velocities up to the velocity of sound (1082 ft. a sec.) the formula

$$R = cv^2 \tag{2}$$

gives a sufficient approximation for many of the cases that arise in practice. We shall speak of other formulas in the next paragraph.

Problem 1. A man is rowing in still water at the rate of 3 miles an hour, when he ships his oars. Determine the subsequent motion of the boat.

Here Newton's Second Law gives us:

$$m\frac{dv}{dt} = -av. \tag{3}$$

Hence

$$dt = -\frac{m}{a}\frac{dv}{v},$$

$$t = \frac{m}{a}\log\frac{v_0}{v}, \tag{4}$$

where v_0 is the initial velocity, nearly $4\frac{1}{2}$ ft. a sec.

To solve (4) for v, observe that

$$\frac{at}{m} = \log\frac{v_0}{v}, \qquad \text{or} \qquad e^{\frac{at}{m}} = \frac{v_0}{v}.$$

Hence

$$v = v_0 e^{-\frac{at}{m}}. \tag{5}$$

It might appear from (5) that the boat would never come to rest, but would move more and more slowly, since

$$\lim_{t=\infty} e^{-\frac{at}{m}} = 0.$$

We warn the student, however, against such a conclusion. For the approximation we are using, $R = av$, holds only for a limited time, and even for that time is at best an *approximation.* It will probably not be many minutes before the boat is drifting sidewise, and the value of a for this aspect of the boat would be quite different, — if indeed the approximation $R = av$ could be used at all.

To determine the distance travelled, we have from (3):

$$mv\frac{dv}{ds} = -\,av,$$

and consequently:

$$v = v_0 - \frac{a}{m}s. \tag{6}$$

Hence, even if the above law of resistance held up to the limit, the boat would not travel an infinite distance, but would approach a point distant

$$S = \frac{mv_0}{a}$$

feet from the starting point, the distance traversed thus being proportional to the initial momentum.

Finally, to get a relation between s and t, integrate (5):

$$\frac{ds}{dt} = v_0\,e^{-\frac{at}{m}},$$

$$s = \frac{mv_0}{a}\left(1 - e^{-\frac{at}{m}}\right). \tag{7}$$

From this result is also evident that the boat will never cover a distance of S ft. while the above approximation lasts.

EXERCISE

If the man and the boat together weigh 300 lbs. and if a steady force of 3 lbs. is just sufficient to maintain a speed of 3 miles an hour in still water, show that when the boat has gone 20 ft., the speed has fallen off by a little less than a mile an hour.

Problem 2. A drop of rain falls from a cloud with an initial velocity of v_0 ft. a sec. Determine the motion.

We assume that the drop is already of its final size, — not gathering further moisture as it proceeds, — and take as the law of resistance:

$$R = cv^2.$$

The forces which act are *i*) the force of gravity, mg, downward, and *ii*) the resistance of the air, cv^2, upward. As the coordinate of the particle we will take the distance AP, Figure 60, which it has fallen. Then, Newton's Second Law becomes:

$$m\frac{dv}{dt} = mg - cv^2.$$

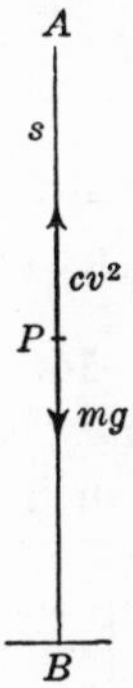

Hence

$$v\frac{dv}{ds} = \frac{mg - cv^2}{m},$$

$$ds = \frac{mv\,dv}{mg - cv^2},$$

$$s = -\frac{m}{2c}\log(mg - cv^2) + C,$$

$$0 = -\frac{m}{2c}\log(mg - cv_0^2) + C,$$

and thus finally

$$s = \frac{m}{2c}\log\frac{mg - cv_0^2}{mg - cv^2}. \tag{8}$$

FIG. 60

Solving for v we have

$$e^{\frac{2cs}{m}} = \frac{mg - cv_0^2}{mg - cv^2},$$

$$v^2 = \frac{mg}{c} - \frac{mg - cv_0^2}{c}e^{-\frac{2cs}{m}}. \tag{9}$$

When s increases indefinitely, the last term approaches 0 as its limit, and hence the velocity v can never exceed (or quite equal) $\bar{v} = \sqrt{mg/c}$ ft. a sec. This is known as the *limiting velocity*. It is independent of the height and also of the initial velocity, and is practically attained by the rain as it falls, for a rain drop is not moving sensibly faster when it reaches the ground than it was at the top of a high building.

EXERCISES

1. Work Problem 2, taking as the coordinate of the rain drop its height above the ground.

2. Find the time in terms of the velocity and the velocity in terms of the time in Problem 2.

3. Show that, if a charge of shot be fired vertically upward, it will return with a velocity about $3\frac{1}{3}$ times that of rain drops of the same size; and that if it be fired directly downward from a balloon two miles high, the velocity will not be appreciably greater.

4. Determine the height to which the shot will rise in Question 3, and show that the time to the highest point is

$$t = \sqrt{\frac{m}{gc}} \tan^{-1}\left(v_0\sqrt{\frac{c}{mg}}\right),$$

where v_0 is the initial velocity.

14. Graph of the Resistance. The resistance which the atmosphere or water opposes to a body of a given size and shape can in many cases be determined experimentally with a reasonable degree of precision and thus the graph of the resistance:

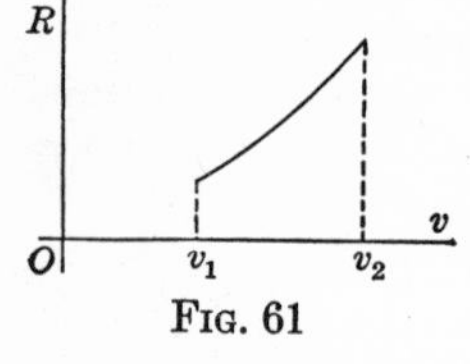

FIG. 61

$$R = f(v)$$

can be plotted. The mathematical problem then presents itself of representing the curve with sufficient accuracy by means of a simple function of v. In the problem of vertical motion in the atmosphere, Problem 2, § 13,

$$m\frac{dv}{dt} = mg \pm f(v),$$

according as the body is going up or coming down, s being measured positively downward. Now if we approximate to $f(v)$ by means of a quadratic polynomial or a fractional linear function,

$$a + bv + cv^2 \qquad \text{or} \qquad \frac{\alpha + \beta v}{\gamma + \delta v},$$

we can integrate the resulting equation readily. And it is obvious that we can so approximate, — at least, for a restricted range of values for v.

Another case of interest is that in which the resistance of the medium is the only force that acts, as in Problem 1:

$$m\frac{dv}{dt} = -f(v).$$

A convenient approximation for the purposes of integration is

$$f(v) = av^b.$$

Here a and b are merely arbitrary constants, enabling us to impose two arbitrary conditions on the curve, — for example, to make it go through two given points, — and are to be determined so as to yield a good approximation to the physical law. Sometimes the simple values $b = 1, 2, 3$ can be used with advantage. But we must not confuse these approximate formulas with similarly appearing formulas that represent exact physical laws. Thus, in geometry, the areas of similar surfaces and the volumes of similar solids are proportional to the squares or cubes of corresponding linear dimensions. This law expresses a fact that holds to the finest degree of accuracy of which physical measurements have shown themselves to be capable and with no restriction whatever on the size of the bodies. But the law $R = av^2$ or $R = cv^3$ ceases to hold, *i.e.* to interpret nature within the limits of precision of physical measurements, when v transcends certain restricted limits, and the student must be careful to bear this fact in mind.

EXERCISES

Work out the relations between v and s, and those between v and t, if the only force acting is the resistance of the medium, which is represented by the formula:

1. $R = a + bv + cv^2.$ **2.** $R = \dfrac{\alpha + \beta v}{\gamma + \delta v}.$ **3.** $R = av^b.$

4. Show that it would be feasible mathematically to use the formulas of Questions 1 and 2 in the case of the falling rain drop.

5. A train weighing 300 tons, inclusive of the locomotive, can just be kept in motion on a level track by a force of 3 pounds to the ton. The locomotive is able to maintain a speed of 60 miles an hour, the horse power developed being reckoned as 1300. Assuming that the frictional resistances are the same at high speeds as at low ones and that the resistance of the air is proportional to the square of the velocity, find by how much the speed of the train will have dropped off in running half a mile if the steam is cut off with the train at full speed.

6. A man and a parachute weigh 150 pounds. How large must the parachute be that the man may trust himself to it at any height, if 25 ft. a sec. is a safe velocity with which to reach the ground? Given that the resistance of the air is as the square of the velocity and is equal to 2 pounds per square foot of opposing surface for a velocity of 30 ft. a sec.

Ans. About 12 ft. in diameter.

7. A toboggan slide of constant slope is a quarter of a mile long and has a fall of 200 ft. Assuming that the coefficient of friction is $\frac{3}{100}$, that the resistance of the air is proportional to the square of the velocity and is equal to 2 pounds per square foot of opposing surface for a velocity of 30 ft. a sec., and that a loaded toboggan weighs 300 pounds and presents a surface of 3 sq. ft. to the resistance of the air; find the velocity acquired during the descent and the time required to reach the bottom.

Find the limit of velocity that could be acquired by a toboggan under the given conditions if the hill were of infinite length.

Ans. (*a*) 68 ft. a sec.; (*b*) 30 secs.; (*c*) 74 ft. a sec.

8. The ropes of an elevator break and the elevator falls without obstruction till it enters an air chamber at the bottom of the shaft. The elevator weighs 2 tons and it falls from a height of 50 ft. The cross-section of the well is 6×6 ft. and its depth is 12 ft. If no air escaped from the well, how far would the elevator sink in? What would be the maximum weight of a man of 170 pounds? Given that the pressure and the volume of air when compressed without gain or loss of heat follow the law:

$$pv^{1.41} = \text{const.},$$

and that the atmospheric pressure is 14 pounds to the square inch.

9. In the early days of modern ballistics the resistance of the atmosphere to a common ball was determined as follows. A number of parallel vertical screens were set up at equal distances, the ball was shot through them (with a practically horizontal trajectory), and the time recorded (through the breaking of an electric circuit) at which it cut each screen. Explain the theory of the experiment, and show how points on the graph of the resistance as a function of the velocity could be obtained.

15. Motion in a Plane and in Space. *Vector Velocity.* When a point P moves in a plane or in space, its position at any instant can be represented by its Cartesian coordinates:

$$(1) \qquad x = f(t), \qquad y = \varphi(t), \qquad z = \psi(t),$$

where the functions are continuous, together with any derivatives we shall have occasion to use.

The velocity of P has been defined as ds/dt. For, hitherto, we have regarded the path as given, and it was a question merely of the speed and sense of description of the path. But now we need more. We need to put into evidence the *direction and sense* of the motion, and so we extend the idea, defining velocity more broadly as a *vector*. Lay off on the tangent to the path, in the sense of the motion, a directed line segment whose length is the speed of the point, and let the vector thus determined be defined as the *vector velocity* of the point P.

Composition and Resolution of Velocities. A mouse runs across the floor of a freight car. To determine the velocity of the mouse in space, if the velocity of the car is u, and the velocity of the mouse relative to the car is v.

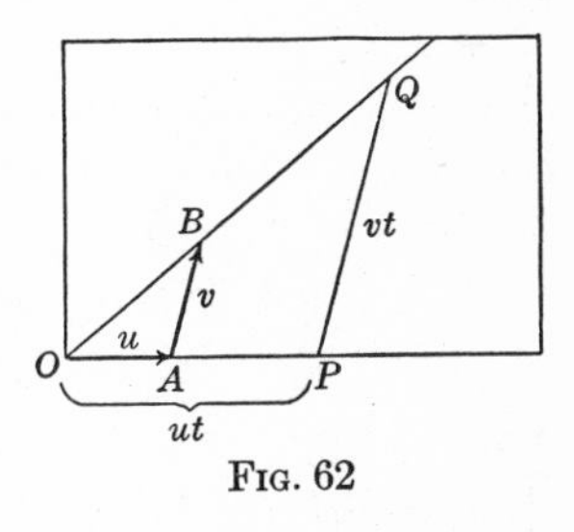

Fig. 62

Let the mouse start from a point P on one side of the car and run across the floor in a straight line with constant velocity, v, relative to the car. Let Q be the point she has reached at the end of t seconds. Then

$$PQ = vt.$$

Let the velocity, u, of the train be constant, and let O be the initial position of P. Then

$$OP = ut.$$

In Figure 62, the line OA represents the vector velocity $\mathbf{u}$ of the train, and AB represents the vector velocity $\mathbf{v}$ of the mouse relative to the freight car. Their geometric, or vector, sum is represented by OB. From similar triangles it appears that the path of the mouse in space is the right line through O and B, and that her velocity in space is the vector OB, or

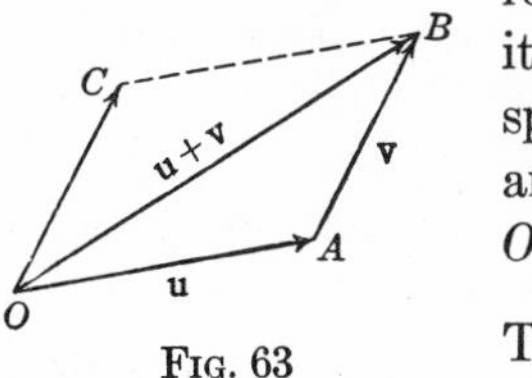

Fig. 63

$$\mathbf{u} + \mathbf{v}.$$

Thus her velocity in space may be described, from analogy with the parallelogram of forces, as the *resultant* of the two *component* velocities, $\mathbf{u}$ along the direction of OA and $\mathbf{v}$ along the direction OC through O parallel to AB.

Similarly, any vector velocity may be *resolved* into two component velocities along any two directions complanar with the given velocity; Fig. 63.

The extension to space is obvious. Any three non-complanar vector velocities can be composed into a single velocity by the parallelepiped law. And conversely any given vector velocity can be decomposed into three component vector velocities along any three non-complanar directions.

The General Case. Returning now to the general case of motion in a plane or in space, we may define the *average vector velocity* for the Δt seconds succeeding a given instant as the vector * (PP') divided by Δt, or the vector (PQ):

$$\frac{(PP')}{\Delta t} = (PQ).$$

When Δt approaches 0 as its limit, the length of this vector, namely, the chord $\overline{PP'}$, divided by Δt, approaches the speed of the point at P; or,

$$\lim_{\Delta t = 0} \frac{\overline{PP'}}{\Delta t} = \lim_{\Delta t = 0} \frac{\Delta s}{\Delta t} = D_t s = v, \quad \text{numerically.}$$

Moreover the direction of the variable vector (PQ) approaches a fixed direction as its limit. And so the variable vector (PQ) approaches a fixed vector, $\mathbf{v}$, as its limit, or

$$\lim_{\Delta t = 0} \frac{(PP')}{\Delta t} = \lim_{\Delta t = 0} (PQ) = \mathbf{v}.$$

This vector, $\mathbf{v}$, is defined as the *vector velocity* of the point P.

Cartesian Coordinates. To prove that the above limit actually exists, consider the components of (PP') and (PQ) along the axes. These are:

$$\Delta x, \ \Delta y, \ \Delta z \qquad \text{and} \qquad \frac{\Delta x}{\Delta t}, \ \frac{\Delta y}{\Delta t}, \ \frac{\Delta z}{\Delta t}.$$

The last three variables approach limits:

$$\lim_{\Delta t = 0} \frac{\Delta x}{\Delta t} = D_t x, \qquad \lim_{\Delta t = 0} \frac{\Delta y}{\Delta t} = D_t y, \qquad \lim_{\Delta t = 0} \frac{\Delta z}{\Delta t} = D_t z.$$

Hence (PQ) approaches a limit, $\mathbf{v}$, and the components of $\mathbf{v}$ along the axes are:

* When it is not feasible to represent vectors by bold face type, the ( ) notation may be used, as: (PP') or, later, (α).—The student should draw the figure which represents the vectors (PP') and (PQ).

$$v_x = \frac{dx}{dt}, \qquad v_y = \frac{dy}{dt}, \qquad v_z = \frac{dz}{dt}.$$

These equations admit the following physical interpretation. Consider the projections, L, M, N, of the point P on the axes of coordinates. The velocities with which these points are moving along the axes are precisely dx/dt, dy/dt, and dz/dt. And so we can say: *The projections of the vector velocity* $\mathbf{v}$ *along the axes are equal respectively to the velocities of the projections.*

Finally, observe that, just as the average vector velocity approaches the actual vector velocity as its limit, so the projections of the average vector velocity approach the projections of the actual vector velocity as their limits.

Remark. The student may raise the question: If v_x, v_y, and v_z are the components of the vector velocity, $\mathbf{v}$, are they not, therefore, themselves vectors, and should they not be written as such, $\mathbf{v}_x$, $\mathbf{v}_y$, $\mathbf{v}_z$? Yes, this is correct. But it does not conflict with the other view of v_x, v_y, and v_z as *directed line segments* on the axes of x, y, and z. For, a system of vectors whose direction (but not sense) is fixed, constitute a system of *one-dimensional vectors,* and these are equivalent to *directed line segments,* since the two systems stand in a one-to-one relation to each other. One-dimensional vectors can be represented arithmetically by the ordinary real numbers, positive, negative, and zero.

EXERCISES

1. Show that, if polar coordinates in the plane are used, the component velocities along and orthogonal to the radius vector are respectively:

$$v_r = \frac{dr}{dt}, \qquad v_\theta = r\frac{d\theta}{dt}.$$

2. A point moves on the surface of a sphere. Show that

$$v_\theta = a\frac{d\theta}{dt}, \qquad v_\varphi = a\sin\theta\frac{d\varphi}{dt},$$

where θ and φ denote respectively the co-latitude and the longitude.

3. A point moves in space. Show that

$$v_r = \frac{dr}{dt}, \qquad v_\theta = r\frac{d\theta}{dt}, \qquad v_\varphi = r\sin\theta\frac{d\varphi}{dt},$$

where r, θ, φ are the spherical coordinates of the point.

16. Vector Acceleration. Let a point describe a path, as in § 15. By the *vector change* in its velocity is meant the vector

$$(1) \qquad \Delta \mathbf{v} = \mathbf{v}' - \mathbf{v},$$

cf. Fig. 65, p. 96. The *average vector acceleration* is defined as the vector

$$\frac{\Delta \mathbf{v}}{\Delta t}.$$

When Δt approaches 0, the average vector acceleration approaches a limiting value, and this limiting vector is defined as the *vector acceleration* of the point:

$$(\alpha) = \lim_{\Delta t = 0} \frac{\Delta \mathbf{v}}{\Delta t}.$$

Cartesian Coordinates. The components of the vector acceleration along the Cartesian axes, α_x, α_y, and α_z, are readily computed. For, the components of the vector (1) along the axes are respectively:

$$\frac{v'_x - v_x}{\Delta t}, \qquad \frac{v'_y - v_y}{\Delta t}, \qquad \frac{v'_z - v_z}{\Delta t}.$$

As in the case of velocities, the components of the limiting vector and the limits approached by the components of the variable vector are respectively equal.* Hence

$$\alpha_x = \lim_{\Delta t = 0} \frac{\Delta v_x}{\Delta t} = D_t v_x, \quad \alpha_y = \lim_{\Delta t = 0} \frac{\Delta v_y}{\Delta t} = D_t v_y, \quad \alpha_z = \lim_{\Delta t = 0} \frac{\Delta v_z}{\Delta t} = D_t v_z,$$

or:

$$\alpha_x = \frac{d^2 x}{dt^2}, \qquad \alpha_y = \frac{d^2 y}{dt^2}, \qquad \alpha_z = \frac{d^2 z}{dt^2}.$$

Osculating Plane and Principal Normal.† Let a vector $\mathbf{r}$ be drawn from an arbitrary fixed point O of space to the variable point P that is tracing out the curve (1), § 15. Then

$$\mathbf{v} = \frac{d\mathbf{r}}{dt} = \dot{\mathbf{r}}.$$

Let s be the arc, measured in the sense of the motion; and let

$$\frac{d\mathbf{r}}{ds} = \mathbf{r}'.$$

* This theorem is true of any vector which approaches a limit, as the student can readily verify.

† Cf. the Author's *Advanced Calculus*, p. 304, § 8.

Then $\mathbf{r}'$ is a unit vector lying along the tangent and directed in the sense of the motion. Furthermore,

$$\frac{d\mathbf{r}'}{ds} = \mathbf{r}''$$

is a vector drawn along the principal normal, toward the concave side of the projection of the curve on the osculating plane, and its length is the curvature, κ, at P.

On the other hand, the acceleration

$$(\alpha) = \frac{d\mathbf{v}}{dt} \quad \text{and} \quad \mathbf{v} = \frac{ds}{dt}\frac{d\mathbf{r}}{ds} = v\,\mathbf{r}'.$$

Hence

$$(\alpha) = \frac{d^2 s}{dt^2}\mathbf{r}' + \left(\frac{ds}{dt}\right)^2 \mathbf{r}''.$$

EXERCISES

1. A point describes a circle with constant velocity. Show that the vector acceleration is normal to the path and directed toward the centre of the circle, and that its magnitude is

$$\frac{v^2}{r}, \quad \text{or} \quad \omega^2 r.$$

2. Show that, when a point is describing an arbitrary plane path, the components of the vector acceleration along the tangent and normal are:

$$\alpha_t = \frac{d^2 s}{dt^2}, \qquad \alpha_n = \frac{v^2}{\rho},$$

where ρ denotes the radius of curvature, and the component α_n is directed toward the concave side of the curve.

3. A point describes a cycloid, the rolling circle moving forward with constant velocity. Show that the acceleration is constant in magnitude and always directed toward the centre of the circle.

4. Prove by vector methods that, in the case of motion in a plane,

$$\alpha_r = \frac{d^2 r}{dt^2} - r\left(\frac{d\theta}{dt}\right)^2, \qquad \alpha_\theta = \frac{1}{r}\frac{d}{dt}\left(r^2\frac{d\theta}{dt}\right),$$

where α_r, α_θ denote the components of the acceleration along and perpendicular to the radius vector.

Use the system of ordinary complex numbers, $a + bi$, where $i = \sqrt{-1}$, and set

$$\mathbf{r} = r e^{\theta i}.$$

5. Obtain the same results by geometric methods.

17. Newton's Second Law. Let a particle move under the action of any forces, and let **F** be their resultant. Let (α) be its vector acceleration. Then Newton's Second Law of Motion asserts that *the mass times the vector acceleration is proportional to the vector force*, or, if the absolute unit of force is adopted,

$$m(\alpha) = \mathbf{F}. \tag{1}$$

In Cartesian coordinates the law becomes:

$$\left\{\begin{aligned} m\frac{d^2x}{dt^2} &= X, \\ m\frac{d^2y}{dt^2} &= Y, \\ m\frac{d^2z}{dt^2} &= Z. \end{aligned}\right. \tag{2}$$

If, in particular, X, Y, Z are continuous functions of x, y, z, dx/dt, dy/dt, dz/dt, and t, it then follows from the theory of differential equations that the path is uniquely determined by the initial conditions; *i.e.* if the particle is projected from a point (x_0, y_0, z_0) with a velocity whose components along the axes are (u_0, v_0, w_0), the path is completely determined. This remark is striking when one considers that the corresponding theorem is not true if one determines the motion by means of the principle of Work and Energy; cf. the Author's *Advanced Calculus*, p. 351, Singular Solutions. The essential point here is that Equations (2) never admit a singular solution, whereas the equations of Work and Energy do.

In the more general cases it is also seen that the path is uniquely determined by the initial conditions. This statement is confirmed in the case of each of the examples considered below. For a general treatment, cf. Appendix A.

Osculating Plane. The force, **F**, always lies in the osculating plane of the path. For, from § 16, and Equation (1) above,

$$m\frac{d^2s}{dt^2}\mathbf{r}' + mv^2\mathbf{r}'' = \mathbf{F}.$$

Hence we can resolve **F** into a component T along the path and a component N along the principal normal, and we shall then have:

$$m\frac{d^2s}{dt^2} = T, \qquad \frac{mv^2}{\rho} = N,$$

where $\rho = 1/\kappa$.

EXERCISE

Show that
$$(\mathbf{r}' \times \mathbf{r}'') \cdot \mathbf{F} = 0.$$

Hence, in Cartesian coordinates,
$$(y'\,z'')\,X + (z'\,x'')\,Y + (x'\,y'')\,Z = 0,$$

where
$$(y'\,z'') = y'\,z'' - z'\,y'', \qquad \text{etc.}$$

and
$$x' = \frac{dx}{ds}, \qquad x'' = \frac{d^2x}{ds^2}, \qquad \text{etc.}$$

18. Motion of a Projectile. *Problem.* To find the path of a projectile acted on only by the force of gravity.

The degree of accuracy of the approximation to the true motion obtained in the following solution depends on the projectile and on the velocity with which it moves. For a cannon ball it is crude, though suggestive, whereas for the 16 lb. shot, used in putting the shot, it is decidedly good.

Hitherto we have known the path of the body; here we do not. The path will obviously be a plane curve, and so Newton's Second Law of Motion becomes:

$$(1) \qquad \begin{cases} m\dfrac{d^2x}{dt^2} = X, \\[2ex] m\dfrac{d^2y}{dt^2} = Y, \end{cases}$$

where X, Y are the components of the resultant force along the axes, measured in absolute units.

In the present case $X = 0$, $Y = -\,mg$, and we have

$$(2) \qquad \begin{cases} m\dfrac{d^2x}{dt^2} = 0, \\[2ex] m\dfrac{d^2y}{dt^2} = -\,mg. \end{cases}$$

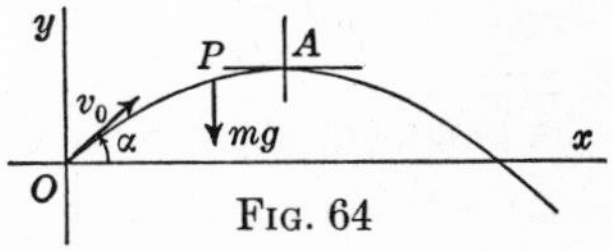

FIG. 64

If we suppose the body projected from O with velocity v_0 at an angle α with the horizontal, the integration of these equations gives:

$$\frac{dx}{dt} = C = v_0 \cos\alpha, \qquad\qquad x = v_0\,t\cos\alpha;$$

$$\frac{dy}{dt} = v_0 \sin\alpha - gt, \qquad\qquad y = v_0\,t\sin\alpha - \tfrac{1}{2}gt^2.$$

Eliminating t we get:

(3) $$y = x \tan \alpha - \frac{gx^2}{2v_0^2 \cos^2 \alpha}.$$

The curve has a maximum at the point $A : (x_1, y_1)$,

$$x_1 = \frac{v_0^2 \sin \alpha \cos \alpha}{g}, \qquad y_1 = \frac{v_0^2 \sin^2 \alpha}{2g}.$$

Transforming to a set of parallel axes through A, we have:

$$x = x' + x_1, \qquad y = y' + y_1,$$

(4) $$y' = -\frac{gx'^2}{2v_0^2 \cos^2 \alpha}.$$

This curve is a parabola with its vertex at A. The height of its directrix above A is $v_0^2 \cos^2 \alpha/2g$, and hence the height above O of the directrix of the parabola represented by (3) is

$$\frac{v_0^2 \sin^2 \alpha}{2g} + \frac{v_0^2 \cos^2 \alpha}{2g} = \frac{v_0^2}{2g}.$$

The result is independent of the angle of elevation α, and so it appears that all the parabolas traced out by projectiles leaving O with the same velocity have their directrices at the same level, the distance of this level above O being the height to which the projectile would rise if shot perpendicularly upward.

EXERCISES

1. Show that the range on the horizontal is

$$R = \frac{v_0^2}{g} \sin 2\alpha,$$

and that the maximum range $\bar{R}$ is attained when $\alpha = 45°$:

$$\bar{R} = \frac{v_0^2}{g}.$$

The height of the directrix above O is half this latter range.

2. A projectile is launched with a velocity of v_0 ft. a sec. and is to hit a mark at the same level and within range. Show that there are two possible angles of elevation and that one is as much greater than 45° as the other is less.

3. Find the range on a plane inclined at an angle β to the horizon and show that the maximum range is

$$R_\beta = \frac{v_0^2}{g} \frac{1}{1 + \sin \beta}.$$

4. A small boy can throw a stone 100 ft. on the level. He is on top of a house 40 ft. high. Show that he can throw the stone 134 ft. from the house. Neglect the height of his hand above the levels in question.

5. The best collegiate record for putting the shot was, at one time, 46 ft. and the amateur and world's record was 49 ft. 6 in.

If a man puts the shot 46 ft. and the shot leaves his hand at a height of 6 ft. 3 in. above the ground, find the velocity with which he launches it, assuming that the angle of elevation α is the most advantageous one. *Ans.* $v_0 = 35.87$.

6. How much better record can the man of the preceding question make than a shorter man of equal strength and skill, the shot leaving the latter's hand at a height of 5 ft. 3 in.?

7. Show that it is possible to hit a mark $B: (x_b, y_b)$, provided

$$y_b + \sqrt{x_b^2 + y_b^2} \leqq \frac{v_0^2}{g}.$$

8. A revolver can give a bullet a muzzle velocity of 200 ft. a sec. Is it possible to hit the vane on a church spire a quarter of a mile away, the height of the spire being 100 ft.?

9. It has been assumed that the path of the projectile is a plane curve. Prove this assumption to be correct by using all three Equations (2), § 17.

19. Constrained Motion. Let a particle be constrained to move in a given curve, like a smooth bead that slides on a wire. Consider first the case of a plane curve. Let the component of the resultant of all the forces along the tangent be T and along the normal be N. Then Newton's Second Law of Motion, § 17, gives the following equations:

$$(1) \qquad \begin{cases} m\dfrac{d^2 s}{dt^2} = T, \\[2ex] \dfrac{mv^2}{\rho} = N. \end{cases}$$

The proof given in § 17 was based on vector analysis. We will give one for the plane case without the use of vector methods.

Geometric Proof. Compute the components of the vector acceleration along the tangent and along the normal. Let φ

be the angle which the tangent has turned through in passing from P to P'. Then the component of $\Delta\mathbf{v}$ along the tangent will be

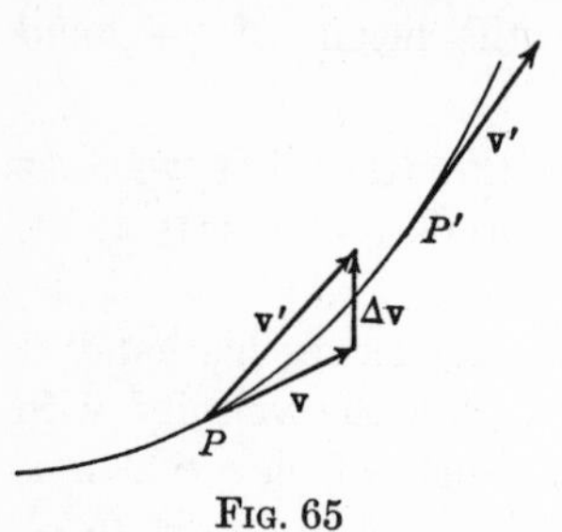

FIG. 65

$$v' \cos \varphi - v = (v + \Delta v) \cos \varphi - v$$
$$= \Delta v \cos \varphi - v\,(1 - \cos \varphi).$$

By the definition of curvature,

$$\kappa = \lim \frac{\varphi}{\widehat{PP'}}, \qquad \rho = \lim \frac{\widehat{PP'}}{\varphi}.$$

Now, the component of the average acceleration along the tangent is

$$\frac{v' \cos \varphi - v}{\Delta t} = \frac{\Delta v}{\Delta t} \cos \varphi - v\,\frac{1 - \cos \varphi}{\Delta t}.$$

Let Δt approach 0. Then φ approaches 0, and the limit of the first term on the right is

$$\left(\lim_{\Delta t = 0} \frac{\Delta v}{\Delta t}\right)\left(\lim_{\varphi = 0} \cos \varphi\right) = D_t v.$$

To evaluate the limit of the second term, write

$$\frac{1 - \cos \varphi}{\Delta t} = \frac{1 - \cos \varphi}{\varphi} \cdot \frac{\varphi}{\Delta s} \cdot \frac{\Delta s}{\Delta t}.$$

The first factor approaches 0, and the second and third factors remain finite, since each approaches a limit. Hence the limit of the right hand side is 0.

We have proved, then, that

$$\lim_{\Delta t = 0} \frac{v' \cos \varphi - v}{\Delta t} = D_t v,$$

and thus the first of Equations (1) is established.

To obtain the second of Equations (1), consider the component of the average acceleration along the normal, or

$$\frac{v' \sin \varphi}{\Delta t}.$$

This can be written as

$$v' \frac{\sin \varphi}{\varphi} \frac{\varphi}{\Delta s} \frac{\Delta s}{\Delta t},$$

where s is assumed to increase with t. The limit of this product is seen to be:

$$v \times 1 \times \frac{1}{\rho} \times v = \frac{v^2}{\rho},$$

and this proves the theorem.

The component N measures the reaction of the curve. It is the centripetal force due to the motion.

The foregoing analysis yields the first of Equations (1) for twisted curves.

EXERCISE

Use the present geometric method to obtain the formulas:

$$\alpha_r = \frac{d^2 r}{dt^2} - r\left(\frac{d\theta}{dt}\right)^2, \qquad \alpha_\theta = \frac{1}{r}\frac{d}{dt}\left(r^2 \frac{d\theta}{dt}\right),$$

where α_r, α_θ denote respectively the components of the vector acceleration along and perpendicular to the radius vector.

20. Simple Pendulum Motion. Consider the simple pendulum. Here

$$m\frac{d^2 s}{dt^2} = -mg \sin\theta,$$

and since $s = l\,\theta$,

$$(1) \qquad \frac{d^2\theta}{dt^2} = -\frac{g}{l}\sin\theta.$$

This differential equation is characteristic for *Simple Pendulum Motion.* We can obtain a first integral by the method of § 7:

$$2\frac{d\theta}{dt}\frac{d^2\theta}{dt^2} = -\frac{2g}{l}\sin\theta\frac{d\theta}{dt},$$

$$\frac{d\theta^2}{dt^2} = -\frac{2g}{l}\int \sin\theta\, d\theta = \frac{2g}{l}\cos\theta + C,$$

$$0 = \frac{2g}{l}\cos\alpha + C,$$

where α is the initial angle; hence

$$(2) \qquad \frac{d\theta^2}{dt^2} = \frac{2g}{l}(\cos\theta - \cos\alpha).$$

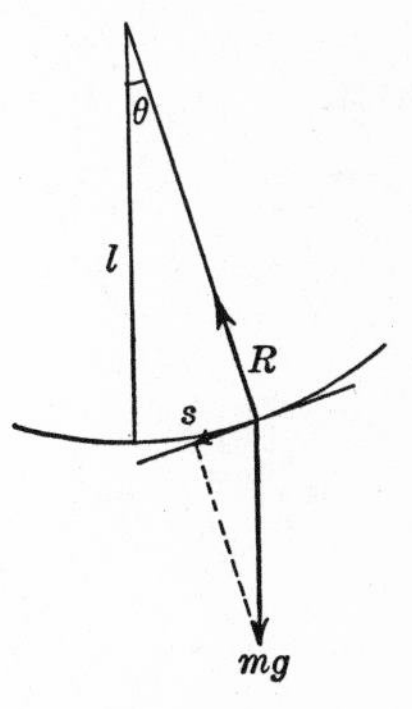

Fig. 66

The velocity in the path at the lowest point is l times the angular velocity for $\theta = 0$, or $\sqrt{2gl(1 - \cos\alpha)}$, and is the same that would have been acquired if the bob had fallen freely under the force of gravity through the same difference in level. Equation (2) is virtually the Integral of Energy.

If we attempt to obtain the time by integrating Equation (2), we are led to the equation:

$$t = \sqrt{\frac{l}{2g}} \int \frac{d\theta}{\sqrt{\cos\theta - \cos\alpha}}.$$

This integral cannot be expressed in terms of the functions at present at our disposal. It is an Elliptic Integral.* When θ, however, is small, $\sin\theta$ differs from θ by only a small percentage of either quantity, and hence we may expect to obtain a good approximation to the actual motion if we replace $\sin\theta$ in (1) by θ:

$$(3) \qquad \frac{d^2\theta}{dt^2} = -\frac{g}{l}\theta.$$

This latter equation is of the type of the differential equation of Simple Harmonic Motion, § 7, A), n^2 having here the value g/l. Hence, when a simple pendulum swings through a small amplitude, its motion is approximately harmonic and its period is approximately

$$T = 2\pi\sqrt{\frac{l}{g}}.$$

The Tautochrone. A question that interested the mathematicians of the eighteenth century was this: In what curve should a pendulum swing in order that the period of oscillation may be absolutely independent of the amplitude? It turns out that the cycloid has this property. For, the differential equation of motion is

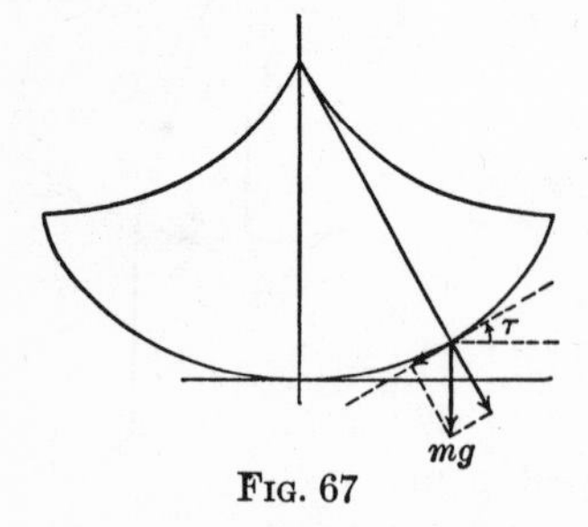

Fig. 67

$$m\frac{d^2s}{dt^2} = -mg\sin\tau,$$

where s is measured from the lowest point, and since

$$s = 4a\sin\tau,$$

we have
$$\frac{d^2s}{dt^2} = -\frac{g}{4a}s.$$

* Cf. the author's *Advanced Calculus*, Chapter IX, page 195, where this integral is reduced to the normal form.

This is the differential equation of Simple Harmonic Motion, § 7, A), and hence the period of the oscillation,

$$T = 2\pi\sqrt{\frac{4a}{g}} = 4\pi\sqrt{\frac{a}{g}},$$

is independent of the amplitude.

A cycloidal pendulum may be constructed by causing the cord of the pendulum to wind on the evolute of the path. The resistances due to the stiffness of the cord as it winds up and unwinds would thus be slight; but in time they would become appreciable.

21. Motion on a Smooth Curve. Let a bead slide on a smooth wire under the force of gravity. Consider first the plane case. Choosing the axes as indicated, we have:

$$(1) \qquad m\frac{d^2s}{dt^2} = mg\cos\tau = mg\frac{dx}{ds}.$$

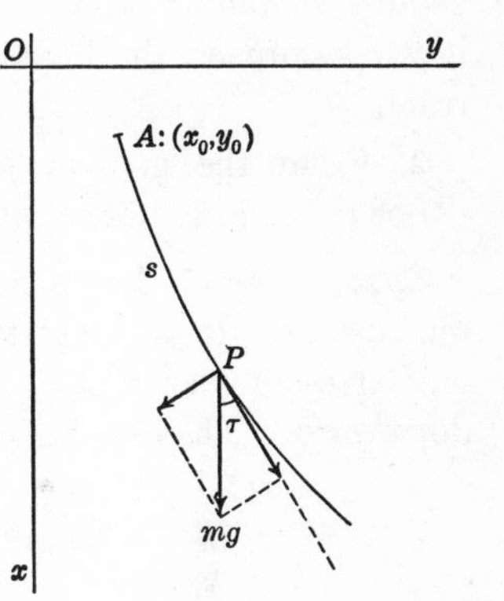

FIG. 68

Hence

$$2\frac{ds}{dt}\frac{d^2s}{dt^2} = 2g\frac{dx}{ds}\frac{ds}{dt} = 2g\frac{dx}{dt}.$$

Integrating this equation with respect to t, we find:

$$\frac{ds^2}{dt^2} = 2gx + C.$$

If we suppose the bead to start from rest at A, then

$$0 = 2gx_0 + C,$$

$$(2) \qquad v^2 = \frac{ds^2}{dt^2} = 2g(x - x_0).$$

But the velocity that a body falling freely from rest a distance of $x - x_0$ attains is expressed by precisely the same formula.

In the more general case that the bead passes the point A with a velocity v_0 we have:

$$v_0^2 = 2gx_0 + C,$$

$$(3) \qquad v^2 - v_0^2 = 2g(x - x_0).$$

Thus it is seen that the velocity at P is the same that the bead would have acquired at the second level if it had been projected vertically from the first with velocity v_0.

The theorem also asserts that the change in kinetic energy is equal to the work done on the bead; cf. § 10.

If the bead starts from rest at A, it will continue to slide till it reaches the end of the wire or comes to a point A' at the same level as A. In the latter case it will in general just rise to the point A' and then retrace its path back to A. But if the tangent to the curve at A' is horizontal, the bead may approach A' as a limiting position without ever reaching it.

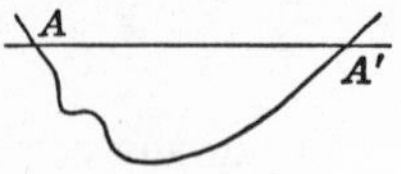

FIG. 69

EXERCISES

1. A bead slides on a smooth vertical circle. It is projected from the lowest point with a velocity equal to that which it would acquire in falling from rest from the highest point. Show that it will approach the highest point as a limit which it will never reach.

2. From the general theorem (2) deduce the first integral (2) of the differential equation (1), § 20.

Space Curves. The same treatment applies to space of three dimensions. It is interesting, however, to give a solution based on Cartesian coordinates. Choose the axis of x as before positive downward. Then we have:

$$(4) \qquad \begin{cases} m\dfrac{d^2x}{dt^2} = mg + R_x, \\[2ex] m\dfrac{d^2y}{dt^2} = R_y, \\[2ex] m\dfrac{d^2z}{dt^2} = R_z, \end{cases}$$

where R_x, R_y, R_z are the components of the reaction $\mathbf{R}$ of the wire along the axes. Since $\mathbf{R}$ is normal to the curve, we have:

$$(5) \qquad R_x\frac{dx}{ds} + R_y\frac{dy}{ds} + R_z\frac{dz}{ds} = 0.$$

To integrate Equations (4) multiply through respectively by dx/dt, dy/dt, dz/dt and add. We thus find, with the aid of (5):

$$(6) \qquad m\left(\frac{dx}{dt}\frac{d^2x}{dt^2} + \frac{dy}{dt}\frac{d^2y}{dt^2} + \frac{dz}{dt}\frac{d^2z}{dt^2}\right) = mg\frac{dx}{dt}.$$

But

$$v^2 = \left(\frac{dx}{dt}\right)^2 + \left(\frac{dy}{dt}\right)^2 + \left(\frac{dz}{dt}\right)^2.$$

Hence (6) reduces to

$$\tfrac{1}{2}\, m\, d\,(v^2) = mg\, dx. \tag{7}$$

On integrating this equation, we find:

$$\frac{mv^2}{2} - \frac{mv_0^2}{2} = mg\,(x - x_0). \tag{8}$$

This is precisely the Equation of Energy. It could have been written down at the start from the Principle of Work and Energy. It is the generalization of (2) for space curves.

EXERCISE

A bead slides on a smooth wire in the form of a helix, axis vertical. Determine the reaction of the wire in magnitude and direction.

22. Centrifugal Force. When a particle of mass m describes a circle with constant velocity, the acceleration is directed toward the centre, and its magnitude is

$$\frac{v^2}{r}, \qquad \text{or} \qquad \omega^2 r.$$

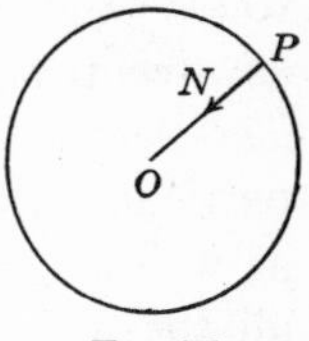

The force which holds the particle in its path is, therefore, normal to the path and directed inward. Its magnitude is

$$N = \frac{mv^2}{r} = m\omega^2 r.$$

FIG. 70

Why, then, the term "centri*fugal* force" — the force that "flees the centre"? The explanation is a confusion of ideas. If the mass is held in its path by a string fastened to a peg at the centre, O, does not the string tug at O in the direction OP away from the centre and is not this force exerted by the particle in its attempt, or tendency, to fly away from the centre? The answer to the first question is, of course, "Yes." Now one of the standard methods of the sophists is to begin with a question on a non-controversial point, conceded without opposition in their favor, and then to confuse the issue in their second question — "and is not this force exerted by the particle?"

Matter cannot exert force, for a *force* is a *push* or a *pull,* and matter can neither push nor pull; it is inert. The particle does not pull on the string, the string pulls on the particle. But even this statement will be accepted only half-heartedly, if at all, by people who have not yet grasped the basic idea of the science of Mechanics — the study of the motion of matter under the action of forces. What comes first is *a material system* — solid bodies, particles, laminae and material surfaces, wires, any combination of these things, including even deformable media (hydrodynamics, elasticity) — and then this system is acted on by forces.

ISOLATE THE SYSTEM

The man who first uttered these words deserves a *monumentum aere.* In the present case there are *two* systems, each of which can be isolated: (1) the particle; (2) whatever the peg is attached to — think of a smooth table, the particle going round and round in a horizontal circle and being held in its path by a string whose other end is attached to a peg at a point O of the table. In the case of the first system, the force that acts is the pull of the string toward the centre, and this force is what is now-a-days described as "centripetal" force — the force that "seeks the centre." The second system has nothing to do with the particle. In particular, this system may be the table. In that case, the floor, as well as gravity, exerts certain forces, and under the action of all the forces, the table stays at rest. The force of the string, varying in direction, causes the forces of the floor to vary.

And now, after all is said and done, comes the rejoinder: "But the particle *did* pull on the string, for otherwise the string would not have pulled on the peg." There is no answer to these people. Some of them are good citizens. They vote the ticket of the party that is responsible for the prosperity of the country; they belong to the only true church; they subscribe to the Red Cross drive — but they have no place in the Temple of Science; they profane it.

Example 1. A bullet weighing 1 oz. is shot into a sling, consisting of a string 5 ft. long with one end fastened at O, the other end carrying a leather cup. If the velocity of the bullet is 600 ft. a sec., how strong must the string be, not to break?

The tension in the string will be

$$\frac{mv^2}{r},$$

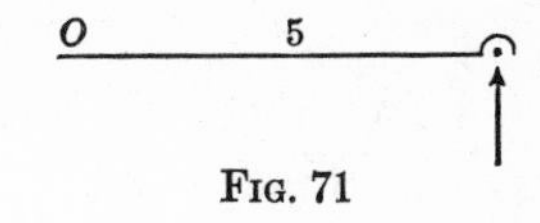

FIG. 71

where $m = \frac{1}{16}$, $v = 600$, $r = 5$; or

$$\frac{600^2}{16 \times 5} = 4500;—$$

4500 what? pounds? No, for the force is measured in *absolute units*, or *poundals*, and so, to get the answer in pounds, we must divide by 32. The tension, then, that the string must be able to withstand is 141 lbs.

Example 2. A railroad train rounds a curve of 1000 ft. radius at 30 m. an hour. How high should the outer rail be raised, if the flanges of the wheels are not to press against either track? Standard gauge, 4 ft. $8\frac{1}{2}$ in.

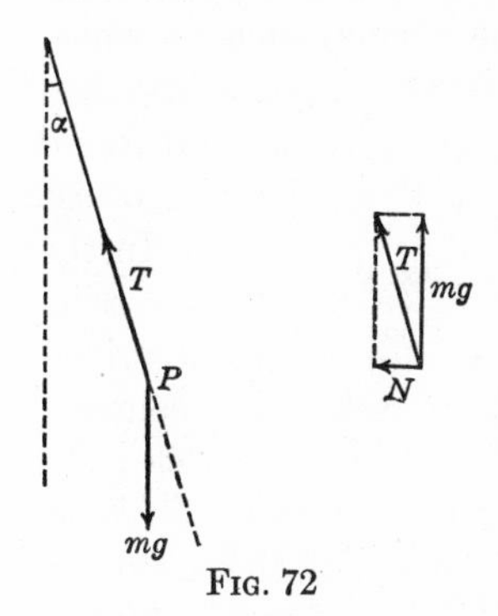

FIG. 72

If a plumb bob is hung up in a car, and does not oscillate, then it should be at right angles to the axles of the wheels. It will describe its circular path in space under the action of two forces, namely, gravity, mg, downward, and the tension, T, of the string. Let the string make an angle α with the vertical. Then the vertical component of T just balances gravity, for there is no vertical motion of the bob. Hence

$$T \cos \alpha = mg.$$

The horizontal component of T yields the normal force N which keeps the bob in its circular path, or

$$T \sin \alpha = N = \frac{mv^2}{r}.$$

Hence

$$\tan \alpha = \frac{v^2}{gr} = \frac{44^2}{32 \times 1000} = .0605.$$

Since the distance between the rails is 4 ft. $8\frac{1}{2}$ in., it follows that the outer rail must be raised 3.42 in.

EXERCISES

1. A particle weighing 4 oz. is attached to a string which passes through a small hole, O, in a smooth table and carries a weight W at its other end. If the first weight is projected along the table from a point P at a distance of 2 ft. from O with a velocity of 50 ft. a second in a direction at right angles to OP, the string being taut and the part below the table vertical, how great must W be, that the 4 oz. weight may describe a circular path? *Ans.* 9 lbs. 12 oz.

2. A boy on a bicycle rounds a corner on a curve of 60 ft. radius at the rate of 10 m. an hour, and his bicycle slips out from under him. What is the greatest value μ could have had? *Ans.* Not quite $\frac{1}{8}$.

3. A conical pendulum is like a simple pendulum, only it is projected so that it moves in a horizontal circle instead of in a vertical one. Show that

$$l\omega^2 = g \sec \alpha.$$

4. If the earth were gradually to stop rotating, how much would Bunker Hill Monument be out of plumb? Given, that the height of the monument is 225 ft. and the latitude of Charlestown is 42° 22′. *Ans.* About $4\frac{1}{2}$ in.

5. An ocean liner of 80,000 tons is steaming east on the equator at the rate of 30 knots an hour. If she puts about and steams west at the same rate, what is the increase in her apparent weight?

6. If the earth were held in her course by steel wires attached to the surface on the side toward the sun and evenly distributed as regards a cross-section by a plane at right angles to them, show that they would have to be as close together as blades of grass. It is assumed that their other ends are guided near the earth's surface.

7. Show that a steel wire one end of which is made fast to the sun and which rotates in a plane with constant velocity, making one rotation in a year, could just about reach to the earth without breaking. Neglect the heat of the sun and all forces of gravitation.

8. A steel wire 1 sq. mm. in cross-section, breaking strength 70 kgs., is strung round the earth along the equator. Show that, if the earth gradually stopped rotating, the wire would snap.

9. What is the smallest latitude such that the wire described in the preceding question, if strung round the earth on that parallel, would not break?

10. If the earth had a satellite close by, how often would the latter rise and set in a day? *Ans.* About 18 times.

11. A boy swings a bucket of water around in a vertical circle without spilling any. Does not the bucket exert a pull on the boy's hand?

Explain the situation by isolating a suitable system, namely: *i*) the bucket of water; *ii*) the boy.

23. The Centrifugal Oil Cup. A device once used for determining the speed of a locomotive consisted of a cylindrical cup containing oil and caused to rotate about its axis, which was vertical, with an angular velocity proportional to the speed of the train. Let us see how it worked.

Suppose the oil to be rotating like a rigid body, with no cross currents or other internal disturbances. What will be the form of the free surface? Imagine a small particle floating on the oil. It will be acted on by the force of gravity, mg, downward and the buoyancy, B, of the oil normal to the surface. The resultant of these two forces must just yield the centripetal force N required to keep the particle in its path. Now

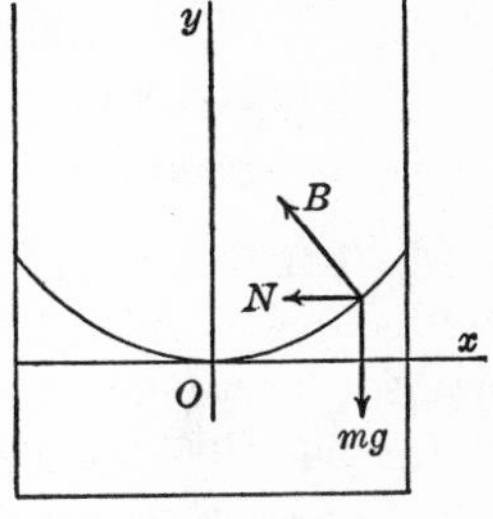

FIG. 73

$$N = m\omega^2 x.$$

On the other hand, the slope of the curve is determined by the fact that the tangent is normal to B. Thus

$$B \cos \tau = mg, \qquad B \sin \tau = N.$$

Hence

$$\tan \tau = \frac{\omega^2 x}{g},$$

or

$$\frac{dy}{dx} = \frac{\omega^2}{g} x.$$

It follows, then, that

$$(1) \qquad y = \frac{\omega^2}{2g} x^2.$$

Thus it appears that the free surface is a *paraboloid of revolution.*

To Graduate the Cup. It is easily shown that the volume of a segment of a paraboloid of revolution is always half the volume of the circumscribing cylinder. If, then, we mark the level of the oil when it is at rest, the height, h, to which it rises above this level when it is in motion will just equal the depth, h, of the lowest point of the surface below this point. From (1) it follows, then, that if a denotes the radius of the cup,

$$2h = \frac{\omega^2}{2g} a^2,$$

or

$$h = \frac{\omega^2}{4g} a^2.$$

EXERCISES

1. A tomato can 4 in. in diameter is filled with water and sealed up. It is placed on a revolving table and caused to rotate about its axis, which is vertical, at the rate of 30 rotations a sec. Find the pressure on the top of the can.

Ans. The weight of a column of water 4 ft. high (nearly) and standing on top of the can.

2. How great is the tendency of the can to rip along the seam?

24. The Centrifugal Field of Force. It is possible to view the mechanical situation in the oil cup from a statical standpoint. Imagine very tiny insects crawling slowly round on the surface of the oil. To them the oil and all they could see of the walls and top of the cup would appear stationary, and they would refer their motion to the rotating space as if it were at rest.

We can reproduce the situation, so far as statical problems are concerned, in a space that is actually at rest by creating a *field of force*, in which the force which acts on a particle of mass m distant r from a fixed vertical axis is the resultant of the force of gravity, mg, vertical and downward, and a force $m\omega^2 r$ directed *away from* the axis, where ω is a constant. Thus the magnitude of the force would be

$$\sqrt{(mg)^2 + (m\omega^2 r)^2} = mg\sqrt{1 + \frac{\omega^4}{g^2} r^2},$$

and it would make an angle φ with the downward vertical, where

$$\tan \varphi = \frac{\omega^2 r}{g}.$$

To bring the mechanical situation nearer to our human intuition, we might think of a large round cup, 500 ft. across at the top, constructed with the flooring in the form of the paraboloid in question and rotating with the suitable angular velocity. There would be a small opening at the vertex, through which observers could enter and leave. The view of all surrounding objects would be cut off, and the mechanical construction would be so nearly perfect that, when we were inside the cup, we should not perceive its motion. Suppose, for example, that the slope of the floor along the rim were 45°. Then, since

$$\tan \tau = \frac{\omega^2 x}{g},$$

it follows that

$$1 = \frac{250\,\omega^2}{32},$$

$$\omega = \tfrac{4}{11} \text{ (nearly), or } .36.$$

The time, T, of a complete revolution is given by the equation:

$$\omega = \frac{2\pi}{T},$$

or

$$T = \frac{2\pi}{\omega} = 17 \text{ secs.}$$

Thus the cup would make nearly four revolutions a minute. Since $\omega^2/g = \frac{1}{250}$, the intensity of the field would be

$$mg\sqrt{1 + (0.004r)^2},$$

and upon the rim of the cup, this would amount to $mg\sqrt{2}$, or 41 per cent greater than gravity on the fixed surface of the earth — roughly, two-fifths more. A movie actress who was maintaining her weight in Hollywood, would tip the scales at, — well, how much?

What we have said applies, however, only to bodies that are at rest in the field. When a body moves, still other forces enter, and these will be considered in the chapter on Relative Motion. Nevertheless, we can describe the motion of a projectile directly, since it would be a parabola in the fixed space we started with. Imagine a tennis court laid out with its centre at the lowest point of the bowl. Lob the ball from the back line to the back line, and watch the slice!

One may reasonably inquire concerning the engineering problems of the construction. There will be a tendency of the cup to burst — to fly apart, due to the "centrifugal force." Can it be held together by reinforcing it with steel bands round the outer rim, or will these have all they can do to hold themselves together? It turns out that only one-seventieth of the breaking strength will be needed to hold the band together, thus leaving sixty-nine seventieths for reinforcing.

But since at the rim the "centrifugal force" is as great as the force of gravity, any unbalanced load will cause the cup to tug on its anchorage unmercifully. A hundred men weigh approximately 8 tons, and if they were bunched at a point of the rim, the reaction on the anchorage would be 8 tons. The student will find it interesting to compute the reaction in case a racing car were driven along the rim at 100 miles an hour.

25. Central Force. Let a particle be acted on by a force directed toward a fixed point, O, and depending only on the distance from O, not on the direction. Newton's Second Law of Motion, § 17, then becomes:

$$(1) \qquad \begin{cases} m\left(\dfrac{d^2r}{dt^2} - r\dfrac{d\theta^2}{dt^2}\right) = R, \\[2ex] \dfrac{m}{r}\dfrac{d}{dt}\left(r^2\dfrac{d\theta}{dt}\right) = 0, \end{cases}$$

where R is a continuous function of r.

Law of Areas. The second equation admits a first integral:

$$(2) \qquad r^2\frac{d\theta}{dt} = h.$$

This equation admits a striking interpretation. Consider the area, A, swept out by the radius vector drawn from O to the particle. Then

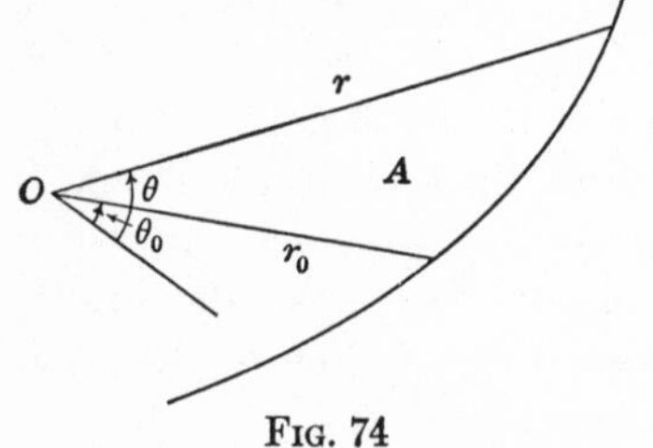

FIG. 74

$$A = \int_{\theta_0}^{\theta} r^2 d\theta,$$

$$\frac{dA}{dt} = r^2\frac{d\theta}{dt} = h,$$

and hence

$$A = h(t - t_0), \tag{3}$$

or, *equal areas are swept out in equal times.*

We have tacitly assumed that $h \neq 0$. If $h = 0$, then (2) reduces to $d\theta = 0$, and the path is a straight line through O.

Work and Energy. The kinetic energy of the particle is

$$\frac{mv^2}{2} = \frac{m}{2}\left(\frac{dr^2}{dt^2} + r^2\frac{d\theta^2}{dt^2}\right). \tag{4}$$

By virtue of (2) this becomes:

$$\frac{mv^2}{2} = \frac{mh^2}{2}\left[\left(\frac{1}{r^2}\frac{dr}{d\theta}\right)^2 + \frac{1}{r^2}\right]. \tag{5}$$

On the other hand, the work, cf. Chap. VII, § 3.

$$W = \int_{r_0}^{r} R\,dr. \tag{6}$$

Hence

$$\left(\frac{1}{r^2}\frac{dr}{d\theta}\right)^2 + \frac{1}{r^2} = \frac{2}{mh^2}\int_{r_0}^{r} R\,dr + \frac{v_0^2}{h^2}. \tag{7}$$

This is a differential equation of the first order connecting r and θ, and its integral gives the form of the path.

The Law of Nature. Newton discovered the Law of Universal Gravitation, which says that any two particles in the universe attract each other with a force proportional to their masses and inversely proportional to the square of the distance between them. This law is often referred to as the Law of Nature.

In the present case, then, the particle is attracted toward O with a force proportional to $1/r^2$, and so

$$|R| \propto \frac{1}{r^2}, \qquad R = -\frac{\lambda}{r^2}. \tag{8}$$

Thus

$$W = \int_{r_0}^{r} -\frac{\lambda}{r^2}\,dr = \lambda\left(\frac{1}{r} - \frac{1}{r_0}\right). \tag{9}$$

The Law of Energy, as expressed in the form of Equation (7), here becomes:

$$\left(\frac{1}{r^2}\frac{dr}{d\theta}\right)^2 + \frac{1}{r^2} = \frac{2\mu}{h^2}\cdot\frac{1}{r} + C, \tag{10}$$

where $\lambda = m\mu$, and C is a constant depending on the initial conditions.

The form of this equation suggests a simplification consisting in substituting for r its reciprocal:

$$u = \frac{1}{r}. \tag{11}$$

Thus (10) becomes:

$$\frac{du^2}{d\theta^2} + u^2 = \frac{2\mu}{h^2}u + C. \tag{12}$$

This equation admits further reduction. Write:

$$\frac{du^2}{d\theta^2} + \left(u - \frac{\mu}{h^2}\right)^2 = C + \frac{\mu^2}{h^4}. \tag{13}$$

Since the left-hand side can never be negative, the right-hand side can be written as B^2, and B itself may be chosen as either one of the square roots. Finally, set

$$x = u - \frac{\mu}{h^2}.$$

Then (13) goes over into:

$$\frac{dx^2}{d\theta^2} + x^2 = B^2. \tag{14}$$

The general integral of this differential equation can be written in the form:

$$x = B\cos(\theta - \gamma), \tag{15}$$

where γ is the constant of integration. When $B = 0$, the truth of this statement is obvious, for then (14) reduces to

$$\frac{dx^2}{d\theta^2} + x^2 = 0,$$

and the only solution of this differential equation is *

$$x = 0.$$

If $B^2 \neq 0$, then (14) yields:

$$d\theta = \frac{dx}{\pm\sqrt{B^2 - x^2}},$$

$$\theta = \int \frac{dx}{\pm\sqrt{B^2 - x^2}} = \pm \cos^{-1}\frac{x}{B} + \gamma,$$

* We have here an example of a differential equation of the first order, handed to us by physics, whose general integral does not depend on an arbitrary constant, but consists of a unique function of θ alone.

where, however, the two $\pm$ signs are not necessarily the same. But in all cases this last equation leads to (15).*

We set out to integrate Equation (12), and we have arrived at the result:

$$u = \frac{\mu}{h^2} + B \cos (\theta - \gamma). \tag{16}$$

This equation can be thrown into familiar form by taking B as the negative radical and setting

$$B = -\frac{\mu e}{h^2},$$

where e now is the constant of integration. Thus (16) yields:

$$r = \frac{h^2}{\mu} \frac{1}{1 - e \cos (\theta - \gamma)}. \tag{17}$$

The Orbit. The path of the particle is given by Equation (17). This is the equation of a conic referred to a focus as pole and having the eccentricity e.

The Case $e < 1$. If $e < 1$, the conic is an ellipse, and the length of the transverse axis is

$$\frac{2h^2}{\mu (1 - e^2)}.$$

Denoting the length of the semi-axes by a and b, we have:

$$a = \frac{h^2}{\mu (1 - e^2)}, \qquad b = \frac{h^2}{\mu\sqrt{1 - e^2}}. \tag{18}$$

The distance between the foci is

$$c = \frac{h^2 e}{\mu (1 - e^2)}. \tag{19}$$

The area of the ellipse is

$$\pi ab = \frac{\pi h^4}{\mu^2 (1 - e^2)^{\frac{3}{2}}}. \tag{20}$$

The periodic time T is connected with the area A by the relation:

$$A = \tfrac{1}{2} hT.$$

Hence

$$T^2 = 4\pi^2 \frac{a^3}{\mu}. \tag{21}$$

* It is worth the student's while to follow through these multiple-valued functions, that he may secure a firmer hold on the Calculus, even though the final result — Equation (15) — is simple.

Determination of the Constants of Integration. Let the body be projected from the point $(r, \theta) = (a, 0)$ with an initial velocity v_0 in a direction making an angle β with the prime direction $\theta = 0$. To determine the orbit.

We will mention first a general formula. Let ψ be the angle from the radius vector produced to the tangent. Then

$$r\frac{d\theta}{dt} = v \sin \psi,$$

since each side represents the component v_θ of the vector velocity, $\mathbf{v}$, perpendicular to the radius vector. By virtue of (2) this becomes:

$$h = vr \sin \psi, \tag{22}$$

and this is the formula we had in mind.

To determine the constants in (17), then, write the equation in the form:

$$u = \frac{\mu}{h^2}\big(1 - e \cos (\theta - \gamma)\big). \tag{23}$$

Hence

$$\frac{1}{a} = \frac{\mu}{h^2}(1 - e \cos \gamma), \qquad e \cos \gamma = 1 - \frac{h^2}{a\mu}. \tag{24}$$

Furthermore,

$$\frac{du}{d\theta} = \frac{\mu e}{h^2} \sin (\theta - \gamma).$$

Since

$$\frac{du}{d\theta} = -\frac{1}{r^2}\frac{dr}{d\theta} = -\frac{1}{h}\frac{dr}{dt} = -\frac{v \cos \psi}{h}, \tag{25}$$

we have initially:

$$-\frac{\mu e}{h^2} \sin \gamma = -\frac{v_0 \cos \beta}{h}, \qquad e \sin \gamma = \frac{v_0 h \cos \beta}{\mu}. \tag{26}$$

From (22),

$$h = v_0 a \sin \beta, \qquad \cos^2 \beta = 1 - \frac{h^2}{v_0^2 a^2}. \tag{27}$$

Squaring the second equation in (24) and (26), and adding, we find by the aid of (27):

$$e^2 = 1 - \frac{2h^2}{a\mu} + \frac{v_0^2 h^2}{\mu^2}. \tag{28}$$

The evaluation is now complete. By means of (27), h is determined; (28) then gives e, and (24) and (26) yield γ.

From (28) we infer that

$$1 - e^2 = \frac{h^2}{\mu^2}\left(\frac{2\mu}{a} - v_0^2\right), \tag{29}$$

and this equation contains the interesting result that the orbit will be the following conic:

i) ellipse, if $v_0^2 < \frac{2\mu}{a}$;

ii) parabola, if $v_0^2 = \frac{2\mu}{a}$;

iii) hyperbola, if $v_0^2 > \frac{2\mu}{a}$,

irrespective of the direction, β, in which the body is launched.

For motion in a circle, $e = 0$. From (24)

$$\frac{1}{a} = \frac{\mu}{h^2}, \qquad h^2 = \mu a. \tag{30}$$

Moreover, from (26) we see that $\beta = \pm \pi/2$, and so we infer from (27) that

$$h^2 = v_0^2 a^2.$$

Hence, by the aid of (30),

$$v_0^2 = \frac{\mu}{a}. \tag{31}$$

Conversely, conditions (30) and (31) are sufficient, that the path be a circle. For from (27) follows that $\cos^2 \beta = 0$, and (29) gives $e = 0$. The result checks with the fact that the numerical value of R, or $m\mu/a^2$, is equal to the centripetal force, or mv_0^2/a.

EXERCISES

1. Show that if

$$r^2 \frac{d\theta}{dt} = h \qquad \text{and} \qquad u = \frac{1}{r},$$

then

$$\frac{d^2 r}{dt^2} - r \frac{d\theta^2}{dt^2} = - h^2 u^2 \left(\frac{d^2 u}{d\theta^2} + u\right).$$

2. It has been assumed that the orbit is a plane curve. Prove this to be the case by means of a constraint, consisting of a smooth plane through O, the point of projection, and the tangent to the path at that point. Use Newton's Equations, § 17, (2), and show that the force of the constraint is 0.

26. The Two Body Problem. If two particles of masses m, m', attracting each other according to the law of nature, and acted on by no other forces, be projected in any manner, their centre of gravity, G, will describe a right line, with constant velocity, or remain permanently at rest; cf. Chapter IV, § 1. Consider the latter case. Let G be the fixed point O, and let the distances of the particles from O be r, r'. Then the force of their mutual attraction is

$$f = K\frac{mm'}{(r + r')^2},$$

where K is the gravitational constant. On the other hand,

$$mr = m'r'.$$

Hence

$$r + r' = \frac{m + m'}{m'}r,$$

and so

$$f = K\frac{mM}{r^2}, \qquad M = \left(\frac{m'}{m + m'}\right)^2 m'.$$

Thus the particle m is attracted toward O with the force that would be exerted by a mass M fixed at O, and so the orbit of m is determined by the work of § 25. In particular, if m describes an ellipse, m' will describe a similar ellipse with the same focus, being turned through an angle of 180°.

27. The Inverse Problem — to Determine the Force. Let a particle move in a plane according to the Law of Areas. Then

$$r^2\frac{d\theta}{dt} = h,$$

and the component of the force perpendicular to the radius vector, Θ, is nil. Hence the particle is acted on by a central force, R, either attractive or repulsive. From Exercise 1, § 25, we have:

$$-\,mh^2u^2\left(\frac{d^2u}{d\theta^2} + u\right) = R.$$

Example. Let the path be an ellipse (or, more generally, any conic) with the centre of force at a focus. Then

$$u = \frac{1 - e\cos(\theta - \gamma)}{p}, \qquad p = \text{const.},$$

$$\frac{d^2u}{d\theta^2} + u = \frac{1}{p},$$

and

$$R = -\frac{mh^2}{p}\frac{1}{r^2}.$$

The force is, therefore, an attractive force, inversely proportional to the square of the distance from the centre, when r lies between its extreme values for this ellipse. But an arbitrary range of values, $0 < \alpha < \gamma < \beta$, can be included in such an ellipse, and so the result is general.

EXERCISE

Show that if the path is an ellipse with the centre of force at the centre, the force is proportional to the distance from the centre.

28. Kepler's Laws. From observations made by Tycho Brahe, Kepler deduced the laws which govern the motion of the planets.

1. *The planets describe plane curves about the sun according to the law of areas;*

2. *The curves are ellipses with the sun at a focus;*

3. *The squares of the periodic times of revolution are proportional to the cubes of the major axes of the ellipses.*

Newton's Inferences. From Kepler's laws Newton drew the following inferences. Consider a particular planet. From the first law it follows that the force acting on it is a central force, since the component Θ at right angles to the radius vector is nil.

From the second law, combined with the first, it follows from § 27 that the force is inversely proportional to the square of the distance from the centre, or

$$P = \frac{m\mu}{r^2}.$$

It has been shown in § 25, (21) that

$$T^2 = 4\pi^2 \frac{a^3}{\mu},$$

where T denotes the periodic time, and a is the semi-axis major. For a second planet,

$$T'^2 = 4\pi^2 \frac{a'^3}{\mu'}.$$

Kepler's third law gives, then, that $\mu' = \mu$, or that μ is the same for all the planets.

To sum up, then, Newton inferred that the planets are attracted toward the sun with a force proportional to their masses and inversely proportional to the square of their distances from the sun.

From here it is but a step to the Law of Universal Gravitation. If the sun attracts the planets, so must, by the principal of action and reaction, the planets attract the sun. Let M denote the mass of the sun, thought of as at rest.* Then

$$P = K \frac{mM}{r^2}.$$

Thus the Law of Universal Gravitation is evolved: Any two bodies (particles) in the universe attract each other with a force proportional to their masses and inversely proportional to the square of the distance between them, or

$$f \propto \frac{mm'}{r^2}, \qquad f = K \frac{mm'}{r^2}.$$

The factor K is called the *gravitational constant.* Its value in c.g.s. units is

$$K = 6.5 \times 10^{-8};$$

cf. Appell, l.c., pp. 390–405.

EXERCISES

1. Show that the first of the equations (1), § 25:

$$\frac{d^2 r}{dt^2} - r \frac{d\theta^2}{dt^2} = -\frac{\mu}{r^2},$$

* For a more detailed treatment cf. Appell, *Mécanique rationnelle,* vol. 1, 3d ed., 1909, § 229 et seq.

on making the transformation (11):

$$u = \frac{1}{r},$$

and employing (2):

$$r^2 \frac{d\theta}{dt} = h,$$

goes over into the equation:

$$\frac{d^2 u}{d\theta^2} + u = \frac{\mu}{h^2}.$$

Hence obtain (16):

$$u = \frac{\mu}{h^2} + B \cos (\theta - \gamma).$$

2. Prove that

$$v = \frac{h}{p},$$

where p denotes the distance from O to the tangent to the path.

3. Show that the earth's orbit, assumed circular, would become parabolic if half the sun's mass were suddenly annihilated, the sun being assumed to be at rest.

4. A smooth tube revolves around one end in a fixed plane with constant angular velocity. A particle is free to move in the tube. Determine the motion.

5. If, in the preceding question, an elastic string is made fast to the particle and attached to the end of the tube, determine the motion.

6. A particle is attracted toward a fixed centre with a force proportional to the distance. Show that the path is a plane curve, and that it can be represented by the equations:

$$x = A \cos (nt + \alpha), \qquad y = B \sin (nt + \alpha).$$

Is it an ellipse?

7. Show that a comet describing a parabolic path cannot remain within the earth's orbit, assumed circular, for more than the $\left(\frac{2}{3\pi}\right)$-th part of a year, or nearly 76 days.

8. A shell is describing an elliptical orbit under a central attractive force. Prove that, if it explodes, all the pieces will meet again at the same moment; and that after half the interval between the explosion and the collision, each piece will be moving with the same velocity as at the instant of explosion, but in the opposite direction.

9. Show that a particle, moving under the action of a central force, cannot have more than two apsidal distances; cf. Appendix B.

10. Find the law of force when a particle describes a circle, the centre of force being situated on the circumference.

Ans. The inverse fifth power.

11. If two spheres, each one foot in diameter and of density equal to the mean density of the earth (5.6) were released from rest in interstellar space with their surfaces $\frac{1}{4}$ inches apart, how long would it take them to come together?

How great would the error be if their mutual attraction were taken as constant?

12. A cannon ball is fired vertically upward from the Equator with a muzzle velocity of 1500 ft. a sec. How far west of the cannon would it fall, if the earth had no atmosphere?

13. Show that a particle acted on by a central repulsive force varying according to the inverse square, will in general describe a branch of a hyperbola with the centre of force at that focus which lies on the convex side of the branch. What is the exceptional case?

29. On the Notion of Mass. Matter is inert. It cannot exert a force; it cannot push or pull. It yields to force, acquiring velocity in the direction in which the force acts — we are thinking of a particle. By virtue of its inertness it possesses *mass*, which may be described as the quantity of matter which a body contains.

Mass is measured by the effect which force produces on the motion of a body. We assume that force may be measured by a spring balance. If a force, constant in magnitude and direction, be applied to a body initially at rest, the body will acquire a certain velocity in a given time. If the same force be applied to another body, and if the second body acquire the same velocity in the same time, the two bodies shall be said to have the same mass. Thus different substances can be compared as to their masses and on adopting an arbitrary mass as the unit in the case of one substance, the unit can be determined in the case of other substances.

It was proved experimentally by Newton that the forces with which gravity attracts two masses equal according to the above definition, are equal. And so one is led to infer the physical law that the weight of a body is proportional to its mass. This law affords a convenient means of measuring masses, namely, by weighing.

In abstract dynamics, however (to quote from Maxwell), matter is considered under no other aspect than that under which it can have its motion changed by the application of force. Hence any two bodies are of equal mass if equal forces applied to these bodies produce, in equal times, equal changes of velocity. This is the only definition of equal masses which can be admitted in dynamics, and it is applicable to all material bodies, whatever they may be made of.*

In Engineering it has become customary to *define* masses as equal when their weights are equal. We have here a question of a sense of values, and Maxwell has gone on record as declaring unequivocally for the inertia property. To use weight to define mass is like saying that two lengths are equal when the rods by which we measure them have the same weight. Just as space and time stand above mass and force, so, in its elementary importance, the inertia property towers above the law of gravitation.

* Maxwell, *Matter and Motion*, Art. XLVI.

CHAPTER IV

DYNAMICS OF A RIGID BODY

1. Motion of the Centre of Gravity. Let a system of particles be acted on by any forces whatever. The latter may be divided into two classes: *i*) the internal forces; *ii*) the external forces.

By *i*) we mean that the particle m_2 exerts on m_1 a force $\mathbf{F}_{12}$ which may have any magnitude and any direction whatever, or in particular not be present at all, $\mathbf{F}_{12} = 0$. The particle m_1 exerts a force on m_2, which is denoted by $\mathbf{F}_{21}$. And now we assume the physical law that *action and reaction are equal and opposite; i.e.* that the vector $\mathbf{F}_{21}$ is equal and opposite to the vector $\mathbf{F}_{12}$, or

$$\mathbf{F}_{12} + \mathbf{F}_{21} = 0.$$

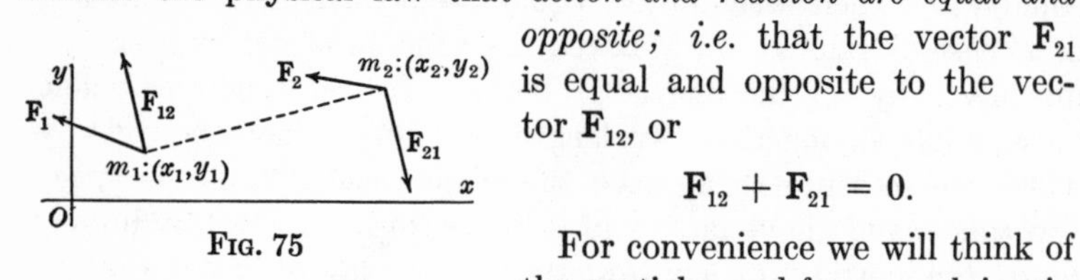

FIG. 75

For convenience we will think of the particles and forces as lying in a plane. The transition to space of three dimensions is immediate.

Denote the components of a vector force $\mathbf{F}$ along the axes of coordinates by X, Y. Then

$$X_{12} + X_{21} = 0, \qquad Y_{12} + Y_{21} = 0.$$

Suppose there are three particles. Then Newton's Second Law of Motion gives for the first of them the equations:

$$m_1 \frac{d^2 x_1}{dt^2} = X_1 + X_{12} + X_{13},$$

$$m_1 \frac{d^2 y_1}{dt^2} = Y_1 + Y_{12} + Y_{13}.$$

There are in all three such pairs of equations, those in x being the following:

$$m_1 \frac{d^2 x_1}{dt^2} = X_1 + X_{12} + X_{13}, \qquad m_2 \frac{d^2 x_2}{dt^2} = X_2 + X_{23} + X_{21},$$

$$m_3 \frac{d^2 x_3}{dt^2} = X_3 + X_{31} + X_{32}.$$

On adding these three equations together, the components X_{ij} on the right, arising from the internal forces, annul one another in pairs, and only the sum of the X_i remains:

$$m_1 \frac{d^2 x_1}{dt^2} + m_2 \frac{d^2 x_2}{dt^2} + m_3 \frac{d^2 x_3}{dt^2} = X_1 + X_2 + X_3.$$

In a similar manner we infer, by writing down the three equations in y and adding, that

$$m_1 \frac{d^2 y_1}{dt^2} + m_2 \frac{d^2 y_2}{dt^2} + m_3 \frac{d^2 y_3}{dt^2} = Y_1 + Y_2 + Y_3.$$

Precisely the same reasoning shows that if, instead of three, we have any number, n, of particles, the internal forces annul one another in pairs, and thus we obtain the result:

$$(1) \qquad \left\{ \begin{aligned} \sum_k m_k \frac{d^2 x_k}{dt^2} &= \sum_k X_k, \\ \sum_k m_k \frac{d^2 y_k}{dt^2} &= \sum_k Y_k. \end{aligned} \right.$$

Coordinates of the Centre of Mass. The left-hand sides of these equations admit a simple interpretation in terms of the motion of the centre of mass of the system. The coordinates, $(\bar{x}, \bar{y})$, of the centre of mass are given by the equations:

$$(2) \qquad \left\{ \begin{aligned} \bar{x} &= \frac{m_1 x_1 + \cdots + m_n x_n}{m_1 + \cdots + m_n} = \frac{\Sigma m_k x_k}{\Sigma m_k}, \\ \bar{y} &= \frac{m_1 y_1 + \cdots + m_n y_n}{m_1 + \cdots + m_n} = \frac{\Sigma m_k y_k}{\Sigma m_k}. \end{aligned} \right.$$

If we denote the total mass by M, then

$$\sum_k m_k x_k = M\bar{x}, \qquad \sum_k m_k y_k = M\bar{y}.$$

Hence we have:

$$\sum_k m_k \frac{d^2 x_k}{dt^2} = M \frac{d^2 \bar{x}}{dt^2}, \qquad \sum_k m_k \frac{d^2 y_k}{dt^2} = M \frac{d^2 \bar{y}}{dt^2},$$

and thus Equations (1) can be written in the form:

$$(3) \qquad \left\{ \begin{aligned} M \frac{d^2 \bar{x}}{dt^2} &= \sum_k X_k, \\ M \frac{d^2 \bar{y}}{dt^2} &= \sum_k Y_k. \end{aligned} \right.$$

These equations are precisely Newton's Second Law of Motion for a particle of mass M, acted on by the given external forces, each transferred to the particle. We can state the result as follows.

THEOREM. *The centre of mass of any system of particles moves as if all the mass were concentrated there and all the external forces acted there.*

In the case of particles in space, there is a third equation, (3) being superseded now by

$$(4)\quad M\frac{d^2\bar{x}}{dt^2} = \sum_k X_k, \qquad M\frac{d^2\bar{y}}{dt^2} = \sum_k Y_k, \qquad M\frac{d^2\bar{z}}{dt^2} = \sum_k Z_k.$$

Remark. There is one detail in the statement of the theorem that requires explicit consideration. We have written down the differential equations of the motion, but we have not integrated them. If we do not start the particle of mass M in coincidence with the initial position of the centre of mass, it obviously cannot describe the same path. More than this, we must give it the same initial velocity (*i.e.* vector velocity). Is this enough to insure its always remaining in coincidence with the centre of mass? The answer to this question is a categorical Yes; cf. Chapter III, § 17 and Appendix B.

Generalized Theorem. We have proved the theorem of the motion of the centre of mass for a system of particles. In the case of a rigid body, we can think of the body as divided up into a large number of cells, each of small maximum diameter; the mass of each cell as then concentrated at one of its points, and the n particles thus resulting as connected by massless rods, after the manner of a truss.* To this auxiliary system of particles the theorem as above developed applies. And now it is intuitionally evident, or plausible, that the system of particles will move in a manner closely similar to that of the rigid body, when the cells are taken very small. One is tempted to say

* It is often necessary to use a truss, at some of whose vertices there are no masses. We may think of minute masses attached at these points and acted on by gravity or by no external forces at all. The effect of these small masses is to modify slightly the value of M in Equations (4). And now it follows from the theory of differential equations that the integrals of (4) are thereby also modified only slightly. Hence the physical assumption is made, that Equations (4) hold even when there are no masses at the vertices in question.

that the motion of the actual body is the limit approached by the motion of the system of particles as n grows large and the cells small. And this is, in fact, true. But this is not a mathematical inference — far from it — it is a *new physical postulate.* We thus extend the theorem and elevate it to a Principle.*

PRINCIPLE OF THE MOTION OF THE CENTRE OF MASS. *The centre of mass of any material system whatsoever moves as if all the mass were concentrated there, and all the external forces acted there:*

$$\text{A)} \qquad \left\{ \begin{aligned} M\frac{d^2\bar{x}}{dt^2} &= \sum_k X_k, \\ M\frac{d^2\bar{y}}{dt^2} &= \sum_k Y_k, \\ M\frac{d^2\bar{z}}{dt^2} &= \sum_k Z_k. \end{aligned} \right.$$

2. Applications. *The Glass of Water.* Suppose a glass of water is thrown out of a third-story window. As the water falls, it takes on most irregular forms, breaking first into large pieces, and these into smaller ones. The forces that act are gravity and the resistance of the atmosphere, the latter spread out all over the surfaces of the pieces. And now the Principle of the last paragraph tells us that the centre of gravity moves as if all the mass were concentrated there and all these forces transferred bodily (*i.e.* as vectors) to that point.

The Falling Chain. Let a chain hang at rest, the lower end just touching a table, and let it be released. To determine the pressure, F, on the table.

We idealize the chain as a uniform flexible string, of length l and density ρ (hence of mass $M = \rho l$), and think of it as impinging always at the same fixed point, O, of the table. Let s be the distance the chain has fallen and let $\bar{x}$ be the height of the centre of gravity above the table. Then the Principle of the Motion of the Centre of Mass gives the equation:

* A "Principle" in Mechanics is well described in the words of Professor Koopman (cf. the Author's *Advanced Calculus*, p. 436): "According to the usage of the present day the word *principle* in physics has lost its metaphysical implication, and now denotes a physical truth of a certain importance and generality. Like all physical truths, it rests ultimately on experiment; but whether it is taken as a physical law, or appears as a consequence of physical laws already laid down, does not matter."

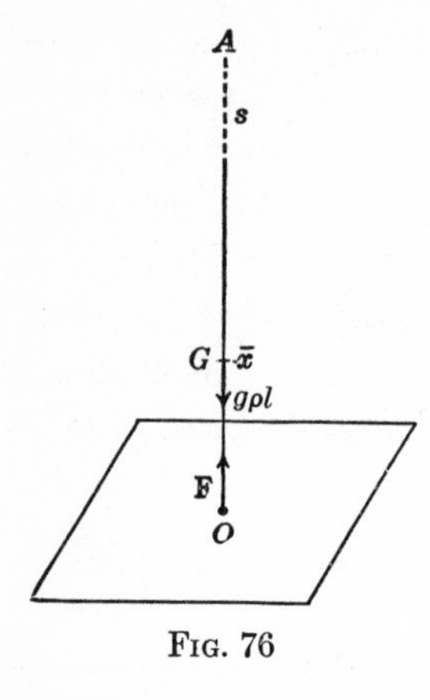

Fig. 76

(1) $$\rho l \frac{d^2 \bar{x}}{dt^2} = F - g\rho l.$$

Now, $$\bar{x} = \frac{(l-s)^2}{2l},$$

$$\frac{d^2 \bar{x}}{dt^2} = \frac{s-l}{l}\frac{d^2 s}{dt^2} + \frac{1}{l}\frac{ds^2}{dt^2}.$$

Moreover, from the laws of freely falling bodies,

$$\frac{d^2 s}{dt^2} = g, \qquad \frac{ds^2}{dt^2} = v^2 = 2gs.$$

On substituting those values in (1), we obtain:

(2) $$g\rho\,(s - l) + \rho v^2 = F - g\rho l.$$

Hence

(3) $$F = g\rho s + \rho v^2,$$

or $$F = 3g\rho s.$$

This means that the pressure of the chain on the table is always just three times the weight of that part of the chain which has already come to rest on the table.

It appears, then, that F is made up of two parts, *i*) the pressure $g\rho s$ on the table, of that part of the chain already at rest; and *ii*) a pressure

(4) $$P = \rho v^2,$$

due to the impact of the chain against the table.

A Stream of Water, Impinging on a Wall. Suppose a hose is turned on a wall (or a convict!). To determine the pressure.

We idealize the motion by thinking of the stream as hitting the wall at right angles, the water spattering in all directions along the wall and thus giving up all its velocity in the line of motion of the stream.

Dynamically, this is precisely the same case as that of the chain falling on the table, so far as the impact is concerned, and hence the pressure is given by (4):

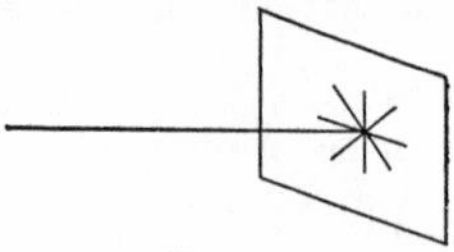

Fig. 77

$$P = \rho v^2.$$

Example. A fire engine is able to send a 2 in. stream to a vertical height of 200 ft. Find the pressure if the stream is played directly on a door. *Ans.* 541 lbs.

The Crew on the River. The crew is out for practice. Observe the cut-water of the shell and describe how it moves, and why it moves as it does. What system do you decide to isolate? the shell? or the shell, oars, and crew?

EXERCISES

1. If a man were placed on a perfectly smooth table, how could he get off?

2. If a shell were fired from a gun on the moon and exploded in its flight, what could you say about the motion of the pieces?

3. A goose is nailed up in an air tight box which rests on platform scales. The goose flies up. Will the scales register more or less or the same?

4. A pail filled with water is placed on some scales. A cork is held submerged by a string tied to the bottom of the pail. The string breaks. Do the scales register more or less or the same?

5. A man, standing in the stern of a row boat at rest, walks forward to the prow. What can you say about the motion of the boat?

6. When the man stops at the prow of the boat, boat and man will be moving forward with a small velocity. Explain why.

7. A uniform flexible heavy string is laid over a smooth cylinder, axis horizontal, and kept from slipping by holding one end, A, fast, the part of the string from A up to the cylinder being vertical. The part of the string on the other side of the cylinder is, of course, also vertical, its lower end, B, being below the level of A, and the whole string lies in a vertical plane perpendicular to the axis of the cylinder. The string is released from rest. Determine the motion, there being a smooth guard which prevents the string from leaving the upper side of the cylinder.

8. If, in Question 7, the difference in level between A and B is 2 ft., and if the distance from A up to the cylinder is 8 ft., compute the velocity of the string when the upper end reaches the cylinder, correct to three significant figures.

9. Find how long it takes the upper end of the string to reach the cylinder.

10. The sporting editor of a leading newspaper recently reported a new stroke which a certain coach had developed, the advantage of which was that it gave an even motion to the shell and avoided the jerkiness of the old-fashioned strokes. Examine this news item.

3. The Equation of Moments. Recall the formula for the moment of a force **F** about the origin, namely,

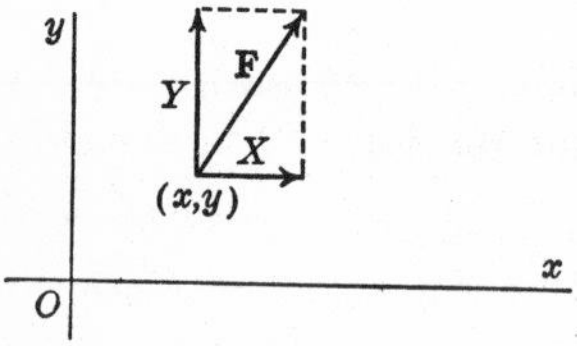

$$(1) \qquad xY - yX.$$

Fig. 78

Consider a system of particles acted on by any external forces whatever, and interacting on one another by forces that are equal and opposite, but are now assumed each time to lie in the line joining the two particles in question. Moreover, the particles shall lie in a fixed plane. Begin with the case of three particles, as in § 1, and write down the six equations that express Newton's Second Law of Motion for these particles.* Next, form the expression:

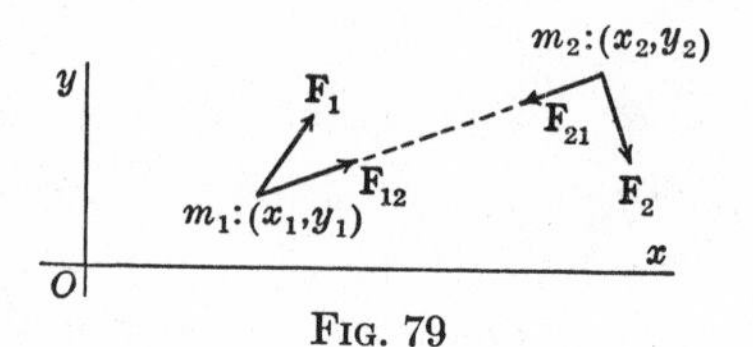

Fig. 79

$$m_1\left(x_1\frac{d^2y_1}{dt^2} - y_1\frac{d^2x_1}{dt^2}\right),$$

and compute its value from the equations in question, namely,

$$(x_1 Y_1 - y_1 X_1) + (x_1 Y_{12} - y_1 X_{12}) + (x_1 Y_{13} - y_1 X_{13}).$$

The parentheses represent respectively the moments of $\mathbf{F}_1$, $\mathbf{F}_{12}$, $\mathbf{F}_{13}$ about the origin.

Now, do the same thing for the particle m_2, and finally, for m_3. On adding these three equations together, it is seen that the

* It is important that the student *do* this, and do it *neatly*, and not merely gaze at the three equations printed in § 1 and try to imagine the three not printed. He should write out the full equation derived below from these, neatly on a single line, and then write the other two under this one.

moments of the internal forces about the origin destroy one another, and there remains on the right-hand side only the sum of the moments of the applied forces.

If there are $n > 3$ particles, $m_1, m_2, \cdots, m_n$, the procedure is the same, and we are thus led to the

THEOREM OF MOMENTS:

$$\text{B)} \qquad \sum_k m_k \left(x_k \frac{d^2 y_k}{dt^2} - y_k \frac{d^2 x_k}{dt^2} \right) = \sum_k (x_k Y_k - y_k X_k).$$

We refrain from writing down the corresponding theorem in three dimensions, because we shall have no need of it for the present.

4. Rotation about a Fixed Axis under Gravity. Let the system of particles of § 3 be rigidly connected, and let one point, O, of the truss-work be at rest, so that the system rotates about O as a pivot. For example, take the case of a uniform rod, one end of which is held fast, and which is released from rest under gravity. Divide the rod into n equal parts, and concentrate the mass of each part, for definiteness, at its most remote point. We thus have a system of n particles, and we connect them rigidly by a massless truss-work as shown in the figure.*

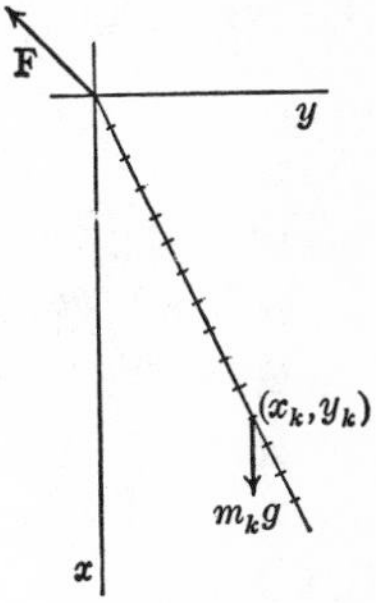

F
y
(x_k, y_k)
$m_k g$
x

FIG. 80

We are now ready to compute each side of Equation B) for the auxiliary system of n particles. Let r be the distance from O to any point fixed in the rod. Then

$$(1) \qquad x = r \cos \theta, \qquad y = r \sin \theta,$$

where θ varies with the time, t, but r is constant with respect to t. Hence

$$(2) \qquad \frac{dx}{dt} = -r \sin \theta \frac{d\theta}{dt}, \qquad \frac{dy}{dt} = r \cos \theta \frac{d\theta}{dt}.$$

We observe next that, in all generality, by mere differentiation, *i.e.* purely mathematically,

$$(3) \qquad \frac{d}{dt}\left(x \frac{dy}{dt} - y \frac{dx}{dt} \right) = x \frac{d^2 y}{dt^2} - y \frac{d^2 x}{dt^2},$$

* Cf. the footnote, § 1.

and we proceed to compute the parenthesis by means of Equations (1) and (2). We find:

(4)
$$x\frac{dy}{dt} - y\frac{dx}{dt} = r^2\frac{d\theta}{dt}.$$

In the present case we have:

$$\frac{d}{dt}\left(x_k\frac{dy_k}{dt} - y_k\frac{dx_k}{dt}\right) = r_k^2\frac{d^2\theta}{dt^2},$$

for r_k does not change with the time, and so $dr_k/dt = 0$. Hence

$$\sum_k m_k\left(x_k\frac{d^2y_k}{dt^2} - y_k\frac{d^2x_k}{dt^2}\right) = \sum_k m_k r_k^2\frac{d^2\theta}{dt^2}.$$

The sum which here appears is the *moment of inertia** of the system about O:

$$I = \sum_k m_k r_k^2.$$

Thus the left-hand side of the Equation of Moments reduces to the expression:

(5)
$$I\frac{d^2\theta}{dt^2}.$$

Turning now to the right-hand side of B) we see that the kth particle, m_k, yields a moment about O equal to the quantity $-m_k g r_k \sin\theta$, and so the sum in question becomes

$$\sum_k - m_k g r_k \sin\theta, \qquad \text{or} \qquad -\left(\sum_k m_k r_k\right) g \sin\theta.$$

But
$$\sum_k m_k r_k = Mh,$$

where h is the distance from O to the centre of gravity, G, of the system of particles. Hence, finally,

(6)
$$I\frac{d^2\theta}{dt^2} = -Mgh\sin\theta.$$

This is substantially the equation of Simple Pendulum Motion, Chapter III, § 20:

(7)
$$\frac{d^2\theta}{dt^2} = -\frac{g}{l}\sin\theta.$$

* Moments of inertia for such bodies as interest us here are treated in the Author's *Introduction to the Calculus*, p. 323.

Hence the system of n particles *oscillates like a simple pendulum* of length

$$(8_1) \qquad l = \frac{I}{Mh}$$

or

$$(8_2) \qquad l = \frac{k^2}{h}, \qquad \text{where} \qquad I = Mk^2,$$

k denoting the *radius of gyration.*

More precisely, what we mean by the last statement is this. Let a simple pendulum be supported at O, let its length be k^2/h, and let it be placed alongside the rod, the bob being at a point distant l from O. If now both be released from rest at the same instant, they will oscillate side by side, though not touching each other.

The Actual Rod. As n grows larger and larger, the massless rod weighted with the n particles comes nearer and nearer to the actual rod, dynamically. This is not a mathematical statement. It expresses our feeling from physics for the situation — our *intuition.* And so when we say that the motion of the actual rod is the limit approached by the motion of the auxiliary rod, we are stating a new physical postulate. The result is, that the actual rod oscillates like a simple pendulum of length

$$\frac{k^2}{h} = \frac{\frac{1}{3}l^2}{\frac{1}{2}l} = \frac{2}{3}\,l.$$

EXERCISES

Apply the *method* set forth in the text, introducing each time an auxiliary set of particles, and proceeding to the limit. Do not try short cuts by attempting to use in part the *result* of the exercise worked in the text.

1. A rod 10 ft. long and weighing 30 lbs. carries a 20 lb. weight at one end and a 30 lb. weight at the other. It is supported at its middle point. Find the length of the equivalent simple pendulum. *Ans.* 30 ft.

2. Equal masses are fixed at the vertices of an equilateral triangle and the latter is supported at one of the vertices. If it be allowed to oscillate in a vertical plane, find the length of the equivalent simple pendulum.

3. A rigid uniform circular wire * 6 in. in diameter and weighing 12 lbs. has a 4 lb. weight fastened at one of its points and is free to oscillate about its centre in its own plane. Find the length of the equivalent simple pendulum.

4. Equal particles are placed at the vertices of a regular hexagon and connected rigidly by a weightless truss. The system is pivoted at one of the particles and allowed to oscillate in a vertical plane under gravity. Find the length of the equivalent simple pendulum.

5. Generalize to the case of n equal particles placed at the vertices of a regular n-gon.

5. The Compound Pendulum. Consider an arbitrary lamina, or plane plate of variable density. Let it be supported at a point O and allowed to oscillate freely in its own plane, assumed vertical, under gravity. This is essentially the most general *compound pendulum*. To determine the motion.

Divide the lamina up in any convenient manner into small pieces and concentrate the mass of each piece at one of its points. Connect these particles with one another and with the support at O by a truss-work. The auxiliary system can be dealt with by the Principle of Moments. Set

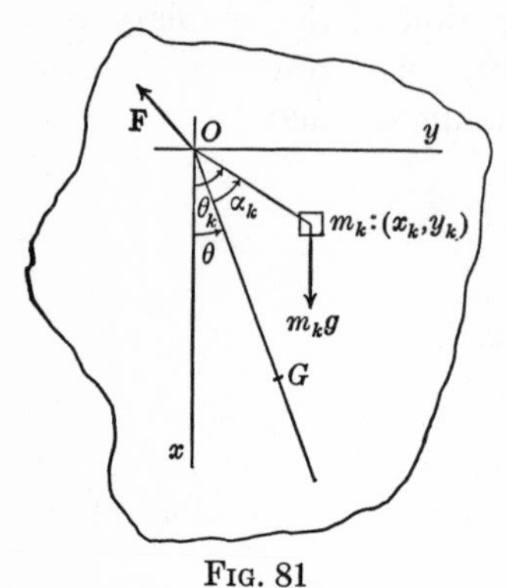

FIG. 81

$$x_k = r_k \cos \theta_k, \qquad y_k = r_k \sin \theta_k.$$

Then

$$x_k \frac{dy_k}{dt} - y_k \frac{dx_k}{dt} = r_k^2 \frac{d\theta_k}{dt}.$$

Now, draw a line in the lamina, — for example, the line through O and the centre of gravity, G, of the particles, — and denote the angle it makes with the axis of x by θ. Then

$$\theta_k = \theta + \alpha_k,$$

where α_k varies with k, but is constant as regards the time. Hence

$$\frac{d\theta_k}{dt} = \frac{d\theta}{dt}, \qquad \frac{d^2\theta_k}{dt^2} = \frac{d^2\theta}{dt^2}.$$

* By a *wire* is always meant a *material curve.*

Thus the left-hand side of the Equation of Moments, § 3, becomes

$$(1) \qquad \sum_k m_k r_k^2 \frac{d^2\theta}{dt^2} = I \frac{d^2\theta}{dt^2},$$

where I denotes the moment of inertia of the system of particles about O.

The right-hand side of B), § 3, can be written

$$(2) \qquad \sum_k - m_k g y_k = - g \sum_k m_k y_k.$$

The last sum has the value $M\bar{y}$, where the coordinates of G are denoted by $(\bar{x}, \bar{y})$. Let the distance from O to G be h. Then

$$\bar{y} = h \sin \theta$$

and (2) becomes

$$(3) \qquad - Mgh \sin \theta.$$

On equating (1) and (3) to each other, we have

$$(4) \qquad I \frac{d^2\theta}{dt^2} = - Mgh \sin \theta.$$

This is the Equation of Simple Pendulum Motion, and

$$(5) \qquad l = \frac{I}{Mh}.$$

It appears, then, that the auxiliary system of particles oscillates like a simple pendulum. As we allow n to increase without limit, the maximum diameter of the little pieces approaching 0, it seems plausible that the motion will approximate more and more closely to that of the actual compound pendulum, and this consideration leads us to lay down the physical law, or *postulate*, that Equation (4) holds for the compound pendulum, where I and h now refer to the latter body.

Remark. We have thought of the mass of the compound pendulum as two-dimensional, or lying in a plane. But this is obviously an unnecessary restriction. Conceive a block of granite, blasted from the quarry — as irregular and jagged as you please. Mount it on two knife-edges, so it can swing about a horizontal axis. Now this block will obviously oscillate exactly as a plane lamina perpendicular to the axis would, if the mass of the actual block were projected parallel to the axis on a plane at right angles to the axis.

The above "obviously" is not to be taken mathematically, but is a new physical law, or postulate. It is true that when we come to treat the general case of motion in three dimensions, this postulate will be merged in more general ones.

EXERCISES

Find the length of the equivalent simple pendulum when the compound pendulum is one of the following.

1. A uniform circular disc, free to rotate in its own plane about a point in its circumference. *Ans.* $l = \frac{3}{2}r$.

2. A circular wire, about a point of the wire. *Ans.* $l = 2r$.

3. Question 1, when the axis is tangent to the disc.
Ans. $l = \frac{5}{4}r$.

4. Question 2, when the axis is tangent to the wire.
Ans. $l = \frac{3}{2}r$.

5. A rectangular lamina, about a side.

6. A square lamina, about a vertex.

7. A triangle, about a vertex.

6. Continuation. Discussion of the Point of Support. Let

$$I_0 = Mk^2$$

be the moment of inertia of the compound pendulum about a parallel axis through the centre of gravity, G. By the theorem of § 10 the moment of inertia about the actual axis will be:

$$I = M(k^2 + h^2),$$

and the length of the equivalent simple pendulum is seen from (5), § 5, to be:

$$l = \frac{h^2 + k^2}{h}. \tag{6}$$

The question arises: What other points of support, O (*i.e.* what other parallel axes), yield the same period of oscillation?

Clearly they are those, and only those, whose distance, x, from G satisfies the equation,

$$l = \frac{x^2 + k^2}{x},$$

$$x^2 - lx + k^2 = 0, \tag{7}$$

where k and l are given, and where, moreover, (6) is true, or

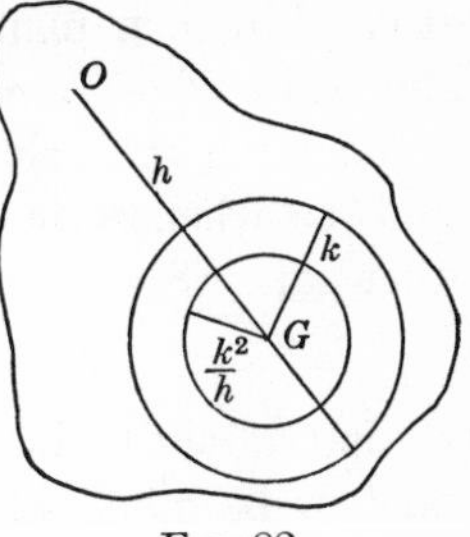

FIG. 82

$$(8) \qquad h^2 - lh + k^2 = 0.$$

One root of Equation (7) is $x_1 = h$. The other is seen to be

$$x_2 = l - h = \frac{k^2}{h}.$$

We can state the result as a theorem.

THEOREM. *The locus of the points O, for which the time of oscillation is the same, consists of two concentric circles with their centre at G, their radii being*

$$h \textit{ and } \frac{k^2}{h}.$$

EXERCISES

1. Draw two concentric circles about G, of radii h and k^2/h. Show that the length, l, of the equivalent simple pendulum corresponding to an axis through a point O on one of these circles is obtained by drawing a line from O through G, and terminating it where it meets the other circle.

This theorem is due to Huygens.

2. Show that the locus of the points of support, for which the time of oscillation is least, form a circle with G as centre and of radius k.

7. Kater's Pendulum. The experiment for determining the value of g, the acceleration of gravity, by means of a simple pendulum and the formula

$$T = 2\pi\sqrt{\frac{l}{g}}$$

is familiar to all students of physics and mathematics. The chief error in the result arises from the error in determining l. The bob is not sensibly a particle and the string stretches.

To attain greater accuracy, Kater made use of Huygens's Theorem, § 6, Ex. 1, constructing a compound pendulum that could be reversed. It consists essentially of a massive rod, or bar, provided with two sets of adjustable knife-edges. These edges lie in two parallel lines, and the centre of gravity, G, is situated in their plane, at unequal distances, h and h', from them. The knife-edges are now so adjusted experimentally that the period when the pendulum oscillates about the one pair is the same as when it is reversed and allowed to oscillate about the other pair. Since

$$l = h + h',$$

the determination of the length of the equivalent simple pendulum can now be made with great accuracy by measuring the distance between the knife-edges. Indeed, the accuracy in thus determining g is now so great that very small errors, like those due to the buoyancy of the air, the changes in the pendulum due to changes in temperature, and the give of the supports have to be considered. For an elaborate and interesting account, cf. Routh, *Elementary Rigid Dynamics*, § 98 *et seq.*

8. Atwood's Machine. An Atwood's Machine consists of a pulley free to rotate about a horizontal axis, and a string passing over the pulley and carrying weights, M and $M + m$, at its two ends. It may be used to measure the acceleration of gravity.

Our problem is to determine the motion of the system. The "system" which we choose to isolate is the complete system of pulley and weights, the mass of the string being assumed negligible. This is not a rigid system, but still, if we replace the pulley by a system of particles rigidly connected, the internal forces of the complete auxiliary system will satisfy the hypothesis * of § 3, and thus the Equation of Moments will hold.

For the auxiliary system of particles due to the wheel the contribution to the left-hand side of the Equation of Moments, B), § 3, becomes as in the case of the compound pendulum:

$$(1) \qquad I\frac{d^2\theta}{dt^2},$$

where I denotes the moment of inertia of this system about the axis, and θ is the angle through which the wheel has rotated.

* Consider a short interval of time in the duration of the motion. In the auxiliary system, let each vertical segment of the string be fastened to a particle near the point of tangency of the string in the actual case. Then it is plausible physically that the motion of the auxiliary system during this short interval differs but slightly from that of the actual system. Hence we may assume that the force of the string always acts at the points of tangency with the wheel, and neglect the rest of the string which is in contact with the wheel. But this is a new physical law.

Let the radius of the wheel (more precisely, of the groove in which the string lies) be a. Observe, too, that

$$y_1 = \text{const.} + a\theta, \qquad y_2 = \text{const.} - a\theta.$$

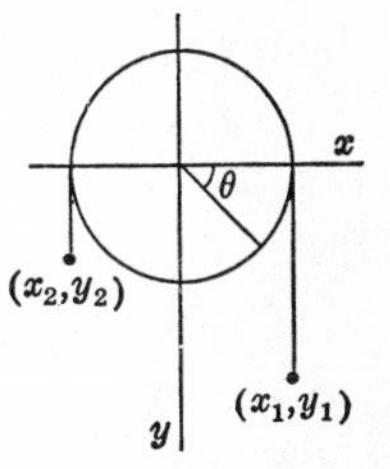

FIG. 83

Thus the remaining contributions to the left-hand side of B), § 3, will be

$$(2) \qquad (M + m)\, a^2 \frac{d^2\theta}{dt^2} + Ma^2 \frac{d^2\theta}{dt^2}.$$

The right-hand side of B) reduces to

$$(3) \qquad (M + m)\, ga - Mga = mga.$$

Thus B) becomes:

$$(4) \qquad [I + (2M + m)\, a^2] \frac{d^2\theta}{dt^2} = mga.$$

This, for the auxiliary system of particles. And now we assume, *physically*, that the limit approached by the motion of the auxiliary system is the motion of the actual system; *i.e.* that Equation (4) holds for the actual system.

Let s denote the distance the weight and rider have descended. Then $s = a\theta$, and from (4) it follows that

$$(5) \qquad \frac{d^2 s}{dt^2} = \frac{mga^2}{I + (2M + m)\, a^2}.$$

On integrating this equation we have, in particular, that

$$(6) \qquad s = \frac{\frac{1}{2} mga^2 t^2}{I + (2M + m)\, a^2}.$$

Corresponding values of s and t can be observed experimentally. Thus Equation (6) is equivalent to a linear equation in the two unknowns, I/a^2 and g:

$$\frac{I}{a^2} - \tfrac{1}{2} m \frac{t^2}{s} g + (2M + m) = 0.$$

If M is held fast and m is given different values, it is clear that the coefficient of g will take on different values, and so we shall have two independent linear equations for determining the unknown physical constants, I/a^2 and g.

EXERCISES

In working these exercises use the *method*, not the *result*, of the text. Begin each time by drawing a figure.

1. Suppose that the wheel is a uniform circular disc weighing 10 lbs., and that 5 lb. weights are fastened to the two ends of the string. What will be the acceleration due to a 1 oz. rider?

2. Work the case in which the wheel is a hoop, *i.e.* a uniform circular wire, the masses of the spokes being negligible; and show that the acceleration of the rider does not depend on the radius, but only on the mass of the hoop, and M and m.

3. Determine the tensions in the string in the general case.

4. Find the reaction on the axis.

5. Prove the assertion in the text about the coefficient of g's taking on different values when m is varied.

6.* How rough must the string be in the general case, in order not to slip?

9. The General Case of Rotation about a Point. Consider an arbitrary rigid body in two dimensions, acted on by any forces in its plane, and free to rotate about a point O, *i.e.* about an axis through O perpendicular to the plane. Then, I say, its motion is determined by the Principle of Moments,

$$\text{B)} \qquad I\frac{d^2\theta}{dt^2} = \sum \text{Moments about } O.$$

The Principle is rendered plausible by dividing the actual distribution into small pieces, as in the example of the compound pendulum and the Atwood's machine, and observing that the Principle is true for the auxiliary system. The limit approached by the motion of the auxiliary system is the motion defined by Equation B) of the present paragraph. And thus we are led to lay down the *physical postulate* that this is the motion of the actual system. Equation B), then, is an independent physical law, made plausible by the mathematical considerations set forth above, but not following mathematically from them.

The Effect of Gravity. Whenever gravity acts, the contribution of this force to the right-hand side of Equation B) can

* This problem is more difficult than the others, and is essentially a problem in the Calculus; cf. the author's *Advanced Calculus*, Chapter 14, § 8.

always be written as the moment of a single force, that force being the attraction of gravity on a single particle of mass equal to the mass of the entire body and situated at the centre of gravity of the body. This is true in the most general case of motion, when no point of the body is permanently at rest. Here, again, we have a new physical postulate.

EXERCISES

1. A turn table consisting of a uniform circular disc is free to rotate without friction about its centre. A man walks along the rim of the table. Find the ratio of the angle turned through by the table to the angle described by the man, if man and table start from rest.

2. The same problem when the man walks in along a radius of the table, — the system not being, however, initially at rest.

10. Moments of Inertia. The moment of inertia of the simpler and more important distributions of matter are determined by the methods of the Integral Calculus; cf. for example the author's *Introduction to the Calculus*, p. 323, and the *Advanced Calculus*, pp. 58, 79, 88.

1. A uniform * rod of length l about one end: $\frac{Ml^2}{3}$.

2. A rod of length $2a$ about its midpoint: $\frac{Ma^2}{3}$.

3. A circular disc about its centre: $\frac{Mr^2}{2}$.

4. A circular disc about a diameter: $\frac{Mr^2}{4}$.

5. A square about its centre: $\frac{2}{3}Ma^2$.

6. A square about a side; cf. Example 1.

7. A scalene triangle about a side: $\frac{Mh^2}{6}$, where h denotes the altitude.

8. A sphere about a diameter: $\frac{2Mr^2}{5}$.

9. A cube about a line through the centre parallel to an edge; cf. Example 5.

* It will henceforth be understood that the distribution is uniform unless the contrary is stated.

A General Theorem. *The moment of inertia of any distribution of matter whatever, about an arbitrary axis, is equal to the moment of inertia about a parallel axis through the centre of gravity, increased by* Mh^2:

$$I = I_0 + Mh^2 = M(h^2 + k^2),$$

where h denotes the distance between the axes.

We will begin by proving the theorem for a system of particles. Let the first axis be taken as the axis of z in a system of Cartesian coordinates, (x, y, z); and let the second axis be the axis of z' in a system of parallel axes. Then

$$I = \sum_k m_k (x_k^2 + y_k^2), \qquad I_0 = \sum_k m_k (x_k'^2 + y_k'^2).$$

Since

$$x = x' + \bar{x}, \qquad y = y' + \bar{y},$$

it follows that

$$\sum_k m_k (x_k^2 + y_k^2) = \sum_k m_k (x_k'^2 + y_k'^2) + \sum_k m_k (\bar{x}^2 + \bar{y}^2)$$
$$+ 2\bar{x} \sum_k m_k x_k' + 2\bar{y} \sum_k m_k y_k'.$$

The last two terms vanish because O' is the centre of gravity, and hence

$$\sum_k m_k x_k' = 0, \qquad \sum_k m_k y_k' = 0.$$

It remains merely to interpret the terms that are left, and thus the theorem is proved for a system of particles.

If we have a body consisting of a continuous distribution of matter, we divide it up into small pieces, concentrate the mass of each piece at its centre of gravity, form the above sums, and take their limits. We shall have as before $\Sigma\, m_k x_k' = 0$, $\Sigma\, m_k y_k' = 0$, and hence

$$\sum m_k (x_k^2 + y_k^2) = \sum m_k (x_k'^2 + y_k'^2) + Mh^2,$$

$$\lim_{n=\infty} \sum_k m_k (x_k^2 + y_k^2) = \lim_{n=\infty} \sum_k m_k (x_k'^2 + y_k'^2) + Mh^2,$$

or

$$I = I_0 + Mh^2,$$

since these limits are by definition the moments of inertia for the continuous distribution.

Example. To find the moment of inertia of a uniform circular disc about a point in its circumference. Here, $I_0 = \frac{1}{2}Mr^2$ and $h = r$. Hence

$$I = \tfrac{3}{2}Mr^2.$$

11. The Torsion Pendulum. Let a rod be clamped at its mid-point to a steel wire and suspended, the rod horizontal and the wire vertical. Let the rod be displaced slightly in its horizontal plane, the wire remaining vertical, and then released. To determine the motion.

The forces acting on the rod amount to a couple, due to the torsion of the wire, and the moment of the couple is proportional to the angle through which the rod is displaced — such is the law of elasticity. Thus the Principle of Moments, § 9, yields in this case the differential equation,

$$I\frac{d^2\theta}{dt^2} = -\kappa\theta, \tag{1}$$

where $I = \dfrac{Ma^2}{3}$ is the moment of inertia of the rod, and κ is the constant of the wire.

Equation (1) is the equation of Simple Harmonic Motion, and thus the period of oscillation,

$$T = 2\pi\sqrt{\frac{I}{\kappa}}, \tag{2}$$

is the same, no matter what the initial displacement may have been, provided merely that the distortion of the wire is not so great as to impair the physical law above stated, and provided damping is neglected.

12. Rotation of a Plane Lamina, No Point Fixed. Let a rigid plane lamina be acted on by any forces in its plane, and let it move in its plane. To determine the motion.

The centre of gravity will move as if all the mass were concentrated there and all the forces were transferred to that point; § 1. It remains to consider the rotation.

Principle of Moments. *The lamina rotates as if the centre of gravity were held fast and the same forces acted on the lamina as those applied in the actual case,*

$$I\frac{d^2\theta}{dt^2} = \sum \text{Moments about } G, \tag{1}$$

where I denotes the moment of inertia about the centre of gravity, G; θ is the angle that a line fixed in the lamina makes with a line fixed in the plane, and the right-hand side is the sum of the moments of the forces about G.

Proof. Consider first a system of particles rigidly connected. Let (x, y) be axes fixed in the plane, and (ξ, η) parallel axes whose origin is at G. Then

$$x = \xi + \bar{x}, \qquad y = \eta + \bar{y}, \tag{2}$$

and

$$\sum_k m_k \left(x_k \frac{dy_k}{dt} - y_k \frac{dx_k}{dt} \right) = \tag{3}$$

$$\sum_k m_k \left(\xi_k \frac{d\eta_k}{dt} - \eta_k \frac{d\xi_k}{dt} \right) + \sum_k m_k \left(\bar{x} \frac{d\bar{y}}{dt} - \bar{y} \frac{d\bar{x}}{dt} \right),$$

the omitted terms vanishing for the reason that

$$\sum_k m_k \, \xi_k = 0, \qquad \sum_k m_k \, \eta_k = 0,$$

and hence, too,

$$\sum_k m_k \frac{d\xi_k}{dt} = 0, \qquad \sum_k m_k \frac{d\eta_k}{dt} = 0.$$

Remembering that

$$\frac{d}{dt}\left(x \frac{dy}{dt} - y \frac{dx}{dt} \right) = x \frac{d^2 y}{dt^2} - y \frac{d^2 x}{dt^2},$$

we see that Equation B), § 3, here becomes:

$$\sum_k m_k \left(\xi_k \frac{d^2 \eta_k}{dt^2} - \eta_k \frac{d^2 \xi_k}{dt^2} \right) \tag{4}$$

$$+ M \left(\bar{x} \frac{d^2 \bar{y}}{dt^2} - \bar{y} \frac{d^2 \bar{x}}{dt^2} \right) = \sum_k \left(x_k Y_k - y_k X_k \right).$$

Because

$$x_k = \xi_k + \bar{x}, \qquad y_k = \eta_k + \bar{y},$$

the right-hand side of Equation (4) becomes:

$$\sum_k (\xi_k Y_k - \eta_k X_k) + \bar{x} \sum_k Y_k - \bar{y} \sum_k X_k.$$

Since

$$M \frac{d^2 \bar{x}}{dt^2} = \sum_k X_k, \qquad M \frac{d^2 \bar{y}}{dt^2} = \sum_k Y_k,$$

it follows that

$$M\left(\bar{x}\frac{d^2\bar{y}}{dt^2} - \bar{y}\frac{d^2\bar{x}}{dt^2}\right) = \bar{x}\sum_k Y_k - \bar{y}\sum_k X_k.$$

On subtracting this equation from (4), there remains:

$$(5) \qquad \sum_k m_k\left(\xi_k\frac{d^2\eta_k}{dt^2} - \eta_k\frac{d^2\xi_k}{dt^2}\right) = \sum_k\left(\xi_k Y_k - \eta_k X_k\right).$$

In this equation is contained the proof of the theorem for a system of n particles. For, the left-hand side reduces to the left-hand side of (1), since the distance of the point (ξ_k, η_k) from the centre of gravity, G, does not change with t; and the right-hand side expresses precisely the sum of the moments of the applied forces about G.

Finally, we pass to a continuous distribution of matter in the usual way, laying down a new physical postulate to the effect that Equation (1) shall hold for all rigid distributions of matter in a plane.

13. Examples. A hoop* rolls down a rough inclined plane without slipping. Determine the motion.

The forces are: the force of gravity and the reaction of the plane. Let the latter force be resolved into a normal component, R, and the tangential force of friction, F, acting up the plane. Then, for the motion of the centre of gravity, we shall have:

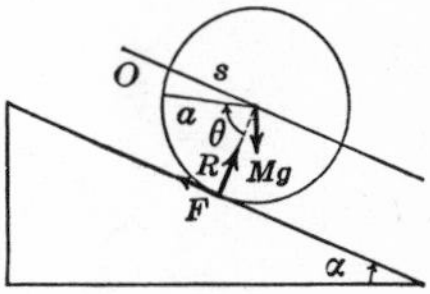

FIG. 84

$$(1) \qquad M\frac{d^2 s}{dt^2} = Mg \sin\alpha - F.$$

The second equation for the motion of the centre of gravity merely tells us that

$$(2) \qquad R = Mg\cos\alpha,$$

a fact that we could have guessed, since the centre of gravity always remains at the same distance from the plane. However, let us formulate the second equation, and prove our guess right. Let $\bar{y}$ denote the distance of the centre of gravity from the plane.

* A pipe, the thickness of which is negligible, when placed on the plane with its axis horizontal, would move in the same way. The two problems are dynamically identical.

Then

$$M\frac{d^2\bar{y}}{dt^2} = R - Mg\cos\alpha. \tag{3}$$

But $\bar{y} = a$, the radius of the hoop, and so the left-hand side of this equation is 0.

Turning now to the rotation of the hoop, we write down Equation (1) of the Theorem, § 12:

$$I\frac{d^2\theta}{dt^2} = aF, \qquad I = Ma^2. \tag{4}$$

Since there is no slipping,

$$s = a\theta, \tag{5}$$

where, for convenience, we take as θ the angle that the radius drawn to the point of contact with the plane at the start has turned through, s being also 0 at the start.

Equations (1) and (4) can now be written in the form:

$$\begin{cases} Ma\dfrac{d^2\theta}{dt^2} = Mg\sin\alpha - F, \\[2ex] Ma^2\dfrac{d^2\theta}{dt^2} = aF. \end{cases} \tag{6}$$

On eliminating F between these equations, we find:

$$\frac{d^2\theta}{dt^2} = \frac{g}{2a}\sin\alpha, \tag{7}$$

or

$$\frac{d^2 s}{dt^2} = \frac{g}{2}\sin\alpha. \tag{8}$$

Hence it appears that the centre of the hoop moves down the plane with just half the acceleration it would have if the plane were smooth.

Equation (2) appears to have played no part in the solution. But we have assumed that there is no slipping, and so F cannot be greater than μR:

$$F \leqq \mu R. \tag{9}$$

To ascertain what this condition means for the coefficient of friction, μ, and the steepness of the plane, α, solve Equations (6) for F and substitute:

$$F = \frac{Mg\sin\alpha}{2},$$

$$\frac{Mg \sin \alpha}{2} \leqq \mu Mg \cos \alpha,$$

$$\tan \alpha \leqq 2\mu. \tag{10}$$

Hence it appears that α may not exceed $\tan^{-1} 2\mu$.

EXERCISES

1. Show that, if the hoop be released from rest,

$$v = \frac{gt}{2} \sin \alpha, \qquad s = \frac{gt^2}{4} \sin \alpha,$$

$$v^2 = gs \sin \alpha.$$

2. Show furthermore that

$$\omega = \frac{gt}{2a} \sin \alpha, \qquad \theta = \frac{gt^2}{4a} \sin \alpha,$$

$$\omega^2 = \frac{g\theta}{a} \sin \alpha.$$

3. Solve the problem studied in the text for a sphere. Show that

$$\frac{d^2\theta}{dt^2} = \frac{5g}{7a} \sin \alpha, \qquad \frac{d^2s}{dt^2} = \frac{5g}{7} \sin \alpha.$$

4. Prove that the sphere will slip unless

$$\tan \alpha \leqq \tfrac{7}{5}\mu.$$

5. Make a complete study of a disc, or solid cylinder.

14. Billiard Ball, Struck Full. A billiard ball is struck full by the cue. To determine the motion.

The forces are: the force of gravity, acting downward at the centre of gravity, and the reaction of the billiard table, which yields a vertical component, R, and a horizontal component, F. Let s be the space described by the centre of the ball, and θ, the angle through which the ball has turned.*

The Principle of the Motion of the Centre of Gravity, § 1, yields the equations:

$$\left\{\begin{aligned} M\frac{d^2s}{dt^2} &= -F \\ R &= Mg \end{aligned}\right. \tag{1}$$

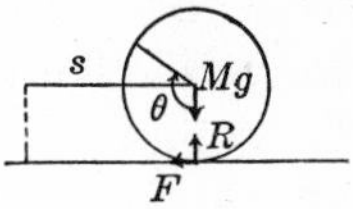

FIG. 85

* It is of prime importance that the student begin each new problem, as here, by drawing a figure showing the forces and the coordinates used in setting up the differential equations of the motion. It is well, too, to note at the same time any auxiliary relations, as in the present instance, $F = \mu R$.

The Principle of Rotation about the Centre of Mass, § 12, yields the equation:

$$(2) \qquad I\frac{d^2\theta}{dt^2} = aF, \qquad I = \frac{2Ma^2}{5}.$$

Finally, so long as there is slipping,

$$(3) \qquad F = \mu R.$$

From Equations (1), (2), and (3) it appears that

$$(4_1) \qquad \frac{d^2 s}{dt^2} = -\mu g,$$

$$(4_2) \qquad \frac{d^2\theta}{dt^2} = \frac{5\mu g}{2a}.$$

The integrals of these equations are as follows:

$$(5_1) \qquad \begin{cases} v = v_0 - \mu g t, \qquad s = v_0 t - \tfrac{1}{2}\mu g t^2, \\ v^2 = v_0^2 - 2\mu g s, \end{cases}$$

and

$$(5_2) \qquad \begin{cases} \omega = \dfrac{5\mu g t}{2a}, \qquad \theta = \dfrac{5\mu g t^2}{4a}, \\ \omega^2 = \dfrac{5\mu g\theta}{a}. \end{cases}$$

Thus as the ball advances, its centre moves more and more slowly, while the speed of rotation steadily increases. Finally, pure rolling will set in. This takes place when the velocity of the point of the ball in contact with the table is nil. Now, the velocity of this point of the ball is made up of two velocities, namely, *i*) the velocity of translation, or the velocity the point would have if the ball were not rotating, *i.e.* v, as given by (5_1); and *ii*) the velocity due to rotation, or the velocity the point would have if the ball were spinning about its centre, thought of as at rest. The latter is a velocity of $a\omega$ in the direction opposite to the motion of the centre, and is given by (5_2). Thus the velocity forward of the point of the ball in contact with the table is

$$(6) \qquad v - a\omega.$$

Slipping continues so long as this expression is positive, and ceases when it vanishes:

$$(7) \qquad v - a\omega = 0.$$

The time is given by the equation

$$v_0 - \mu g t = \tfrac{5}{2}\mu g t,$$

or

$$t_1 = \frac{2v_0}{7\mu g}. \tag{8}$$

The corresponding value of s is seen to be:

$$s_1 = \frac{12v_0^2}{49\mu g}. \tag{9}$$

The angle through which the ball turns is

$$\theta_1 = \frac{5v_0^2}{49\mu g a}. \tag{10}$$

Finally,

$$v_1 = \frac{5}{7}v_0, \qquad \omega_1 = \frac{5v_0}{7a}. \tag{11}$$

EXERCISES

1. Solve the same problem in case the table is slightly tipped and the ball is projected straight down the plane.

2. Work the last problem with the modification that the ball is projected straight up the plane.

15. Continuation. The Subsequent Motion. At the end of the stage of the motion just discussed, the ball has both a motion of translation and one of rotation, the point of the ball in contact with the table being at rest. If from now on the force exerted by the table on the ball consists solely of an upward component R and a tangential component F, the latter force will vanish, and the ball will continue to roll without slipping. For, suppose the table is rough enough to prevent slipping. Then $s = a\theta$, and since equations (1) and (2) still hold, we have:

$$Ma\frac{d^2\theta}{dt^2} = -F, \qquad \tfrac{2}{5}Ma\frac{d^2\theta}{dt^2} = F. \tag{12}$$

Hence F vanishes, and the angular and linear accelerations are both 0, too.

But in practice the ball *will* slow up. How is this to be accounted for, if the resistance of the air is negligible? The answer is, that the reaction of the table is not merely a force, with components R and F, but, in addition, a *couple*, the moment of which we will denote by C. This couple has no influence on the motion of the centre of gravity; thus Equations (1), § 14, remain as before. But Equation (2) now becomes

(13) $$I\frac{d^2\theta}{dt^2} = aF - C.$$

FIG. 86

Furthermore,

(14) $$s = a\theta.$$

Hence

(15) $$\frac{d^2s}{dt^2} = -\frac{5C}{7Ma}, \qquad \frac{d^2\theta}{dt^2} = -\frac{5C}{7Ma^2}, \qquad F = \frac{5C}{7a}.$$

Since C is small, the ball slows up gradually.

EXERCISES

1. If the centre of the ball was moving initially at the rate of 6 ft. a sec. and if the ball stops after rolling 18 ft., show that

$$C = \tfrac{7}{5}Ma.$$

2. If the initial velocity of the centre was v_0 and if the ball rolled l ft., show that C is proportional to the initial kinetic energy and inversely proportional to the distance rolled.

16. Further Examples. *i*) HOOP ON ROUGH STEEPLY INCLINED PLANE. Suppose, in the Example studied in the text of § 13, that α does exceed $\tan^{-1} 2\mu$. What will the motion then be, the hoop being released from rest?

Equations (1), (2), and (4) will be as before. But now (5) is replaced by the equation:

(1) $$F = \mu R,$$

all the friction now being called into play. On eliminating F and R, we find:

(2) $$\begin{cases} \dfrac{d^2s}{dt^2} = g(\sin\alpha - \mu\cos\alpha), \\[2ex] \dfrac{d^2\theta}{dt^2} = \dfrac{\mu g}{a}\cos\alpha. \end{cases}$$

The integrals of these differential equations can be written down at once. In particular, it is seen that the ratio of s to θ is constant, if the hoop starts from rest:

$$\frac{s}{\theta} = \frac{a(\sin\alpha - \mu\cos\alpha)}{\mu\cos\alpha} = a(\tan\alpha\cot\lambda - 1).$$

The last parenthesis has the value 1 when $\tan \alpha = 2\mu$, and is > 1 when α is larger. Thus the motion is one in which a circle of radius

$$r = a(\tan \alpha \cot \lambda - 1)$$

and centre at the centre of the hoop rolls without slipping on a line parallel to the plane and beneath it. We have here an illustration of the general theorem that any motion of a lamina in its own plane can be realized by the rolling without slipping of a curve drawn in the lamina on a curve drawn in the plane; cf. Chapter V, § 4.

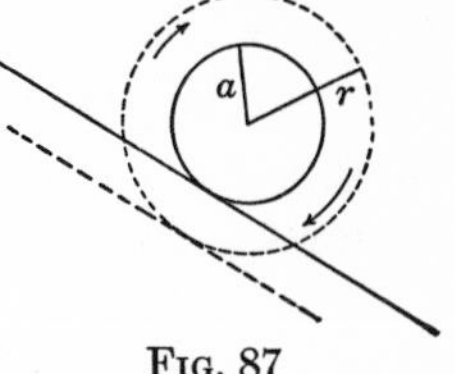

FIG. 87

ii) LADDER SLIDING DOWN A SMOOTH WALL. First, draw a figure representing the forces and the coordinates. The three equations of motion thus become:

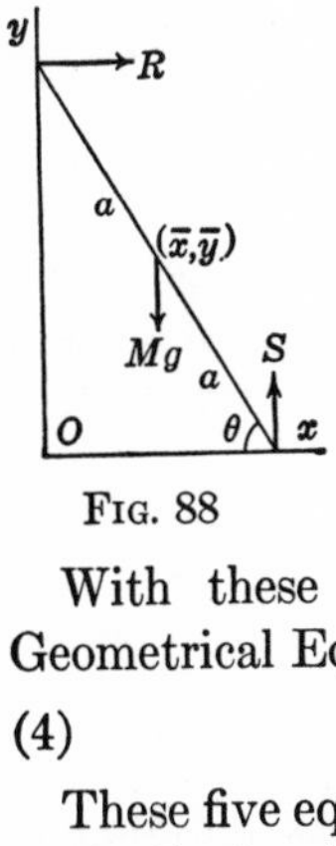

FIG. 88

$$(3)\quad \begin{cases} M\dfrac{d^2\bar{x}}{dt^2} = R, \\[2ex] M\dfrac{d^2\bar{y}}{dt^2} = S - Mg, \\[2ex] I\dfrac{d^2\theta}{dt^2} = aR\sin\theta - aS\cos\theta, \quad I = \tfrac{1}{3}Ma^2. \end{cases}$$

With these three Dynamical Equations are associated two Geometrical Equations:

$$(4)\qquad \bar{x} = a\cos\theta, \qquad \bar{y} = a\sin\theta.$$

These five equations determine the five unknown functions $\bar{x}$, $\bar{y}$, θ, R, S, the time being the independent variable; or they determine five of the variables $\bar{x}$, $\bar{y}$, θ, R, S, t as functions of the sixth. Eliminate R, S between the first three equations:

$$(5)\qquad I\frac{d^2\theta}{dt^2} = Ma\sin\theta\frac{d^2\bar{x}}{dt^2} - Ma\cos\theta\frac{d^2\bar{y}}{dt^2} - Mga\cos\theta.$$

From the Geometrical Equations follows:

$$\frac{d\bar{x}}{dt} = -a\sin\theta\frac{d\theta}{dt}, \qquad \frac{d\bar{y}}{dt} = a\cos\theta\frac{d\theta}{dt},$$

$$\frac{d^2\bar{x}}{dt^2} = -a\sin\theta\frac{d^2\theta}{dt^2} - a\cos\theta\left(\frac{d\theta}{dt}\right)^2,$$

$$\frac{d^2\bar{y}}{dt^2} = a\cos\theta\frac{d^2\theta}{dt^2} - a\sin\theta\left(\frac{d\theta}{dt}\right)^2.$$

Combining these with (5) and reducing we obtain:

$$\frac{d^2\theta}{dt^2} = -\frac{3g}{4a}\cos\theta. \tag{6}$$

This differential equation can be integrated by the device of multiplying through by $2d\theta/dt$ and then integrating each side with respect to t:

$$2\frac{d\theta}{dt}\frac{d^2\theta}{dt^2} = -\frac{3g}{2a}\cos\theta\frac{d\theta}{dt},$$

$$\int 2\frac{d\theta}{dt}\frac{d^2\theta}{dt^2}dt = -\frac{3g}{2a}\int\cos\theta\, d\theta.$$

Since

$$d\left(\frac{d\theta}{dt}\right)^2 = 2\frac{d\theta}{dt}d\left(\frac{d\theta}{dt}\right) = 2\frac{d\theta}{dt}\frac{d^2\theta}{dt^2}dt,$$

it follows that

$$\left(\frac{d\theta}{dt}\right)^2 = -\frac{3g}{2a}\sin\theta + C.$$

The constant of integration, C, is determined by the initial conditions. If the ladder is released from rest, making an angle α with the horizontal, then $d\theta/dt = 0$ and $\theta = \alpha$ initially, and so

$$0 = -\frac{3g}{2a}\sin\alpha + C.$$

Hence, finally,

$$\left(\frac{d\theta}{dt}\right)^2 = \frac{3g}{2a}\left(\sin\alpha - \sin\theta\right). \tag{7}$$

To find where the ladder will leave the wall. This question is answered by computing R and setting it $= 0$:

$$R = M\frac{d^2\bar{x}}{dt^2} = -Ma\sin\theta\frac{d^2\theta}{dt^2} - Ma\cos\theta\left(\frac{d\theta}{dt}\right)^2,$$

$$R = \tfrac{3}{4}Mg\cos\theta\,(3\sin\theta - 2\sin\alpha). \tag{8}$$

Hence $R = 0$ when

$$3\sin\theta - 2\sin\alpha = 0.$$

Let β be the root of this equation:

$$\beta = \sin^{-1}\left(\tfrac{2}{3}\sin\alpha\right).$$

Observe that $\cos\theta$ cannot vanish when $0 \leqq \theta < \dfrac{\pi}{2}$.

The intuitional evidence is here complete: — the ladder leaves the wall and slides along with the lower end in contact with the floor. But suppose a person is unwilling to trust his intuition and says: — "Ah, you have not proven your point in merely showing that $R = 0$ for a certain value of θ. The ladder might still remain in contact with the wall, R increasing as the ladder continues to slide." The logic of this objection is valid. The objection can be met as follows.

Think of the upper end of the ladder as provided with a ring that slides on a smooth vertical rod. Then the ladder will not leave the wall. How about R in this case? Formula (8) now holds clear down to the floor; but $R < 0$ when $\theta < \sin^{-1}(\frac{2}{3}\sin\alpha)$, and so the vertical rod has to pull on the ladder instead of pushing. This proves that our intuition was correct.

The Time. From Equation (7) it appears that

$$t = -\sqrt{\frac{2a}{3g}}\int\frac{d\theta}{\sqrt{\sin\alpha - \sin\theta}}. \tag{9}$$

This integral cannot be evaluated in terms of the elementary functions. On making the substitution

$$x = \sin\theta,$$

the integral goes over into an Elliptic Integral of the First Kind, and can be treated by well-known methods; cf. the Author's *Advanced Calculus*, Chapter IX.

iii) Coin on Smooth Table. A coin is released from rest with one point of the rim touching a smooth horizontal table. To determine the motion.

The forces acting are: Gravity, Mg, down, and the reaction, R, of the table upward. Thus the centre of gravity of the coin descends in a right line. Let its height above the table be denoted by $\bar{y}$. Then the further Dynamical Equations become:

$$\begin{cases} M\dfrac{d^2\bar{y}}{dt^2} = R - Mg, \\ Mk^2\dfrac{d^2\theta}{dt^2} = -\,aR\cos\theta. \end{cases} \tag{10}$$

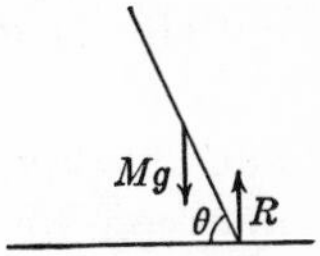

Fig. 89

The Geometrical Equation is:

$$\bar{y} = a\sin\theta. \tag{11}$$

On eliminating R and $\bar{y}$ we find:

$$(12)\quad (k^2 + a^2 \cos^2 \theta)\frac{d^2\theta}{dt^2} - a^2 \sin\theta \cos\theta \left(\frac{d\theta}{dt}\right)^2 = -\, ag \cos\theta.$$

This differential equation comes under a general class, namely, those in which one of the variables fails to appear explicitly. The general plan of solution in such cases is to introduce a new variable,

$$p = \frac{d\theta}{dt}.$$

And this can be done here. But in the present case there is a short cut, due to the special form of the differential equation. It is observed that, on multiplying the equation through by $2d\theta/dt$, the left-hand side becomes the derivative of a certain function with respect to t, so that the equation takes on the form:

$$(13)\qquad \frac{d}{dt}\left\{(k^2 + a^2 \cos^2 \theta)\left(\frac{d\theta}{dt}\right)^2\right\} = -\, 2ag \cos\theta \frac{d\theta}{dt}.$$

On integrating each side of this equation with respect to t, we find:

$$(k^2 + a^2 \cos^2 \theta)\left(\frac{d\theta}{dt}\right)^2 = -\, 2ag \sin\theta + C.$$

To determine C make use of the initial conditions, $d\theta/dt = 0$ and $\theta = \alpha$. Thus

$$(14)\qquad (k^2 + a^2 \cos^2 \theta)\left(\frac{d\theta}{dt}\right)^2 = 2ag\,(\sin\alpha - \sin\theta).$$

The angular velocity, ω, of the coin when it falls flat on the table is given by the equation:

$$(15)\qquad \omega_1^2 = \frac{2ag \sin\alpha}{k^2 + a^2}.$$

But here is an assumption, namely, that the point of the coin initially in contact with the table remains in contact till $\theta = 0$. This is plausible enough physically; but in this guess, is there not an appreciable admixture of unimaginativeness and the question which the moron so frequently asks: "Why shouldn't it?" The angular velocity $d\theta/dt$ of the coin is steadily increasing, as we see both intuitionally and from (14). May it not increase to such an extent that the lowest point in the coin may kick up and leave the table before the centre comes clear down? The moron certainly cannot answer this objection by physical intuition.

It is here that mathematics sits as judge over the situation. Replace the actual problem by one equivalent during the early stage of the motion, and see whether this stage lasts through to the end. Let the lowest point of the coin be provided with a ring that slides on a smooth horizontal rod. Then the coin *will* fall as we guessed. Compute now the reaction, R. The test is: Does R remain positive throughout the motion? We leave it to the student to find out.

EXERCISES ON CHAPTER IV

1. A homogeneous solid cylinder is placed on a rough inclined plane and released from rest. Will it slip as it rolls, or will it roll without slipping?

Ans. It will slip if the angle of inclination of the plane is greater than $\tan^{-1}(\frac{3}{2}\mu)$.

2. The same problem for a homogeneous spherical shell (material surface).

3. A billiard ball is set spinning about a horizontal axis and is released, just touching the cloth of the billiard table. How far will it go before pure rolling sets in?

4. A circular disc has a string wound round its circumference. The free end of the string is fastened to a peg, A, and the disc is released from rest in a vertical plane with its centre below the level of A, and the string taut and vertical. Show that the centre of the disc will descend in a vertical right line with two-thirds the acceleration of gravity.

5. The disc of the preceding problem is laid flat on a smooth horizontal table; the string is carried over a smooth pulley at the edge of the table, and a weight equal to the weight of the disc is attached to the end of the string. The system is released from rest, the string being taut and the weight hanging straight down. Show that the acceleration of the weight is three-fourths that of gravity.

6. Find the tension of the string in the last question.

7. Solve the problem of Question 3 with the modification that the table is inclined at an angle α to the horizon.

Discuss in full the case that the rotation of the ball is in such a sense that the ball moves down the plane faster than it would if it had not been rotating.

8. Study the problem of the last question when the rotation is in the opposite sense.

9. A billiard ball is placed on a billiard table inclined to the horizontal at an angle α, and is struck full by the cue, so that it starts off straight down the plane without any initial rotation. Study the motion.

10. The same problem when the ball is so struck that it starts straight up the plane.

11. If a man were placed on a perfectly smooth table, how could he turn round?

12. A plank can rotate about one end, on a smooth horizontal table. A man, starting from the other end, walks toward the pivot. Determine the motion.

13. A smooth tube, the weight of which may be neglected, can turn freely about one end. A rod is placed in the tube and the system is released from rest with the rod horizontal. Determine the motion.

14. A spindle consists of two equal discs connected rigidly with an axle, which is a solid cylinder. The spindle is placed on a rough horizontal table, and a string is wound round the axle and carried over a smooth pulley above the edge of the table. A weight is attached to the lower end of the string and the system is released from rest. Determine the motion.

Consider first the case in which the string leaves the axle from the top; then, the case that the string leaves the axle from the bottom. In each case, the segment of the string between the axle and the pulley shall be horizontal and at right angles to the axis, and the part below the pulley, vertical.

15. The centre of gravity of a four-wheeled freight car is 5 ft. above the track and midway between the axles, which are 8 ft. apart. The coefficient of friction between the wheels (when they are locked) and the track is $\frac{1}{8}$. If the car is running at the rate of 30 m. an h., in how short a distance can it be stopped by applying the brakes to the rear wheels only? How far, if the brakes are applied to the front wheels only?

16. A uniform rod is suspended in a horizontal position by two vertical strings attached to its ends. One string is cut. Find the initial tension in the other one.

17. A hoop is hung up on a peg and released. Find whether it will slip.

18. A uniform circular disc, of radius 1 ft. and weight 10 lbs., can rotate freely about its centre, its plane being vertical. There is a particle weighing 1 lb. fixed in the rim, and a fine inextensible weightless string, wound round the rim of the disc, has a weight of P lbs. fastened to it. The system is released from rest with the 1 lb. weight at the lowest point and the other weight hanging freely at the same level. How great may P be, if the 1 lb. weight is not to be pulled over the top?

19. A billiard ball rolls in a punch bowl. Determine the motion.

20. A solid sphere is placed on top of a rough cylinder of revolution, axis horizontal, and slightly displaced, under the action of gravity. Find where it will leave the cylinder.

21. A uniform rod is released from rest, inclined at an angle, with its lower end in contact with a rough horizontal plane. Will it slip at the start? Determine the motion.

22. A packing box is sliding over an icy side walk. It comes to bare ground. Will it tip up?

CHAPTER V

KINEMATICS IN TWO DIMENSIONS

1. The Rolling Wheel. When a wheel rolls over a level road without slipping, the nature of the motion is particularly accessible to our intuition, for the points of the wheel low down move slowly, the point in contact with the ground actually being at rest for the instant, and it is much as if the whole wheel were pivoted at this point and rotating about it as an axis. This is, in fact, precisely the case, the velocity of each point of the wheel *at the instant* being the same as if the wheel were rotating *permanently* about that point.

If the wheel is skidding, it is not so easy to see that a similar situation exists, — and yet it does. No matter how the wheel is moving, provided it is rotating at all, there is *at each instant* a definite point (far away it may be), about which the wheel rotates *at that instant.* This point is called the *instantaneous centre.*

To prove this assertion, we will begin by giving a general formulation of the problem of the motion of any plane lamina in its plane. It makes the problem more concrete to think of an actual lamina, like a disc or a triangle or a finite surface, S. But we are really dealing with the motion of the *whole plane,* thought of as rigid.

The motion may be described mathematically as follows. Draw a pair of Cartesian axes in the moving plane; *i.e.* think

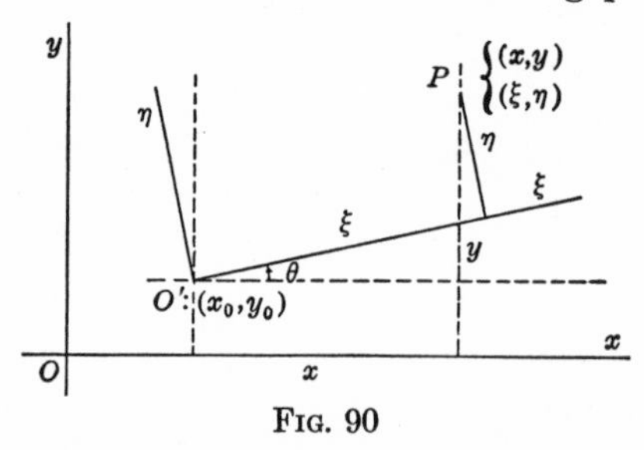

FIG. 90

of this plane as a sheet of paper, and draw the (ξ, η)-axes in red ink on the paper. Assume further a system of axes fixed in space — the (x, y)-axes. Then the (ξ, η)-coordinates of an arbitrary point P of the moving plane are connected with the (x, y)-coordinates of the same point by the relations:

$$\text{A)} \qquad \begin{cases} x = x_0 + \xi \cos\theta - \eta \sin\theta, \\ y = y_0 + \xi \sin\theta + \eta \cos\theta. \end{cases}$$

The position of the moving plane is known when one point, as O', is known and the orientation, as given by θ, is known. The motion may, therefore, be completely described by stating how x_0, y_0, θ vary with the time; *i.e.* by saying what functions x_0, y_0, θ are of t:

$$(1) \qquad x_0 = f(t), \qquad y_0 = \varphi(t), \qquad \theta = \psi(t).$$

We shall assume at the outset that these functions are continuous and possess continuous derivatives of the first order. Later, it will be desirable to restrict them further by requiring that they have continuous derivatives of the second order.

EXERCISE

Express ξ and η in terms of x and y, *i*) geometrically, by reading the result off from the figure; *ii*) analytically, by solving Equations A) for ξ, η. The formulas are:

$$\text{A}^{-1}) \qquad \begin{cases} \xi = \quad (x - x_0) \cos\theta + (y - y_0) \sin\theta, \\ \eta = -(x - x_0) \sin\theta + (y - y_0) \cos\theta. \end{cases}$$

2. The Instantaneous Centre. Let P be a point fixed in the moving plane — mark it with a dot of red ink on the sheet of paper. Let the coordinates of P be (x, y). Then they are determined as functions of t by Equations A), (ξ, η) being the coordinates of P with reference to the moving axes. Of course, ξ and η are constants with respect to the time, for the red ink dot does not move in the paper — it moves in space.

The vector velocity, $\mathbf{v}$, of P in space can be determined by means of its components along the axes of x and y, which are fixed in space, Chapter III, § 15:

$$v_x = \frac{dx}{dt}, \qquad v_y = \frac{dy}{dt}.$$

These derivatives can be computed in terms of the known functions (1), namely, x_0, y_0, θ, and of their derivatives, by means of Equations A). Thus

$$(2)\qquad \begin{cases} \dfrac{dx}{dt} = \dfrac{dx_0}{dt} + (-\,\xi \sin\theta - \eta \cos\theta)\,\dfrac{d\theta}{dt}, \\[2ex] \dfrac{dy}{dt} = \dfrac{dy_0}{dt} + (\quad \xi \cos\theta - \eta \sin\theta)\,\dfrac{d\theta}{dt}. \end{cases}$$

The parentheses that here enter are seen from Equations A) to have the values:

$$-\,(y - y_0), \qquad\qquad x - x_0.$$

Hence

$$(3)\qquad \begin{cases} \dfrac{dx}{dt} = \dfrac{dx_0}{dt} - (y - y_0)\,\dfrac{d\theta}{dt}, \\[2ex] \dfrac{dy}{dt} = \dfrac{dy_0}{dt} + (x - x_0)\,\dfrac{d\theta}{dt}. \end{cases}$$

These equations express the components of the vector velocity $\mathbf{v}$ of the point P along the axes fixed in space, in terms of the coordinates (x, y) of P and the known functions (1).

New Notation. Since derivatives with respect to the time occur frequently in the work which follows, the Newtonian notation with the dot is expedient:

$$(4)\qquad \dot{x} = \frac{dx}{dt}, \qquad \ddot{x} = \frac{d^2x}{dt^2}, \qquad \text{etc.}$$

Thus the formula for the components of the velocity assumes the final form:

$$\text{B)}\qquad \begin{cases} \dot{x} = \dot{x}_0 - (y - y_0)\,\dot{\theta}, \\ \dot{y} = \dot{y}_0 + (x - x_0)\,\dot{\theta}. \end{cases}$$

The Instantaneous Centre. We now inquire what point or points (if any) of the body are at rest at a given instant. A point is "at rest" if its velocity is 0. Hence the condition is, that $\dot{x} = 0$ and $\dot{y} = 0$, or:

$$(5)\qquad \begin{cases} 0 = \dot{x}_0 - (y - y_0)\,\dot{\theta}, \\ 0 = \dot{y}_0 + (x - x_0)\,\dot{\theta}. \end{cases}$$

These equations yield a unique solution for the unknown x and y when, and only when, $\dot{\theta} \neq 0$:

$$\text{C)} \qquad \begin{cases} x_1 = x_0 - \dfrac{\dot{y}_0}{\dot{\theta}}, \\[2ex] y_1 = y_0 + \dfrac{\dot{x}_0}{\dot{\theta}}. \end{cases}$$

THEOREM. *At any instant at which $d\theta/dt = \dot{\theta}$ is not 0, there is one and only one point of the body at rest.*

This point is called the instantaneous centre, and its coordinates, (x_1, y_1), are given by Equations C).

If $\dot{\theta} = 0$, no point of the body is at rest, or else all points are; there is never a single point at rest, to the exclusion of all others.

When $\dot{\theta} = 0$, $\dot{x}_0$ and $\dot{y}_0$ not both vanishing, all points of the body are moving in the same direction with the same speed, and we have a motion of *translation*.

EXERCISES

1. Show that the coordinates (ξ_1, η_1) of the instantaneous centre, referred to the moving axes, are given by the equations:

$$\text{(C)} \qquad \begin{cases} \xi_1 = \dfrac{\dot{x}_0 \sin\theta - \dot{y}_0 \cos\theta}{\dot{\theta}}, \\[2ex] \eta_1 = \dfrac{\dot{x}_0 \cos\theta + \dot{y}_0 \sin\theta}{\dot{\theta}}. \end{cases}$$

2. A circle rolls on a line without slipping. Show that the point of contact is at rest.

3. A billiard ball is struck full by the cue. Find the instantaneous centre during the subsequent motion.

3. Rotation about the Instantaneous Centre. The very name "instantaneous centre" implies that the motion of the body is one of rotation about that point. Let us make this statement precise.

Suppose the body is rotating about the origin, O, with angular velocity $\dot{\theta} = \omega$. What will be the vector velocity of an arbitrary point $P: (x, y)$? The answer is given by Equations B), where O' is taken at O, and thus $x_0 = y_0 = \dot{x}_0 = \dot{y}_0 = 0$. Hence

$$(1) \qquad \begin{cases} \dot{x} = -\, y\omega, \\ \dot{y} = x\omega. \end{cases}$$

The result checks, for these are the components of a vector at right angles with the radius vector r drawn from O to P and having the sense of the increasing angle θ, its length being

$$\sqrt{x^2 + y^2}\,\dot{\theta} = r\omega, \qquad 0 < \dot{\theta}.$$

If $\dot{\theta} < 0$, its sense is reversed.

It is the form of Equations (1) that is important. We say that any motion of the points of the (x, y)-plane such that, at a given instant, the velocity of each point is given by (1), is one of *rotation of the plane as a rigid body about O*. The velocities of the points in the actual motion before and after the instant in question may be different from those of the rigid body that is rotating permanently about O. But for a short space of time before and after the instant, the discrepancy will be small because of the continuity of the motion, and at the one instant, the velocities will all tally exactly.

If the point about which the body is permanently rotating had been the point (a, b) instead of the origin, Equations (1) would have been the following:

$$(2) \qquad \begin{cases} \dot{x} = -\,(y - b)\dot{\theta}, \\ \dot{y} = (x - a)\dot{\theta}. \end{cases}$$

We are now ready to state and prove the following theorem.

THEOREM. *The motion of the actual body at an arbitrary instant t, at which $\dot{\theta} \neq 0$, is one of rotation about the instantaneous centre.*

To prove the theorem we have to show that, at the instant t,

$$(3) \qquad \begin{cases} \dot{x} = -\,(y - y_1)\dot{\theta}, \\ \dot{y} = (x - x_1)\dot{\theta}, \end{cases}$$

where (x_1, y_1) are given by Equations C), § 2, and $(\dot{x}, \dot{y})$ are given by Equations B) of the same paragraph. To do this, eliminate x_0 and y_0 between Equations B) and C). This can be done most conveniently by writing Equations C) in the form (5) of § 2:

$$0 = \dot{x}_0 - (y_1 - y_0)\dot{\theta},$$
$$0 = \dot{y}_0 + (x_1 - x_0)\dot{\theta},$$

and then, in this form, subtracting them respectively from Equations B). The result is Equations (3) of this paragraph, and the theorem is proved.

Translation and Rotation. From the foregoing result a new theorem about the motion of the plane can be derived at once. Let A be an arbitrary point, and let its vector velocity be denoted by $\mathbf{V}$. Impress on each point of the plane, as it moves under the given law, a vector velocity equal and opposite to $\mathbf{V}$. Then A is reduced to rest, and the new motion is one of rotation about A with the same angular velocity as before. We thus have the

THEOREM. *The field of vector velocities is the vector sum of the fields consisting i) of the translation field due to the vector velocity of an arbitrary point, A; and ii) of the rotational field with A as centre.*

In other words, the given motion consists of rotation about an arbitrary point, A, plus the translation of A.

The theorem also follows immediately from Equations B), if we take the point O' at A.

4. The Centrodes. We are familiar with the motion of a circular disc when it rolls without slipping on a right line or a curve — a wheel rolling on the ground. Consider, more generally, the motion of a lamina when an arbitrary curve drawn in it rolls without slipping on an arbitrary curve fixed in space. We may think of a brass cylinder, or cam, as cut with its face corresponding to the first curve, and attached to the body; a second such cam, with its face corresponding to the second curve, being fixed in space. And now the first cam is allowed to roll without slipping on the second cam.

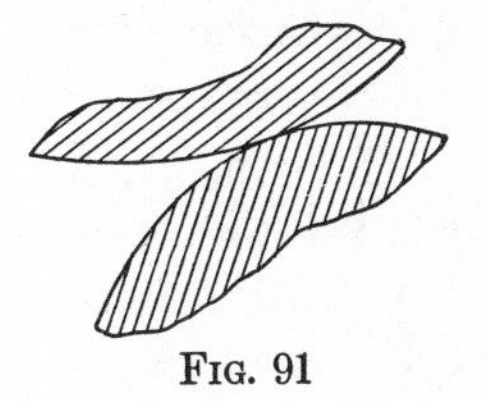

FIG. 91

Thus a great variety of motions of the lamina can be realized, and now the remarkable fact is that *all* motions can be generated in this way, with the single exception of the translations, — provided that the functions (1) of § 1 have continuous derivatives of the second order, and the space centrode is traced out by the instantaneous centre with non-vanishing velocity.

A necessary condition for the truth of this statement is evident from intuition, namely: — the point of contact of the two cams must be the instantaneous centre of the actual motion. This fact suggests the proof — the faces of the cams, *i.e.* the curves, must be the loci of the instantaneous centres in the body and in space.

Definition. The locus of the instantaneous centre in the body is called the *body centrode,* and the locus of the instantaneous centre in space is called the *space centrode.*

THEOREM. *Any motion of a rigid lamina which is not translation can be generated by the rolling of the body centrode (without slipping) on the space centrode, provided the space centrode is traced out by the instantaneous centre with non-vanishing velocity; the functions* (1) *of* § 1 *having continuous derivatives of the second order.*

Before we can prove the theorem, we must make clear to ourselves how to formulate mathematically the rolling of one curve without slipping on a second curve. As the independent variable, the time most naturally suggests itself; but it is better at the outset not to choose it, but to take, rather, a variable λ which merely corresponds to the fact that, for an arbitrary (*i.e.* variable) value of λ, the curves meet in a (variable) point P. And now we shall demand further:

i) that the curves be tangent to each other at P;

ii) that the arc of the one curve corresponding to any two different values of λ, namely, λ_1 and λ_2; and the arc of the other curve corresponding to the same values of λ, have the same length.

Thus, in particular, both curves may be moving — a more general case than the one that interests us here.

Let the equation of the one curve, C, referred to a system of Cartesian axes, (x, y), be:

$$x = g(\lambda), \qquad y = h(\lambda), \tag{1}$$

where the functions $g(\lambda)$, $h(\lambda)$ are continuous together with their first derivatives, and the latter do not vanish simultaneously:

$$0 < g'(\lambda)^2 + h'(\lambda)^2. \tag{2}$$

Let the second curve, Γ, referred to a second system of Cartesian axes, (ξ, η), be represented by similar equations,

$$\xi = g_1(\lambda), \qquad \eta = h_1(\lambda), \tag{3}$$

$$0 < g_1'(\lambda)^2 + h_1'(\lambda)^2. \tag{4}$$

The coordinates of any point of the plane, referred to the one set of axes, are connected with the coordinates of the same point, referred to the other set of axes, by the equations:

$$(5)\qquad \begin{cases} x = x_0 + \xi \cos\theta - \eta \sin\theta, \\ y = y_0 + \xi \sin\theta + \eta \cos\theta. \end{cases}$$

And now we require that x_0, y_0, θ be functions of λ which have continuous first derivatives:

$$(6)\qquad x_0 = f(\lambda), \qquad y_0 = \varphi(\lambda), \qquad \theta = \psi(\lambda),$$

where $f'(\lambda)$, $\varphi'(\lambda)$, $\psi'(\lambda)$ exist and are continuous.

Since C and Γ always meet in a point P, whose coordinates are expressed by the equations (1) and (3), it follows that Equations (5) will hold identically in λ if the values of x, y from (1), and those of ξ, η from (3), be substituted therein.

The vector **v** whose components are

$$v_x = \frac{dx}{d\lambda}, \qquad v_y = \frac{dy}{d\lambda}$$

is tangent to C at P and its length is

$$\sqrt{v_x^2 + v_y^2} = \sqrt{\frac{dx^2}{d\lambda^2} + \frac{dy^2}{d\lambda^2}} = \frac{ds}{d\lambda} > 0.$$

The vector **u** whose components are

$$u_\xi = \frac{d\xi}{d\lambda}, \qquad u_\eta = \frac{d\eta}{d\lambda}$$

is tangent to Γ at P and its length is

$$\sqrt{u_\xi^2 + u_\eta^2} = \sqrt{\frac{d\xi^2}{d\lambda^2} + \frac{d\eta^2}{d\lambda^2}} = \frac{d\sigma}{d\lambda} > 0.$$

The requirements *i*) and *ii*) demand that these two vectors be identical. This condition is both necessary and sufficient. The analytical formulation of the condition is as follows:

$$(7)\qquad \begin{cases} v_x = u_\xi \cos\theta - u_\eta \sin\theta, \\ v_y = u_\xi \sin\theta + u_\eta \cos\theta. \end{cases}$$

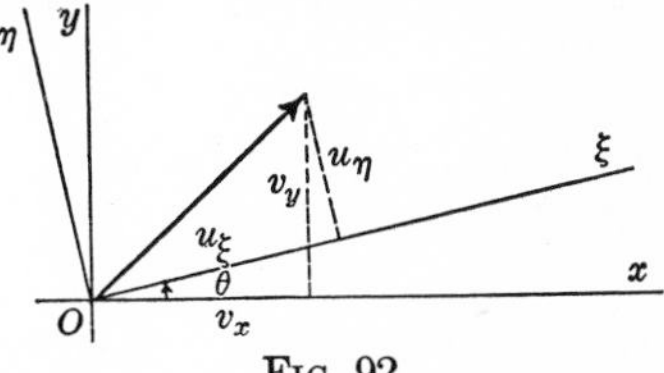

FIG. 92

We now have all the material out of which to construct the proof. From Equations (5) it follows that

$$\frac{dx}{d\lambda} = \frac{d\xi}{d\lambda}\cos\theta - \frac{d\eta}{d\lambda}\sin\theta$$

$$+ \frac{dx_0}{d\lambda} - (\xi\sin\theta + \eta\cos\theta)\frac{d\theta}{d\lambda},$$

$$\frac{dy}{d\lambda} = \frac{d\xi}{d\lambda}\sin\theta + \frac{d\eta}{d\lambda}\cos\theta$$

$$+ \frac{dy_0}{d\lambda} + (\xi\cos\theta - \eta\sin\theta)\frac{d\theta}{d\lambda}.$$

The first line in these equations is nothing more or less than the first of Equations (7), and the latter equations we have set out to prove. Hence the second line must vanish, if the equation is to be true, and so, by the aid of (5), we obtain the first of Equations (8):

$$(8)\qquad \begin{cases} \dfrac{dx_0}{d\lambda} - (y - y_0)\dfrac{d\theta}{d\lambda} = 0, \\[2ex] \dfrac{dy_0}{d\lambda} + (x - x_0)\dfrac{d\theta}{d\lambda} = 0. \end{cases}$$

The second equation is obtained in a similar manner from the second of the above equations.

Equations (8) represent a new form of necessary and sufficient condition for the fulfilment of Conditions *i*) and *ii*).

5. Continuation. Proof of the Fundamental Theorem. It is now easy to prove the theorem of § 4. The two curves, C and Γ, are here the space centrode and the body centrode, and we will now take as our parameter λ, the time t. Equations (1), § 4, thus represent the coordinates, x_1 and y_1, of the instantaneous centre in space, and in Equations (8), the (x, y) are the coordinates of this same point, (x_1, y_1). The other quantities that enter into (8) are the functions (6) that determine the position of the moving body; and $\lambda = t$. Thus Equations (8) go over into the following:

$$(9)\qquad \begin{cases} \dot{x}_0 - (y_1 - y_0)\,\dot{\theta} = 0, \\ \dot{y}_0 + (x_1 - x_0)\,\dot{\theta} = 0. \end{cases}$$

But these are precisely Equations (5) of § 2, which determine the instantaneous centre. Equations (8) are thus shown to be true.

Discussion of the Result. From Equations (8) § 4, it appears that a necessary condition for the truth of the theorem is, that the coordinates of a point P of C satisfy the equations:

$$(10)\qquad \begin{cases} x = x_0 - \dfrac{dy_0}{d\lambda} \Big/ \dfrac{d\theta}{d\lambda}, \\[2ex] y = y_0 + \dfrac{dx_0}{d\lambda} \Big/ \dfrac{d\theta}{d\lambda}. \end{cases}$$

But these conditions are not sufficient, since the functions x and y thus defined will not in general admit derivatives.

To meet this latter requirement we demand, therefore, that the functions (6) possess, furthermore, continuous second derivatives. But this is not enough, even if the case that x and y reduce to constants is excluded (rotation about a fixed point). It is, however, sufficient when we add the hypothesis of (2), § 4, and so demand that

$$(11)\qquad \frac{d}{d\lambda}\left(x_0 - \frac{dy_0}{d\lambda} \Big/ \frac{d\theta}{d\lambda}\right), \qquad \frac{d}{d\lambda}\left(y_0 + \frac{dx_0}{d\lambda} \Big/ \frac{d\theta}{d\lambda}\right)$$

be not both 0 ($d\theta/d\lambda$ being, of course, $\neq 0$). In other words, the equations

$$(12)\qquad \begin{cases} \dfrac{d\theta}{d\lambda}\dfrac{d^2x_0}{d\lambda^2} - \dfrac{d^2\theta}{d\lambda^2}\dfrac{dx_0}{d\lambda} + \dfrac{d\theta^2}{d\lambda^2}\dfrac{dy_0}{d\lambda} = 0, \\[2ex] \dfrac{d\theta}{d\lambda}\dfrac{d^2y_0}{d\lambda^2} - \dfrac{d^2\theta}{d\lambda^2}\dfrac{dy_0}{d\lambda} - \dfrac{d\theta^2}{d\lambda^2}\dfrac{dx_0}{d\lambda} = 0 \end{cases}$$

shall never hold simultaneously. This excluded case includes the case in which an ordinary cusp occurs; but it also includes more complicated singularities.

If, in particular, the functions (6) are analytic in the neighborhood of a point, $\lambda = \lambda_0$, and if the case of permanent rotation about a fixed point be excluded, the curve C will at most have a cusp and otherwise be smooth in the neighborhood of the point; and the same will be true of Γ.

Acceleration of the Point of Contact. Let the point (x_0, y_0), at a given instant, t, be taken at the point of contact of C and Γ. Then it follows from (10) — or, more simply, from (8) — since $x = x_0$ and $y = y_0$, that

$$\frac{dx_0}{d\lambda} = 0, \qquad \frac{dy_0}{d\lambda} = 0.$$

Let the origin, furthermore, be taken at this point (x_0, y_0), and let C be tangent to the x-axis here. Now, the derivatives of x and y in (10) cannot both vanish. On computing them it is seen that they reduce respectively to

$$-\frac{d^2y_0}{d\lambda^2}\bigg/\frac{d\theta}{d\lambda}, \qquad \frac{d^2x_0}{d\lambda^2}\bigg/\frac{d\theta}{d\lambda}.$$

The second, $dy/d\lambda$, has the value 0, since C is tangent to the axis of x at the origin. Hence we infer that

$$\text{(13)} \qquad \frac{d^2x_0}{d\lambda^2} = 0, \qquad \frac{d^2y_0}{d\lambda^2} \neq 0.$$

If λ is the time, t, these derivatives become the components along the axes of the acceleration of the point of contact, thought of as a point fixed in the moving body. From (13) it appears that this acceleration is never 0, but is a vector orthogonal to the centrodes at their point of contact. The reader can verify this result in the case of the cycloid.

Example. A billiard ball is projected along a smooth horizontal table with an initial spin about the horizontal diameter which is perpendicular to the line of motion of the centre. Determine the two centrodes.

Take the path described by the centre of the ball as the axis of x, and the centre of the ball as (x_0, y_0). Then

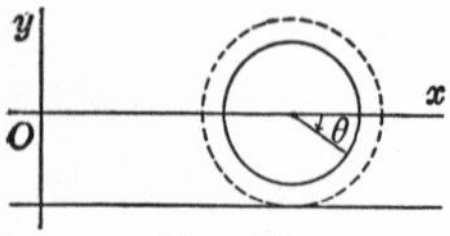

$$x_0 = ct, \qquad y_0 = 0, \qquad \theta = -\omega t.$$

Equations (10) give:

$$x = x_0, \qquad y = -\frac{c}{\omega}.$$

FIG. 93

Hence the space centrode is a horizontal straight line at a distance c/ω below the centre of the ball, and the instantaneous centre is always beneath the centre of the ball. This means that the ball rolls without slipping on a right line distant c/ω below the centre. Hence the body centrode is a circle of radius c/ω about the centre of the ball.

EXERCISE

A billiard ball is struck full by the cue. Determine the space centrode and the body centrode during the stage of slipping; cf. Chapter IV, § 14.

The coordinates being chosen as in the Example, the equations of the space centrode are:

$$\begin{cases} x = x_0 = ct - \frac{1}{2}\mu g t^2, \\ y = \dfrac{2a}{5} - \dfrac{2ac}{5\mu g}\dfrac{1}{t}, \end{cases}$$

where a denotes the radius of the ball, and c the initial velocity of its centre. The time that elapses during the stage of slipping is $2c/7\mu g$ seconds. The space centrode meets the billiard table at the angle

$$\tan^{-1}\frac{343\mu a g}{50c^2}.$$

The equations of the body centrode, referred to suitable polar coordinates, are:

$$\begin{cases} \rho = -y = \dfrac{2ac}{5\mu g}\dfrac{1}{t} - \dfrac{2a}{5}, \\ \varphi = \dfrac{\pi}{2} - \theta = \dfrac{\pi}{2} + \dfrac{5\mu g}{4a}t^2. \end{cases}$$

6. The Dancing Tea Cup. When an empty tea cup is set down on a saucer, the cup sometimes will dance for a long time before coming to rest. Two features of this phenomenon attract attention; first, that the energy, obviously slight, is not earlier dissipated by damping, and secondly, that we can hear a noise in which so little energy is involved. The second point can be disposed of easily because of the physical fact that the energy of sound waves is surprisingly small.

To examine the first critically we need more light on the nature of the motion. The results which we have obtained in this chapter furnish the clue. The following example is highly suggestive.

Consider the motion of a lamina, in which the body centrode is a right line making a small (variable) angle with the horizontal. For the space centrode take a curve suggested by the figure. Such a curve can be defined suggestively as follows. Begin with the curve

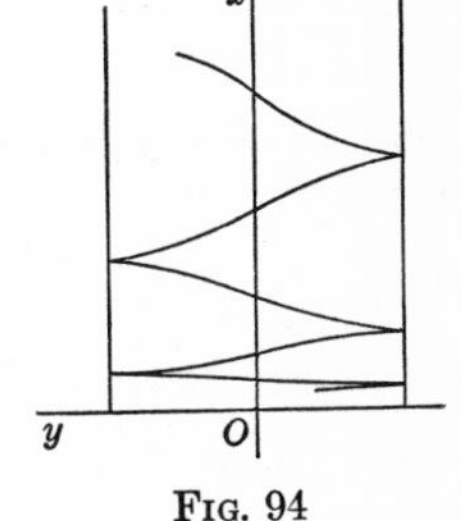

FIG. 94

$$(1) \qquad y = \sin\frac{1}{x}.$$

This curve gives satisfactorily the part of the figure not too near the lines $y = \pm 1$, but it is tangent to these lines, whereas it should have cusps on them.

The desired modification is simple. For example, to convert the curve

$$y = f(x) = x^2$$

from one which is tangent to the axis of x into one which has a cusp on the axis, it is enough to replace $f(x)$ by $[f(x)]^{\frac{1}{3}}$:

$$y = (x^2)^{\frac{1}{3}} = x^{\frac{2}{3}}.$$

Apply this idea to the curve (1). It will suffice to set

$$(2) \qquad y = \left(1 + \sin\frac{1}{x}\right)^{\frac{1}{3}} - \left(1 - \sin\frac{1}{x}\right)^{\frac{1}{3}},$$

as the reader can easily verify.

Now allow the body centrode — the tangent line — to descend according to a reasonable law. We have here a picture of what goes on as the tea cup dances. The line oscillates through smaller and smaller angles as its point of intersection with the axis of x descends.

Tyndall,* in his popular lectures, showed an experiment with a coal shovel illustrating the same phenomenon. The all-metal shovel was heated near its centre of gravity and laid across two thin lead plates clamped in a vise, with their edges horizontal. As the shovel bore more heavily on one of the plates, the latter expanded with the heat, throwing the shovel onto the other plate. Then the process was reversed. Thus vibrations like those of the tea cup arose, and died down.

7. The Kinetic Energy of a Rigid System. The kinetic energy of any system of particles is defined as

$$T = \tfrac{1}{2}\sum_k m_k v_k^2, \qquad v_k^2 = \dot{x}_k^2 + \dot{y}_k^2 + \dot{z}_k^2.$$

We restrict ourselves to two dimensions, and thus

$$v_k^2 = \dot{x}_k^2 + \dot{y}_k^2.$$

Suppose, now, that the particles are rigidly connected. In Equations A), § 1, let the point (x_0, y_0) be taken at the centre of gravity, $(\bar{x}, \bar{y})$. Thus Equations (2), § 2, become:

* Tyndall, *Heat Considered as a Mode of Motion*, Lecture IV.

$$\dot{x}_k = \dot{\bar{x}} - (\xi_k \sin\theta + \eta_k \cos\theta)\,\dot{\theta},$$

$$\dot{y}_k = \dot{\bar{y}} + (\xi_k \cos\theta - \eta_k \sin\theta)\,\dot{\theta}.$$

On squaring and adding, multiplying by m_k, and then adding with respect to k, we find:

$$\sum_k m_k(\dot{x}_k^2 + \dot{y}_k^2) = \sum_k m_k(\dot{\bar{x}}^2 + \dot{\bar{y}}^2) + \sum_k m_k(\xi_k^2 + \eta_k^2)\,\dot{\theta}^2.$$

For, each of the remaining terms involves as a factor one of the quantities

$$\sum_k m_k \xi_k, \qquad \sum_k m_k \eta_k,$$

and each of these is 0, since the centre of gravity is at the origin of the (ξ, η)-axes. Hence it follows that

$$(1) \qquad T = \tfrac{1}{2}MV^2 + \tfrac{1}{2}I\Omega^2,$$

where V denotes the velocity of the centre of gravity and I is the moment of inertia about the centre of gravity, Ω being the angular velocity.

Second Proof. The result may also be obtained by means of the instantaneous centre, O. For the motion, so far as the velocities that enter into the definition of T are concerned, is one of rotation about O. Hence

$$(2) \qquad T = \tfrac{1}{2}I'\Omega^2,$$

where I' denotes the moment of inertia about O. Now,

$$I' = I + Mh^2,$$

where h is the distance from O to the centre of gravity, and

$$(3) \qquad V = h\Omega.$$

On substituting this value of I' in (2) and then making use of (3), T takes on the form (1), and this completes the proof.

Generalization. The most general rigid bodies with which we are concerned are made up of particles and material distributions spread out continuously along curves, over surfaces, and throughout regions of space. When such a body rotates about an axis, the kinetic energy is defined by Equation (1). We can state the result in the form:

The kinetic energy of any rigid material system which is rotating about an axis, is given by the formula:

$$T = \tfrac{1}{2}MV^2 + \tfrac{1}{2}I\Omega^2.$$

Remark. The formula holds even for the most general case of motion of any rigid distribution of matter in space. For, such motion is *helical*, *i.e.* due to the composition of two vector fields of velocity, *i*) a field corresponding to rotation about an axis; and *ii*) a field of translation along that axis; cf. § 12 below.

EXERCISES

1. A ball rolls down a rough plane without slipping. Determine the kinetic energy in terms of the velocity of its centre.

2. A ladder slides down a wall, the lower end sliding on the floor. Find the kinetic energy in terms of the angular velocity.

3. A uniform lamina in the form of an ellipse is rotating in its plane about a focus. Compute the kinetic energy.

4. A homogeneous cube is rotating about one edge. Determine the kinetic energy.

8. Motion of Space with One Point Fixed. Consider any motion of rigid space, one point, O, being fixed. We shall show that there is an *instantaneous axis*, *i.e.* a line through O, the velocity of each point of which is 0; and that the velocities of all the points of the moving space, considered at an arbitrary instant, form a vector field which coincides with the vector field arising from the permanent rotation of space about this axis.

We give first a geometrical proof which appeals strongly to the intuition. The refinements which a critical examination of the details calls for are best given through a new proof by vector methods.

Let Q be a point of the fixed space, distinct from O. If its velocity is 0, then the velocity of every point of the indefinite right line through O and Q is 0, since a variable right line is evidently at rest if two of its points are at rest.

If, on the other hand, Q is moving, pass a sphere, with centre at O, through Q and consider the field of vector velocities corresponding to the points of this sphere. The vectors are evidently all tangent to the sphere, and they vary continuously, together with their first derivatives, for we are not concerned with discontinuous motions.

Pass a great circle, C, through Q perpendicular to the vector velocity of Q. Let P be a point of C near Q. Then the vector velocity of P will also be at right angles to the plane of C and on the same side of C as the vector at Q. For, since the vector velocity of Q is at right angles to the chord QP, the vector velocity of P must lie in the plane through P perpendicular to QP. But it also lies in the tangent plane to the sphere at P. And now I say, *there must be a point A of C* (and hence *two* points) *whose velocity is* 0. For, otherwise, all the vectors that represent the velocities of the points of C would be directed toward the same side of C. In particular, then, the point Q' diametrically opposite Q would have such a vector velocity. But that would mean that the mid-point of the diameter $Q'Q$, *i.e.* the centre O of the sphere, is not at rest. From this contradiction follows the truth of the assertion that there is a point A of C which is at rest. Hence the whole indefinite line through O and A is at rest, and the existence of an instantaneous axis, I, is established.

Rotation about the Instantaneous Axis. It remains to prove that the vector field of the actual velocities coincides with the field of the vector velocities due to a rotation about I. Consider an arbitrary point, P, not on I. Then P cannot be at rest, unless all space is at rest. For, if three points, not in a line, of moving space are at rest, all points must be at rest. Pass a plane, M, through P and the axis. Then the vector velocity of P must be perpendicular to M. For let Q be any point of I. Since Q is at rest, the vector velocity of P must lie in a plane through P perpendicular to QP.

Consider next the circle, C, through P with I as its axis. The vector velocity of P is tangent to C. For it is perpendicular to any line joining P with a point of I. Moreover, the vector velocities of all points of C are of the same length. For otherwise two points of C would be approaching each other, or receding from each other.*

Lastly, the magnitude of the vector velocity of P is proportional to its distance from I. Let M be the plane determined by I and P. Consider two points, P_1 and P_2, in M but not on I, distant h_1 and h_2 respectively from I. Let ωh_1 be the magnitude of the vector velocity of P_1. Then ωh_2 is the magnitude of the vector velocity of P_2. For otherwise P_2 would issue from the rigid plane M.† This completes the proof.

* Exercise 4 below. † Exercise 5 below.

EXERCISES

1. Give a rigorous analytic proof that if two points of a moving straight line are at rest, every point of the line is at rest.

2. A point Q is moving in any manner, and a second point, P, is so moving that, at a given instant, it is neither approaching Q nor receding from Q. Give a rigorous analytic proof that the vector velocity of P is orthogonal to the line QP, if the vector velocity of Q is orthogonal to that line.

3. If three points of space are at rest, and if these points do not lie on a line, all space is at rest. Prove rigorously analytically.

4. Prove analytically the statement of the text which refers to this Exercise.

5. The same for this statement.

9. Vector Angular Velocity. Let space rotate as a rigid body about a fixed axis, L, with angular velocity ω. Let P be an arbitrary point fixed in the moving space. Then the velocity of P will be represented by a vector $\mathbf{v}$ perpendicular to the plane determined by P and the line L, and of length $h\omega$, where h is the distance of P from L.

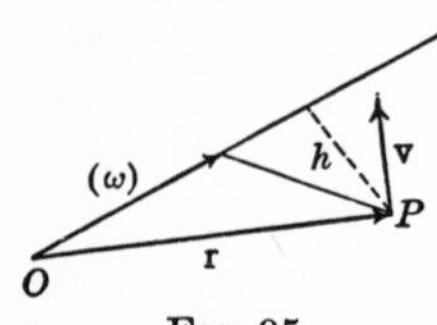

Fig. 95

Let O be any point of L. Lay off from O along L a vector of length ω and denote this vector by (ω). Let $\mathbf{r}$ be the vector drawn from O to P. Then the vector velocity $\mathbf{v}$ of P is represented by the *vector product* of (ω) and $\mathbf{r}$:

$$\mathbf{v} = (\omega) \times \mathbf{r}; \tag{1}$$

cf. Appendix A.

Let a system of Cartesian axes (x, y, z) be assumed with O as origin, and let $\mathbf{i}, \mathbf{j}, \mathbf{k}$ be unit vectors along these axes. Write

$$(\omega) = \omega_x \mathbf{i} + \omega_y \mathbf{j} + \omega_z \mathbf{k}. \tag{2}$$

Then

$$\mathbf{v} = \begin{vmatrix} \mathbf{i} & \mathbf{j} & \mathbf{k} \\ \omega_x & \omega_y & \omega_z \\ x & y & z \end{vmatrix}. \tag{3}$$

The components of $\mathbf{v}$ along the axes are thus seen to be:

$$(4) \qquad \begin{cases} v_x = z\,\omega_y - y\,\omega_z \\ v_y = x\,\omega_z - z\,\omega_x \\ v_z = y\,\omega_x - x\,\omega_y. \end{cases}$$

Composition of Angular Velocities. Consider two rotations about axes which pass through O. Let them be represented by the vectors (ω) and (ω'). An arbitrary point P of space has a vector velocity $\mathbf{v}$ given by (1):

$$\mathbf{v} = (\omega) \times \mathbf{r},$$

due to the first rotation, and a vector velocity $\mathbf{v}'$:

$$\mathbf{v}' = (\omega') \times \mathbf{r},$$

due to the second rotation.

Let these vectors, $\mathbf{v}$ and $\mathbf{v}'$, be added. Then a third vector field results — one in which to the point P is assigned the vector $\mathbf{v} + \mathbf{v}'$. It is not obvious that this third vector field can be realized by a motion of rigid space — far less, then, that it is precisely the field of velocities due to the angular velocity represented by the vector

$$(5) \qquad (\Omega) = (\omega) + (\omega').$$

That this is in fact true — that is the *Law of the Composition of Angular Velocities.*

The proof is immediate. We have:

$$(6) \qquad \mathbf{v} + \mathbf{v}' = (\omega) \times \mathbf{r} + (\omega') \times \mathbf{r}.$$

Now, the vector, or outer, product is distributive:

$$(7) \qquad \{(\omega) + (\omega')\} \times \mathbf{r} = (\omega) \times \mathbf{r} + (\omega') \times \mathbf{r}.$$

Hence

$$(8) \qquad \mathbf{v} + \mathbf{v}' = \{(\omega) + (\omega')\} \times \mathbf{r} = (\Omega) \times \mathbf{r},$$

and we are through. The result can be formulated as the following theorem.

THEOREM. *Angular velocities can be compounded by the Law of Vector Addition.*

EXERCISE

Prove the law of composition for angular velocities by means of Equations (4).

10. Moving Axes. Proof of the Theorem of § 8. Let space be moving as a rigid body with one point, O, fixed. Let **i**, **j**, **k** be three mutually orthogonal unit vectors drawn from O and fixed in space, and let α, β, γ be a second set of such vectors fixed in the body. The scheme of their direction cosines shall be the following:

(1)

	α	β	γ
i	l_1	l_2	l_3
j	m_1	m_2	m_3
k	n_1	n_2	n_3

	ξ	η	ζ
x	l_1	l_2	l_3
y	m_1	m_2	m_3
z	n_1	n_2	n_3

Thus

$$\alpha = l_1\mathbf{i} + m_1\mathbf{j} + n_1\mathbf{k},$$

with similar expressions for β, γ, where the direction cosines are *any* functions of the time, t, continuous with their first (and for later purposes their second or even third) derivatives, and satisfying the familiar identities; cf. Appendix A. Observe that

$$(2)\qquad \begin{cases} \alpha^2 = 1, \quad \beta^2 = 1, \quad \gamma^2 = 1; \\ \beta\gamma = 0, \quad \gamma\alpha = 0, \quad \alpha\beta = 0; \\ \alpha\dot{\alpha} = 0, \quad \text{etc.} \\ \beta\dot{\gamma} + \gamma\dot{\beta} = 0, \quad \text{etc.} \end{cases}$$

We are now in a position to prove analytically the *existence of an instantaneous axis*. Let P be an arbitrary point fixed in the body, and let **r** be the vector drawn from the fixed point O to P. Then

$$(3)\qquad \mathbf{r} = \xi\alpha + \eta\beta + \zeta\gamma.$$

Since P is fixed in the body, ξ, η, ζ are constant with respect to the time, and so

$$(4)\qquad \dot{\mathbf{r}} = \xi\dot{\alpha} + \eta\dot{\beta} + \zeta\dot{\gamma}.$$

A necessary and sufficient condition that P be at rest is, that the projections of $\dot{\mathbf{r}}$ on three non-complanar axes all vanish. Hence, in particular, the condition that P be at rest can be expressed in the form:

$$(5)\qquad \alpha\dot{\mathbf{r}} = 0, \qquad \beta\dot{\mathbf{r}} = 0, \qquad \gamma\dot{\mathbf{r}} = 0.$$

Applying this condition to the vector (4), we find the three ordinary equations:

$$(6) \qquad \begin{cases} \qquad\qquad \eta\alpha\dot{\beta} + \zeta\alpha\dot{\gamma} = 0 \\ \xi\beta\dot{\alpha} + \qquad\qquad \zeta\beta\dot{\gamma} = 0 \\ \xi\gamma\dot{\alpha} + \eta\gamma\dot{\beta} \qquad\qquad = 0 \end{cases}$$

Let

$$(7) \qquad a = \gamma\dot{\beta}, \qquad b = \alpha\dot{\gamma}, \qquad c = \beta\dot{\alpha}.$$

From (2) it follows that

$$a = -\beta\dot{\gamma}, \qquad b = -\gamma\dot{\alpha}, \qquad c = -\alpha\dot{\beta}.$$

Equations (6) are now seen to admit the particular solution:

$$\xi = a, \qquad \eta = b, \qquad \zeta = c.$$

These cannot all be 0 unless the body is at rest, since the vanishing of the above scalar products would mean that

$$\beta\dot{\alpha} = 0, \qquad \gamma\dot{\alpha} = 0;$$

and of course $\alpha\dot{\alpha} = 0$. Thus the vector α would be at rest, and likewise, each of the other vectors, β and γ.

The general solution of the equations (6) is given by the equations:

$$(8) \qquad \xi = \lambda a, \qquad \eta = \lambda b, \qquad \zeta = \lambda c, \qquad -\infty < \lambda < \infty.$$

These points, and these only, are at rest. They form the *instantaneous axis,* and it remains to show that the latter deserves its name.

Instantaneous Axis. Let a vector (ω) be defined as follows:

$$(9) \qquad \omega_\xi = \gamma\dot{\beta}, \qquad \omega_\eta = \alpha\dot{\gamma}, \qquad \omega_\zeta = \beta\dot{\alpha};$$

$$(10) \qquad (\omega) = \omega_\xi\alpha + \omega_\eta\beta + \omega_\zeta\gamma.$$

Then (ω) is collinear with the instantaneous axis, whose equations (8) can now be written in the form:

$$(11) \qquad \frac{\xi}{\omega_\xi} = \frac{\eta}{\omega_\eta} = \frac{\zeta}{\omega_\zeta}.$$

We have seen that the vector velocity $\mathbf{v}$ of an arbitrary point fixed in the body is given by (4). The components of $\mathbf{v}$ along the (ξ, η, ζ)-axes can be written in the form:

$$\begin{aligned} v_\xi &= \alpha\dot{\mathbf{r}} = \xi\alpha\dot{\alpha} + \eta\alpha\dot{\beta} + \zeta\alpha\dot{\gamma} \\ v_\eta &= \beta\dot{\mathbf{r}} = \xi\beta\dot{\alpha} + \eta\beta\dot{\beta} + \zeta\beta\dot{\gamma} \\ v_\zeta &= \gamma\dot{\mathbf{r}} = \xi\gamma\dot{\alpha} + \eta\gamma\dot{\beta} + \zeta\gamma\dot{\gamma} \end{aligned}$$

Hence

$$(12) \qquad \begin{cases} v_\xi = \zeta\,\omega_\eta - \eta\,\omega_\zeta \\ v_\eta = \xi\,\omega_\zeta - \zeta\,\omega_\xi \\ v_\zeta = \eta\,\omega_\xi - \xi\,\omega_\eta \end{cases}$$

From (12) it follows that

$$(13) \qquad \mathbf{v} = \begin{vmatrix} \alpha & \beta & \gamma \\ \omega_\xi & \omega_\eta & \omega_\zeta \\ \xi & \eta & \zeta \end{vmatrix} = (\omega) \times \mathbf{r},$$

and so we see that the actual vector velocity $\mathbf{v}$ of P is the same as the vector velocity which P would have if rigid space were rotating about the instantaneous axis with angular velocity ω.

Thus the actual field of vector velocities of the points P coincides with the field of vector velocities due to rotation about the instantaneous axis represented by the vector angular velocity (ω), and the proof is complete.

11. Space Centrode and Body Centrode. The locus of the instantaneous axis in fixed space is called the *space centrode*, and its locus in the moving space, the *body centrode*. The actual motion consists of the rolling of the one cone (the body centrode) without slipping on the other cone (the space centrode).

To prove this statement consider the path traced out by a specified point in the instantaneous axis. Take, for instance, the terminal point of the vector (ω), the initial point being at O. The locus of this point is a certain curve C on the space centrode:

$$(\omega) = \omega_x\mathbf{i} + \omega_y\mathbf{j} + \omega_z\mathbf{k},$$

and a certain curve Γ of the body centrode:

$$(\omega) = \omega_\xi\alpha + \omega_\eta\beta + \omega_\zeta\gamma.$$

It is sufficient to show that these curves are tangent and that corresponding arcs are equal. This will surely be the case if $d(\omega)/dt$ for C is equal to $d(\omega)/dt$ for Γ. Now, the first vector has the value

$$(\dot\omega) = \dot\omega_x\mathbf{i} + \dot\omega_y\mathbf{j} + \dot\omega_z\mathbf{k}.$$

The value of the second vector is:

$$\begin{aligned} &\dot\omega_\xi\alpha + \dot\omega_\eta\beta + \dot\omega_\zeta\gamma \\ &+ \omega_\xi\dot\alpha + \omega_\eta\dot\beta + \omega_\zeta\dot\gamma. \end{aligned}$$

The last line vanishes because it represents the velocity of the point (ω) fixed in the body, this point lying on the instantaneous axis. The first line is the vector ($\dot{\omega}$). This completes the proof.

EXERCISE

Treat the motion of the plane by analogous vector methods. Let

$$\zeta = x_0 + y_0 i, \qquad i = \sqrt{-1},$$

be the vector drawn from the fixed to the moving origin, and let ρ, σ be unit vectors drawn along the positive axes of ξ and η. Let $\mathbf{r}$ be the vector from the fixed origin to an arbitrary point P. Then

$$\mathbf{r} = \zeta + \xi\rho + \eta\sigma.$$

The vector velocity in space of a point P fixed in the plane is given by the vector equation:

$$\dot{\mathbf{r}} = \dot{\zeta} + \xi\dot{\rho} + \eta\dot{\sigma}.$$

The instantaneous centre is given by setting $\dot{\mathbf{r}} = 0$.

On the other hand,

$$\rho = e^{\theta i}, \qquad \sigma = e^{\left(\theta + \frac{\pi}{2}\right)i}.$$

$$\dot{\rho} = i\,e^{\theta i}\,\dot{\theta} = \sigma\dot{\theta}, \qquad \dot{\sigma} = i\,e^{\left(\theta + \frac{\pi}{2}\right)i}\,\dot{\theta} = -\,\rho\dot{\theta}.$$

The complete treatment can now be worked out without difficulty.

12. Motion of Space. General Case. Let rigid space be moving in any manner, subject to the ordinary assumptions about continuity. Reduce a point A to rest by impressing on all space a motion of translation whose vector is equal and opposite to the vector velocity of A. The vector field of the velocities in the original motion is compounded by the parallelogram law of vector addition out of the two vector fields *i*) of translation and *ii*) of rotation about the instantaneous axis, I.

Let the vector that represents the translation be resolved into two vectors, one, $\mathbf{T}$, collinear with I, the other, $\mathbf{A}$, at right angles to I. The velocity of any point, P, distant h from the axis, is, in the case of pure rotation, $h\omega$; its direction is at right angles to the plane through P and the axis, and its sense is a definite one of the two possible senses. Hence it is seen

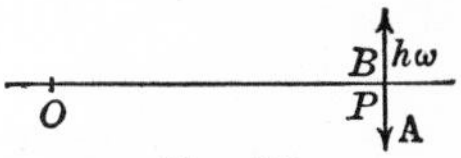

FIG. 96

that it is possible to find a point, B, whose vector velocity due to the rotation is equal and opposite to the vector velocity **A**. (Draw a line from a point O of the axis, perpendicular to I and **A**, and measure off on it, in the proper direction, a distance $h = A/\omega$.) All points in the line L through B parallel to I will also be at rest. It thus appears that the original motion is one of rotation about L compounded by the law of vector addition with a motion of translation parallel to L and represented by the vector **T**.

This vector field is, in general, the same as that of the vector velocities of the points of a nut which moves along a fixed machine screw (or of the points of a machine screw which moves through a fixed nut). The two exceptional cases are those of rotation, corresponding to a pitch 0 of the threads, and translation, the limiting case, as the pitch becomes infinite.

13. The Ruled Surfaces. We have seen in § 12 that the vector field of velocities, in the general case of the motion of rigid space, is the sum of two vector fields — one, rotation about an axis, L; the other, translation parallel to L. The locus of L in space is a ruled surface S, the *space centrode*, and the locus of L in the moving space is also a ruled surface, Σ, the *body centrode.* From analogy with the rolling cones we should anticipate the theorem governing the present case.

THEOREM. *The surface Σ is tangent to S along L, and it rolls and slides on S.*

An intuitional proof can be given as follows. First of all, it is clear from the very definition of L that Σ slides on S along L. So it is necessary to prove only the tangency of the two surfaces. Let L_0 be the line L at time $t = t_0$, and let P_0 be a point of L_0. Pass a plane through P_0 perpendicular to L_0, cutting S in the curve C, and let P be the point in which L at time $t = t_0 + \Delta t$ cuts C. Let Q be the point fixed in Σ, which will coincide with P at time $t_0 + \Delta t$. The vector velocity of Q at the instant $t = t_0$ has a component, c, parallel to L_0 and a component $h\omega$ at right angles to the plane through L_0 and Q. Obviously h is infinitesimal with Δt. In time Δt the point Q will, then, have been displaced, save as to an infinitesimal of higher order, parallel to L_0 by a distance $c\,\Delta t$. But it will have reached P.

The proof is now clear. The plane through L_0 and Q makes an infinitesimal angle with the tangent plane to Σ at P_0 because it contains a point Q of Σ infinitely near to P_0, but not on L_0. The plane through L_0 and P makes an infinitesimal angle with the tangent plane to S at P_0 because it contains a point P of S infinitely near to P_0, but not on L_0. And these two planes make an infinitesimal angle with each other, because when Q is displaced parallel to L_0 by a distance $c\Delta t$, its distance from P is an infinitesimal of higher order than the distance of P from P_0.

Instead of developing the details needed to make the intuitive proof rigorous, we will treat the whole question by vector methods. First, however, a digression on relative velocities.

14. Relative Velocities. Let a point P move in any manner in space, and let its motion be referred to a system of moving axes.

Consider, first, the case that the moving axes have a fixed origin, O. Let a system of axes (x, y, z), fixed in space, with origin at O be chosen; let the moving axes be denoted by (ξ, η, ζ), and referred to the fixed axes by the scheme of direction cosines of § 10. Let $\mathbf{r}$ be the vector drawn from O to P:

$$\mathbf{r} = \xi\alpha + \eta\beta + \zeta\gamma. \tag{1}$$

Then

$$\frac{d\mathbf{r}}{dt} = \frac{d\xi}{dt}\alpha + \frac{d\eta}{dt}\beta + \frac{d\zeta}{dt}\gamma + \xi\dot{\alpha} + \eta\dot{\beta} + \zeta\dot{\gamma} \tag{2}$$

or

$$\mathbf{v} = \mathbf{v}_r + \mathbf{v}_e, \tag{3}$$

where the terms on the right have the following meanings. The vector,

$$\mathbf{v}_r = \frac{d\xi}{dt}\alpha + \frac{d\eta}{dt}\beta + \frac{d\zeta}{dt}\gamma, \tag{4}$$

represents the velocity of P *relative* to the moving axes; *i.e.* what its absolute velocity would be if the (ξ, η, ζ)-axes were fixed and P moved relative to them just as it actually does move.

Secondly, the vector

$$\mathbf{v}_e = \xi\dot{\alpha} + \eta\dot{\beta} + \zeta\dot{\gamma} \tag{5}$$

represents the velocity in space of that point fixed in the body, which at the instant t coincides with P. To say the same thing in other words: — Let us consider the point P at an arbitrary instant of time, $t = t$. Let Q be the point fixed in the body, which at this one instant coincides with P. Then $\mathbf{v}_e$ is the vector velocity of Q. It is the *vitesse d'entraînement*, the velocity with which the point Q is being transported by the body at the instant t.

The analytic expression for $\mathbf{v}_e$ we know all about. In vector form it is:

$$\mathbf{v}_e = (\omega) \times \mathbf{r} \tag{6}$$

or

$$\mathbf{v}_e = \begin{vmatrix} \alpha & \beta & \gamma \\ \omega_\xi & \omega_\eta & \omega_\zeta \\ \xi & \eta & \zeta \end{vmatrix}. \tag{7}$$

Its components along the axes, if we write $\mathbf{v}' = \mathbf{v}_e$, are:

$$\begin{cases} \mathbf{v}'_\xi = \zeta\,\omega_\eta - \eta\,\omega_\zeta \\ \mathbf{v}'_\eta = \xi\,\omega_\zeta - \zeta\,\omega_\xi \\ \mathbf{v}'_\zeta = \eta\,\omega_\xi - \xi\,\omega_\eta \end{cases} \tag{8}$$

Thus we have as the final solution of our problem this: *The components of the vector velocity of P along the axes of ξ, η, ζ are:*

$$\begin{cases} v_\xi = \dfrac{d\xi}{dt} + \zeta\,\omega_\eta - \eta\,\omega_\zeta \\ v_\eta = \dfrac{d\eta}{dt} + \xi\,\omega_\zeta - \zeta\,\omega_\xi \\ v_\zeta = \dfrac{d\zeta}{dt} + \eta\,\omega_\xi - \xi\,\omega_\eta \end{cases} \tag{9}$$

The General Case. Let the axes of (x, y, z) be fixed in space. Let (ξ, η, ζ) be the moving axes, whose origin, O', has the coordinates (x_0, y_0, z_0). Then

$$\mathbf{r} = \mathbf{r}_0 + \mathbf{r}'. \tag{10}$$

Hence

$$\mathbf{v} = \mathbf{v}_0 + \mathbf{v}'. \tag{11}$$

Here,

$$\mathbf{v}_0 = \frac{dx_0}{dt}\mathbf{i} + \frac{dy_0}{dt}\mathbf{j} + \frac{dz_0}{dt}\mathbf{k}, \tag{12}$$

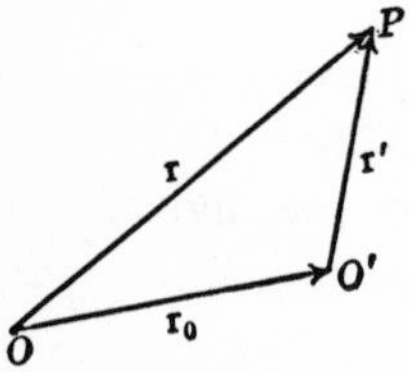

FIG. 97

and $\mathbf{v}'$ is given by (9). The $\mathbf{v}'$ of (11) is, of course, not the $\mathbf{v}'$ of (8).

EXERCISE

Denoting the components of $\mathbf{v}_0$ along the (ξ, η, ζ)-axes by v_ξ^0, v_η^0, v_ζ^0, show that the components of v along these axes are:

$$(13) \qquad \begin{cases} v_\xi = v_\xi^0 + \dfrac{d\xi}{dt} + \zeta\,\omega_\eta - \eta\,\omega_\zeta \\[2ex] v_\eta = v_\eta^0 + \dfrac{d\eta}{dt} + \xi\,\omega_\zeta - \zeta\,\omega_\xi \\[2ex] v_\zeta = v_\zeta^0 + \dfrac{d\zeta}{dt} + \eta\,\omega_\xi - \xi\,\omega_\eta \end{cases}$$

Here,

$$v_\xi^0 = \alpha \dot{\mathbf{r}}_0, \qquad v_\eta^0 = \beta \dot{\mathbf{r}}_0, \qquad v_\zeta^0 = \gamma \dot{\mathbf{r}}_0.$$

15. Proof of the Theorem of § 12. Let a system of Cartesian axes fixed in space, (x, y, z), with origin in O be assumed. Let O': (x_0, y_0, z_0) be a point fixed in the body, the motion of O' being known:

$$(1) \qquad x_0 = f(t), \qquad y_0 = \varphi(t), \qquad z_0 = \psi(t).$$

Finally, let P be any point fixed in the body. Then

$$(2) \qquad \mathbf{r} = \mathbf{r}_0 + \mathbf{r}',$$

cf. § 14, (10) with the specialization that here

$$(3) \qquad \frac{d\xi}{dt} = 0, \qquad \frac{d\eta}{dt} = 0, \qquad \frac{d\zeta}{dt} = 0.$$

Then the components of the absolute velocity of P (*i.e.* its velocity in fixed space) along the axes of (ξ, η, ζ) are given by the formulas of §§ 14, 13:

$$(4) \qquad \begin{cases} v_\xi = \alpha \dot{\mathbf{r}}_0 + \zeta\,\omega_\eta - \eta\,\omega_\zeta \\ v_\eta = \beta \dot{\mathbf{r}}_0 + \xi\,\omega_\zeta - \zeta\,\omega_\xi \\ v_\zeta = \gamma \dot{\mathbf{r}}_0 + \eta\,\omega_\xi - \xi\,\omega_\eta \end{cases}$$

We can formulate the problem as follows: To find a point P: (ξ_1, η_1, ζ_1) fixed in the moving space whose absolute vector velocity is collinear with the vector (ω), or is 0:

$$(5) \qquad \dot{\mathbf{r}}_1 = k\,(\omega).$$

Here, (ω) is the vector angular velocity of the moving space, whose rotation is defined by the direction cosines of § 10.

By virtue of (4) the vector equation (5) is equivalent to the three ordinary equations:

$$
(6) \qquad \begin{cases} \alpha \mathfrak{k}_0 \qquad\qquad - \eta_1 \omega_\zeta + \zeta_1 \omega_\eta = k\,\omega_\xi \\ \beta \mathfrak{k}_0 + \xi_1 \omega_\zeta \qquad\qquad - \zeta_1 \omega_\xi = k\,\omega_\eta \\ \gamma \mathfrak{k}_0 - \xi_1 \omega_\eta + \eta_1 \omega_\xi \qquad\qquad = k\,\omega_\zeta \end{cases}
$$

Since

$$
(7) \qquad \begin{vmatrix} 0 & -\omega_\zeta & \omega_\eta \\ \omega_\zeta & 0 & -\omega_\xi \\ -\omega_\eta & \omega_\xi & 0 \end{vmatrix} = 0,
$$

a further necessary condition is:

$$
(8) \qquad \omega_\xi \alpha \mathfrak{k}_0 + \omega_\eta \beta \mathfrak{k}_0 + \omega_\zeta \gamma \mathfrak{k}_0 = k\,\omega^2.
$$

We can dispose at once of the case $\omega = 0$; for then the space in which O' is at rest, is stationary, and so the motion of the given space is translation (unless it be at rest). Thus all lines parallel to the vector that represents the translation are axes such as we seek.

If $\omega \neq 0$, we obtain from (8) a unique determination of k. On substituting this value in (6), two of these equations, suitably chosen, determine uniquely two of the three unknowns ξ_1, η_1, ζ_1 as linear functions of the third, and then the remaining equation (6) is true because of (8).

If

$$
\xi_1 = a_1, \qquad \eta_1 = b_1, \qquad \zeta_1 = c_1
$$

be a particular solution of (6), then an arbitrary solution, ξ_1, η_1, ζ_1, will satisfy the equations:

$$
\begin{cases} \qquad\qquad - (\eta_1 - b_1)\omega_\zeta + (\zeta_1 - c_1)\omega_\eta = 0 \\ (\xi_1 - a_1)\omega_\zeta \qquad\qquad - (\zeta_1 - c_1)\omega_\xi = 0 \\ -(\xi_1 - a_1)\omega_\eta + (\eta_1 - b_1)\omega_\xi \qquad\qquad = 0 \end{cases}
$$

Hence

$$
(9) \qquad \frac{\xi_1 - a_1}{\omega_\xi} = \frac{\eta_1 - b_1}{\omega_\eta} = \frac{\zeta_1 - c_1}{\omega_\zeta}
$$

and thus ξ_1, η_1, ζ_1 is seen to be any point of the line through (a_1, b_1, c_1) collinear with (ω). This line we define as L. These conditions are sufficient as well as necessary.

The locus of L in space is the ruled surface S; its locus in the body (*i.e.* the moving space) is the surface Σ. These surfaces have the line L in common. We wish to show that they are tangent along L, and that Σ slides over S in the direction of L. The last fact is clear from the definition of L.

The point (ξ_1, η_1, ζ_1) is not uniquely determined by the time, but may be any point of L. We will, for our purposes, select it as follows. Let L_0 be a particular L, and let P_0 be an arbitrary point of L_0, once chosen and then held fast. Pass a plane M through P_0 orthogonal to L_0. Then (x_1, y_1, z_1) shall be the intersection of the variable line L with M, and its locus shall be denoted by C. The point (ξ_1, η_1, ζ_1) shall be the point of Σ which coincides with (x_1, y_1, z_1) at time $t = t$. Its locus in Σ shall be denoted by Γ. This curve can be represented in the form:

$$\Gamma: \qquad \xi_1 = F(t), \qquad \eta_1 = \Phi(t), \qquad \zeta_1 = \Psi(t).$$

Its tangent vector at an arbitrary point is

$$\frac{d\xi_1}{dt}\alpha + \frac{d\eta_1}{dt}\beta + \frac{d\zeta_1}{dt}\gamma, \tag{10}$$

provided this vector $\neq 0$.

To show that two surfaces which intersect at a point P_0 are tangent it is sufficient to show *i*) that they have a common tangent vector, $\mathbf{t}$; and *ii*) that a tangent vector $\mathbf{t}_1$ to the one surface and a tangent vector $\mathbf{t}_2$ to the other surface, neither collinear with $\mathbf{t}$, are complanar with $\mathbf{t}$, all three vectors, emanating from P_0.

The surfaces S and Σ satisfy *i*) because they are both tangent to L. Secondly, consider the vector $\mathbf{r}_1$ drawn from O to the point of intersection of C and Γ at time $t = t$. Its derivative is a vector

$$\mathbf{t}_1 = \frac{d\mathbf{r}_1}{dt},$$

tangent to C, provided it $\neq 0$.

On the other hand, consider the point (ξ_1, η_1, ζ_1) of Γ, for which $t = t$. Let the vector drawn from O' to this point be denoted by $\mathbf{r}_1'$. Then

$$\mathbf{r}_1 = \mathbf{r}_0 + \mathbf{r}_1',$$

where $\mathbf{r}_0$ is given by (1), and

$$\mathbf{r}_1' = \xi_1\alpha + \eta_1\beta + \zeta_1\gamma.$$

Hence

$$\frac{d\mathbf{r}_1}{dt} = \frac{d\xi_1}{dt}\alpha + \frac{d\eta_1}{dt}\beta + \frac{d\zeta_1}{dt}\gamma \tag{11}$$

$$+ \dot{\mathbf{r}}_0 + \xi_1\dot{\alpha} + \eta_1\dot{\beta} + \zeta_1\dot{\gamma}.$$

This last line is precisely the vector velocity of that point *fixed in* Σ, which at the instant in question, $t = t$, coincides with (x_1, y_1, z_1). This vector, $\mathbf{t}$, let us call it, lies along L because of (5), — unless it be 0.

The other vector on the right of (11) is the vector (10); *i.e.* a vector $\mathbf{t}_2$ tangent to Γ at (ξ_1, η_1, ζ_1) or (x_1, y_1, z_1). Equation (11) thus says that

$$\mathbf{t}_1 = \mathbf{t}_2 + \mathbf{t}.$$

Now, the vector $d\mathbf{r}_1/dt = \mathbf{t}_1 \neq 0$ will not lie along L. Hence $\mathbf{t}_2$ will not, either. Consequently Condition *ii*) is satisfied, and the surfaces S and Γ are tangent along L. The case $t = 0$ is included; it does not lead to an exception.

EXERCISE

Let a cylinder of revolution roll and slide on a second cylinder which is fixed, the first cylinder always being tangent along an element, and there being no slipping oblique to the element. Choose the point (x_0, y_0, z_0) in the axis of the moving cylinder, and discuss the whole problem by the method of this paragraph.

16. Lissajou's Curves. In one dimension, or with one degree of freedom, the most important periodic motion is Simple Harmonic Motion. It can be represented analytically in the form:

$$x = a \cos (nt + \gamma), \tag{1}$$

where

$$T = \frac{2\pi}{n} \tag{2}$$

is the period, where a is the amplitude, and where γ is determined by the phase.

In two dimensions, or with two degrees of freedom, an important case of oscillatory motion about a fixed point is that in which the projections of the moving point on two fixed axes at right angles to each other, execute, each by itself, simple harmonic motion:

$$(3)\qquad \begin{cases} x = a\cos(nt+\gamma) \\ y = b\cos(mt+\epsilon) \end{cases}$$

It is possible to generalize at once to n dimensions:

$$(4)\qquad x_k = a_k \cos(n_k t + \gamma_k), \qquad k = l, \cdots, n.$$

Let us study first the two-dimensional case, beginning with some simple examples. We may set $\gamma = 0$ if, as usually happens, the instant from which the time is measured is unimportant.

Example 1: $m = n$. Dynamically, this case can be realized approximately by the small oscillations of a *spherical pendulum*. Let $\gamma = 0$,

$$\varphi = nt.$$

Then

$$mt + \epsilon = \varphi + \epsilon,$$

$$(5)\qquad \begin{cases} x = a\cos\varphi \\ y = A\cos\varphi - B\sin\varphi \end{cases}$$

$$A = b\cos\epsilon, \qquad B = b\sin\epsilon.$$

Assume that neither a nor b vanishes, since otherwise we should be thrown back on right line motion along one of the axes. We will take $a > 0$, $b > 0$.

In general, $B \neq 0$. The path of the moving point is then an ellipse with its centre at the origin. For,

$$(6)\qquad \cos\varphi = \frac{x}{a}, \qquad \sin\varphi = \frac{A}{aB}x - \frac{1}{B}y.$$

On squaring and adding we find:

$$(7)\qquad B^2x^2 + (Ax - ay)^2 = a^2B^2,$$

and this equation represents a central conic which does not reach to infinity, *i.e.* an ellipse.

The axes can be found by the methods of analytic geometry, or computed directly by making the function

$$x^2 + y^2 = (a^2 + A^2)\cos^2\varphi - 2AB\cos\varphi\sin\varphi + B^2\sin^2\varphi$$

a maximum or a minimum.*

We have omitted the special case: $B = 0$. Here, $\epsilon = 0$ or π, and the motion is rectilinear, along the line:

$$\frac{x}{a} = \frac{y}{A}. \tag{8}$$

In all cases, the path is confined within the rectangle:

$$x = \pm a, \qquad y = \pm b,$$

and continually touches all four sides, — sometimes being a diagonal, but, in general, an ellipse inscribed in the rectangle.

Example 2. $m = n + h$, *where* h *is small.* If we write the equations in the form:

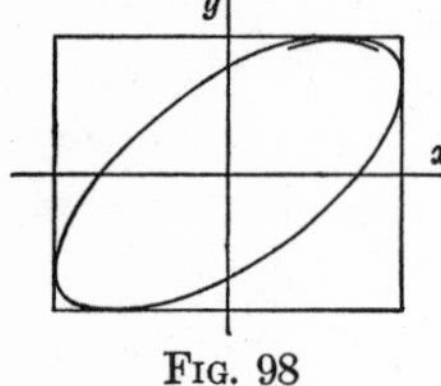

Fig. 98

$$(9) \qquad \begin{cases} x = a\cos nt \\ y = b\cos(nt + \overline{ht + \epsilon}) \end{cases}$$

then, for the duration of time $T = 2\pi/n$, $ht + \epsilon$ is nearly constant, and the path is nearly an ellipse — which, however, does not quite close. And now, in the next interval of time, the path again will be a near-ellipse, but in a slightly different orientation — its points of tangency with the circumscribing rectangle will be slightly advanced or retarded, depending on whether h is positive or negative.

Thus a succession of near-ellipses will be described, all inscribed in the same fixed rectangle $x = \pm a$, $y = \pm b$. The motion can be realized approximately experimentally as follows.

Blackburn's Pendulum. By this is meant the mechanical system that consists of an ordinary pendulum, the upper end of

* Their directions are determined, in either way, by the formula:

$$\cos 2\varphi = \frac{a^2 + b^2\cos 2\epsilon}{-b^2\sin 2\epsilon} \qquad \text{or} \qquad \cos 2\gamma = \frac{a^2 - b^2}{2ab\cos\epsilon},$$

where γ denotes the angle from the axis of x to an axis of the conic. The lengths of the axes are found to be:

$$\sqrt{2(a^2 + b^2 + \Delta)}, \qquad \sqrt{2(a^2 + b^2 - \Delta)},$$

where

$$\Delta^2 = a^4 + 2a^2b^2\cos 2\epsilon + b^4.$$

which is made fast at the mid-point of an inextensible string whose two ends are fastened at the same level. When the bob oscillates in the vertical plane through the supports, the second string remains at rest, and we have simple pendulum motion, the length of the pendulum being l, the length of the first string.

Secondly, let the bob oscillate in a vertical plane at right angles to the line through the points of support, and mid-way between these. Again, we have simple pendulum motion; but the length is now $l' = l + d$, where d denotes the sag in the second string. For small oscillations, the coordinates of the bob will evidently be given approximately by Equations (3), and by suitably choosing l and d, we can realize an arbitrary choice of m and n.

*The Sand Tunnel.** If the bob of the pendulum is a tunnel of small opening, filled with fine sand, the sand, as it issues from the tunnel, will trace out a curve on the floor which shows admirably the whole phenomenon of the Lissajou's Curves. In particular, if the second string is drawn as taut as is feasible, so that d is small, the two periods will be nearly, but not quite, equal; and it is possible to observe the near-ellipses steadily advancing, flashing through near-right lines (the diagonals of the fixed rectangle).

Example 3. $m = 2n$. Begin with the case $\gamma = 0, \epsilon = 0$, and set $\varphi = nt$:

$$x = a \cos \varphi, \qquad y = b \cos 2\varphi. \tag{10}$$

Hence

$$y = \frac{2b}{a^2} x^2 - b \tag{11}$$

and the curve is an arc of a parabola, passing through the vertices (a, b), $(-a, b)$ of the circumscribing rectangle and tangent to the opposite side at the mid-point. The sand pendulum may be released from rest at the point (a, b), and it then traces repeatedly the parabolic arc. In the general case,

$$x = a \cos \varphi, \qquad y = A \cos 2\varphi - B \sin 2\varphi; \tag{12}$$

$$A = b \cos \epsilon, \qquad B = b \sin \epsilon.$$

* This experiment should be shown in the course. It is not necessary to have a physical laboratory. A tunnel can be bought at the Five and Ten, and string is still available, even in this age of cellophane and gummed paper.

When ϵ is small, the curve runs along near to the parabola; cf. Fig. 100. It is symmetric in the axis of y, since φ and $\varphi' = \varphi + \pi$ give $x' = -x$, $y' = y$. It is tangent once to each of the sides $x = a$, $x = -a$ of the circumscribing rectangle, and twice to each of the other two sides. When ϵ has increased to $\pi/2$, $A = 0$ and

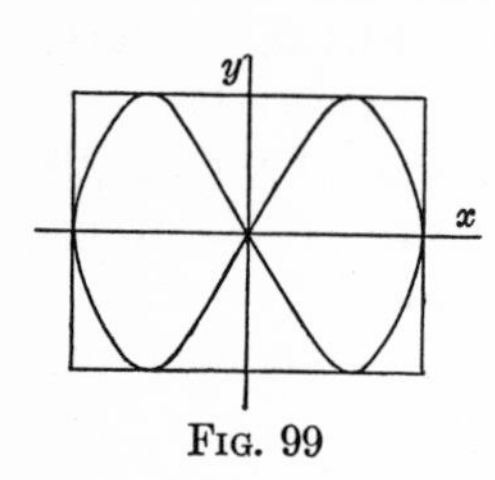

FIG. 99

$$(13) \qquad x = a \cos \varphi, \qquad y = -b \sin 2\varphi$$

or

$$(14) \qquad \left(\frac{y}{b}\right)^2 = 4\left(\frac{x}{a}\right)^2\left[1 - \left(\frac{x}{a}\right)^2\right].$$

This curve is obtained at once by affine transformations from the curve

$$(15) \qquad y^2 = 4x^2(1 - x^2),$$

which is readily plotted.

When ϵ has reached the value π, we have again an arc of a parabola — the former arc, turned upside down. As ϵ continues to increase, the new curves are the mirrored images of the old in the axis of x, for $\epsilon' = \epsilon + \pi$ reverses the signs of A and B. All these curves except the arcs of parabolas are quartics, inscribed in the fixed rectangle, and having symmetry in the axis of y.

Example 4. *$m = 2n + h$, where h is small.* Here,

$$(16) \qquad \begin{cases} x = a \cos nt \\ y = b \cos [2nt + \overline{ht + \epsilon}] \end{cases}$$

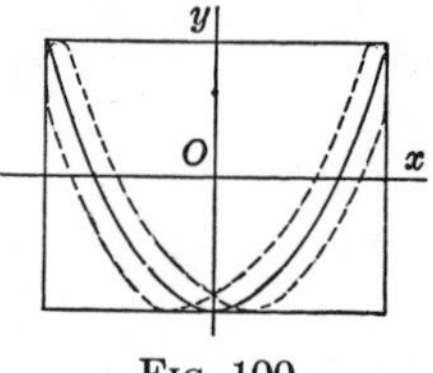

FIG. 100

and for a single excursion, ht is nearly 0. Thus the new curve runs along close to an old curve for a suitable fixed ϵ, but as time elapses, the suitable ϵ advances, too.

The student can readily trace these curves with the sand tunnel. If he does his best to make $d = l$, there will be enough discrepancy to provide for a small h.

17. Continuation. The General Case. The Commensurable Case. Periodicity. Let m and n be commensurable,

$$\frac{n}{m} = \frac{p}{q},$$

where p and q are natural numbers prime to each other. Then

$$n = \alpha p, \qquad m = \alpha q.$$

Let $\varphi = \alpha t$, $\gamma = 0$. Then

$$(17) \qquad \begin{cases} x = a \cos p\varphi, \\ y = b \cos (q\varphi + \epsilon). \end{cases}$$

These functions are periodic with the primitive periods $2\pi/p$ and $2\pi/q$, and evidently have the common period 2π. The smallest positive value of ω for which

$$a \cos p(\varphi + \omega) = a \cos p\varphi$$

$$b \cos \{q(\varphi + \omega) + \epsilon\} = b \cos \{q\varphi + \epsilon\}$$

is $\omega = 2\pi$. For, if ω is to be a period of the first function, then

$$\omega = \lambda \frac{2\pi}{p}.$$

And if ω is to be a period of the second function, then

$$\omega = \mu \frac{2\pi}{q}.$$

Hence

$$\frac{\lambda}{p} = \frac{\mu}{q}, \qquad \lambda q = \mu p,$$

and the smallest values of λ, μ in natural numbers which satisfy this equation are $\lambda = p$, $\mu = q$.

From the periodicity of the functions it appears that the curve is closed, and thus, as t increases, the curve is traced out repeatedly. For a non-specialized value of ϵ, the curve is tangent to each of the sides $x = a, -a$ of the circumscribing rectangle p times, corresponding to the solutions of the equations $\cos p\varphi = 1, -1$; and q times to each of the sides $y = b, -b$. A line $x = x'$, $-a < x' < a$, cuts the curve in $2p$ points; a line $y = y'$, $-b < y' < b$, in $2q$ points.

These curves are all algebraic, and *rational*, or *unicursal.* For, on setting $\xi = \tan \frac{1}{2}\varphi$, the variables x and y appear, by de Moivre's Theorem, as rational functions of ξ. The curves are all symmetric in the axis of y.

The Incommensurable Case. Aperiodic. If on the other hand n/m is incommensurable, the curve never closes. It courses every region contained within the rectangle. If P be an arbitrary point of the rectangle, the curve will not in general pass through P; but it will come indefinitely near to P — not merely once, but infinitely often; possibly, occasionally passing through P.

The proof can be given as follows. Consider a circle and the angle φ at the centre. Let $\varphi = \xi = 2\pi\alpha$ be an angle which is incommensurable with 2π; *i.e.* let α be irrational. Then the points of the circle which correspond to $\xi, 2\xi, 3\xi, \cdots$ (denote them by $P_1, P_2, \cdots$) are all distinct. Hence they must have at least one point of condensation, P. But from this follows that *every* point must be a point of condensation. For, let P_n and P_m be two points near P. Then the point corresponding to $n\xi - m\xi$ must be near the point corresponding to $\varphi = 0$. Having thus obtained an arc of arbitrarily small length, we have but to take multiples of it, *i.e.* to construct the points $P_{k(n-m)}$, $k = 1, 2, 3, \cdots$, to come arbitrarily near to any point on the circumference.

Turning now to the equations of the curve, let

$$\varphi = nt, \qquad \frac{m}{n} = \alpha.$$

Then

$$(18) \qquad \begin{cases} x = a \cos \varphi, \\ y = b \cos (\alpha\varphi + \eta). \end{cases}$$

Let $-a \leqq x' \leqq a$, and let $\varphi = \varphi'$ be a root of the equation

$$x' = a \cos \varphi.$$

The curve cuts the line $x = x'$ in the points for which

$$y = b \cos \{\alpha(\varphi' + 2k\pi) + \eta\}, \quad b \cos \{\alpha(-\varphi' + 2k\pi) + \eta\}.$$

And now, since the angles $2k\alpha\pi$ lead to points on the circle which are everywhere dense, the corresponding values of the cosine factor are also everywhere dense between -1 and $+1$.

It is of interest to study the multiple points of the curve. These occur when

i) $$\varphi' = k\pi + \frac{l\pi}{\alpha}, \qquad l \neq 0;$$

ii) $$\varphi' = k\pi + \frac{l\pi}{\alpha} - \frac{\eta}{\alpha}, \qquad k \neq 0;$$

provided

$$(19) \qquad \eta \neq (l_0 + k_0\alpha)\pi.$$

When the inequality (19) holds, there is a one-to-one correspondence between the values of φ and the points of the curve, provided the multiple points (which are always double points with distinct tangents) are counted multiply. If, however,

$$\eta = (l_0 + k_0\alpha)\pi, \tag{20}$$

then

$$\alpha\varphi + \eta = \alpha(\varphi + k_0\pi) + l_0\pi.$$

Set

$$\theta = \varphi + k_0\pi. \tag{21}$$

Then the Equations (18) become:

$$\begin{cases} x = a' \cos \theta \\ y = b' \cos \alpha\theta, \end{cases} \tag{22}$$

where $a' = a$ or $-a$, and likewise $b' = b$ or $-b$. Let

$$0 \leqq \theta < \infty.$$

Then there is a one-to-one correspondence between the values of θ and the points on the curve. The point for which $\theta = 0$:

$$x = a', \qquad y = b',$$

is an end-point of the curve. It is simple, no other branch going through it. The double points correspond to the values

$$\theta' = k\pi + \frac{l\pi}{\alpha} > 0,$$

where

$$k\pi - \frac{l\pi}{\alpha} > 0, \qquad l \neq 0;$$

or where

$$-k\pi + \frac{l\pi}{\alpha} > 0, \qquad k \neq 0.$$

Three and n-Dimensions. In the case of motion with three degrees of freedom, the equations can be reduced to the form:

$$\begin{cases} x = a \cos \varphi \\ y = b \cos (\alpha\varphi + \eta) \\ z = c \cos (\beta\varphi + \zeta) \end{cases} \tag{23}$$

The case that α, β are both commensurable can be discussed as before. The curve closes, the motion is periodic. When α and β are both irrational, and their ratio is also irrational, it can happen that the curve courses every region, however small, of the parallelepiped:

$$-a \leqq x \leqq a, \qquad -b \leqq y \leqq b, \qquad -c \leqq z \leqq c,$$

and has no multiple points, the correspondence between the points of the curve and the values of φ when $-\infty < \varphi < \infty$ being one-to-one without exception Whether the former property is present for all such values of α and β, provided furthermore that α, β, and β/α are not connected by a linear non-homogeneous equation with integral coefficients, and that η, ζ are not specialized, I cannot say, though I surmise it to be. The latter property, however, can be established.

The same statements hold in the general case,

$$x_k = a_k \cos(\alpha_k \varphi + \eta_k), \qquad k = 1, \cdots, n.$$

It may happen, in a dynamical system with n degrees of freedom and coordinates $q_1, \cdots, q_n$, that only a sub-set, $q_1, \cdots, q_m$, $1 \leqq m < n$, execute a Lissajou's motion. Thus a Blackburn's Pendulum suspended in a moving elevator will have its projection on a horizontal plane executing a Lissajou's motion, whereas the vertical motion is not periodic at all.

The late Professor Wallace Clement Sabine drew mechanically some very beautiful curves, which are here reproduced in half-tone. I still have the half-tone which Dr. Sabine gave me. So far as I have been able to ascertain, the curves were never published. The figures here shown were made from lantern slides in possession of the Jefferson Physical Laboratory, and it is through the courtesy of the Laboratory that I have been enabled to reproduce them here.

Lissajou's Curves —

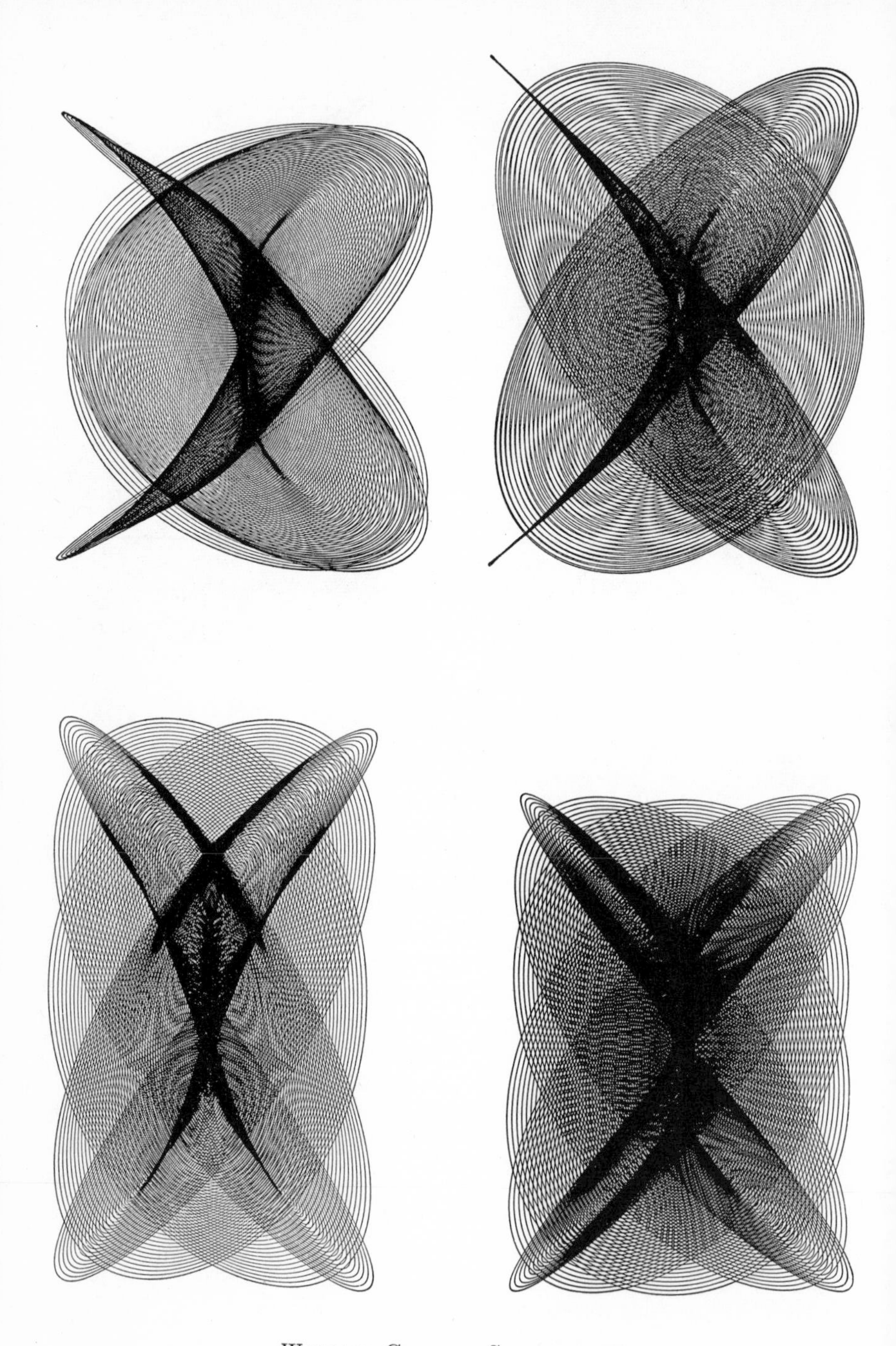

Wallace Clement Sabine fecit

CHAPTER VI

ROTATION

1. Moments of Inertia. The *moment of inertia* of n particles, m_i, with respect to an axis is defined as the sum:

$$I = \sum_{i=1}^{n} m_i r_i^2, \tag{1}$$

where r_i denotes the distance of m_i from the axis; cf. Chapter IV, § 10.

Let O be an arbitrary point of space, and let Cartesian axes with O as origin be assumed. Let the moments of inertia about the axes be denoted as follows:

$$\begin{cases} A = \sum m_i(y_i^2 + z_i^2), \\ B = \sum m_i(z_i^2 + x_i^2), \\ C = \sum m_i(x_i^2 + y_i^2). \end{cases} \tag{2}$$

The *products of inertia* are defined as the sums:

$$D = \sum m_i y_i z_i, \qquad E = \sum m_i z_i x_i, \qquad F = \sum m_i x_i y_i. \tag{3}$$

These definitions are extended in the usual way by the methods of the calculus to continuous distributions.

In terms of the above six constants it is possible to express the moment of inertia about an arbitrary axis through O. Let the direction cosines of the axis be α, β, γ and let $P: (x, y, z)$ be an arbitrary point in space. Then

$$r^2 = \rho^2 - \sigma^2,$$

or

$$r^2 = x^2 + y^2 + z^2 - (\alpha x + \beta y + \gamma z)^2.$$

Since

$$\alpha^2 + \beta^2 + \gamma^2 = 1,$$

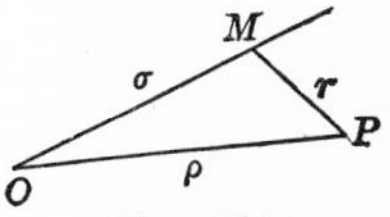

FIG. 101

the last expression for r^2 can be written in the form:

$$(x^2 + y^2 + z^2)(\alpha^2 + \beta^2 + \gamma^2) - (\alpha x + \beta y + \gamma z)^2.$$

Hence

$$r^2 = \alpha^2(y^2 + z^2) + \beta^2(z^2 + x^2) + \gamma^2(x^2 + y^2) - 2\beta\gamma yz - 2\gamma\alpha zx - 2\alpha\beta xy.$$

Thus

$$I = \alpha^2 \sum m_i(y_i^2 + z_i^2) + \beta^2 \sum m_i(z_i^2 + x_i^2) + \text{etc.}$$

or:

$$(4) \qquad I = A\alpha^2 + B\beta^2 + C\gamma^2 - 2D\beta\gamma - 2E\gamma\alpha - 2F\alpha\beta.$$

This is the desired result. The meaning of the formula can be illustrated by the *Ellipsoid of Inertia*. Consider the quadric surface,

$$(5) \qquad Ax^2 + By^2 + Cz^2 - 2Dyz - 2Ezx - 2Fxy = 1.$$

It is known as the *Ellipsoid of Inertia*, and its use is as follows. Let an arbitrary line through O with the direction cosines α, β, γ meet the surface in the point (X, Y, Z), and let ρ be thc length of the segment of the axis included between the centre of the ellipsoid and its surface. Then

$$X = \pm\, \alpha\rho, \qquad Y = \pm\, \beta\rho, \qquad Z = \pm\, \gamma\rho.$$

Since X, Y, Z satisfy (5), it follows that

$$(6) \quad \rho^2(A\alpha^2 + B\beta^2 + C\gamma^2 - 2D\beta\gamma - 2E\gamma\alpha - 2F\alpha\beta) = 1.$$

On combining (4) and (6) we find:

$$(7) \qquad \rho^2 I = 1, \qquad I = \frac{1}{\rho^2},$$

and I is seen to be the square of the reciprocal of ρ. From this property it appears that the Ellipsoid of Inertia is invariant of the choice of the coordinate axes.

If all n particles m_i lie on a line, Equation (5) no longer represents an ellipsoid. Let the axis of z be taken along this line. Then $A = B$ and all the other coefficients vanish. Thus (4) becomes

$$I = A(\alpha^2 + \beta^2).$$

Here, $A \neq 0$ except in the single case that $i = 1$ and m_1 lies at the origin. In all cases but this one, the quadric surface (5) still exists, being the cylinder of revolution

$$(8) \qquad A(x^2 + y^2) = 1,$$

and the theorem embodied in (7) is still true.

Suppose, conversely, that (5) fails to represent a true (*i.e.* non-degenerate) ellipsoid. If all the coefficients A, $B, \cdots,$ F are 0, the system of particles evidently reduces to a single particle situated at O. In all other cases, (5) represents a central quadric surface, S. If this is not a true ellipsoid, then there is a line, L, which meets S at infinity; *i.e.* which does not meet S in any proper point, but is such that a suitably chosen variable line L' always meets S, the points of intersection receding indefinitely as L' approaches L. The moment of inertia about L' is given by (7) and approaches 0 as L' approaches L. Hence the moment of inertia about L is 0. But if the moment of inertia of a system of particles about a given axis is 0, it is obvious that all the particles must lie on this axis.

We see, then, that (5) represents a true ellipsoid in all cases except the one in which the particles lie on a line, and that (7) holds in all the latter cases, too, except the one in which the system reduces to a single particle situated at O.

Parallel Axes. We recall, finally, the theorem relative to parallel axes; Chapter IV, § 10:

THEOREM. *The moment of inertia, I, about any axis, L, is equal to the moment of inertia, I_0, about a parallel axis, L_0, through the centre of gravity, plus the total mass times the square of the distance, h, between the axes:*

$$I = I_0 + Mh^2.$$

EXERCISE

Show that the moment of inertia about any line, L, in space is given by the formula:

$$\begin{aligned} I = \{A + M(y_1^2 + z_1^2)\}\,\alpha^2 &+ \{B + M(z_1^2 + x_1^2)\}\,\beta^2 \\ &+ \{C + M(x_1^2 + y_1^2)\}\,\gamma^2 - 2(D + My_1z_1)\,\beta\gamma \\ &- 2(E + Mz_1x_1)\,\gamma\alpha - 2(F + Mx_1y_1)\,\alpha\beta, \end{aligned}$$

where the origin of coordinates is at the centre of gravity and x_1, y_1, z_1 are the coordinates of any point on L, and where α, β, γ are the direction cosines of L.

For an arbitrary system of axes, replace x_1, y_1, z_1 respectively by

$$x_1 - \bar{x}, \qquad y_1 - \bar{y}, \qquad z_1 - \bar{z},$$

where $\bar{x}$, $\bar{y}$, $\bar{z}$ are the coordinates of the centre of gravity, and x_1, y_1, z_1 are the coordinates of any point on L — all referred to the new axes.

2. Principal Axes of a Central Quadric. Let a quadric surface be given by the equation:

$$(1) \qquad Ax^2 + By^2 + Cz^2 + 2Dyz + 2Ezx + 2Fxy = 1,$$

where the coefficients are arbitrary subject to the sole restriction that they shall not all vanish. The problem is, so to rotate the axes that the new equation contains only the square terms. Let

$$(2) \quad F(x, y, z) = Ax^2 + By^2 + Cz^2 + 2Dyz + 2Ezx + 2Fxy,$$

$$(3) \qquad \Phi(x, y, z) = x^2 + y^2 + z^2.$$

Consider the value of the function $F(x, y, z)$ on the surface of the sphere

$$(4) \qquad x^2 + y^2 + z^2 = a^2, \qquad \text{or} \qquad \Phi(x, y, z) = a^2.$$

Since $F(x, y, z)$ is continuous and the sphere is a closed surface, the function must attain a maximum value there, and also a minimum.

Let the axes be so rotated that the maximum value is assumed on the axis of z, in the point $(0, 0, \zeta)$, where $\zeta = \pm a$. We think of Equations (2) and (3) now as referring to the new axes.

In accordance with the Method of Lagrange * we form the function

$$F + \lambda\Phi,$$

the independent variables being x, y, z, with λ as a parameter; and we then set each of the first partial derivatives equal to 0:

$$(5) \quad F_1 + \lambda\Phi_1 = 0, \qquad F_2 + \lambda\Phi_2 = 0, \qquad F_3 + \lambda\Phi_3 = 0.$$

These three equations, combined with (4), form a *necessary* condition on the four unknowns x, y, z, λ for a maximum:

$$(6) \qquad \begin{cases} Ax + Fy + Ez + \lambda x = 0 \\ Fx + By + Dz + \lambda y = 0 \\ Ex + Dy + Cz + \lambda z = 0 \end{cases}$$

But we know that the point $(0, 0, \zeta)$, $\zeta \neq 0$, yields a maximum. Hence

$$D = 0, \qquad E = 0,$$

* Lagrange's Multipliers, *Advanced Calculus*, Chapter VII, § 5.

and the new $F(x, y, z)$ has the form:

$$F(x, y, z) = Ax^2 + 2Fxy + By^2 + Cz^2.$$

If the coefficient of the term in xy does not vanish, it can be made to do so by a suitable rotation of the axes about the axis of z; cf. *Analytic Geometry*, Chap. XII, § 2. Thus $F(x, y, z)$ is reduced finally by at most two rotations (these may be combined into a single rotation, but that is unessential) to the desired form:

$$F(x, y, z) = Ax^2 + By^2 + Cz^2. \tag{7}$$

Here, A, B, C may be any three numbers, — positive, negative, or 0, — except that we have excluded as trivial the case that all three vanish. The original equation (1) will obviously represent an ellipsoid if and only if the new coefficients A, B, C in (7) are all positive. We have thus established the following theorem.

THEOREM. *An arbitrary homogeneous quadratic function $F(x, y, z)$ can be reduced by a suitable rotation of the axes of coordinates to a sum of squares. The new coefficients of x', y', z' may be any numbers, positive, negative, or* 0.

EXERCISE

Show, by the method of mathematical induction and Lagrange's Multipliers, that an arbitrary homogeneous quadratic function in n variables,

$$F(x_1, \cdots, x_n) = A_1 x_1^2 + \cdots + A_n x_n^2 +$$
$$2A_{12}\, x_1 x_2 + 2A_{13}\, x_1 x_3 + \cdots + 2A_{n-1,\,n} x_{n-1} x_n,$$

can be reduced to a sum of squares by a suitable rotation.

By a *rotation* is meant a linear transformation:

$$\begin{cases} x_1' = a_{11} x_1 + \cdots + a_{1n} x_n \\ x_2' = a_{21} x_1 + \cdots + a_{2n} x_n \\ \cdots\cdots\cdots\cdots\cdots\cdots \\ x_n' = a_{n1} x_1 + \cdots + a_{nn} x_n \end{cases}$$

such that, for any two corresponding points $(x_1, \cdots, x_n)$ and $(x_1', \cdots, x_n')$, the relation holds:

$$x_1'^2 + \cdots + x_n'^2 = x_1^2 + \cdots + x_n^2,$$

and the determinant of the transformation,

$$\Delta = \Sigma \pm a_{11} \cdots a_{nn},$$

which necessarily has the value ± 1, is equal to $+ 1$.

It is easy to write down the conditions that must hold between the coefficients of the transformation, but these conditions are not needed for our present purpose. Obviously, the result of any two rotations is a rotation.

3. Continuation. Determination of the Axes. In the foregoing paragraph we have been content to show the *existence* of at least *one* rotation, whereby the given function is reduced to a sum of squares. We have not computed the values of the new coefficients, nor have we determined the lengths of the axes. Now, any rotation of the axes carries the second function, Φ, over into itself:

$$\Phi'(x', y', z') \equiv \Phi(x', y', z') = \Phi(x, y, z).$$

The function:

$$\Omega = F + \lambda\Phi,$$

goes over into the function:

$$F' + \lambda\Phi',$$

where λ remains unchanged. Now, the condition:

$$\frac{\partial\Omega}{\partial x'} = 0, \qquad \frac{\partial\Omega}{\partial y'} = 0, \qquad \frac{\partial\Omega}{\partial z'} = 0,$$

is equivalent to the condition:

$$\frac{\partial\Omega}{\partial x} = 0, \qquad \frac{\partial\Omega}{\partial y} = 0, \qquad \frac{\partial\Omega}{\partial z} = 0,$$

since the determinant of the linear transformation does not vanish. Hence Equations (6) of the preceding paragraph will be of the same form for the transformed functions. When

$$F'(x', y', z') = A'x'^2 + B'y'^2 + C'z'^2,$$

the equation for determining λ reduces to the following:

$$(A' + \lambda)(B' + \lambda)(C' + \lambda) = 0.$$

Thus the three roots of the determinant of Equations (6),

$$\text{(8)} \qquad \Delta \equiv \begin{vmatrix} A+\lambda & F & E \\ F & B+\lambda & D \\ E & D & C+\lambda \end{vmatrix}$$

are seen to be the negatives of the coefficients A', B', C', and so the axes of the quadric are found. If the roots of the determinant (8) are denoted by λ_1, λ_2, λ_3, the lengths of the semi-axes are

$$\frac{1}{\sqrt{|\lambda_1|}}, \qquad \frac{1}{\sqrt{|\lambda_2|}}, \qquad \frac{1}{\sqrt{|\lambda_3|}}, \qquad \lambda_i \neq 0.$$

In case a $\lambda_i = 0$, the quadric reduces to a cylinder, or more specially, to two planes. All three λ_i's will vanish if and only if the original $F(x, y, z)$ vanishes identically.

When the λ_i have once been determined, Equations (6) give the equations of the axes of the quadric. In general, the three λ_i are distinct, and Equations (6) then represent a right line for each λ_i.

4. Moment of Momentum. Moment of a Localized Vector. Let **A** be a vector whose initial point, P, is given, and let O be any point of space. Let **r** be the vector drawn from O to P. By the *moment of* **A** *with respect to O* is meant the vector, or outer product: *

$$\mathbf{r} \times \mathbf{A}. \tag{1}$$

We have met this idea in Statics, where the *moment of a force* **F**, acting at a point P, *with respect to a point O* was defined as the vector

$$\mathbf{M} = \mathbf{r} \times \mathbf{F}.$$

The *moment of momentum of a particle with respect to a point* is defined as the vector †

$$\sigma = \mathbf{r} \times m\mathbf{v}, \tag{2}$$

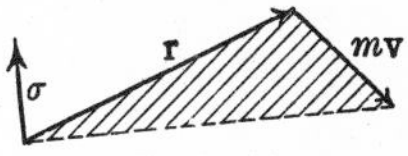

where **r** is the vector drawn from the point to the particle, and **v** is the vector velocity of the particle.

FIG. 102

The *moment of momentum of a system of particles with respect to a point* is defined as the vector

$$\sigma = \sum_{k=1}^{n} \mathbf{r}_k \times m_k \mathbf{v}_k, \tag{3}$$

* Cf. Appendix A.

† Contrary to the usual notation of writing vectors in boldface, as **a**, **x**, **i**, etc., or by parentheses, as (ω), it seems here expedient to denote the vector moment of momentum by σ, the vector momentum by ρ, and the vector angular velocity by $\tilde{\omega}$.

where $\mathbf{r}_k$ is drawn from the point in question to m_k, and $\mathbf{v}_k$ is the vector velocity of m_k.

In the case of a continuous distribution of matter the extension of the definition is made in the usual way by definite integrals.

In Cartesian form σ has the value, for a single particle:

$$\sigma = \begin{vmatrix} \mathbf{i} & \mathbf{j} & \mathbf{k} \\ x & y & z \\ m\frac{dx}{dt} & m\frac{dy}{dt} & m\frac{dz}{dt} \end{vmatrix} \tag{4}$$

$$= m\left(y\frac{dz}{dt} - z\frac{dy}{dt}\right)\mathbf{i} + m\left(z\frac{dx}{dt} - x\frac{dz}{dt}\right)\mathbf{j} + m\left(x\frac{dy}{dt} - y\frac{dx}{dt}\right)\mathbf{k},$$

the origin being at O. And so, for a system of particles, the components of σ along the axes are:

$$\left\{\begin{aligned} \sigma_x &= \sum_{k=1}^{n} m_k\left(y_k\frac{dz_k}{dt} - z_k\frac{dy_k}{dt}\right) \\ \sigma_y &= \sum_{k=1}^{n} m_k\left(z_k\frac{dx_k}{dt} - x_k\frac{dz_k}{dt}\right) \\ \sigma_z &= \sum_{k=1}^{n} m_k\left(x_k\frac{dy_k}{dt} - y_k\frac{dx_k}{dt}\right) \end{aligned}\right. \tag{5}$$

Rate of Change of σ. Since

$$\frac{d}{dt}\left(x\frac{dy}{dt} - y\frac{dx}{dt}\right) = x\frac{d^2y}{dt^2} - y\frac{d^2x}{dt^2}, \tag{6}$$

it is seen from Equations (5) that

$$\left\{\begin{aligned} \frac{d\sigma_x}{dt} &= \sum_{k=1}^{n} m_k\left(y_k\frac{d^2z_k}{dt^2} - z_k\frac{d^2y_k}{dt^2}\right) \\ \frac{d\sigma_y}{dt} &= \sum_{k=1}^{n} m_k\left(z_k\frac{d^2x_k}{dt^2} - x_k\frac{d^2z_k}{dt^2}\right) \\ \frac{d\sigma_z}{dt} &= \sum_{k=1}^{n} m_k\left(x_k\frac{d^2y_k}{dt^2} - y_k\frac{d^2x_k}{dt^2}\right) \end{aligned}\right. \tag{7}$$

These equations, in vector form, become:

$$\frac{d\sigma}{dt} = \sum_{k=1}^{n} m_k\mathbf{r}_k \times \mathbf{a}_k, \tag{8}$$

where $\mathbf{a}_k$ denotes the vector acceleration of the k-th particle.

The result, Equation (8), could have been obtained at once from (3). If we differentiate Equation (2), we find:

$$\frac{d\sigma}{dt} = m\mathbf{r} \times \frac{d\mathbf{v}}{dt} + m\frac{d\mathbf{r}}{dt} \times \mathbf{v}.$$

Now,

$$\frac{d\mathbf{r}}{dt} = \mathbf{v} \qquad \text{and} \qquad \mathbf{v} \times \mathbf{v} = 0.$$

Hence

$$\frac{d\sigma}{dt} = m\mathbf{r} \times \frac{d\mathbf{v}}{dt} = m\mathbf{r} \times \mathbf{a},$$

where $\mathbf{a}$ denotes the vector acceleration. Similarly, from (3) we derive (8).

5. The Fundamental Theorem of Moments. In Chap. IV, § 3, it was shown that, in the case of any system of particles in a plane such that the internal forces between any two particles are equal and opposite and lie along the line through the particles, the moments of the internal forces annul each other, and the equation of rotation becomes:

$$(1) \qquad \sum_{k=1}^{n} m_k \left(x_k \frac{d^2 y_k}{dt^2} - y_k \frac{d^2 x_k}{dt^2} \right) = \sum_{k=1}^{n} (x_k Y_k - y_k X_k).$$

The theorem and its proof can be generalized at once to space of three dimensions. Newton's Second Law of Motion is expressed for the particle m_k by the equations

$$\left\{ \begin{aligned} m_k \frac{d^2 x_k}{dt^2} &= X_k + \sum_j X_{kj} \\ m_k \frac{d^2 y_k}{dt^2} &= Y_k + \sum_j Y_{kj} \\ m_k \frac{d^2 z_k}{dt^2} &= Z_k + \sum_j Z_{kj} \end{aligned} \right.$$

where $\mathbf{F}_{kj}$ denotes the internal force which is exerted on the particle m_k from the particle m_j. Multiplying the third of these equations by y_k, the second by $-z_k$ and adding, and observing that the moments of the internal force cancel in pairs, since the forces $\mathbf{F}_{jk}$ and $\mathbf{F}_{kj}$ are equal and opposite and have the same line of action, the first of the following three equations is obtained. The other two are deduced in a similar manner.

$$(2)\quad \begin{cases} \sum_{k=1}^{n} m_k \left(y_k \frac{d^2 z_k}{dt^2} - z_k \frac{d^2 y_k}{dt^2} \right) = \sum_{k=1}^{n} (y_k Z_k - z_k Y_k) \\ \sum_{k=1}^{n} m_k \left(z_k \frac{d^2 x_k}{dt^2} - x_k \frac{d^2 z_k}{dt^2} \right) = \sum_{k=1}^{n} (z_k X_k - x_k Z_k) \\ \sum_{k=1}^{n} m_k \left(x_k \frac{d^2 y_k}{dt^2} - y_k \frac{d^2 x_k}{dt^2} \right) = \sum_{k=1}^{n} (x_k Y_k - y_k X_k) \end{cases}$$

These equations express the Fundamental Theorem of Moments. In vector form it is:

$$(3)\qquad \frac{d\sigma}{dt} = \sum_{k=1}^{n} \mathbf{r}_k \times \mathbf{F}_k$$

or:

$$\frac{d\sigma}{dt} = \sum \text{(Moments of the Applied Forces about } O).$$

Equation (3) can be deduced more simply by vector methods. Write Newton's Second Law in the vector form:

$$(4)\qquad m_k \mathbf{a}_k = \mathbf{F}_k + \sum_j \mathbf{F}_{kj}.$$

Next, form the vector product,

$$(5)\qquad m_k \mathbf{r}_k \times \mathbf{a}_k = \mathbf{r}_k \times \mathbf{F}_k + \sum_j \mathbf{r}_k \times \mathbf{F}_{kj},$$

and add. The sum on the left is equal to $d\sigma/dt$ by § 4, (8). On the right, the vector moments of the internal forces cancel in pairs, and there remains the right-hand side of (3).

Fundamental Theorem of Moments. *The rate of change of the vector moment of momentum of any system of particles is equal to the vector moment of the applied forces, provided that the internal forces between each pair of particles are equal and opposite and in the line through the particles:*

$$\frac{d\sigma}{dt} = \sum_{k=1}^{n} \mathbf{r}_k \times \mathbf{F}_k,$$

or, in Cartesian form, Equations (2).

The foregoing result applies to the most general system of particles, subject merely to internal forces of the very general nature indicated. By the usual physical postulate of continuity we extend the theorem to the case of continuous distributions of matter, or to any material point set. For example, our solar system is a case in point, and we will speak of it in detail in § 7.

6. Vector Form for the Motion of the Centre of Mass. Let O be an arbitrary fixed point in space, and let $\bar{\mathbf{r}}$ be the vector drawn from O to the centre of gravity, G, of a material system. Let $\mathbf{F}_1, \cdots, \mathbf{F}_n$ be the forces that act; *i.e.* the applied, or external, forces. Then the Principle of the Motion of the Centre of Mass is expressed by the equation:

$$(1) \qquad M\frac{d\bar{\mathbf{v}}}{dt} = \sum_k \mathbf{F}_k,$$

where $\bar{\mathbf{v}} = d\bar{\mathbf{r}}/dt$. Equation (1) is merely the vector form of Equations A), Chapter IV, § 1. It can be derived by vector methods, by adding Equations (4), § 5, and observing that

$$M\bar{\mathbf{r}} = \sum_{=1}^{n} m_k \mathbf{r}_k.$$

Let ρ denote the momentum,

$$\rho = M\bar{\mathbf{v}}.$$

Equation (1) now takes on the form:

$$(2) \qquad \frac{d\rho}{dt} = \sum_{k=1}^{n} \mathbf{F}_k.$$

Thus we have for any system of particles, rigid or deformable, and even for rigid bodies and fluids, the two equations of momentum:

THE EQUATION OF LINEAR MOMENTUM:

$$\text{A)} \qquad \frac{d\rho}{dt} = \sum_{k=1}^{n} \mathbf{F}_k.$$

THE EQUATION OF MOMENT OF MOMENTUM:

$$\text{B)} \qquad \frac{d\sigma}{dt} = \sum_{k=1}^{n} \mathbf{r}_k \times \mathbf{F}_k.$$

7. The Invariable Line and Plane. In case no external forces act,

$$(1) \qquad \frac{d\sigma}{dt} = 0,$$

and the vector σ remains constant. The line through O collinear with σ is called the *invariable line with respect to* O, and a plane perpendicular to it, the *invariable plane with respect to* O.

The solar system is a case in point, if we may neglect any force the stars may exert. We may consider the actual distribution of matter and velocities, and then, on choosing a fixed point, O, the corresponding value of σ will be constant.

Or we may replace the sun and each planet by an equal mass concentrated at its centre of gravity, and consider this system. Again, the vector σ corresponding to a given point O will be constant, and obviously nearly equal to the former σ.

Let us choose one of these cases arbitrarily and discuss it further. The vector σ depends on the choice of O. Can we normalize this choice? The centre of mass of the solar system is not at rest, and so, since we are neglecting any force exerted by the stars, the momentum of the system, $\bar{\rho} = M\bar{\mathbf{v}}$, is constant and $\neq 0$. The direction of this vector, $\bar{\rho}$ or $\bar{\mathbf{v}}$, does not depend on the choice of O. *The point O' can be so chosen that σ' is collinear with $\bar{\rho}$.*

We shall show in the next paragraph, Equation (5), that

$$\sigma = \sigma' + M\mathbf{r}_0 \times \bar{\mathbf{v}}, \tag{2}$$

where σ, σ' are referred to O, O' respectively. If σ is not already collinear with $\bar{\rho}$, let σ be resolved into two components; one, collinear with ρ, the other, σ_0, at right angles. We wish, then, so to determine $\mathbf{r}_0$ that

$$M\mathbf{r}_0 \times \bar{\mathbf{v}} = \sigma_0, \qquad \sigma_0 \neq 0. \tag{3}$$

Since σ_0 and $\bar{\mathbf{v}}$ are perpendicular to each other, this can be done. The point O' will be any point of a line collinear with $\bar{\rho}$. This is known as the *invariable line of the solar system.* For a further discussion, cf. Routh, *Rigid Dynamics*, Vol. I, p. 242.

8. Transformation of σ. Let O be a point fixed in space, and let O' be a second point, moving or fixed. Let P be the position of a particle of the system. Then

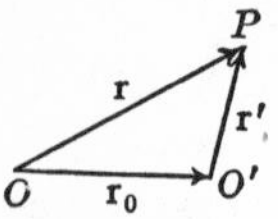

Fig. 103

$$\text{(1)} \qquad \begin{cases} \mathbf{r} = \mathbf{r}' + \mathbf{r}_0; \\ \mathbf{v} = \mathbf{v}' + \mathbf{v}_0, \end{cases}$$

where $\mathbf{v}'$ expresses the velocity of P relative to O', and $\mathbf{v}_0$ is the velocity of O'.

For a single particle, the moment of momentum with respect to O is the vector

$$\sigma = \mathbf{r} \times m\mathbf{v} = m\mathbf{r}' \times (\mathbf{v}' + \mathbf{v}_0) + m\mathbf{r}_0 \times \mathbf{v},$$

or

$$\sigma = m\mathbf{r}' \times \mathbf{v}' + m\mathbf{r}_0 \times \mathbf{v} + m\mathbf{r}' \times \mathbf{v}_0. \tag{2}$$

The first term on the right has the value

$$\sigma_r' = \mathbf{r}' \times m\mathbf{v}',$$

or the relative moment of momentum, referred to O' as a moving point.

For a system of particles we infer that

$$\sigma = \sum_k m_k \mathbf{r}_k' \times \mathbf{v}_k' + \sum_k m_k \mathbf{r}_0 \times \mathbf{v}_k + \sum_k m_k \mathbf{r}_k' \times \mathbf{v}_0$$

or

$$\sigma = \sigma_r' + M\mathbf{r}_0 \times \bar{\mathbf{v}} + M\bar{\mathbf{r}}' \times \mathbf{v}_0, \tag{3}$$

where σ_r' is the relative moment of momentum referred to O' as a moving point; $\bar{\mathbf{v}}$ is the velocity of the centre of mass; and $\bar{\mathbf{r}}'$ is the vector drawn from O' to the centre of mass.

The second term on the right,

$$M\mathbf{r}_0 \times \bar{\mathbf{v}} = \mathbf{r}_0 \times M\bar{\mathbf{v}},$$

can be interpreted as the moment of momentum, relative to O, of the total momentum, $M\bar{\mathbf{v}}$, of the system, thought of as a mass, M, concentrated at O' and moving with the velocity $\bar{\mathbf{v}}$.

The third term,

$$M\bar{\mathbf{r}}' \times \mathbf{v}_0 = \bar{\mathbf{r}}' \times M\mathbf{v}_0,$$

is the moment of momentum, relative to O', of the total mass, M, concentrated at the centre of gravity and moving with velocity $\mathbf{v}_0$.

If, in particular, O' be taken at G, then $\bar{\mathbf{r}}' = 0$, $\mathbf{v}_0 = \bar{\mathbf{v}}$, and

$$\sigma = \sigma_r' + M\bar{\mathbf{r}} \times \bar{\mathbf{v}}, \tag{4}$$

where σ_r' denotes the relative moment of momentum, referred to G as a moving point, and

$$M\bar{\mathbf{r}} \times \bar{\mathbf{v}} = \bar{\mathbf{r}} \times M\bar{\mathbf{v}}$$

is the moment of momentum, $M\bar{\mathbf{v}}$, of the total mass, concentrated at G and moving with the velocity of G, referred to O.

Let σ' denote the value of σ referred to the point O' as a fixed point; *i.e.*

$$\sigma' = \sum_k m_k \mathbf{r}'_k \times \mathbf{v}_k.$$

Then

$$\sigma' = \sum_k m_k \mathbf{r}'_k \times \mathbf{v}'_k + \sum_k m_k \mathbf{r}'_k \times \mathbf{v}_0$$

or

$$\sigma' = \sigma'_r + M\bar{\mathbf{r}}' \times \mathbf{v}_0.$$

Thus Equation (3) goes over into:

$$\sigma = \sigma' + M\mathbf{r}_0 \times \bar{\mathbf{v}}. \tag{5}$$

This amounts to setting $\mathbf{v}_0 = 0$ in (3).

9. Moments about the Centre of Mass. We have the Fundamental Equation of Moments, § 5:

$$\frac{d\sigma}{dt} = \sum_k \mathbf{r}_k \times \mathbf{F}_k. \tag{1}$$

And we have the Equation of Transformation, § 8, (4):

$$\sigma = \sigma'_r + M\bar{\mathbf{r}} \times \bar{\mathbf{v}}, \tag{2}$$

where σ'_r is the relative moment of momentum, referred to G as a moving point.

Differentiate this last equation, observing that

$$\frac{d\bar{\mathbf{r}}}{dt} \times \bar{\mathbf{v}} = 0,$$

since $d\bar{\mathbf{r}}/dt = \bar{\mathbf{v}}$ is either 0 or else collinear with $\bar{\mathbf{v}}$. Thus we find:

$$\frac{d\sigma}{dt} = \frac{d\sigma'_r}{dt} + M\bar{\mathbf{r}} \times \frac{d\bar{\mathbf{v}}}{dt}. \tag{3}$$

On the other hand,

$$\mathbf{r} = \mathbf{r}' + \bar{\mathbf{r}},$$

where $\mathbf{r}$, $\bar{\mathbf{r}}$ are drawn from O; $\mathbf{r}'$ from G. Thus

$$\sum_k \mathbf{r}_k \times \mathbf{F}_k = \sum_k \mathbf{r}'_k \times \mathbf{F}_k + \bar{\mathbf{r}} \times \sum_k \mathbf{F}_k. \tag{4}$$

Substituting in Equation (1) the values found in Equations (3) and (4), we obtain the result:

$$\frac{d\sigma'_r}{dt} + M\bar{\mathbf{r}} \times \frac{d\bar{\mathbf{v}}}{dt} = \sum_k \mathbf{r}'_k \times \mathbf{F}_k + \bar{\mathbf{r}} \times \sum_k \mathbf{F}_k. \tag{5}$$

Recall the Equation of the Motion of the Centre of Mass, § 6:

$$(6) \qquad M\frac{d\bar{\mathbf{v}}}{dt} = \sum_k \mathbf{F}_k.$$

From it follows that

$$M\bar{\mathbf{r}} \times \frac{d\bar{\mathbf{v}}}{dt} = \bar{\mathbf{r}} \times \sum_k \mathbf{F}_k.$$

Thus these terms cancel in (5) and there remains:

$$(7) \qquad \frac{d\sigma_r'}{dt} = \sum_k^n \mathbf{r}_k' \times \mathbf{F}_k.$$

In this equation is embodied the result which may be described as the

PRINCIPLE OF MOMENTS WITH RESPECT TO THE CENTRE OF MASS. *The rate of change of the relative vector moment of momentum, referred to the centre of mass G regarded as a moving point, is equal to the sum of the vector moments of the applied forces with respect to G:*

$$\frac{d\sigma_r'}{dt} = \sum_{k=1}^n \mathbf{r}_k' \times \mathbf{F}_k.$$

EXERCISE

Show that a rigid body is dynamically equivalent, in general, to a pair of equal masses on the axis of x, a second pair on the axis of y, and a third pair on the axis of z, each pair being situated symmetrically with respect to the origin, and all six distances from the origin being the same; it being assumed that the principal axes of inertia lie along the coordinate axes. Discuss the exceptional cases. Use the results of § 12.

10. Moments about an Arbitrary Point. Consider the most general transformation, § 8, (3):

$$(1) \qquad \sigma = \sigma_r' + M\mathbf{r}_0 \times \bar{\mathbf{v}} + M\bar{\mathbf{r}}' \times \mathbf{v}_0,$$

and differentiate:

$$(2) \quad \frac{d\sigma}{dt} = \frac{d\sigma_r'}{dt} + M\mathbf{r}_0 \times \frac{d\bar{\mathbf{v}}}{dt} + M\frac{d\mathbf{r}_0}{dt} \times \bar{\mathbf{v}} + M\bar{\mathbf{v}}' \times \mathbf{v}_0 + M\bar{\mathbf{r}}' \times \frac{d\mathbf{v}_0}{dt}.$$

On the other hand,

$$(3) \qquad \sum_k \mathbf{r}_k \times \mathbf{F}_k = \sum_k \mathbf{r}_k' \times \mathbf{F}_k + \sum_k \mathbf{r}_0 \times \mathbf{F}_k.$$

From the Equation of Linear Momentum, § 6, (1) follows that

$$M\mathbf{r}_0 \times \frac{d\bar{\mathbf{v}}}{dt} = \sum_k \mathbf{r}_0 \times \mathbf{F}_k. \tag{4}$$

Moreover,

$$M\frac{d\mathbf{r}_0}{dt} \times \bar{\mathbf{v}} + M\bar{\mathbf{v}}' \times \mathbf{v}_0 = 0.$$

For,

$$\bar{\mathbf{v}} = \bar{\mathbf{v}}' + \mathbf{v}_0;$$

hence

$$\mathbf{v}_0 \times \bar{\mathbf{v}} = \mathbf{v}_0 \times \bar{\mathbf{v}}' + \mathbf{v}_0 \times \mathbf{v}_0 = \mathbf{v}_0 \times \bar{\mathbf{v}}',$$

and

$$\mathbf{v}_0 \times \bar{\mathbf{v}}' + \bar{\mathbf{v}}' \times \mathbf{v}_0 = 0.$$

Substituting, then, in the Equation of Moment of Momentum, § 6, B):

$$\frac{d\sigma}{dt} = \sum_k \mathbf{r}_k \times \mathbf{F}_k,$$

and reducing, we find:

$$\frac{d\sigma_r'}{dt} + M\bar{\mathbf{r}}' \times \frac{d\mathbf{v}_0}{dt} = \sum_k \mathbf{r}_k' \times \mathbf{F}_k. \tag{5}$$

This equation is general, covering all cases of taking moments about a moving point O', relative to that point. When, however, one uses the expression: "taking moments about a point O'" the meaning ordinarily attached to these words is, that the equation

$$\frac{d\sigma_r'}{dt} = \sum_k \mathbf{r}_k' \times \mathbf{F}_k \tag{6}$$

shall be true. Hence we must have

$$\bar{\mathbf{r}}' \times \frac{d\mathbf{v}_0}{dt} = 0 \tag{7}$$

for every value of t.

Let $t = \tau$ be an arbitrary instant. Let O' be a point which describes a certain path,

$$\mathbf{r}_0 = \mathbf{r}_0(t, \tau). \tag{8}$$

Consider this as the vector $\mathbf{r}_0$ of the foregoing treatment. Then

$$\mathbf{v}_0 = \frac{\partial \mathbf{r}_0}{\partial t}, \qquad \frac{d\mathbf{v}_0}{dt} = \frac{\partial^2 \mathbf{r}_0}{\partial t^2}. \tag{9}$$

At the instant $t = \tau$,

$$(10) \qquad \left(\frac{d\mathbf{v}_0}{dt}\right)_{t=\tau} = \left(\frac{\partial^2 \mathbf{r}_0}{\partial t^2}\right)_{t=\tau}.$$

Then Equation (7) is to hold for this vector $\mathbf{r}_0$ at the one instant $t = \tau$.

Thus we have in general, not a single curve traced out by O' and (7) considered for a variable point of that curve, as in the case of § 8, where $\bar{\mathbf{r}}' = 0$, — but a one-parameter family of curves, and Equation (7) considered for one point of each curve; cf. for example, the next paragraph.

11. Moments about the Instantaneous Centre. Consider the motion of a lamina, *i.e.* a rigid plane system, in its own plane. Let Q be the instantaneous centre at a given instant, $t = \tau$. Then σ, referred to the point Q, is a vector perpendicular to the plane, and its length is

$$I\frac{d\theta}{dt},$$

where I is the moment of inertia of the lamina about Q. What does it mean to "take moments about Q"? From the foregoing it means to take moments about a point O' describing a curve

$$\mathbf{r}_0 = \mathbf{r}_0(t, \tau)$$

which at the instant $t = \tau$ passes through Q.

There is an unlimited set of such curves. Let us select, in particular, the curve C which is the path of that point fixed in the lamina, which passes through Q at the instant $t = \tau$. Observe that this is an arbitrary choice of C. This curve C is known in terms of the rolling of the body centrode on the space centrode. The velocity of O' at Q is 0, but its acceleration, if Q is an ordinary point, is normal to the centrodes at Q and does not vanish; Chapter V, § 5. If, then, Equation (7), § 10 is to be satisfied, the centre of gravity, G, must lie in the normal to the centrodes. In particular, the normal to the body centrode must pass through the centre of gravity. Hence the body centrode must be a circle with the centre of gravity at the centre, if the condition is to be permanently satisfied. The equation of moments now becomes:

$$I\frac{d^2\theta}{dt^2} = \sum \text{(Moments about Inst. Centre)}.$$

The only case, then, of motion in a plane, in which we may permanently take moments about the instantaneous centre, thought of as a point fixed in the moving body, is that in which a circle rolls on an arbitrary curve, the centre of gravity being at the centre of the circle; and the limiting case, namely, that the point Q is permanently at rest. This last case corresponds to the identical vanishing of $d\mathbf{v}_0/dt$.

Moments about an Arbitrary Point. Consider now an arbitrary point O' fixed in the body. Let it be at Q at the instant $t = \tau$, and let C be the curve,

$$\mathbf{r}_0 = \mathbf{r}_0(t, \tau),$$

which it is describing. Take moments about Q with reference to this point, O'. Then

$$\frac{d\sigma'_\tau}{dt} = I\frac{d^2\theta}{dt^2}.$$

If, furthermore, Equation (7), § 10 is satisfied, the equation of moments becomes:

$$I\frac{d^2\theta}{dt^2} = \sum \text{(Moments about } Q\text{)}.$$

Equation (7) here means, in general, that the acceleration of O' is collinear with the line determined by Q and G. In particular, the equation is satisfied if the acceleration of O' is 0 at Q; or if Q coincides with G.*

EXERCISE

A billiard ball is struck full by the cue. Consider the motion while there is slipping. Show that it is not possible to take moments about the instantaneous centre.

Find the points of zero acceleration and verify the fact that it is possible to take moments about them, explaining carefully what you mean by these words.

Show that the points whose acceleration passes through the centre of the ball lie on a circle through the centre of the ball, of radius one-fifth that of the ball, the centre being directly above the centre of the ball.

12. Evaluation of σ for a Rigid System; One Point Fixed. Consider a rigid system of particles with one point, O, fixed.

* Edward V. Huntington has discussed this question, *Amer. Math. Monthly*, vol. XXI (1914) p. 315.

The motion is then one of rotation about an axis passing through O; cf. Chapter V, § 8. Let the vector angular velocity be denoted by $\tilde{\omega}$, and let α, β, γ be a system of mutually perpendicular unit vectors lying along Cartesian axes with the origin at O. When we wish these axes to be fixed, we shall use the coordinates (x, y, z) and replace α, β, γ by $\mathbf{i}$, $\mathbf{j}$, $\mathbf{k}$. In the general case, the coordinates shall be ξ, η, ζ.

Let P be any point fixed in the body, and let $\mathbf{r}$ be the vector drawn from O to P:

$$\mathbf{r} = \xi\alpha + \eta\beta + \zeta\gamma. \tag{1}$$

The velocity of P,

$$\mathbf{v} = \frac{d\mathbf{r}}{dt}, \tag{2}$$

is expressed in terms of the vector $\tilde{\omega}$ as follows (Chapter V, § 9):

$$\mathbf{v} = \tilde{\omega} \times \mathbf{r}. \tag{3}$$

In Cartesian form,

$$\mathbf{v} = \begin{vmatrix} \alpha & \beta & \gamma \\ \omega_\xi & \omega_\eta & \omega_\zeta \\ \xi & \eta & \zeta \end{vmatrix} \tag{4}$$

or

$$\begin{cases} v_\xi = \zeta\,\omega_\eta - \eta\,\omega_\zeta \\ v_\eta = \xi\,\omega_\zeta - \zeta\,\omega_\xi \\ v_\zeta = \eta\,\omega_\xi - \xi\,\omega_\eta \end{cases} \tag{5}$$

For a single particle, then, σ has the value:

$$\sigma = \mathbf{r} \times m\mathbf{v}, \tag{6}$$

$$\sigma = m \begin{vmatrix} \alpha & \beta & \gamma \\ \xi & \eta & \zeta \\ v_\xi & v_\eta & v_\zeta \end{vmatrix} \tag{7}$$

Hence

$$\begin{cases} \sigma_\xi = m(\eta\,v_\zeta - \zeta\,v_\eta) \\ \sigma_\eta = m(\zeta\,v_\xi - \xi\,v_\zeta) \\ \sigma_\zeta = m(\xi\,v_\eta - \eta\,v_\xi) \end{cases} \tag{8}$$

These formulas lead in turn to the following:

$$(9)\quad \begin{cases} \sigma_\xi = m\left[(\eta^2 + \zeta^2)\,\omega_\xi - \xi\eta\,\omega_\eta - \xi\zeta\,\omega_\zeta\right] \\ \sigma_\eta = m\left[-\eta\xi\,\omega_\xi + (\zeta^2 + \xi^2)\,\omega_\eta - \eta\zeta\,\omega_\zeta\right] \\ \sigma_\zeta = m\left[-\zeta\xi\,\omega_\xi - \zeta\eta\,\omega_\eta + (\xi^2 + \eta^2)\,\omega_\zeta\right] \end{cases}$$

For a system of particles they become:

$$\sigma_\xi = \sum_k m_k\left[(\eta_k{}^2 + \zeta_k{}^2)\,\omega_\xi - \xi_k\eta_k\,\omega_\eta - \xi_k\zeta_k\,\omega_\zeta\right], \quad \text{etc.}$$

and so, finally,

$$(10)\quad \begin{cases} \sigma_\xi = \quad A\omega_\xi - F\omega_\eta - E\omega_\zeta \\ \sigma_\eta = -F\omega_\xi + B\omega_\eta - D\omega_\zeta \\ \sigma_\zeta = -E\omega_\xi - D\omega_\eta + C\omega_\zeta \end{cases}$$

These are the formulas which give the components of σ along the axes of ξ, η, ζ when the origin is fixed. It is obviously immaterial whether the axes are fixed or moving.

13. Euler's Dynamical Equations. Consider the case of a rigid body, one point O of which is fixed. The Equation of Moments,

$$(1)\quad \frac{d\sigma}{dt} = \sum (\text{Moments about } O),$$

referred to this point, admits a simple expression in terms of the angular velocity, $\bar{\omega}$, of the body. Let the (ξ, η, ζ)-axes be fixed in the body, and let P be a point which moves according to any law. Let

$$\mathbf{r} = \xi\alpha + \eta\beta + \zeta\gamma,$$

where $\mathbf{r}$ is the vector drawn from O to P. Then we have seen (Chapter V, § 14):

$$\frac{d\mathbf{r}}{dt} =$$

$$\left(\frac{d\xi}{dt} + \zeta\,\omega_\eta - \eta\,\omega_\zeta\right)\alpha + \left(\frac{d\eta}{dt} + \xi\,\omega_\zeta - \zeta\,\omega_\xi\right)\beta + \left(\frac{d\zeta}{dt} + \eta\,\omega_\xi - \xi\,\omega_\eta\right)\gamma.$$

This result applies to the vector:

$$\sigma = \sigma_\xi\,\alpha + \sigma_\eta\,\beta + \sigma_\zeta\,\gamma,$$

and thus gives us the left-hand side of (1).

On the right-hand side of (1) let

$$\sum_k \mathbf{r}_k \times \mathbf{F}_k = L\alpha + M\beta + N\gamma.$$

Thus we have:

$$(2) \qquad \begin{cases} \dfrac{d\sigma_\xi}{dt} + \omega_\eta\, \sigma_\zeta - \omega_\zeta\, \sigma_\eta = L, \\[2ex] \dfrac{d\sigma_\eta}{dt} + \omega_\zeta\, \sigma_\xi - \omega_\xi\, \sigma_\zeta = M, \\[2ex] \dfrac{d\sigma_\zeta}{dt} + \omega_\xi\, \sigma_\eta - \omega_\eta\, \sigma_\xi = N. \end{cases}$$

On substituting for σ_ξ, σ_η, σ_ζ their values from (10), § 12, the equations known as *Euler's Dynamical Equations* result. In particular, if the (ξ, η, ζ)-axes are laid along the principal axes of inertia, then

$$\sigma_\xi = A\omega_\xi, \qquad \sigma_\eta = B\,\omega_\eta, \qquad \sigma_\zeta = C\omega_\zeta$$

and Equations (2) assume the form:

$$(3) \qquad \begin{cases} A\dfrac{dp}{dt} - (B - C)\, qr = L, \\[2ex] B\dfrac{dq}{dt} - (C - A)\, rp = M, \\[2ex] C\dfrac{dr}{dt} - (A - B)\, pq = N, \end{cases}$$

where

$$p = \omega_\xi, \qquad q = \omega_\eta, \qquad r = \omega_\zeta.$$

When the axes of coordinates do not coincide with the principal axes of inertia, Euler's Equations take the general form:

$$(4) \quad A\frac{d\omega_\xi}{dt} - F\frac{d\omega_\eta}{dt} - E\frac{d\omega_\zeta}{dt}$$
$$- (E\omega_\eta - F\omega_\zeta)\,\omega_\xi - D(\omega_\eta^2 - \omega_\zeta^2) + (C - B)\,\omega_\eta\omega_\zeta = L,$$

and two others obtained by advancing the letters cyclically.

Euler's Dynamical Equations also apply to the rotation of a rigid body about its centre of mass; § 9. Here, there is no restriction whatsoever on the motion.

14. Motion about a Fixed Point. Let the body move under the action of no forces, save the reaction at O. Then Euler's Equations become the following:

$$(1)\qquad \begin{cases} A\dfrac{dp}{dt} - (B - C)\,qr = 0 \\[2ex] B\dfrac{dq}{dt} - (C - A)\,rp = 0 \\[2ex] C\dfrac{dr}{dt} - (A - B)\,pq = 0 \end{cases}$$

A first integral is obtained by multiplying the equations respectively by p, q, and r, and adding:

$$Ap\frac{dp}{dt} + Bq\frac{dq}{dt} + Cr\frac{dr}{dt} = 0,$$

$$(2)\qquad Ap^2 + Bq^2 + Cr^2 = h.$$

This is the Equation of Energy, Chapter VII, §§ 5, 6.

A second integral is found by multiplying Equations (1) respectively by Ap, Bq, and Cr, and adding:

$$A^2p\frac{dp}{dt} + B^2q\frac{dq}{dt} + C^2r\frac{dr}{dt} = 0,$$

$$(3)\qquad A^2p^2 + B^2q^2 + C^2r^2 = l.$$

This equation corresponds to the fact that

$$\frac{d\sigma}{dt} = 0,$$

and so

$$\sigma = Ap\,\alpha + Bq\,\beta + Cr\,\gamma$$

is constant.

From Equations (2) and (3), two of the variables, as p^2 and q^2, can, in general, be determined in terms of the third, and then, on substituting in the third equation (1), a differential equation for r alone is found. It is seen that t is expressed as an elliptic integral of the first kind in r. Thus, p, q, and r are found as functions of t.

Exercise. Let $A = 3$, $B = 2$, $C = 1$; and let p, q, r all have the initial value 1. Work out the value of t in terms of the integral.

The Body Cone. On multiplying (2) by l, (3) by h, and subtracting, we find:

$$(4) \qquad A(l - Ah)p^2 + B(l - Bh)q^2 + C(l - Ch)r^2 = 0.$$

The equations of the instantaneous axis are:

$$(5) \qquad \frac{\xi}{p} = \frac{\eta}{q} = \frac{\zeta}{r},$$

when p, q, r are the above functions of t. Hence the locus of the instantaneous axis in the body is the quadric cone:

$$(6) \qquad A(l - Ah)\xi^2 + B(l - Bh)\eta^2 + C(l - Ch)\zeta^2 = 0.$$

More explicitly, let p, q, r satisfy (2) and (3) and hence (4). Then any point (ξ, η, ζ) of (5) satisfies (6), and hence lies on the quadric cone. Conversely, let (ξ, η, ζ) be a point of the quadric cone, (6). If $(\xi, \eta, \zeta) \neq (0, 0, 0)$, determine p, q, r, μ by the four equations

$$p = \mu\xi, \qquad q = \mu\eta, \qquad r = \mu\zeta,$$

$$\mu^2(A^2\xi^2 + B^2\eta^2 + C^2\zeta^2) = l.$$

Thus (3) is satisfied. And (4) holds, too. Hence (2) is true. Consequently, (ξ, η, ζ) lies on an instantaneous axis.

The Space Cone. Poinsot obtained an elegant determination of the space centrode. Consider the ellipsoid of inertia, Σ. It is a surface fixed in the body. Let m be the point in which the ray drawn from O and collinear with $\bar{\omega}$ cuts Σ. Then the tangent plane M to Σ at m is a plane fixed in space, the same for all points, m. The motion is seen to be one of rolling of the surface Σ on the plane M without slipping.

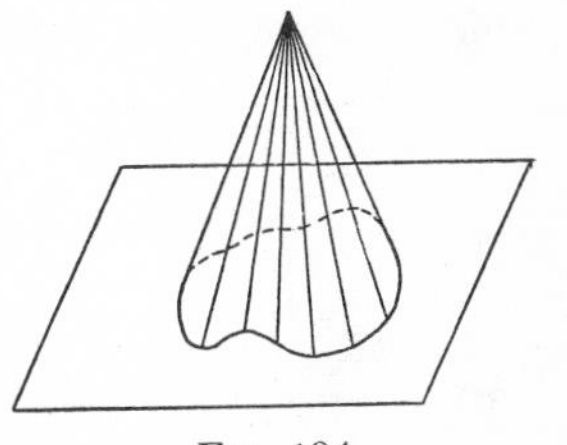

Fig. 104

To prove the first statement, it is sufficient to show that the tangent plane at m is perpendicular to σ, and that its distance from O does not depend on m. The equation of Σ is:

$$\Sigma: \qquad A\xi^2 + B\eta^2 + C\zeta^2 = 1.$$

The coordinates of m are:

$$\rho p, \qquad \rho q, \qquad \rho r, \qquad \text{where} \qquad \rho = \frac{1}{\sqrt{h}}.$$

Hence the equation of M is

$$M: \qquad \rho A p \xi + \rho B q \eta + \rho C r \zeta = 1.$$

The direction components of its normal are Ap, Bq, Cr. But these are precisely the projections of σ on the axes. Hence M is perpendicular to σ. Moreover, the distance of the plane M from O is

$$\frac{1}{\sqrt{(\rho A p)^2 + (\rho B q)^2 + (\rho C r)^2}} = \frac{1}{\rho \sqrt{l}} = \sqrt{\frac{h}{l}},$$

and so is constant. This completes the proof of the first statement.

To prove the second statement, consider so much of the body cone, (6), as lies in Σ. Let Γ be the curve on Σ which marks the intersection of these two surfaces. Then Γ rolls on M without slipping, and the curve of contact, C, can serve as a directrix of the space centrode. For the body cone, (6), rolls without slipping on the space cone, and the curves Γ, C are two curves on these cones, which curves are always tangent at the point M of the instantaneous axis. The angular velocity, ω, is proportional to the distance Om; for

$$\overline{Om}^2 = (\rho p)^2 + (\rho q)^2 + (\rho r)^2 = \frac{\omega^2}{h}.$$

15. Euler's Geometrical Equations. Euler introduced as coordinates describing the position of a rigid body, one point of which is fixed at O, the three angles, θ, φ, and ψ represented in the figure. Between the components of the angular velocity $\tilde{\omega}$ about the instantaneous axis,

$$\omega_\xi = p, \qquad \omega_\eta = q, \qquad \omega_\zeta = r,$$

and these coordinates and their derivatives, exist the following relations:

$$(1) \qquad \begin{cases} p = -\sin\theta\cos\varphi\dfrac{d\psi}{dt} + \sin\varphi\dfrac{d\theta}{dt} \\[2ex] q = \sin\theta\sin\varphi\dfrac{d\psi}{dt} + \cos\varphi\dfrac{d\theta}{dt} \\[2ex] r = \cos\theta\dfrac{d\psi}{dt} + \dfrac{d\varphi}{dt} \end{cases}$$

These are known as *Euler's Geometrical Equations.*

A geometrical proof can be given by computing the vector velocities of certain suitably chosen points in two ways. Begin

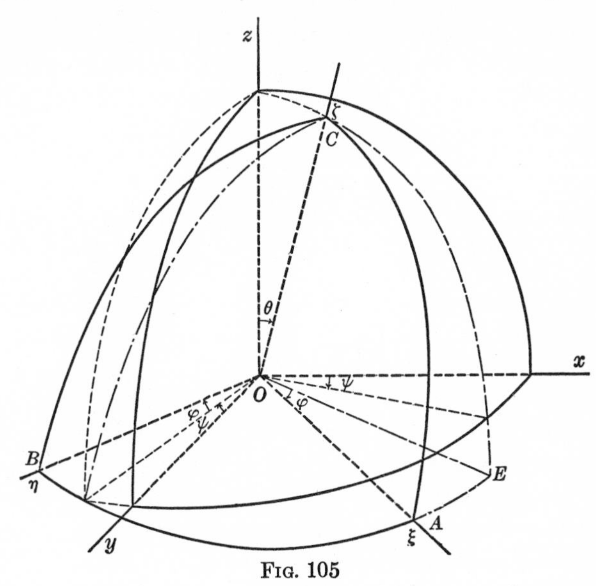

Fig. 105

with the point C in which the positive axis of ζ pierces the surface of the unit sphere. As we look down on the sphere from above this point, it is evident from the figure that

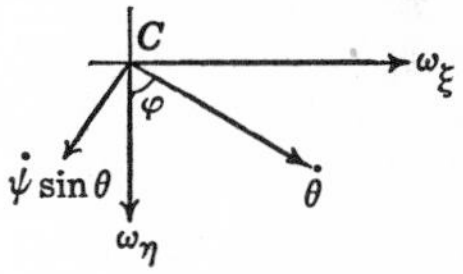

$$\frac{d\theta}{dt} = p \sin \varphi + q \cos \varphi$$

$$\sin \theta \frac{d\psi}{dt} = - p \cos \varphi + q \sin \varphi.$$

Fig. 106

These equations yield the first two of Equations (1).

To obtain the third equation, consider the motion of E. Its velocity is made up of a velocity ω_ζ tangent to the arc EA, and two velocities perpendicular to this arc. On the other hand, its velocity is composed of the velocity $\dot{\varphi}$ tangent to the arc EA; the velocity $\dot{\psi} \cos \theta$, also tangent to EA; and $\dot{\theta}$ perpendicular to EA. Hence

$$r = \cos \theta \frac{d\psi}{dt} + \frac{d\varphi}{dt},$$

and this is the third Equation (1).

If p, q, r have once been determined as functions of the time, Equations (1) yield a system of three simultaneous differential equations of the first order for determining θ, φ, ψ as functions of the time. Thus in the problem of § 14, — the motion of a rigid body under no forces, or acted on by the one force of constraint that holds the point O fixed, — p, q, r were determined explicitly as functions of the time, and the further study of the problem is based on the above Equations (1).

16. Continuation. The Direction Cosines of the Moving Axes. The moving axes are related to the fixed axes by the scheme,

	ξ	η	ζ
x	l_1	l_2	l_3
y	m_1	m_2	m_3
z	n_1	n_2	n_3

and the question is, to express the nine direction cosines in terms of the Eulerian angles. This can be done conveniently by vector methods, if we effect the displacement one step at a time. Let $\mathbf{i}, \mathbf{j}, \mathbf{k}$ be unit vectors along the original axes, and α, β, γ unit vectors along the displaced axes. Let the first displacement be a rotation about the axis of z through the angle ψ, and let $\mathbf{i}, \mathbf{j}$ go over into $\mathbf{i}_1, \mathbf{j}_1$. Then

$$\left\{\begin{aligned} \mathbf{i}_1 &= \mathbf{i}\cos\psi + \mathbf{j}\sin\psi \\ \mathbf{j}_1 &= -\,\mathbf{i}\sin\psi + \mathbf{j}\cos\psi \\ \mathbf{k}_1 &= \mathbf{k}. \end{aligned}\right.$$

Next, rotate about the axis of $\mathbf{j}_1$ through an angle θ, whereby $\mathbf{i}_1$ goes into $\mathbf{i}_2$, and $\mathbf{k}_1$ into $\mathbf{k}_2 = \gamma$:

$$\left\{\begin{aligned} \mathbf{i}_2 &= \mathbf{i}_1\cos\theta - \mathbf{k}_1\sin\theta \\ \mathbf{j}_2 &= \mathbf{j}_1 \\ \mathbf{k}_2 &= \mathbf{i}_1\sin\theta + \mathbf{k}_1\cos\theta. \end{aligned}\right.$$

Finally, rotate about $\mathbf{k}_2$ through an angle φ, whereby $\mathbf{i}_2$ goes over into $\mathbf{i}_3 = \alpha$ and $\mathbf{j}_2$ goes into $\mathbf{j}_3 = \beta$:

$$\left\{\begin{aligned} \mathbf{i}_3 &= \mathbf{i}_2\cos\varphi + \mathbf{j}_2\sin\varphi \\ \mathbf{j}_3 &= -\,\mathbf{i}_2\sin\varphi + \mathbf{j}_2\cos\varphi \\ \mathbf{k}_3 &= \mathbf{k}_2. \end{aligned}\right.$$

From these equations it appears that

$$\begin{cases} l_1 = \cos\theta\cos\varphi\cos\psi - \sin\varphi\sin\psi \\ m_1 = \cos\theta\cos\varphi\sin\psi + \sin\varphi\cos\psi \\ n_1 = -\sin\theta\cos\varphi \end{cases}$$

$$\begin{cases} l_2 = -\cos\theta\sin\varphi\cos\psi - \cos\varphi\sin\psi \\ m_2 = -\cos\theta\sin\varphi\sin\psi + \cos\varphi\cos\psi \\ n_2 = \sin\theta\sin\varphi \end{cases}$$

$$\begin{cases} l_3 = \sin\theta\cos\psi \\ m_3 = \sin\theta\sin\psi \\ n_3 = \cos\theta. \end{cases}$$

17. The Gyroscope. It is now possible to set forth in simplest terms the essential characteristics of the motion of a rotating rigid body, which is the basis of gyroscopic action. By a *gyroscope* is meant a rigid body spinning at high velocity about an axis passing through the centre of gravity, which is at rest, and acted on by a couple whose representative vector is perpendicular to the axis.

Consider, in particular, the following motion. Let $A = B$, $C \neq 0$, and let the axis of ζ be caused to rotate with constant angular velocity, c, in the plane $\psi = 0$. What will be the couple? Here, $d\psi/dt = 0$ and Euler's Geometrical Equations give

$$p = \sin\varphi\frac{d\theta}{dt}, \qquad q = \cos\varphi\frac{d\theta}{dt}, \qquad r = \frac{d\varphi}{dt},$$

where $d\theta/dt = c$. The components L and M are unknown, but $N = 0$. The third of the Dynamical Equations becomes:

$$C\frac{dr}{dt} = 0; \qquad \text{hence} \qquad r = \nu,$$

and ν is a large positive constant. Since here

$$r = \frac{d\varphi}{dt}, \qquad \varphi = \nu t.$$

Hence, from the first two equations,

$$p = c\sin\nu t, \qquad q = c\cos\nu t.$$

On substituting these values in the Dynamical Equations, we find:

$$L = C c \nu \cos \nu t, \qquad M = - C c \nu \sin \nu t.$$

We may think of the couple as made up of a force **F** acting at the point $C : (\xi, \eta, \zeta) = (0, 0, 1)$ in Fig. 105, and an equal and opposite force at O. Then **F** will be tangent to the sphere at C. Let it be resolved into two components, one perpendicular to the (ξ, ζ)-plane; the other, in that plane. The first will have the value L, taken positive in the sense of the negative η-axis; the second will equal M, taken positive in the sense of the ξ-axis. When $t = 0$,

$$L = C c \nu, \qquad M = 0,$$

and at any later time, the result is the same. This can be seen directly from the nature of the problem, since the motion of the axis of ζ in the plane $\psi = 0$ is uniform, and hence the force which the constraint exerts will be the same force relative to the body at one instant as at any other instant.

It is easy to verify analytically the truth of the last statement. For, the force normal to the plane $\psi = 0$ will always be

$$L \cos \varphi - M \sin \varphi = C c \nu;$$

and the force in that plane will always be

$$L \sin \varphi + M \cos \varphi = 0.$$

This result brings out in the simplest form imaginable the essential phenomenon in gyroscopic action, namely, this: *To cause the axis to move in a plane with constant angular velocity, a couple must be applied whose forces act on the axis in a direction at right angles to that plane.*

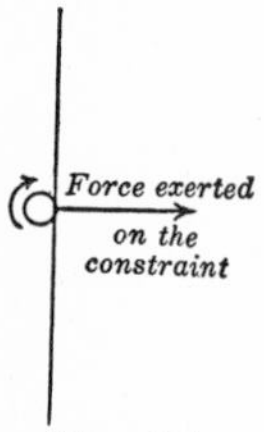

Force exerted on the constraint

Fig. 107

Finally observe that if one thinks of oneself as standing on the gyroscope and moving with it, one's body along the positive axis of ζ and facing in the direction of the motion of the axis, the force L applied to the gyroscope will be directed toward one's left, and hence the *reaction* of the gyroscope on the constraint will be directed *toward the right*, the gyroscope spinning in the *clockwise sense* as one looks down on it. Of course, if the sense of the rotation were reversed, the sense of the reaction would be reversed also.

EXERCISES

1. Show that, if no assumption regarding θ is made, but $\psi = 0$ and $r = \nu$, then

$$L = A \sin \nu t \frac{d^2\theta}{dt^2} + C\nu \cos \nu t \frac{d\theta}{dt}$$

$$M = A \cos \nu t \frac{d^2\theta}{dt^2} - C\nu \sin \nu t \frac{d\theta}{dt}.$$

2. If the axis presses against a rough plane, $\psi = 0$, the tangential force being μ times the normal force, then *

$$A\frac{d^2\theta}{dt^2} - C\mu\nu\frac{d\theta}{dt} = 0,$$

provided $d\theta/dt > 0$ and furthermore the point of the axis in contact with the plane moves backward, *i.e.* in the sense of the decreasing θ.

On the other hand, the axis must have a sufficiently large radius so that the requirement below relating to the motion of the point of contact can be fulfilled. Hence

$$\frac{d\theta}{dt} = c\,e^{\frac{C\mu\nu}{A}t} \qquad \text{and} \qquad \theta = \frac{cA}{C\mu\nu}e^{\frac{C\mu\nu}{A}t},$$

where c denotes the initial value of $d\theta/dt$, and initially $\theta = cA/C\mu\nu$. Moreover,

$$v = \frac{C\mu\nu}{A}s,$$

where $v = d\theta/dt$ and $s = \theta$ refer to the point in which the sphere (of radius 1) is cut by the axis.

3. Prove that, in the problem of the preceding question, the normal reaction of the constraint is

$$cC\nu\, e^{\frac{\mu\nu C}{A}t} = \frac{\mu C^2\nu^2}{A}s.$$

4. Show that, if a (small) constant couple, of moment ϵ, acts on the gyroscope, the vector that represents the couple being at right angles to the plane $\psi = 0$ and directed in the proper

* If we think of the material axis as a cylinder of small radius, there will be a small couple about the axis, tending to reduce r. But as this couple approaches 0 when the radius of the cylinder approaches 0, we may consider the ideal case of an axis that is a material wire of nil cross section, the couple now vanishing.

sense, and if the axis of the gyroscope be constrained to move in the plane $\psi = 0$, the acceleration of θ is constant:

$$A\frac{d^2\theta}{dt^2} = \epsilon.$$

This last equation is true, even when ϵ varies with the time.

5. Prove that, no matter how θ varies, the axis of the gyroscope always being constrained to move in the plane $\psi = 0$, the reaction on the constraining plane $\psi = 0$ is numerically

$$C\nu\frac{d\theta}{dt},$$

its sense being that of the increasing ψ when $d\theta/dt > 0$, but the opposite when $d\theta/dt < 0$.

6. It has been shown that a rigid body is equivalent dynamically to three pairs of particles situated at the six extremities of a three dimensional cross; § 9.

Let the equivalent system move as the gyroscope did in the text, *i.e.* with $\psi = 0$ and $d\theta/dt = c$. Consider, in particular, an instant, at which the moving axes are flashing through the fixed axes; *i.e.* $\theta = \varphi = \psi = 0$. Show, by aid of the expressions for α, β, γ, that the vector acceleration of each of the four particles on the ξ- and the ζ-axes passes through O; but, in the case of each of the other two particles, is parallel to the axis of ζ. Hence explain the reaction of the gyroscope on the constraint.

7. Discuss the problem of Question 2 for the case that the point of contact is allowed to slip forward. Consider also all cases in which $d\theta/dt < 0$ initially.

18. The Top. The top is a rigid body having an axis of material symmetry and, in the case of a fixed peg, supported at a point O of the axis. Let the positive axis of ζ pass through the centre of gravity, G, distant h from O.

The third of Euler's Dynamical Equations becomes, since the applied forces — gravity and the reaction of the peg — both pass through the axis of ζ,

(1) $$C\frac{dr}{dt} = 0.$$

Hence $r = \nu$ (constant).

The equation of energy here becomes:

(2) $$A(p^2 + q^2) + C\nu^2 = H - 2Mgh\cos\theta.$$

Furthermore, the vertical component of the vector σ is constant. For, the applied forces giving a vector moment at O reduce to gravity, which is vertical, and so its vector moment with respect to O is horizontal. Now, the components of σ along the moving axes are Ap, Bq, and Cr. Hence the vertical component of σ is (§ 16):

$$Apn_1 + Bqn_2 + Crn_3.$$

On substituting for n_1, n_2, n_3 their values from § 16 we have:

$$(3) \qquad -Ap \sin\theta \cos\varphi + Aq \sin\theta \sin\varphi + C\nu \cos\theta = K.$$

Turning now to Euler's Geometrical Equations, we find:

$$(4) \qquad \begin{cases} p = -\sin\theta\cos\varphi\dfrac{d\psi}{dt} + \sin\varphi\dfrac{d\theta}{dt} \\[2ex] q = \sin\theta\sin\varphi\dfrac{d\psi}{dt} + \cos\varphi\dfrac{d\theta}{dt} \\[2ex] \nu = \cos\theta\dfrac{d\psi}{dt} + \dfrac{d\varphi}{dt}. \end{cases}$$

On substituting these values of p and q in (2) and (3) we find:

$$(5) \qquad \begin{cases} \sin^2\theta\left(\dfrac{d\psi}{dt}\right)^2 + \left(\dfrac{d\theta}{dt}\right)^2 = \alpha - a\cos\theta \\[2ex] \sin^2\theta\dfrac{d\psi}{dt} = \beta - b\nu\cos\theta \end{cases}$$

$$(6) \qquad a = \frac{2Mgh}{A}, \qquad b = \frac{C}{A},$$

the constants α and β depending on the initial conditions of the motion; *i.e.* they are constants of integration; whereas a and b are constants of the body.

The third Equation (4) determines φ after θ and ψ have been found from (5) as functions of t:

$$(7) \qquad \varphi = \nu t - \int \cos\theta\frac{d\psi}{dt}\,dt.$$

Returning now to Equations (5) and eliminating $d\psi/dt$, we obtain:

$$(8) \qquad \sin^2\theta\left(\frac{d\theta}{dt}\right)^2 = \sin^2\theta(\alpha - a\cos\theta) - (\beta - b\nu\cos\theta)^2.$$

The result is a differential equation for the single dependent variable, θ. It can be improved in form by the substitution

$$(9) \qquad u = \cos\theta:$$

$$(10) \qquad \left(\frac{du}{dt}\right)^2 = (1 - u^2)(\alpha - au) - (\beta - b\nu u)^2 = f(u).$$

Thus $f(u)$ is seen to be a cubic polynomial, which we will presently discuss in detail. But first observe that the second Equation (5) gives:

$$(11) \qquad \frac{d\psi}{dt} = \frac{\beta - b\nu u}{1 - u^2}.$$

Hence ψ is given by a quadrature after u has once been found as a function of t.

Retrospect and Prospect. To sum up, then, we have reduced the problem to the solution of Equation (10) for u as a function of t. Equation (9) gives θ; Equation (11) gives ψ; and Equation (7) gives φ. We may concentrate, then, on the solution of Equation (10).

19. Continuation. Discussion of the Motion. The Polynomial

$$(1) \qquad f(u) = (1 - u^2)(\alpha - au) - (\beta - b\nu u)^2$$

becomes positively infinite for $u = +\infty$. It is negative or 0 for $u = +1, -1$. Hence in general the graph will be as indicated, or

$$0 < f(u), \qquad u_1 < u < u_2;$$

$$f(u_1) = f(u_2) = 0.$$

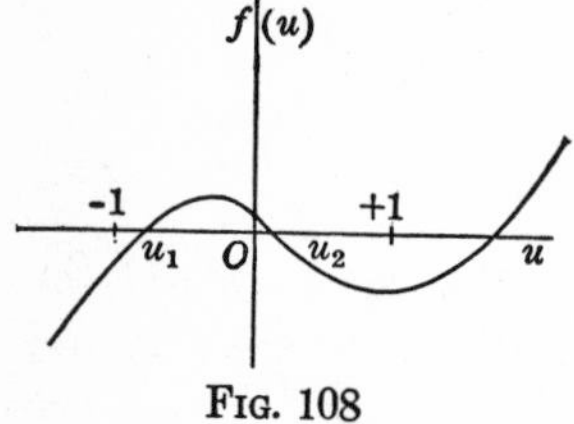

FIG. 108

Moreover, $-1 < u_1 < u_2 < 1$, and $f(u)$ has one root, $u' > 1$. The roots u_1, u_2 will, therefore, be simple roots.

The differential equation

$$(2) \qquad \left(\frac{du}{dt}\right)^2 = f(u)$$

comes under the class discussed in Appendix B. In particular, the solution is a function

$$(3) \qquad u = \varphi(t)$$

single-valued and continuous for all values of t and having the period T, where

(4) $$\tfrac{1}{2}T = \int_{u_1}^{u_2} \frac{du}{\sqrt{f(u)}},$$

or

(5) $$\varphi(t + T) = \varphi(t).$$

Furthermore, if

(6) $$u_1 = \varphi(t_1),$$

then

(7) $$\varphi(t_1 - \tau) = \varphi(t_1 + \tau).$$

And similarly, if

(6′) $$u_2 = \varphi(t_2),$$

then

(7′) $$\varphi(t_2 - \tau) = \varphi(t_2 + \tau).$$

Physical Interpretation. Let a sphere S be placed about O as centre, and let P be the point of intersection of the positive axis of ζ with S. Let C be the curve that P describes on S. The results just obtained show that C lies between the two parallels of latitude corresponding to

(8) $$u = u_1, \qquad u = u_2.$$

For convenience let t be measured from a point on the upper parallel, $u = u_2$. Then there are three cases according as initially

(9) $$\text{I. } \frac{d\psi}{dt} > 0; \qquad \text{II. } \frac{d\psi}{dt} = 0; \qquad \text{III. } \frac{d\psi}{dt} < 0.$$

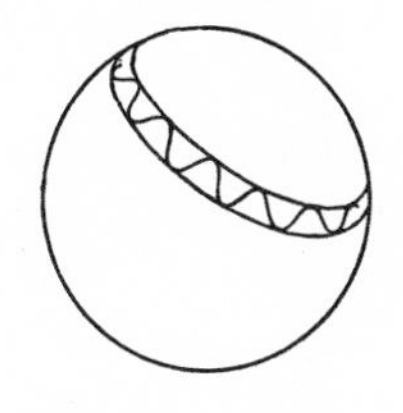
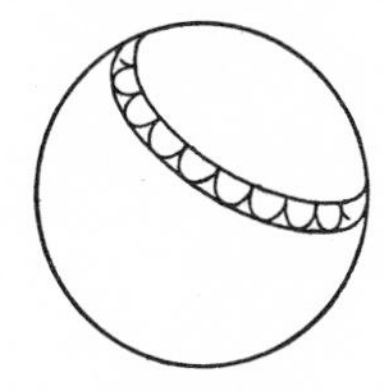
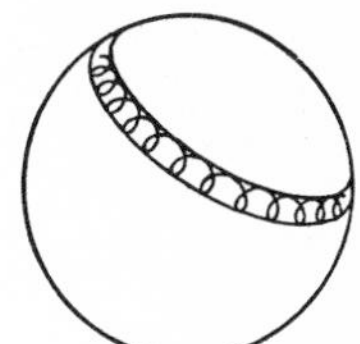

Fig. 109

Case I. Since

$$\frac{d\psi}{dt} = \frac{\beta - b\nu u}{1 - u^2}$$

is positive when u has its greatest value, u_2, $d\psi/dt$ will remain positive, and so ψ will steadily increase with t. Let $\psi = 0$ when $t = 0$. As t increases to $\tfrac{1}{2}T$, ψ will increase to

$$\tfrac{1}{2}\Psi = \int_0^{\frac{1}{2}T} \frac{\beta - b\nu u}{1 - u^2}\, dt,$$

where $u = \varphi(t)$, Equation (3). When $t = T$, ψ will have increased by Ψ, and one complete arch of the curve C will have been described. The arch is symmetric in the plane $\psi = \frac{1}{2}\Psi$. The rest of C is obtained by rotating this arch about the polar axis of S through angles that are multipla of Ψ.

CASE II. Here, $d\psi/dt$ is 0 at the start, and hence

$$\beta - b\nu u_2 = 0.$$

Since u decreases, it follows that in the further course of the motion

$$0 < \beta - b\nu u,$$

and so ψ steadily increases. The curve C has cusps on the upper parallel of latitude.

CASE III. Here

$$\beta - b\nu u < 0$$

at the start, and it is conceivable that this relation should persist forever. But even if this were not the case, it is still conceivable that the value of ψ when P reaches the lower parallel of latitude should be less than or equal to the initial value, $\psi = 0$. That neither of these cases is possible — that the value of ψ corresponding to the first return of P to the upper circle is positive — has been shown by Hadamard.* The curve C has double points in this case, but it proceeds with increasing t in the sense of the advancing ψ, as indicated.

Special Cases. There is still a variety of special cases to be discussed, one of which is that in which $f(u)$ has equal roots lying within the interval:

$$-1 < u_1 = u_2 < 1.$$

Since

$$f(1) \leqq 0$$

in all cases, and since

$$f(+\infty) = +\infty,$$

there must be a third root $u' \geqq 1$. Thus u_1 is a double root and

$$f(u) = (u - u_1)^2\chi(u),$$

where

$$\chi(u) < 0, \qquad -1 < u < 1.$$

* *Bull. des Sci. math.* 1895, p. 228.

The only solution of Equation (2) in this case, which takes on the value u_1 when $t = 0$, is

$$u = u_1.$$

The curve C reduces to a parallel of latitude.

When $u = 1$ is a root, various cases can arise. The point P may pass through the north pole with a velocity ± 0; or it may gradually climb, approaching the north pole as a limit; or the top may permanently rotate about the polar axis. Similarly, when $u = -1$ is a root.

There is a great wealth of literature on the gyroscope and the top. The reader can refer to the article on the *Gyroscope* in the *Encyclopaedia Britannica;* to Webster, *Dynamics;* to Routh, *Elementary Rigid Dynamics;* and to Appell, *Mécanique rationelle*, vol. II.

EXERCISE

Treat the case of a top on a smooth table. Assume that the peg is a surface of revolution. The distance, then, from the centre of gravity to the vertical through the point of contact with the table will be a function of the angle of inclination of the axis.

Assume axes fixed in the body with the origin at the centre of gravity.

Write down *i*) the equation of energy; *ii*) the equation that says that the vertical component of σ is constant.

From this point on the procedure is precisely as before, and the result is again a differential equation of the type treated in Appendix B. Discuss all cases, and show that in general the axis oscillates between two inclinations, both oblique to the vertical.

Begin with the special case that the peg is a point. Having studied this case in detail, proceed to the general case and study it in detail, also. Then derive the special case as a particular case under the general case.

20. Intrinsic Treatment of the Gyroscope.* The most general case of motion of a gyroscope reduces to one in which a single couple acts on the body, and this couple can be broken up into

* The results of this paragraph are contained in a paper by the Author: "On the Gyroscope," *Trans. Amer. Math. Soc.*, vol. 23, April, 1922, p. 240.

two couples — one, represented by a vector at right angles to the axis of the gyroscope; the other, by a vector collinear with the axis. In the most important applications that arise in practice, the latter couple vanishes. But in the general case, it gives rise to the third of the Dynamical Equations in the form:

$$C\frac{dr}{dt} = N. \tag{1}$$

The former couple can be realized by a single force **F** perpendicular to the axis and acting at the point P in which the positive ζ-axis cuts the unit sphere, — the other force of the couple and the resultant force acting at O.*

Definition of the Bending, κ. Let C be the curve described on the unit sphere by P, and let S be the cone which is the locus of the axis of the gyroscope, and of which C is the directrix. Consider the rate at which the tangent plane to S is turning when P describes C with unit velocity. This quantity shall be denoted as the *bending* of the cone and represented by the number κ. It is also the rate at which the terminal point of a unit vector drawn from O at right angles to the tangent plane traces out its path on the unit sphere. κ shall be taken positive when an observer, walking along C, sees C to the left of the tangent plane, and negative, when C is to his right.

It is easy to compute κ. Let V be the angle from the parallel of latitude through P with the sense of the increasing ψ to the tangent to C with the sense of the increasing s. Then it appears form an infinitesimal treatment that

$$\kappa = \frac{dV}{ds} - \frac{d\psi}{ds}\cos\theta. \tag{2}$$

Since

$$\tan V = \frac{d\theta}{d\psi \sin\theta}, \qquad \text{or} \qquad V = \tan^{-1}\frac{\theta'}{\psi' \sin\theta},$$

where accents denote differentiation with respect to s, and since

$$ds^2 = d\theta^2 + d\psi^2 \sin^2\theta, \qquad \text{or} \qquad \theta'^2 + \psi'^2 \sin^2\theta = 1,$$

it follows that

$$\kappa = (\psi'\theta'' - \theta'\psi'')\sin\theta - (1 + \theta'^2)\,\psi'\cos\theta. \tag{3}$$

* The point O need not be the centre of gravity in the following treatment. It may be any point fixed in the axis of material symmetry.

From the definition it follows at once that the bending of a cone of revolution must be constant. To find its value, let the coordinates be so chosen that the equation of the cone is $\theta = \alpha$. Then the length of the arc of C is

$$s = \psi \sin \alpha \qquad \text{and so} \qquad \psi' \sin \alpha = 1.$$

From (3) it now is seen that

$$\kappa = - \cot \alpha. \tag{4}$$

The cone lies to the right of the observer, as he travels along C. If he reverses his sense, the sign of κ will be changed. But both cases are embraced in the single formula (4), the second corresponding to a cone whose angle is $\pi - \alpha$, or again for which s is replaced by $- s$.

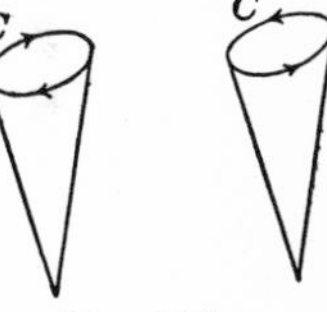

FIG. 110

Conversely, if κ is constant, C is a circular cone. For, the equation (3) can, by eliminating ψ',

$$\psi' = \pm \frac{\sqrt{1 - \theta'^2}}{\sin \theta},$$

be written in the form:

$$\pm \kappa = \frac{\theta''}{\sqrt{1 - \theta'^2}} - \sqrt{1 - \theta'^2} \cot \theta, \tag{5}$$

the $-$ sign holding whenever $\psi' < 0$. Hence

$$\frac{d^2\theta}{ds^2} = \left(1 - \frac{d\theta^2}{ds^2}\right) \cot \theta \pm \kappa \left(1 - \frac{d\theta^2}{ds^2}\right)^{\frac{1}{2}}. \tag{6}$$

If κ is constant, set $\kappa = - \cot \alpha$. Then Equation (6) admits one solution, $\theta = \alpha$, or $\theta = \pi - \alpha$; and, as is shown in the theory of differential equations, this is the only solution which, at a point $s = s_0$, takes on the value α, or $\pi - \alpha$, and whose derivative vanishes there.

*Further Formulas for κ.** From (3) it follows further that

$$\pm \kappa = \frac{\dfrac{d^2\theta}{d\psi^2} \sin \theta - 2 \dfrac{d\theta^2}{d\psi^2} \cos \theta - \sin^2 \theta \cos \theta}{\left(\dfrac{d\theta^2}{d\psi^2} + \sin^2 \theta\right)^{\frac{3}{2}}}, \tag{7}$$

where the $-$ sign holds whenever $\psi' < 0$.

* These results are inserted for completeness. They will not be used in what follows, and the student may pass on without studying them. They are chiefly of interest to the student of Differential Geometry.

If κ is known, or given, as a function of s, then Equation (6) determines θ as a function of s, and ψ is then found by a quadrature:

$$\frac{d\psi}{ds} = \pm \csc\theta \left(1 - \frac{d\theta^2}{ds^2}\right)^{\frac{1}{2}}. \tag{8}$$

The bending, κ, is connected with the curvature, K, of C, regarded as a space curve, by the formula

$$K^2 = \kappa^2 + 1. \tag{9}$$

Furthermore, cf. Fig. 111 below:

$$\mathbf{n} = \mathbf{a} \times \mathbf{t} = \begin{vmatrix} \mathbf{i} & \mathbf{j} & \mathbf{k} \\ x & y & z \\ x' & y' & z' \end{vmatrix}, \tag{10}$$

$$\mathbf{n}' = \kappa \mathbf{t}, \tag{11}$$

hence

$$\begin{cases} \kappa x' = yz'' - zy'' \\ \kappa y' = zx'' - xz'' \\ \kappa z' = xy'' - yx''. \end{cases} \tag{12}$$

Since $|K| = |\mathbf{t}'|$ and

$$\mathbf{t}' = x''\mathbf{i} + y''\mathbf{j} + z''\mathbf{k},$$

$$K^2 = x''^2 + y''^2 + z''^2, \tag{13}$$

formula (9) follows at once from (12) and (13). Moreover, from (12) it follows that

$$\kappa = - \begin{vmatrix} x & y & z \\ x' & y' & z' \\ x'' & y'' & z'' \end{vmatrix}. \tag{14}$$

Finally, the torsion, T, of C is connected with κ by the relation:

$$\frac{d}{ds} \tan^{-1}\kappa = \pm T,$$

the result obtained by Professor Haskins.*

* For the proof of this formula cf. the Author's paper cited above.

21. The Relations Connecting v, F, and κ. The physical phenomenon which it is most important to bring home to one's intuition is the effect of the force **F** on the motion of the gyroscope. Any such explanation must take account of all *three* quantities, **v**, **F**, and κ. But many popular explanations — claiming correctly to be "non-mathematical," but incorrectly to be accurate in their mechanics — fail because they are unaware of κ. Thus, for example, the statement often made that "when a couple is applied to a rotating gyroscope, the forces of the couple intersecting the axis of the gyroscope at right angles, the axis will move in a plane perpendicular to the plane of the forces of the couple" is false. In fact, the axis will begin to move tangentially to this plane, if it starts from rest, and all intermediate cases are possible, according to the initial motion of the axis.

A simple and accurate explanation, in terms of **v**, **F**, and κ, can be given as follows.* First of all, however, the third of Euler's Dynamical Equations, which here becomes:

$$C\frac{dr}{dt} = N, \tag{1}$$

and requires no further comment than *i*) that it is perfectly general, applying to the motion of the gyroscope under any forces whatever; and *ii*) that in the case which most interests us, namely that in which there is only the force **F** (and the reaction at O) we have: $N = 0$, and so $r = \nu$, a constant.

Let **F**, then, be resolved, in the tangent plane, into a component T along the positive tangent, and a component Q, taken positive when directed toward the left of the observer; *i.e.* Q is positive when κ is positive. Then

$$\left\{\begin{aligned} Av\frac{dv}{ds} &= T \\ A\kappa v^2 + Crv &= Q, \end{aligned}\right. \tag{2}$$

where $v = ds/dt$ and s increases in the sense of the motion of P, r being given by Equation (1).

* Cf. the Author's paper "On the Gyroscope" cited above, p. 240.

Proof. Let the unit vector from O to P be denoted by $\mathbf{a}$ (it is the vector γ of the coordinate system); let $\mathbf{t}$ be a unit vector along the positive tangent to C at P; and let $\mathbf{n}$ be a unit vector normal to $\mathbf{a}$ and $\mathbf{t}$ and so oriented with regard to them as β is with regard to γ and α. These are principal axes of inertia, and the moments of inertia about them are:

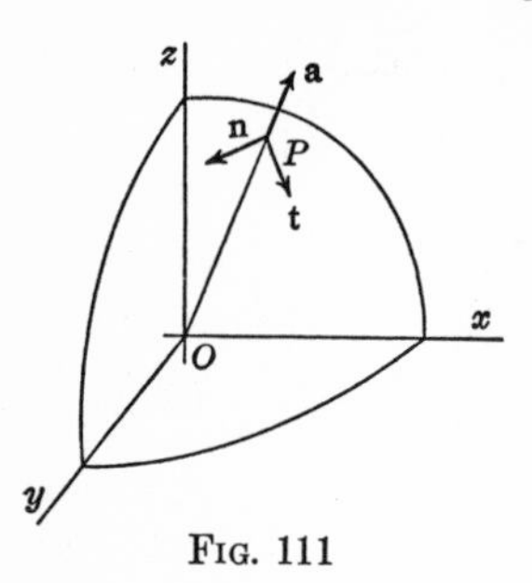

FIG. 111

$$I_t = A, \qquad I_n = A, \qquad I_a = C.$$

The components of the angular velocity $\tilde{\omega}$ about them are:

$$\omega_t = 0, \qquad \omega_n = v, \qquad \omega_a = r.$$

Now, σ has the value:

$$\sigma = I_n\,\omega_n\,\mathbf{n} + I_t\,\omega_t\,\mathbf{t} + I_a\,\omega_a\,\mathbf{a}.$$

Hence

$$\sigma = Av\,\mathbf{n} + Cr\,\mathbf{a}. \tag{3}$$

From this equation we can compute $\frac{d\sigma}{dt}$:

$$\frac{d\sigma}{dt} = A\frac{dv}{dt}\mathbf{n} + C\frac{dr}{dt}\mathbf{a} + Av\frac{d\mathbf{n}}{dt} + Cr\frac{d\mathbf{a}}{dt}.$$

It is clear that

$$\frac{d\mathbf{a}}{dt} = v\,\mathbf{t}. \tag{4}$$

Furthermore, from the definition of the bending, it appears that

$$\frac{d\mathbf{n}}{dt} = \kappa v\,\mathbf{t}. \tag{5}$$

Hence, finally,

$$\frac{d\sigma}{dt} = Av\frac{dv}{ds}\mathbf{n} + (A\kappa v^2 + Crv)\,\mathbf{t} + C\frac{dr}{dt}\mathbf{a}. \tag{6}$$

Let the vector $\mathbf{M}$ which represents the resultant moment of all the applied forces about O be written in the form:

$$\mathbf{M} = M_n\,\mathbf{n} + M_t\,\mathbf{t} + M_a\,\mathbf{a}.$$

Since

$$\frac{d\sigma}{dt} = \mathbf{M},$$

we have:

$$(7) \qquad Av\frac{dv}{ds} = M_n, \qquad A\kappa v^2 + Crv = M_t, \qquad C\frac{dr}{dt} = M_a.$$

Turning now to the case in which we are most interested, namely, that in which a force $\mathbf{F}$ acts at P in a direction at right angles to OP:

$$\mathbf{F} = -Q\mathbf{n} + T\mathbf{t},$$

we see that $M_n = T$, $M_t = Q$, and thus Equations (2) are established. Equation (1) is the third of Equations (7).

We have thus obtained Euler's Dynamical Equations in the form:

$$(8) \qquad \begin{cases} Av\dfrac{dv}{ds} = T \\ A\kappa v^2 + Crv = Q \\ C\dfrac{dr}{dt} = N \end{cases}$$

22. Discussion of the Intrinsic Equations. The first of Equations (8), § 21,

$$Av\frac{dv}{ds} = T,$$

admits a simple interpretation. It shows that the point P describes the curve C exactly as a smooth bead of mass $m = A$ would move along a wire in the form of C if it were acted on by a tangential force T.

The third equation,

$$C\frac{dr}{dt} = N,$$

shows that the component r of the angular velocity $\bar{\omega}$ about the axis of the gyroscope varies exactly as it would if the axis were permanently at rest and the same couple N relative to the axis acted.

The second equation,

$$\text{A)} \qquad A\kappa v^2 + Crv = Q,$$

expresses the sole relation which holds between the four variables κ, v, r, and Q. In the applications, however, r is constant, $r = \nu$, and so the equation

$$\text{A}') \qquad A\kappa v^2 + C\nu v = Q$$

expresses the sole relation between κ, v, and Q.

The Case $\mathbf{F} = 0$. Let us begin with the case that $\mathbf{F}$ vanishes, but the axis is not at rest. Here, $Q = 0$, $T = 0$. Equation A) gives

$$i) \qquad A\kappa v + Cr = 0,$$

or, on introducing the radius of bending, $\rho = 1/|\kappa|$, and choosing $r > 0$:

$$\rho = \frac{Av}{Cr}.$$

If r is a positive constant, $r = \nu > 0$, then

$$\kappa = -\frac{C\nu}{Av}$$

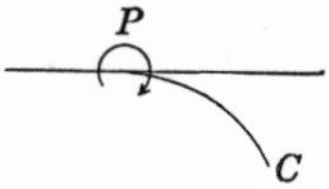

and since v is constant, for

$$Av\frac{dv}{ds} = 0,$$

Fig. 112

κ is also constant, and negative. The axis of the gyroscope is describing a cone of semi-vertical angle α, where

$$\cot \alpha = |\kappa|, \qquad \text{or} \qquad \tan \alpha = \rho,$$

and the sense of the description is such that the observer, walking along C in the positive sense, has the cone on his right.

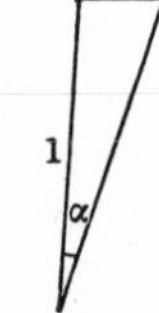

Fig. 113

The Case $\kappa = 0$. Here, the path of P is an arc of a great circle, and

$$Q = Crv, \qquad \text{or} \qquad Q = C\nu v,$$

no matter what T and the motion of P along its path may be. The pressure of the axis against the constraint, in a normal direction, is to the right, and is proportional to r and to v; — or, if r is constant, to v, the coefficient then being $C\nu$. Thus we obtain anew, and with the minimum of effort, the main result of § 17.

General Interpretation of Equation A). We can now give a simple physical interpretation to Equation A):

$$A\kappa v^2 + Crv = Q.$$

The left-hand side is the sum of two terms. The second term expresses the force,

$$Q_2 = Crv,$$

that would be required to cause P to describe a great circle on the sphere; *i.e.* to make the axis move in the plane through O tangent to C. This force, Q_2, is always directed toward the left, for $Q_2 > 0$.

The first term,

$$Q_1 = A\kappa v^2,$$

accounts for the bending. Its numerical value,

$$| Q_1 | = \frac{Av^2}{\rho},$$

can be interpreted as the centripetal force exerted on a particle, of mass $m = A$, to make it describe a circle of radius ρ with velocity v. When κ is positive, this force is positive, and so is directed toward the left; and vice versa.

Consider now the force Q' along the normal $-\mathbf{n}$ at P, which (combined with the smooth constraint of the surface of the sphere) would be required to hold a particle of mass $m = A$ in the path C. Let the vector $\mathbf{a}$ be written in the form:

$$\mathbf{a} = x\mathbf{i} + y\mathbf{j} + z\mathbf{k}.$$

Then

$$\mathbf{t} = x'\mathbf{i} + y'\mathbf{j} + z'\mathbf{k},$$

$$\mathbf{v} = \dot{x}\mathbf{i} + \dot{y}\mathbf{j} + \dot{z}\mathbf{k} = v\mathbf{t},$$

where $x' = dx/ds$, $\dot{x} = dx/dt$, etc. Furthermore,

$$\mathbf{n} = \mathbf{a} \times \mathbf{t} = (yz' - zy')\mathbf{i} + (zx' - xz')\mathbf{j} + (xy' - yx')\mathbf{k}.$$

The acceleration, (α), of P in space is, of course:

$$(\alpha) = \ddot{x}\mathbf{i} + \ddot{y}\mathbf{j} + \ddot{z}\mathbf{k}.$$

Now, the component of the acceleration along the normal $\mathbf{n}$ to the plane of $\mathbf{a}$ and $\mathbf{t}$ is $\mathbf{n}\cdot(\alpha)$, which can be written in the form:

$$\mathbf{n}\cdot(\alpha) = \begin{vmatrix} x & y & z \\ x' & y' & z' \\ \ddot{x} & \ddot{y} & \ddot{z} \end{vmatrix}.$$

Since $\dot{x} = vx'$, it follows that

$$\ddot{x} = v^2x'' + \dot{v}x', \quad \text{etc.},$$

and so

$$\begin{vmatrix} x & y & z \\ x' & y' & z' \\ \ddot{x} & \ddot{y} & \ddot{z} \end{vmatrix} = v^2 \begin{vmatrix} x & y & z \\ x' & y' & z' \\ x'' & y'' & z'' \end{vmatrix} = -\kappa v^2.$$

Thus $m\kappa v^2$ is equal to the force Q' tangent to the sphere and normal to C, which would be required to hold a particle of mass m, describing C, in its path; the component along **t** being $mv\,dv/ds$, and the third component, along **a**, being the reaction normal to the sphere, in which we are not interested.

It is natural to think of the point $\rho\mathbf{n}$ on the line through P along **n** as the *centre of bending*. If we draw the osculating cone of revolution through P, this is the point Q in which that line meets the axis of the cone. An obvious interpretation for this force of $m\kappa v^2$ is the centripetal force of a particle describing a circle of radius ρ, with centre at Q, tangent to C at P, the velocity being v.

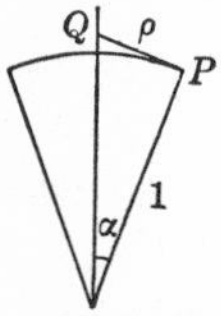

Fig. 114

The force Q_2 can be realized physically as follows. Let an electro-magnetic field of force be generated by a north-pole situated at O, and let the particle m carry a charge, e, of electricity. The force exerted on e by the field will be at right angles to the path and tangent to the sphere, — and, finally, proportional to the velocity, v, of m. Hence e can be so chosen that this force will be precisely equal to $Q_2 = C\nu v$.

In the more general, but less interesting, case that r is variable, the physical interpretation can still be adapted by using a variable charge.*

Summary of the Results. To sum up, then, we can say: The point P, in which the axis of the gyroscope meets the unit sphere about O, moves like a particle of mass $m = A$ constrained to lie on the sphere and carrying a charge of electricity, e. The forces that act on m are supplied by the electromagnetic force of the field, $Q_2 = C\nu v$, acting on e, and a force **F** acting on m, the components of **F** along the tangent and normal at P being T and Q_1 respectively. The case of a variable r can be met by a variable charge, e.

As regards the physical realization of the condition that the particle lie on the surface of the sphere, we may think of a massless rod of unit length, free to turn about one end which is pivoted at O, and carrying the particle at the other end.

* The idea of using the above electro-magnetic field to obtain Q_2 was suggested to me by my colleague, Professor Kemble, to whom I had just communicated the results of the text, down to this point. (Note of Jan. 23, 1933.)

EXERCISES

1. Suppose the axle P of the gyroscope is caused to move in a smooth slot in the form of a meridian circle, which is made to rotate in any manner. The force **F** will then be normal to the meridian, or tangent to the parallel of latitude. Show that

$$\frac{d^2\theta}{dt^2} + \left(\frac{C\nu}{A} - \cos\theta\frac{d\psi}{dt}\right)\sin\theta\frac{d\psi}{dt} = 0,$$

$$F = A\left\{\sin\theta\frac{d^2\psi}{dt^2} + 2\cos\theta\frac{d\theta}{dt}\frac{d\psi}{dt}\right\} - C\nu\frac{d\theta}{dt}.$$

Suggestion: Combine Euler's Geometrical Equations with Euler's Dynamical Equations.

2. Let the components of **F** along the meridian in the sense of the increasing θ and along the parallel of latitude in the sense of the increasing ψ be denoted respectively by Θ and Ψ. Show that

$$A\frac{d^2\theta}{dt^2} + \left(C\nu - A\cos\theta\frac{d\psi}{dt}\right)\sin\theta\frac{d\psi}{dt} = \Theta$$

$$A\left\{\sin\theta\frac{d^2\psi}{dt^2} + 2\cos\theta\frac{d\theta}{dt}\frac{d\psi}{dt}\right\} - C\nu\frac{d\theta}{dt} = \Psi.$$

If Θ and Ψ are known as functions of θ, ψ, t, these equations suffice to determine the path of P.

3. Consider small oscillations of the axis of the gyroscope in the neighborhood of the axis $\theta = \pi/2$, $\psi = 0$. Let

$$\theta = \frac{\pi}{2} + \vartheta, \qquad \vartheta = \theta - \frac{\pi}{2}.$$

Show that the equations of Question 1 lead to the approximate equations:

$$\begin{cases} A\dfrac{d^2\vartheta}{dt^2} + C\nu\dfrac{d\psi}{dt} = \Theta \\[2ex] A\dfrac{d^2\psi}{dt^2} - C\nu\dfrac{d\vartheta}{dt} = \Psi \end{cases}$$

4. Generalize the equations of Question 2 to the case that A, B, C are all distinct.

5. *Intrinsic Equations.* From the equations:

$$Av\frac{dv}{ds} = T,$$

$$A\kappa v^2 + C\nu v = Q,$$

$$\kappa = \frac{\dfrac{d^2\theta}{ds^2}}{\sqrt{1 - \dfrac{d\theta^2}{ds^2}}} - \sqrt{1 - \frac{d\theta^2}{ds^2}}\cot\theta,$$

$$\frac{d\theta^2}{ds^2} + \frac{d\psi^2}{ds^2}\sin^2\theta = 1,$$

the path can be determined if T, Q are known as functions of s and v.

6. *Ship's Stabilizer.* The gyroscope can be used to reduce the rolling of a ship. A massive gyroscope is mounted in a cage, or frame, its axis being fixed with reference to the frame, and vertical. The frame is mounted on trunnions, with axis horizontal and at right angles to the keel, and it is provided with a brake to dampen its oscillations about this axis. Thus the axis of the gyroscope has two degrees of freedom; it can rotate in the plane through the keel and the masts, and this plane rotates with the rolling of the ship.* Isolate the following systems:

i) The ship, exclusive of the gyroscope and frame;
ii) The frame;
iii) The gyroscope.

The rolling of the ship is governed by the equation:

$$I\frac{d^2\psi}{dt^2} = -\kappa\frac{d\psi}{dt} - \lambda\psi - 2hF,$$

where the first term on the right is due to the damping of the water; the second, to the righting moment produced by the buoyancy; and the third, to the force exerted by the trunnions.

The frame may be thought of as rotating about the point, O, regarded as fixed, in which the axis of the gyroscope cuts the

* A picture and an account of the ship's gyroscope is found in the article on the "Gyroscope" in the *Encyclopaedia Britannica* and in Klein-Sommerfeld, *Theorie des Kreisels*, vol. iv., p. 797. For the discussion which follows the reader also needs, however, the theory and practice of Oscillatory Motion with Damping; cf. the Author's *Advanced Calculus*, Chap. XV.

axis of the trunnions. Let Euler's Angles be so chosen that the axis of the sphere: $\theta = 0$, $\psi = 0$, is parallel to the keel, the plane $\psi = 0$ being vertical. Moreover, let θ be replaced by ϑ, where

$$\theta = \frac{\pi}{2} + \vartheta.$$

The motion of the gyroscope about its centre of gravity, the point O, will be governed by the approximate equations of Question 3.

Finally, the motion of the frame is governed by the equations called for in Question 4 above. These equations are modified by the condition $\varphi \equiv 0$, and then reduced still further by setting $\sin\vartheta = 0$, $\cos\vartheta = 1$. Thus

$$\begin{cases} B'\dfrac{d^2\vartheta}{dt^2} = -\kappa'\dfrac{d\vartheta}{dt} - \lambda'\vartheta - \Theta \\[2ex] A'\dfrac{d^2\psi}{dt^2} = 2hF - \Psi \end{cases}$$

where the first term on the right is due to the brake and other damping, and the second, to gravity, since the frame is so constructed that its centre of gravity is appreciably below O.

On combining these five equations and neglecting $A + A'$ in comparison with I we find:

$$\begin{cases} I\dfrac{d^2\psi}{dt^2} + \kappa\dfrac{d\psi}{dt} - C\nu\dfrac{d\vartheta}{dt} + \lambda\psi = 0 \\[2ex] (A + B')\dfrac{d^2\vartheta}{dt^2} + \kappa'\dfrac{d\vartheta}{dt} + C\nu\dfrac{d\psi}{dt} + \lambda'\vartheta = 0 \end{cases}$$

These are the equations which govern the motion. They are discussed at length in Klein-Sommerfeld, l.c.

23. Billiard Ball. Let a billiard ball be projected along the table, with an arbitrary initial velocity of the centre, O, and an arbitrary initial velocity of rotation. To determine the motion.

Let the (x, y)-plane of the axes fixed in space be horizontal. Let moving axes of (ξ, η, ζ) be chosen parallel to (x, y, z), but with the origin at the centre of the ball.

The point of the ball P, in contact with the table, shall be slipping, and the angle from the positive direction of the axis of x or ξ to the direction of its motion shall be ψ.

The forces acting are: gravity, or Mg, downward at O; $R = Mg$ upward at P; and the force of friction, μMg, at P in the sense opposite to that of slipping. Hence, for the motion of the centre of gravity,

$$(1)\qquad \left\{\begin{aligned} M\frac{d^2\bar{x}}{dt^2} &= -\,\mu Mg\cos\psi \\ M\frac{d^2\bar{y}}{dt^2} &= -\,\mu Mg\sin\psi. \end{aligned}\right.$$

The vector momentum σ, referred to the centre of gravity, has for its components along the moving axes:

$$\sigma_\xi = I\omega_\xi, \qquad \sigma_\eta = I\omega_\eta, \qquad \sigma_\zeta = I\omega_\zeta,$$

where

$$I = \tfrac{2}{5}Ma^2.$$

The moment equation,

$$\frac{d\sigma}{dt} = \sum_k \mathbf{r}_k \times \mathbf{F}_k,$$

thus gives:

$$(2)\qquad \left\{\begin{aligned} I\frac{d\omega_\xi}{dt} &= -\,\mu Mga\sin\psi \\ I\frac{d\omega_\eta}{dt} &= \mu Mga\cos\psi \\ I\frac{d\omega_\zeta}{dt} &= 0. \end{aligned}\right.$$

Hence

$$(3)\qquad \omega_\zeta = \text{const.}$$

The angle ψ is unknown. Eliminate it by combining Equations (1) and (2):

$$(4)\qquad \left\{\begin{aligned} \frac{d^2\bar{x}}{dt^2} &= -\,\frac{2a}{5}\frac{d\omega_\eta}{dt} \\ \frac{d^2\bar{y}}{dt^2} &= \frac{2a}{5}\frac{d\omega_\xi}{dt}. \end{aligned}\right.$$

Hence

$$(5)\qquad \left\{\begin{aligned} \frac{d\bar{x}}{dt} &= -\,\frac{2a}{5}\omega_\eta + A \\ \frac{d\bar{y}}{dt} &= \frac{2a}{5}\omega_\xi + B, \end{aligned}\right.$$

where A, B are constants of integration depending on the initial conditions. They may have any values whatever.

Let **V** be the velocity of the lowest point of the ball. Then

$$(6)\quad \begin{cases} V_x = V\cos\psi = \dfrac{d\bar{x}}{dt} - a\omega_\eta \\[2ex] V_y = V\sin\psi = \dfrac{d\bar{y}}{dt} + a\omega_\xi. \end{cases}$$

Combining these equations with (5) we get:

$$(7)\qquad V_x = \frac{7}{2}\left(\frac{d\bar{x}}{dt} + A'\right), \qquad V_y = \frac{7}{2}\left(\frac{d\bar{y}}{dt} + B'\right),$$

where

$$A' = -\tfrac{5}{7}A, \qquad B' = -\tfrac{5}{7}B.$$

Equations (1) now take on the following form. For abbreviation let

$$u = \frac{d\bar{x}}{dt} + A', \quad v = \frac{d\bar{y}}{dt} + B', \quad w = \sqrt{u^2 + v^2} = \tfrac{2}{7}V \neq 0.$$

Then

$$(8)\quad \begin{cases} \dfrac{du}{dt} = -\mu g\dfrac{u}{w} \\[2ex] \dfrac{dv}{dt} = -\mu g\dfrac{v}{w}. \end{cases}$$

Hence

$$v\frac{du}{dt} - u\frac{dv}{dt} = 0,$$

and consequently

$$\beta u = \alpha v,$$

where α, β are constants not both 0. Moreover, u and v are not both 0.

Suppose $u > 0$, $\alpha > 0$. Then

$$\frac{v}{w} = \frac{\alpha v}{\sqrt{\alpha^2 u^2 + \beta^2 u^2}} = \frac{\alpha v}{u\sqrt{\alpha^2 + \beta^2}} = \frac{\beta}{\sqrt{\alpha^2 + \beta^2}},$$

$$\frac{u}{w} = \frac{\alpha}{\sqrt{\alpha^2 + \beta^2}}.$$

The proof of this last equation requires the consideration of the two cases: *i*) $\beta \neq 0$; *ii*) $\beta = 0$. These formulas are general, holding in all cases in which $w \neq 0$.

It thus appears that

$$(9) \qquad \cos\psi = \frac{\alpha}{\sqrt{\alpha^2+\beta^2}}, \qquad \sin\psi = \frac{\beta}{\sqrt{\alpha^2+\beta^2}}.$$

Hence $d^2\bar{x}/dt^2$ and $d^2\bar{y}/dt^2$ are constants, and consequently the centre of the ball describes in general a parabola; in particular, a straight line. The direction, however, in which the point P is slipping, is always the same; cf. Equations (9).

This result comprises the main interest of the problem, so long as there is slipping. Slipping ceases when $V = 0$, or

$$(10) \qquad \left(\frac{d\bar{x}}{dt} + A'\right)^2 + \left(\frac{d\bar{y}}{dt} + B'\right)^2 = 0.$$

The Subsequent Motion. From this instant on the motion is pure rolling — we are, of course, neglecting rolling friction and all other damping. For, at the instant in question, $V = 0$, and the angular velocity is related to the linear velocity of the centre of gravity as follows. Let the centre of the ball be at the origin and let its velocity be directed along the positive axis of x. Then

$$(11) \qquad \begin{cases} \left.\dfrac{d\bar{x}}{dt}\right|_{t=0} = c, \qquad \left.\dfrac{d\bar{y}}{dt}\right|_{t=0} = 0; \\ a\omega_\xi \,|_{t=0} = 0, \qquad a\omega_\eta \,|_{t=0} = c, \qquad \omega_\zeta \,|_{t=0} = \gamma, \end{cases}$$

where γ can have any value, positive, negative, or 0.

Let us consider the motion which consists in pure rolling and pivoting, and see what force at P is necessary. First, we have

$$(12) \qquad \begin{cases} M\dfrac{d^2\bar{x}}{dt^2} = X \\ M\dfrac{d^2\bar{y}}{dt^2} = Y, \end{cases}$$

where X, Y are the components of the unknown reaction at P. Next, taking moments about the centre of gravity, we find:

$$(13) \qquad \begin{cases} \dfrac{d\sigma_\xi}{dt} = I\dfrac{d\omega_\xi}{dt} = aY \\ \dfrac{d\sigma_\eta}{dt} = I\dfrac{d\omega_\eta}{dt} = -\,aX \\ \dfrac{d\sigma_\zeta}{dt} = I\dfrac{d\omega_\zeta}{dt} = 0 \end{cases}$$

Finally,

$$(14)\qquad \begin{cases} V_x = \dfrac{d\bar{x}}{dt} - a\omega_\eta = 0 \\ V_y = \dfrac{d\bar{y}}{dt} + a\omega_\xi = 0. \end{cases}$$

These seven equations, (12), (13), (14), together with the initial conditions (11), formulate the problem completely, and determine the seven unknown functions, $\bar{x}$, $\bar{y}$, ω_ξ, ω_η, ω_ζ, X, Y, as we will now show.

From the second equation (13) it appears that

$$Ma\frac{d\omega_\eta}{dt} = -\tfrac{5}{2}X.$$

Subtracting this equation from the first equation (12), we find:

$$M\left(\frac{d^2\bar{x}}{dt^2} - a\frac{d\omega_\eta}{dt}\right) = \tfrac{7}{2}X.$$

But the left-hand side of this equation vanishes because the first equation (14) is an identity in t. Hence $X = 0$. Similar considerations show that $Y = 0$.

On substituting these values in (12) and (13), these five equations can be solved subject to the five initial conditions (11), and the other condition, that initially $\bar{x} = 0$, $\bar{y} = 0$. The centre of the ball describes the positive axis of x with constant velocity, c. The angular velocity $\tilde{\omega}$ is also constant, its components along the axes being given by their initial values (11). Since γ is arbitrary, $\tilde{\omega}$ may be any vector whatever in the (η, ζ)-plane, whose component along the η-axis is c/a.

The foregoing discussion may be abbreviated by means of the Principle of Work and Energy, Chapter VII.

The motion of pure rolling with pivoting requires, then, no force to be exerted by the table. It is uniquely determined by the initial conditions, and hence it coincides with the actual motion of the billiard ball.

24. Cart Wheels. Consider the forewheels of a cart. Idealized they form two equal discs connected by an axle about which each can turn freely. To determine the motion on a rough inclined plane.

We will begin with a still simpler case — that of a single wheel, or disc, mounted so that it can turn and roll freely, but will always have its plane perpendicular to the plane on which it rolls. The frame which guides it may be thought of as smooth. Its mass can be taken into account, but we will disregard it, in order not to obscure the main points of the problem.

We will choose the coordinates as indicated, the axis of ξ being in the disc and always parallel to the plane; the axis of η, being the axis of the disc, is also parallel to the plane. The axis of y lies in the plane and is horizontal. The axis of x is directed down the plane. Let φ be the angle through which the disc has turned about the axis of η; let s be the arc described by the point of contact, P; and let θ be the angle from the positive axis of x to the positive tangent at P. Let the reaction of the plane be

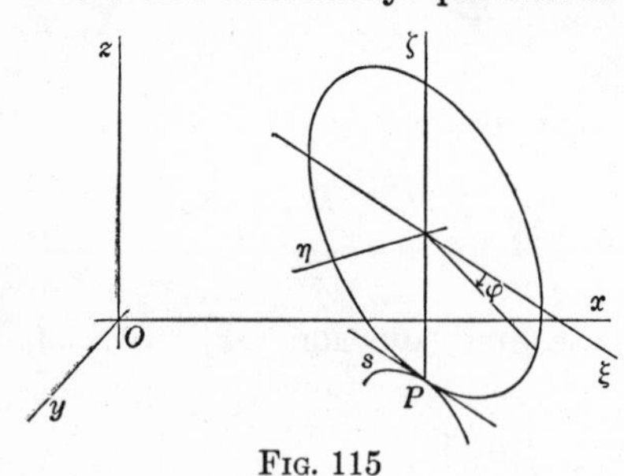

FIG. 115

$$\mathbf{F} = \mathsf{X}\alpha + \mathsf{Y}\beta + \mathsf{Z}\gamma,$$

where α, β, γ are unit vectors along the moving axes. Then $\mathsf{Z} = Mg\cos\epsilon$, where ϵ is the inclination of the plane, and

$$(1)\qquad \begin{cases} M\dfrac{d^2\bar{x}}{dt^2} = \mathsf{X}\cos\theta - \mathsf{Y}\sin\theta + Mg\sin\epsilon \\[2ex] M\dfrac{d^2\bar{y}}{dt^2} = \mathsf{X}\sin\theta + \mathsf{Y}\cos\theta \end{cases}$$

The angular velocity,

$$\bar{\omega} = \omega_\xi\alpha + \omega_\eta\beta + \omega_\zeta\gamma,$$

has the value

$$\bar{\omega} = \dot{\varphi}\beta + \dot{\theta}\gamma.$$

Moreover,

$$\dot{\beta} = -\dot{\theta}\alpha, \qquad \dot{\gamma} = 0.$$

Take moments about the centre of gravity:

$$(2)\qquad \frac{d\sigma}{dt} = \sum_k \mathbf{r}_k \times \mathbf{F}_k,$$

$$\sigma = A\omega_\xi \alpha + B\omega_\eta \beta + C\omega_\zeta \gamma.$$

Since

$$\omega_\xi = 0, \qquad \omega_\eta = \dot{\varphi}, \qquad \omega_\zeta = \dot{\theta},$$

and $A = C = \frac{1}{4}Ma^2$, $B = \frac{1}{2}Ma^2$, we have:

$$\sigma = B\dot{\varphi}\beta + A\dot{\theta}\gamma.$$

Hence

$$\frac{d\sigma}{dt} = B\ddot{\varphi}\beta + A\ddot{\theta}\gamma + B\dot{\varphi}\dot{\beta}$$

or

$$\frac{d\sigma}{dt} = - B\dot{\theta}\dot{\varphi}\alpha + B\ddot{\varphi}\beta + A\ddot{\theta}\gamma. \tag{3}$$

In computing the right-hand side of Equation (2), the couple which keeps the axis of the disc parallel to the plane must be taken into account. The vector which represents it is collinear with the axis of ξ. Hence the couple may be realized by the two forces:

$$\begin{cases} \mathbf{F}_1 = F_1\gamma \quad \text{at} \quad \mathbf{r}_1 = \beta; \\ \mathbf{F}_2 = - F_1\gamma \quad \text{at} \quad \mathbf{r}_2 = - \beta. \end{cases}$$

Thus

$$\sum_k \mathbf{r}_k \times \mathbf{F}_k = \beta \times F_1\gamma + (-\beta) \times (-F_1\gamma) + (-a\gamma) \times \mathbf{F};$$

or, finally:

$$\sum_k \mathbf{r}_k \times \mathbf{F}_k = (2F_1 + a\mathsf{Y})\,\alpha - a\mathsf{X}\beta. \tag{4}$$

Equating, then, the vectors (3) and (4) we find:

$$\begin{cases} -\frac{1}{2}Ma^2 \dfrac{d\theta}{dt}\dfrac{d\varphi}{dt} = 2F_1 + a\mathsf{Y} \\ \frac{1}{2}Ma^2 \dfrac{d^2\varphi}{dt^2} = - a\mathsf{X} \\ \frac{1}{4}Ma^2 \dfrac{d^2\theta}{dt^2} = 0. \end{cases} \tag{5}$$

Finally, the condition of rolling without slipping can be written in the form:

$$\frac{d\bar{x}}{dt} = \bar{v}\cos\theta, \qquad \frac{d\bar{y}}{dt} = \bar{v}\sin\theta,$$

where

$$\bar{v} = a\frac{d\varphi}{dt},$$

and so

$$(6) \qquad \frac{d\bar{x}}{dt} = a\frac{d\varphi}{dt}\cos\theta, \qquad \frac{d\bar{y}}{dt} = a\frac{d\varphi}{dt}\sin\theta.$$

The formulation is now complete. There are seven unknown functions, namely: $\bar{x}$, $\bar{y}$, θ, φ, X, Y, F_1, and seven equations to determine them, namely, Equations (1), (5), (6).

To solve these equations, begin by determining θ from (5):

$$(7) \qquad \frac{d\theta}{dt} = \lambda, \qquad \theta = \lambda t + \mu.$$

Next, eliminate Y in (1):

$$M\left(\frac{d^2\bar{x}}{dt^2}\cos\theta + \frac{d^2\bar{y}}{dt^2}\sin\theta\right) = \mathsf{X} + Mg\sin\epsilon\cos\theta.$$

And now X can be eliminated by (5), and $\bar{x}$, $\bar{y}$ by (6). Thus

$$Ma\frac{d^2\varphi}{dt^2} = -\tfrac{1}{2}Ma\frac{d^2\varphi}{dt^2} + Mg\sin\epsilon\cos\theta,$$

or

$$(8) \qquad \frac{d^2\varphi}{dt^2} = k\cos\theta,$$

where

$$k = \frac{2g\sin\epsilon}{3a}, \qquad \theta = \lambda t + \mu.$$

Hence

$$(9) \qquad \frac{d\varphi}{dt} = \frac{k}{\lambda}(\sin\theta - \sin\mu) + \dot{\varphi}_0,$$

and

$$\varphi = \frac{k}{\lambda^2}(\cos\mu - \cos\theta) + \left(\dot{\varphi}_0 - \frac{k\sin\mu}{\lambda}\right)t + \varphi_0.$$

From (6), $\bar{x}$ and $\bar{y}$ can now be found as functions of t; and finally X, Y, F_1, can be determined from (1) and (5).

The system of Equations (1), (5), (6) is an example of equations called *non-holonomic* by Hertz because some of them, — namely (6), — involve time-derivatives of the first order only and cannot be replaced by geometric equations between the coordinates.

An interesting case of a non-holonomic problem is that of a coin rolling on a rough table. It is studied in detail by Appell,

Mécanique rationelle, Vol. I, p. 242, of the 1904 edition, and an explicit solution is obtained in terms of the hypergeometric function.

EXERCISES

The student should first, without reference to the book, reproduce the treatment just given in the text, arranging in his mind the procedure: *i*) figure, forces, coordinates; *ii*) motion of the centre of gravity; *iii*) moments about the centre of gravity; *iv*) conditions of constraint; *v*) the solution of the equations.

1. Solve the problem of the two wheels mentioned in the text.

2. Coin rolling on a rough table. Read casually Appell, adopting his system of coordinates. Then construct independently the solution, following the method used in the problem of the text.

3. The problem of the text, when the mass of the frame is taken into account. Begin with the case that the bottom of the frame is smooth and its centre of mass is at the centre of the disc.

4. Study the motion of the centre of gravity of the disc treated in the text, by means of the explicit solution of $\bar{x}$, $\bar{y}$ in terms of t.

25. Résumé. In dealing with the motion of a rigid body, there are the two vector equations:

$$\frac{d\rho}{dt} = \sum_k \mathbf{F}_k, \qquad \frac{d\sigma}{dt} = \sum_k \mathbf{r}_k \times \mathbf{F}_k,$$

equivalent to six ordinary equations.

It is always possible to take moments about the centre of gravity.

The Principle of Work and Energy frequently gives a useful integral of the equations of motion.

If the right-hand side of the Moment Equation is a vector lying in a fixed plane, the component of σ normal to this plane is constant, and thus an integral of the equations of motion is obtained.

Sometimes there are conditions which are expressed by equations between time-derivatives of the first order, $t = t$, but which cannot be expressed by equations between the coordinates only.

The first step in solving a problem is to draw the figure, mark the forces, and pass in review each of the items just mentioned; reflecting, in case these are not adequate, on considerations of like nature, which may be germane to the problem.

With the forces and the geometry of the problem in mind, next choose a suitable

Coordinate System. If it is desirable to refer σ to the centre of gravity, a Cartesian system with its origin there is usually the solution. These axes may be fixed in the body, coinciding with the principal axes of inertia. Or their directions may be fixed in space. Or they may move in the body and in space subject to some condition peculiar to the problem in hand.

Final Formulation. It remains to write down the equations arising from each of the above considerations. They must be in number equal to the number of unknown functions. Besides the differential equations of the second order, these may also include differential equations of the first order, not reducible to equations between the coordinates.

The solution of these equations is a purely mathematical problem. Go back frequently over familiar problems and recall the mathematical technique, writing the equations down on paper, neatly, and carrying through all details of the solution. In this way, analytical consciousness is developed; it is composed of experience and common sense.

Further Study. There is a vast fund of interesting problems in Rigid Dynamics, of all orders of difficulty, and two invaluable treatises are Appell, *Mécanique rationelle,* Vols. I and II, and Routh, *Rigid Dynamics,* Vols. I, II. Routh's exposition of the theory is execrable, but his lists of problems, garnered from the old Cambridge Tripos Papers, are capital.

The earth is a top, and the study of the precession and nutation of the polar axis is a good subject for the student to take up next.

Webster's *Dynamics* is also useful in the important applications it contains. The text is hard reading; but the student who once dominates the method as set forth, for example, in the foregoing treatment, can and should construct his own solution of the problem in hand.

Finally, Klein-Sommerfeld, *Theorie des Kreisels,* in four volumes. This is a classic treatment of the subject. The first three volumes treat the theory of the top by modern mathematical methods. The fourth volume, devoted to the applications in engineering, can be studied directly through the theory which we have developed above, without reference to the earlier volumes. There is a detailed study of the gyroscopic effect in the case of railroad wheels, the Whitehead torpedo, the ship's stabilizer, the stability of the bicycle, the gyro-compass, the turbine of Leval, and a large number of further topics.

CHAPTER VII

WORK AND ENERGY

1. Work. In Elementary Physics *work* is defined as the product, *force by distance*:

$$W = Fl, \tag{1}$$

the understanding being that a force **F**, constant in magnitude and direction, acts on a particle, P, or at a point P fixed in a rigid or elastic body, and displaces P a distance l in the direction of the force.

The definition shall now be extended to the case of a variable force, still acting on a particle or at a fixed point of a material body. Let

$$a \leqq x \leqq b$$

be the interval of displacement. Let

$$F = f(x)$$

be the force, where $f(x)$ is a continuous function. Divide the interval into n parts by the points $x_0 = a,\ x_1, \cdots, x_{n-1},\ x_n = b,$ and consider the k-th sub-interval:

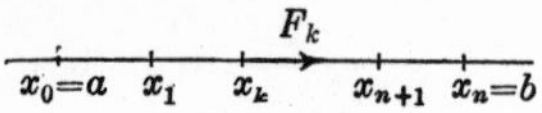

F_k

$x_0=a \quad x_1 \quad x_k \quad x_{n+1} \quad x_n=b$

Fig. 116

$$x_{k-1} \leqq x \leqq x_k, \qquad \Delta x_k = x_k - x_{k-1}.$$

And now we demand that the extended definition of work shall be so laid down that

i) the total work shall be equal to the sum of the partial works:

$$W = \sum_{k=1}^{n} \Delta W_k;$$

ii) the work for any interval shall lie between the work corresponding to the maximum value of the force in that interval, and the work corresponding to the minimum force:

$$F'_k \Delta x_k \leqq \Delta W_k \leqq F''_k \Delta x_k,$$

where

$$F'_k \leqq F \leqq F''_k$$

in the interval in question.

Now, since $f(x)$ is a continuous function, it takes on its minimum value, F'_k, in the interval:

$$F'_k = f(x'_k), \qquad x_{k-1} \leqq x'_k \leqq x_k;$$

and similarly, its maximum value:

$$F''_k = f(x''_k), \qquad x_{k-1} \leqq x''_k \leqq x_k.$$

Hence W lies between the two sums:

$$f(x'_1)\,\Delta x_1 + \cdots + f(x'_n)\,\Delta x_n,$$
$$f(x''_1)\,\Delta x_1 + \cdots + f(x''_n)\,\Delta x_n.$$

But each of these sums approaches a limit as n increases, the longest Δx_k approaching 0, and this limit is the definite integral:

$$\lim_{n=\infty} \sum_{k=1}^{n} f(x'_k)\,\Delta x_k = \int_a^b f(x)\,dx.$$

Hence the requirements, *i.e. physical postulates i*) and *ii*) are sufficient to determine the definition of the work in this case: *

$$(2) \qquad W = \int_a^b f(x)\,dx.$$

The foregoing definition applies to a *negative* force, and also to the case that $b < a$; the *work* now being considered as an algebraic quantity. Thus if a force, instead of overcoming resistance, is itself overcome; *i.e.* yields, it does negative work.

The work which corresponds to a variable displacement, x, where $a \leqq x \leqq b$, is by definition:

$$(3) \qquad W = \int_a^x f(x)\,dx.$$

Hence

$$(4) \qquad \frac{dW}{dx} = F.$$

* Strictly speaking, we have shown that (2) is a *necessary* condition for the definition of *work* according to the postulates *i*) and *ii*). It is seen at once, however, that conversely Equation (2) affords a *sufficient* condition, also.

EXERCISES

1. Show that the work done in stretching an elastic string is proportional to the square of the stretching.

2. Find the work done by the sun on a meteor which falls directly into it.

3. The work corresponding to a variable displacement from x to b, where $a \leqq x \leqq b$, is by definition:

$$W = \int_x^b f(x)\, dx. \tag{5}$$

What is the value of dW/dx?

2. Continuation: Curved Paths. Suppose the particle describes a curved path C in a plane, and that the force, $\mathbf{F}$, varies in magnitude and direction in any continuous manner. What will be the work done in this case?

Suppose the path C is a right line and the force, though oblique to the line, is constant in magnitude and direction; Fig. 117.

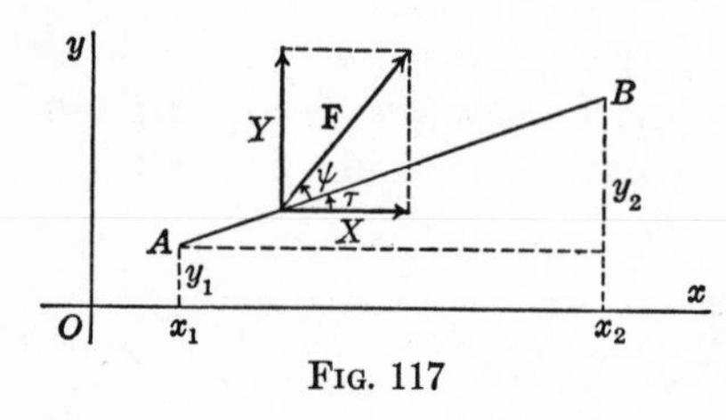

Fig. 117

Resolve the force into its two components along the line and normal to it. Surely, we must lay down our definition of work so that the work done by $\mathbf{F}$ is equal to the sum of the works of the component forces. Now, the work done by the component along the line has already been defined, namely, $Fl \cos \psi$, where $F = |\mathbf{F}|$ is the intensity of the force.

It is an essential part of the idea of *work* that the force overcomes resistance through distance (or is overcome through distance). Now, the normal component does neither; it merely sidles off and sidesteps the whole question. It is natural, therefore, to define it as doing no work. Thus we arrive at our final definition: The work done by $\mathbf{F}$ in the particular case in hand shall be

$$W = Fl \cos \psi. \tag{6}$$

A second form of the expression on the right is as follows. Let X and Y be the components of $\mathbf{F}$ along the axes. Let τ be the angle that the path AB makes with the positive axis of x. Then the projection of **F** on AB is equal to the sum of the projections of X and Y on AB, or

$$F \cos \psi = X \cos \tau + Y \sin \tau.$$

On the other hand,

$$x_2 - x_1 = l \cos \tau, \qquad y_2 - y_1 = l \sin \tau.$$

Hence

$$W = X(x_2 - x_1) + Y(y_2 - y_1). \tag{7}$$

The General Case. If C be any regular curve, divide it into n arcs by the points $s_0 = 0,\ s_1,\ \cdots,\ s_{n-1},\ s_n = l$. Let $\mathbf{F}'_k$ be the value of **F** at an arbitrary point of the k-th arc, and let ψ'_k be the angle from the chord (s_{k-1}, s_k) to the vector $\mathbf{F}'_k$. Then the sum

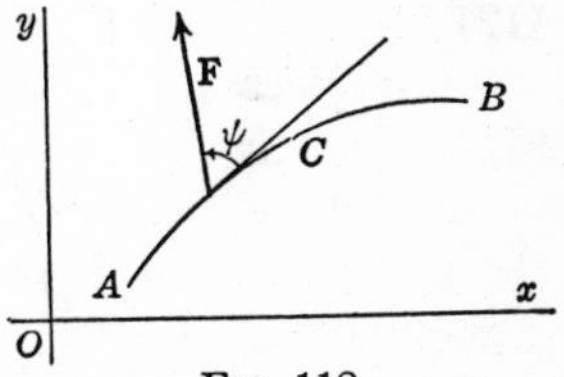

$$\sum_{k=1}^{n} F'_k \cos \psi'_k \, l_k,$$

Fig. 118

where l_k denotes the length of the chord, gives us approximately what we should wish to understand by the work, in view of our physical feeling for this quantity. The limit of this sum, when the longest l_k approaches 0, shall be defined as the work, or

$$W = \lim_{n=\infty} \sum_{k=1}^{n} F'_k \cos \psi'_k \, l_k. \tag{8}$$

Since

$$\lim \frac{l_k}{\Delta s_k} = 1,$$

it is clear that the above limit is the same as*

$$\lim_{n=\infty} \sum_{k=1}^{n} F_k \cos \psi_k \, \Delta s_k = \int_0^l F \cos \psi \, ds.$$

We are thus led to the following definition of *work* in the case of a curved path:

$$W = \int_0^l F \cos \psi \, ds. \tag{9}$$

* Cf. the author's *Advanced Calculus*, p. 217. It is imperative that the student learn thoroughly what is meant by a *line integral*.

A second formula for the work is obtained by means of (7):

$$(10)\quad W = \int_0^l (X\cos\tau + Y\sin\tau)\,ds = \int_0^l \left(X\frac{dx}{ds} + Y\frac{dy}{ds}\right) ds$$

or

$$(11)\qquad W = \int_C X\,dx + Y\,dy.$$

The extension to three dimensions is immediate. The definition (9) applies at once without even a formal change. Formula (11) is replaced by the following:

$$(12)\qquad W = \int_C X\,dx + Y\,dy + Z\,dz$$

or

$$\int_{(a,\,b,\,c)}^{(a',\,b',\,c')} X\,dx + Y\,dy + Z\,dz.$$

Example. To find the work done by gravity on a particle of mass m which moves from an initial point (x_0, y_0, z_0) to a final point (x_1, y_1, z_1) along an arbitrary twisted curve, C.

Let the axis of z be vertical and positive downwards. Then

$$X = 0, \qquad Y = 0, \qquad Z = mg;$$

$$W = \int_C X\,dx + Y\,dy + Z\,dz = \int_{z_0}^{z_1} mg\,dz = mg\,(z_1 - z_0).$$

Hence the work done is equal to the product of the force by the difference in level (taken algebraically), and depends only on the initial and final points, but not on the path joining them.

EXERCISES

1. A well is pumped out by a force pump which delivers the water at the mouth of a pipe which is fixed. Show that the work done is equal to the weight of the water initially in the well, multiplied by the vertical distance of the centre of gravity below the mouth of the pipe.

2. The components of the force which acts on a particle are:
$X = 2x - 3y + 4z - 5,\;\; Y = z - x + 8,\;\; Z = x + y + z + 12.$

Find the work done when the particle describes the arc of the helix

$$x = \cos\theta, \qquad y = \sin\theta, \qquad z = 7\theta,$$

for which $0 \leqq \theta \leqq 2\pi$.

3. If the curve C is represented parametrically:

$$C: \quad x = f(\lambda), \qquad y = \varphi(\lambda), \qquad z = \psi(\lambda), \qquad \lambda_0 \leqq \lambda \leqq \lambda_1,$$

show that the work is given by the integral:

$$(13) \qquad W = \int_{\lambda_0}^{\lambda_1} \left(X \frac{dx}{d\lambda} + Y \frac{dy}{d\lambda} + Z \frac{dz}{d\lambda} \right) d\lambda.$$

3. Field of Force. Force Function. Potential. A particle in the neighborhood of the solar system is attracted by all the other particles of the system with a force $\mathbf{F}$ that varies in magnitude and direction from point to point. Thus $\mathbf{F}$ is a *vector point-function* throughout the region of space just mentioned. Its components along Cartesian axes, namely, X, Y, Z, are ordinary functions of the space coordinates, x, y, z, of the particle. In vector form:

$$(1) \qquad \mathbf{F} = X\mathbf{i} + Y\mathbf{j} + Z\mathbf{k}.$$

The example serves to illustrate the general idea of a *field of force.* We may have an electro-magnetic field, as when a straight wire carries a current. If the north pole, P, of a magnet is brought into the neighborhood of the wire, it will be acted on by a force $\mathbf{F}$ at right angles to any line drawn from P to the wire and of intensity inversely proportional to the distance of P from the wire, the sense of the force depending on the sense of the current.

If the axis of z be taken along the wire, then

$$(2) \qquad X = C\frac{-\sin\theta}{r}, \qquad Y = C\frac{\cos\theta}{r}, \qquad Z = 0,$$

where (r, θ, z) are the cylindrical coordinates of P, and C is a positive or negative constant. Thus in vector form

$$(3) \qquad \mathbf{F} = C\frac{-\sin\theta}{r}\mathbf{i} + C\frac{\cos\theta}{r}\mathbf{j}$$

and

$$F = \frac{|C|}{r}.$$

Force Function. It may happen that there is a function

(4) $$u = \varphi(x, y, z)$$

such that

(5) $$X = \frac{\partial u}{\partial x}, \qquad Y = \frac{\partial u}{\partial y}, \qquad Z = \frac{\partial u}{\partial z}.$$

Such a function, u, is called a *force function.* In vector form:

(6) $$\mathbf{F} = \frac{\partial u}{\partial x}\mathbf{i} + \frac{\partial u}{\partial y}\mathbf{j} + \frac{\partial u}{\partial z}\mathbf{k}.$$

$\mathbf{F}$ can be written in symbolic vector form as follows. Let ∇ be a *symbolic vector operator*, namely:

(7) $$\nabla = \mathbf{i}\frac{\partial}{\partial x} + \mathbf{j}\frac{\partial}{\partial y} + \mathbf{k}\frac{\partial}{\partial z}.$$

Then ∇u is *defined* as:

(8) $$\nabla u = \mathbf{i}\frac{\partial u}{\partial x} + \mathbf{j}\frac{\partial u}{\partial y} + \mathbf{k}\frac{\partial u}{\partial z}.$$

Hence

(9) $$\mathbf{F} = \nabla u.$$

Gravitational Field. In the case of the field generated by a single particle of attracting matter, there is a force function:

(10) $$u = \frac{\lambda}{r},$$

where r is the distance from the given fixed particle to the variable particle, and λ is a positive constant.

In the case of n particles,

(11) $$u = \sum_{k=1}^{n} \frac{m_k}{r_k},$$

provided the units are properly chosen.

Electro-Magnetic Field. For the electro-magnetic field above described,

(12) $$u = C\theta.$$

We may also write:

(13) $$u = C\tan^{-1}\frac{y}{x};$$

but this formula is treacherous, since only certain values of the multiple-valued function are admissible. However, since the wrong values differ from the right ones only by additive constants, we can use the formula for purposes of differentiation, and we shall have:

$$(14)\quad X = \frac{\partial u}{\partial x} = C\frac{-y}{x^2+y^2}, \qquad Y = \frac{\partial u}{\partial y} = C\frac{x}{x^2+y^2}, \qquad Z = 0.$$

Work. When a particle describes an arbitrary path in a field of force, the work done on the particle by the field is given by Equation (12) of § 2. If there is a force function, this formula becomes:

$$(15)\qquad W = \int\limits_C \frac{\partial u}{\partial x}dx + \frac{\partial u}{\partial y}dy + \frac{\partial u}{\partial z}dz = u\Big|_C,$$

i.e. the change which u experiences along the curve C. If the region in which C lies is simply connected, or if u is a single-valued function, then

$$(16)\qquad W = u + \text{const.}$$

Thus W is independent of the path by which the particle arrived at its final destination, and depends only on the starting point and the terminal point:

$$(17)\qquad W = u(x, y, z) - u(a, b, c).$$

For any closed path, $W = 0$.

Such a field of force is called *conservative.* It is true conversely that if the field represented by the vector (1) is conservative, then there always is a force function, u. For then the integral:

$$(18)\qquad u = \int\limits_{(a,b,c)}^{(x,y,z)} X\,dx + Y\,dy + Z\,dz$$

is independent of the path and so defines a function $u(x, y, z)$. Moreover,

$$(19)\qquad \frac{\partial u}{\partial x} = X, \qquad \frac{\partial u}{\partial y} = Y, \qquad \frac{\partial u}{\partial z} = Z.$$

Potential Energy. When a field of force has a force function, u, the negative of u, plus a constant, is defined as the *potential energy*:

$$(20)\qquad \varphi = -u + C.$$

In case, then, a potential φ exists,

$$(21)\qquad X = -\frac{\partial \varphi}{\partial x}, \qquad Y = -\frac{\partial \varphi}{\partial y}, \qquad Z = -\frac{\partial \varphi}{\partial z}.$$

EXERCISES

1. Show that the field of force defined by the vector (3) is not conservative. But if R be any region of space such that an arbitrary closed curve in R can be drawn together continuously to a point not on the axis, without ever meeting the axis, though passing out of R, then the field of force defined in R by (3) is conservative.

2. A meteor, which may be regarded as a particle, is attracted by the sun (considered at rest) and by all the rest of the matter in the solar system. It moves from a point A to a point B. Show that the work done on it by the sun is

$$W = Km\left(\frac{1}{r_1} - \frac{1}{r_0}\right),$$

where r_0 and r_1 represent the distances of A and B, respectively, from the sun, and K is the gravitational constant.

4. Conservation of Energy. Let a particle be acted on by any force whatever. The motion is determined by Newton's Second Law:

$$(1) \qquad m\frac{d^2x}{dt^2} = X, \qquad m\frac{d^2y}{dt^2} = Y, \qquad m\frac{d^2z}{dt^2} = Z.$$

Multiply these equations respectively by dx/dt, dy/dt, dz/dt, and add:

$$(2) \quad m\left(\frac{dx}{dt}\frac{d^2x}{dt^2} + \frac{dy}{dt}\frac{d^2y}{dt^2} + \frac{dz}{dt}\frac{d^2z}{dt^2}\right) = X\frac{dx}{dt} + Y\frac{dy}{dt} + Z\frac{dz}{dt}.$$

The left-hand side of this equation has the value:

$$\frac{m}{2}\frac{d}{dt}v^2,$$

where

$$v^2 = \frac{dx^2}{dt^2} + \frac{dy^2}{dt^2} + \frac{dz^2}{dt^2}.$$

Hence

$$\frac{m}{2}\frac{d}{dt}v^2 = X\frac{dx}{dt} + Y\frac{dy}{dt} + Z\frac{dz}{dt}.$$

Each side of this equation is a function of t, and the two functions are, of course, identical in value. If, then, we integrate each side between any two limits, t_0 and t_1, the results must tally:

$$\int_{t_0}^{t_1} \frac{m}{2}\frac{d}{dt}v^2 dt = \int_{t_0}^{t_1}\left(X\frac{dx}{dt} + Y\frac{dy}{dt} + Z\frac{dz}{dt}\right) dt.$$

The left-hand side of this equation has the value:

$$\frac{mv_1^2}{2} - \frac{mv_0^2}{2}.$$

The right-hand side is nothing more or less than

$$\int_C X\,dx + Y\,dy + Z\,dz,$$

taken over the path of the particle; § 2, (13). But this is precisely the work done on the particle by the force that acts. Hence

$$(3) \qquad \frac{mv_1^2}{2} - \frac{mv_0^2}{2} = W.$$

The quantity

$$\frac{mv^2}{2}$$

is defined as the *kinetic energy* of the particle. We have, then, in Equation (3) the following theorem.

THEOREM. *The change in the kinetic energy of a particle is equal to the work done on the particle.*

If instead of a single particle we have a system of particles, the same result is true. For, from the equations of motion of the individual particles:

$$(4) \qquad m_k\frac{d^2x_k}{dt^2} = X_k, \qquad m_k\frac{d^2y_k}{dt^2} = Y_k, \qquad m_k\frac{d^2z_k}{dt^2} = Z_k,$$

we infer that

$$\frac{d}{dt}\sum_{k=1}^{n}\frac{m_k}{2}\left(\frac{dx_k^2}{dt^2} + \frac{dy_k^2}{dt^2} + \frac{dz_k^2}{dt^2}\right) = \sum_{k=1}^{n}\left(X_k\frac{dx_k}{dt} + Y_k\frac{dy_k}{dt} + Z_k\frac{dz_k}{dt}\right).$$

The kinetic energy of the system is defined as

$$T = \sum_{k=1}^{n}\frac{m_k v_k^2}{2}.$$

On integrating, then, between any limits t_0 and t_1, we have

$$(5) \qquad T_1 - T_0 = \sum_{k=1}^{n} \int_{t_0}^{t_1} \left(X_k \frac{dx_k}{dt} + Y_k \frac{dy_k}{dt} + Z_k \frac{dz_k}{dt} \right) dt.$$

The right-hand side represents the sum of the works done on the individual particles, or the total work done on the system. The result is the Law of Work and Energy in its most general form for a system of particles.

THEOREM. *The change in the kinetic energy of any system of particles is equal to the total work done on the system.*

Conservative Systems. In case the forces are conservative; *i.e.* if there exists a force function U such that

$$(6) \qquad X_k = \frac{\partial U}{\partial x_k}, \qquad Y_k = \frac{\partial U}{\partial y_k}, \qquad Z_k = \frac{\partial U}{\partial z_k},$$

the right-hand side of Equation (5) becomes $U_1 - U_0$, and so

$$(7) \qquad T_1 - T_0 = U_1 - U_0.$$

The potential energy, Φ, is defined as:

$$(8) \qquad \Phi = - U + \text{const.}$$

Hence (7) can be written:

$$(9) \qquad T_1 + \Phi_1 = T_0 + \Phi_0.$$

Let the *total energy* be defined as

$$(10) \qquad E = T + \Phi.$$

We have, then:

$$(11) \qquad E_1 = E_0,$$

or the *total energy remains constant.* This is the Law of the Conservation of Energy in its most general form for a system of particles.

5. Vanishing of the Internal Work for a Rigid System. Consider a set of particles which form a rigid system. Let them be held together by massless rods connecting them in pairs. Thus the internal forces with which any two particles, m_i and m_j, react on each other are equal and opposite:

$$(1) \qquad \mathbf{F}_{ij} + \mathbf{F}_{ji} = 0,$$

and lie along the line joining the particles, and furthermore the distance between the particles is constant; *i.e.*

$$(2) \qquad r_{ij}^2 = (x_i - x_j)^2 + (y_i - y_j)^2 + (z_i - z_j)^2$$

is independent of the time, or

$$(3) \qquad (x_i - x_j)\left(\frac{dx_i}{dt} - \frac{dx_j}{dt}\right) + (y_i - y_j)\left(\frac{dy_i}{dt} - \frac{dy_j}{dt}\right) + (z_i - z_j)\left(\frac{dz_i}{dt} - \frac{dz_j}{dt}\right) = 0.$$

In general, however, if each particle is connected by these rods with all the others, there will be redundant members, so that the stresses in the individual rods will be indeterminate. In that case, let the superfluous rods be suppressed.

Consider the work done on the particle m_i by the rod connecting it with m_j. It is:

$$\int_{C_i} X_{ij}\, dx_i + Y_{ij}\, dy_i + Z_{ij}\, dz_i$$

and can be expressed by means of the parameter t in the form:

$$\int_{t_0}^{t_1} \left(X_{ij}\frac{dx_i}{dt} + Y_{ij}\frac{dy_i}{dt} + Z_{ij}\frac{dz_i}{dt}\right) dt.$$

By the same token, the work done on m_j by m_i is

$$\int_{t_0}^{t_1} \left(X_{ji}\frac{dx_j}{dt} + Y_{ji}\frac{dy_j}{dt} + Z_{ji}\frac{dz_j}{dt}\right) dt.$$

Since

$$X_{ij} + X_{ji} = 0, \qquad Y_{ij} + Y_{ji} = 0, \qquad Z_{ij} + Z_{ji} = 0,$$

the sum of these two works can be written in the form:

$$\int_{t_0}^{t_1} \left\{ X_{ij}\left(\frac{dx_i}{dt} - \frac{dx_j}{dt}\right) + Y_{ij}\left(\frac{dy_i}{dt} - \frac{dy_j}{dt}\right) + Z_{ij}\left(\frac{dz_i}{dt} - \frac{dz_j}{dt}\right)\right\} dt.$$

This last integral vanishes. For, the force $\mathbf{F}_{ij}$ is collinear with the line segment connecting m_i with m_j, or:

$$X_{ij} = \rho(x_i - x_j), \qquad Y_{ij} = \rho(y_i - y_j), \qquad Z_{ij} = \rho(z_i - z_j).$$

Hence the integrand vanishes identically by (3).

We have thus obtained the result that the work done by the internal forces of a rigid system of particles is nil. It follows, then, that the change in the kinetic energy of such a system is equal to the work done by the *external*, or *applied*, forces. Looking backward and also forward we can now state the general

THEOREM. *The change in the kinetic energy of any rigid system whatever is equal to the work done by the applied forces.*

For a system of particles the proof has been given. Before we can extend it to rigid bodies, we must generalize the definitions of kinetic energy and work.

6. Kinetic Energy of a Rigid Body. Consider a rigid body. Let the volume density, ρ, be a continuous function. Denote by v the velocity of a variable point P of the body. Then the *kinetic energy* is defined as

$$T = \tfrac{1}{2} \iiint\limits_{\tau} \rho v^2 \, d\tau \tag{1}$$

extended throughout the region τ of space, occupied by the body.

The vector velocity $\mathbf{v}$ of P is the vector sum *i*) of the velocity $\mathbf{V}$ along the axis of rotation and *ii*) the velocity $\mathbf{v}'$ at right angles to that axis. Hence

$$v^2 = V^2 + r^2\omega^2 \tag{2}$$

where r denotes the distance of P from the axis, and ω is the angular velocity about the axis. Substituting this value in (1) we find:

$$T = \frac{V^2}{2} \iiint\limits_{\tau} \rho \, d\tau + \frac{\omega^2}{2} \iiint\limits_{\tau} \rho r^2 \, d\tau.$$

Hence

$$T = \frac{MV^2}{2} + \frac{I\omega^2}{2}. \tag{3}$$

Let $\bar{v}$ denote the velocity of the centre of gravity, G; and let h be the distance of G from the axis of rotation. Then

$$\bar{v}^2 = V^2 + h^2\omega^2.$$

Moreover,

$$I = I_0 + Mh^2,$$

where I_0 is the moment of inertia about a parallel axis through G. Hence

$$T = \tfrac{1}{2} M \bar{v}^2 + \tfrac{1}{2} I_0 \omega^2. \tag{4}$$

In equations (3) and (4) is contained the following general theorem.

THEOREM. *The kinetic energy of a rigid body is the sum of the kinetic energy of translation along the instantaneous axis and the kinetic energy of rotation about the instantaneous axis.*

It can also be expressed as the sum of the kinetic energy of a particle of like mass, moving with the velocity of the centre of gravity, and the kinetic energy of rotation about an axis through the centre of gravity, parallel to the instantaneous axis.

One Point Fixed. Let a point O of the body be at rest. Let the (ξ, η, ζ)-axes lie along the principal axes of inertia, O being the origin. Then the components of the vector velocity $\mathbf{v}$ of any point fixed in the body are:

$$\begin{cases} v_\xi = \eta\omega_\zeta - \zeta\omega_\eta \\ v_\eta = \zeta\omega_\xi - \xi\omega_\zeta \\ v_\zeta = \xi\omega_\eta - \eta\omega_\xi. \end{cases}$$

Hence

$$T = \tfrac{1}{2}(A\omega_\xi^2 + B\omega_\eta^2 + C\omega_\zeta^2). \tag{5}$$

If the axes of coordinates are not the principal axes of inertia, then

$$T = \tfrac{1}{2}(A\omega_\xi^2 + B\omega_\eta^2 + C\omega_\zeta^2 - 2D\omega_\eta\omega_\zeta - 2E\omega_\zeta\omega_\xi - 2F\omega_\xi\omega_\eta). \tag{6}$$

The General Case. From (4) and (5) we infer that

$$T = \tfrac{1}{2}M\bar{v}^2 + \tfrac{1}{2}(Ap^2 + Bq^2 + Cr^2), \tag{7}$$

where A, B, C are the moments of inertia about the principal axes of inertia through the centre of gravity, and p, q, r are the components of the vector angular velocity $\tilde{\omega}$ along these axes.

7. Final Definition of Work. We have hitherto assumed that the point of application, P, of the force is fixed in the body. Suppose P describes a curve C either in the body or in space. How shall the work now be defined?

Take the time as a parameter. Divide the interval $\tau_0 \leqq t \leqq \tau_1$ into n parts by the points $t_0 = \tau_0 < t_1 < \cdots < t_{n-1} < t_n = \tau_1$. Let Q_k be the point *fixed in the body*, which at time $t = t_k$ will reach C; let Γ_k be its path in space, and let v_k be its velocity in space when it reaches C; cf. Fig. 120. For the interval of time

$$\Delta t_k = t_k - t_{k-1}$$

we may take the force as constant, $\mathbf{F} = \mathbf{F}_k$, the value of $\mathbf{F}$ at the intersection of Γ_k with C, and let $\mathbf{F}_k$ act on the point Q_k throughout the interval Δt_k. Then $\mathbf{F}_k$ will do work equal approximately to

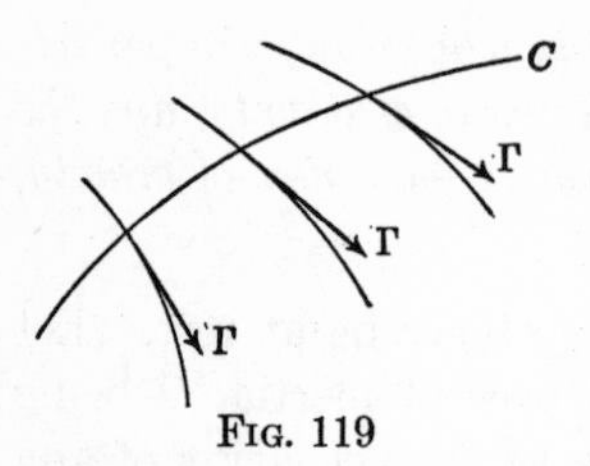

Fig. 119

(1) $$F_k v_k \cos \psi_k \Delta t_k,$$

where ψ_k is the angle from Γ_k to $\mathbf{F}_k$ at P_k. If Q_k is displaced along the tangent to Γ_k a distance $v_k \Delta t_k$, the expression (1) represents the work precisely.

We will now define the work as

$$\lim_{n=\infty} \sum_{k=1}^{n} F_k v_k \cos \psi_k \Delta t_k$$

or, dropping the τ-notation and expressing the interval of time as $t_0 \leqq t \leqq t_1$:

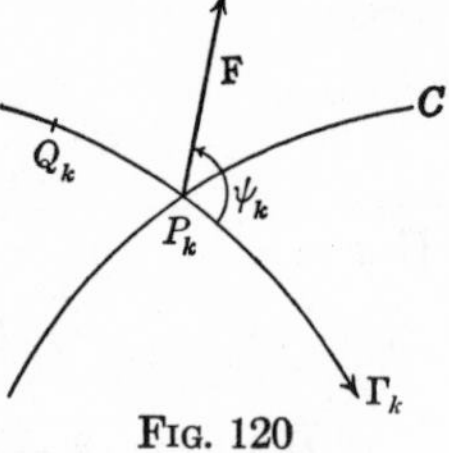

Fig. 120

(2) $$W = \int_{t_0}^{t_1} F v \cos \psi \, dt;$$

cf. Fig. 119. In vector form the work is

(3) $$W = \int_{t_0}^{t_1} \mathbf{F}\mathbf{v} \, dt,$$

where $\mathbf{v}$ is the vector velocity of Q at P, and $\mathbf{F}\mathbf{v}$ is the scalar product of these vectors.

We have used the *time* as the independent variable, or the *parameter*, in terms of which to define the displacement. But the result is in no wise dependent on the time in which the displacement takes place. Any other parameter, λ, would have done equally well, provided $d\lambda/dt$ is continuous and positive (or negative) throughout. For

$$\mathbf{v} \, dt = \frac{d\mathbf{s}}{dt} dt = \frac{d\mathbf{s}}{d\lambda} d\lambda.$$

This formulation of the definition of work in the general case is due to Professor E. C. Kemble.

Example 1. A billiard ball rolls down a rough inclined plane without slipping. Find the work done by the plane.

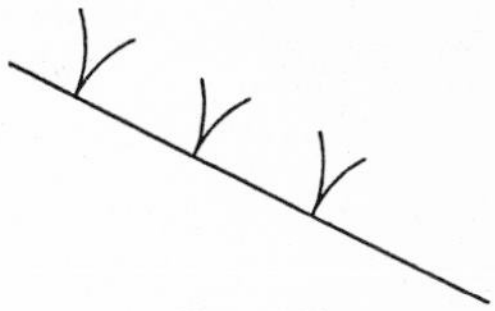

Here, C is either the straight line or the circle; each curve Γ is a cycloid with cusp at P and tangent normal to C; and $v = 0$. Hence

$$W = 0.$$

FIG. 121

Example 2. The same, except that the ball slips.

The curve C shall be taken along the plane. The normal component $R = Mg \cos \alpha$ does no work; the component along the plane,

$$F = \mu Mg \cos \alpha,$$

does. Let s be the distance travelled by the centre of the ball; θ, the angle through which the ball has turned.

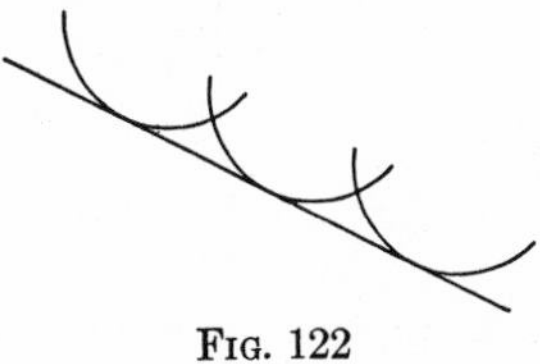

FIG. 122

The curve Γ is a trochoid tangent to C at P. Hence $\psi = 0$ or π,

$$v \cos \psi = \frac{ds}{dt} - a\frac{d\theta}{dt},$$

and

$$W = \int_{t_0}^{t_1} \mu Mg \cos \alpha \left(\frac{ds}{dt} - a\frac{d\theta}{dt}\right) dt$$

$$= \mu Mg \cos \alpha \,\{(s_1 - s_0) - a(\theta_1 - \theta_0)\}.$$

Observe that in the definition, Equation (2), v is positive or 0. It would not, therefore, be right in this example to write

$$v = \frac{ds}{dt} - a\frac{d\theta}{dt}.$$

EXERCISES

1. Check the result in Example 2 by determining the motion of the ball and computing the change in kinetic energy.

2. A train is running at the rate of 40 m. an h. The baggage car is empty, and the small son of the baggage master is disporting himself on the floor. He runs forward, then slides. If he was running at the rate of 6 m. an h. when he began to slide, and slid 5 ft., how much work did he do on the car?

Compute by the definition and check your work by solving for the motion.

3. A rope is frozen to the deck of a ship. The free end is hauled over a smooth pulley at P. It takes a vertical component of $R = 20$ lbs. to free the frozen part. How much work is done?

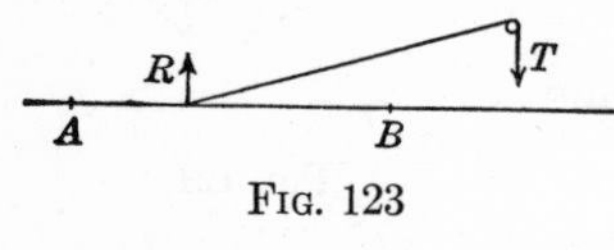

FIG. 123

Take the frozen part as straight, and P in the vertical plane through it.

4. Extend the definition of *work* to a body force, **F**, where **F** is a continuous vector, defined at each point of the body:

$$W = \iiint_{\tau} d\tau \int_{t_0}^{t_1} \mathbf{F}\mathbf{v}\, dt.$$

5. Show that the internal work due to the rope in an Atwood's machine is nil. Would this be the case if the rope stretched?

6. A number of rigid bodies are connected by inextensible cords that can wind and unwind on them in any manner without slipping. Show that the sum of the works done by the cords on the system and the system on the cords is nil. First, extend the definition of *work* so as to include the case of the work done on the system by the part of a cord which is in contact with a body.

8. Work Done by a Moving Stairway. Consider the work which an escalator, or moving stairway, does on a man as he walks up. The forces that act on the man are R, S, and Mg, where R, S are the components of the force which the escalator exerts on his foot, and Mg acts at his centre of gravity. The curve Γ is always a right line lying in the inclined plane, and

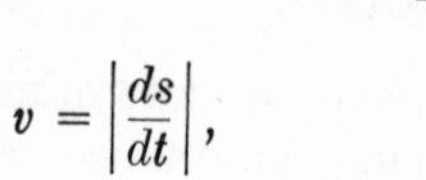

$$v = \left|\frac{ds}{dt}\right|,$$

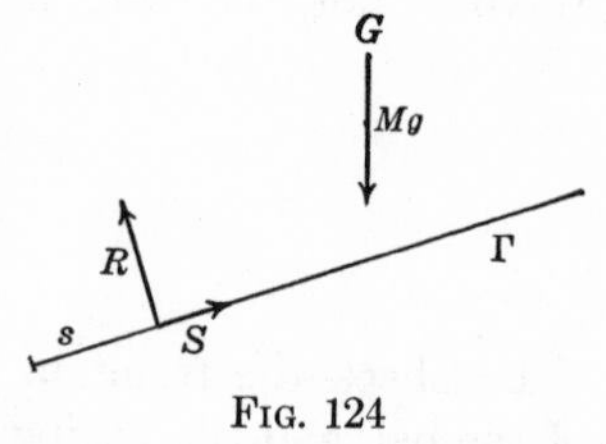

FIG. 124

where s denotes the distance the escalator has moved since the man came aboard.

The force R does no work, since for it $\psi = \pi/2$. The whole work is due to $S = F\cos\psi$, and is:

$$(1) \qquad W = \int_0^t S\frac{ds}{dt}\,dt.$$

The speed of the escalator is constant; denote it by c. Thus

$$(2) \qquad W = c\int_0^{t_1} S\,dt.$$

And

$$(3) \qquad l = ct_1,$$

where l is the distance the escalator has moved while the man is running up.

On the other hand, consider the motion of the centre of gravity of the man. Let the axis of x be taken up the plane. Then

$$M\frac{d^2\bar{x}}{dt^2} = S - Mg\sin\alpha,$$

$$(4) \qquad Mu_1 - Mu_0 = \int_0^{t_1} S\,dt - Mgt_1\sin\alpha,$$

where $u = d\bar{x}/dt$.

It follows, then, from (2) and (4) that

$$(5) \qquad W = c(Mu_1 - Mu_0) + Mgct_1\sin\alpha.$$

If the man steps off with the same velocity with which he stepped on, $u_1 = u_0$, then, with the help of (3),

$$(6) \qquad W = Mgl\sin\alpha.$$

Now

$$h = l\sin\alpha$$

is the vertical distance by which the man would have been raised in the time he was on the escalator if he had not run, but stood still. Hence, finally,

$$(7) \qquad W = Mgh.$$

It makes no difference, then, whether the man runs fast or slowly, up or down. The one thing that counts is *how long* he is on the escalator. Thus when small boys play on the escalator, running up and down, the work the escalator does increases in proportion to the time they are on it, provided they arrive and leave with the same velocity.

9. Other Cases in Which the Internal Work Vanishes.

i) Two Rigid Bodies, Rolling without Slipping. Here, the action and reaction are equal and opposite, though not in general normal to the surfaces. Moreover, the vector velocity of the point of contact, regarded as a point fixed in the one body, must be the same as the vector velocity of the point of contact, regarded as a point fixed in the other body.

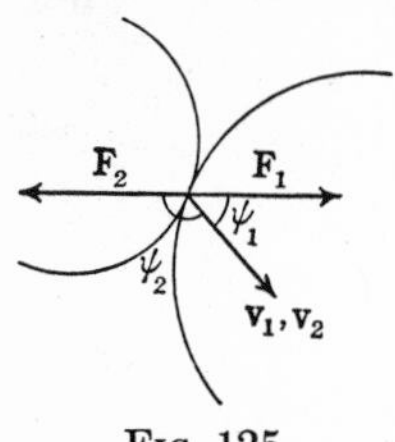

FIG. 125

The works done by the two forces $\mathbf{F}_1$, $\mathbf{F}_2$ on the two bodies are:

$$W_1 = \int_{t_0}^{t_1} F_1 v_1 \cos \psi_1 \, dt, \qquad W_2 = \int_{t_0}^{t_1} F_2 v_2 \cos \psi_2 \, dt.$$

But $F_1 = F_2$, $v_1 = v_2$, $\psi_1 + \psi_2 = \pi$. Hence

$$W_1 + W_2 = 0.$$

ii) Two Smooth Rigid Bodies, Rolling and Slipping. Here the forces $\mathbf{F}_1$ and $\mathbf{F}_2$ are equal and opposite, and normal to the surfaces at the point of contact. The velocities $\mathbf{v}_1$ and $\mathbf{v}_2$ are not equal when there is slipping; but their projections on the normal are equal:

$$v_1 \cos \psi_1 + v_2 \cos \psi_2 = 0,$$

Since furthermore $F_1 = F_2$, we have:

$$W_1 + W_2 = 0.$$

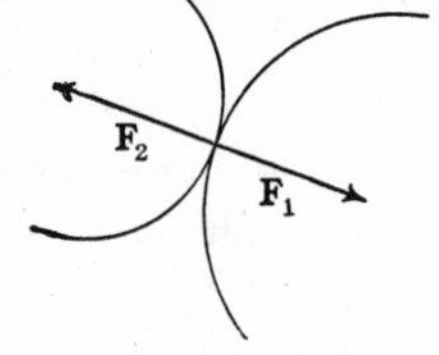

FIG. 126

We have already mentioned the case of rigid bodies on which inextensible massless strings wind and unwind, § 7, Exercise 6; and massless rods were shown in § 5 to do no work. Thus systems of rigid bodies connected by inextensible strings and rods, even though the point of application of the force exerted by the string or rod be variable, show no internal work.

10. Work and Energy for a Rigid Body. THEOREM. *The change in the kinetic energy of a rigid body, acted on by any forces, is equal to the work done by these forces.*

We prove the theorem first for two special cases.

CASE I. *No Rotation.* Here, the change in kinetic energy is

$$(1) \qquad \frac{M\bar{v}_1^2}{2} - \frac{M\bar{v}_0^2}{2},$$

i.e. the change in the kinetic energy of a particle of mass M, moving as the centre of gravity is moving.

On the other hand, consider the work done by one of the forces, $\mathbf{F}$:

$$(2) \qquad W = \int_{t_0}^{t_1} Fv \cos \psi \, dt.$$

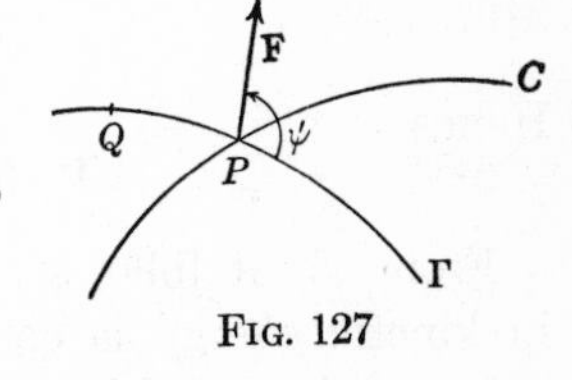

FIG. 127

Since there is no rotation, $\mathbf{v} = \bar{\mathbf{v}}$, $\psi = \bar{\psi}$, and

$$(3) \qquad W = \int_{t_0}^{t_1} F\bar{v} \cos \bar{\psi} \, dt.$$

Hence W is the work done on a particle at the centre of gravity by the same force, and the theorem is true by § 4.

CASE II. *One Point Fixed.* Here, Euler's *Dynamical Equations,* Chapter VI, § 13, determine the motion. Consider a force $\mathbf{F}$ which acts on the body at P. Let $\mathbf{r}$ be the vector drawn from O to P. Then the $\mathbf{v}$ of the definition of work, § 7, (3) is

$$\mathbf{v} = \tilde{\omega} \times \mathbf{r}.$$

Hence

$$(4) \qquad Fv \cos \psi = \mathbf{F} \cdot \mathbf{v} = \mathbf{F} \cdot (\tilde{\omega} \times \mathbf{r}).$$

On the other hand the vector moment of $\mathbf{F}$ about O is

$$\mathbf{M} = \mathbf{r} \times \mathbf{F} = L\alpha + M\beta + N\gamma.$$

From Euler's Equations, l.c., we have:

$$Ap\frac{dp}{dt} + Bq\frac{dq}{dt} + Cr\frac{dr}{dt} = Lp + Mq + Nr.$$

Hence

$$(5) \qquad \tfrac{1}{2}(Ap^2 + Bq^2 + Cr^2)\Big|_{t_0}^{t_1} = \int_{t_0}^{t_1} (Lp + Mq + Nr)\, dt.$$

The left-hand side is the change in kinetic energy. Now

$$(6) \qquad Lp + Mq + Nr = \mathbf{M} \cdot \tilde{\omega} = \tilde{\omega} \cdot (\mathbf{r} \times \mathbf{F}),$$

and so

$$(7) \qquad Fv \cos \psi = Lp + Mq + Nr.$$

For it is true of any three vectors that

$$\mathbf{a} \cdot (\mathbf{b} \times \mathbf{c}) + \mathbf{c} \cdot (\mathbf{b} \times \mathbf{a}) = 0,$$

since

$$\mathbf{a} \cdot (\mathbf{b} \times \mathbf{c}) = \begin{vmatrix} a_1 & a_2 & a_3 \\ b_1 & b_2 & b_3 \\ c_1 & c_2 & c_3 \end{vmatrix}.$$

Moreover,

$$\tilde{\omega} \times \mathbf{r} = -(\mathbf{r} \times \tilde{\omega}).$$

Hence

$$\mathbf{F} \cdot (\tilde{\omega} \times \mathbf{r}) = \tilde{\omega} \cdot (\mathbf{r} \times \mathbf{F}).$$

From (5) it follows, then, that for a single force, the change in kinetic energy is equal to the work done. For the case of n forces the proof is now obvious. The extension to the case of body forces and forces spread out continuously over surfaces or along curves, presents no difficulty.

Remark. We have shown incidentally that the work done on a rigid body with one point fixed is

$$\int_{t_0}^{t_1} \mathbf{M} \tilde{\omega}\, dt,$$

where

$$\mathbf{M} = L\alpha + M\beta + N\gamma$$

is the resultant couple.

The General Case. Consider first a single force, $\mathbf{F}$. The work it does is

$$W = \int_{t_0}^{t_1} \mathbf{F} \mathbf{v}\, dt.$$

Here,

$$\mathbf{v} = \bar{\mathbf{v}} + \mathbf{v}',$$

where $\bar{\mathbf{v}}$ is the velocity of the centre of gravity and $\mathbf{v}'$ is the velocity of the point Q relative to the centre of gravity, as it flashes through P. Hence

$$W = \int_{t_0}^{t_1} \mathbf{F} \bar{\mathbf{v}}\, dt + \int_{t_0}^{t_1} \mathbf{F} \mathbf{v}'\, dt.$$

The first integral has the value

$$\left.\frac{M\bar{v}^2}{2}\right|_{t_0}^{t_1}$$

The second integral is equal to the right-hand side of Equation (5).

Thus the theorem is proved for one force. For a number of forces the proof is now obvious.

EXERCISES

1. A ball is placed on a rough fixed sphere of the same size and slightly displaced near the highest point. Find where it will leave the sphere. Let μ have any value.

2. A weightless tube can turn freely about one end. A smooth rod is inserted in the tube and the system is released from rest with the tube horizontal. How fast will it be turning when it is vertical?

3. A cylindrical can is filled with water and sealed up. It is mounted so that it can rotate freely about an element of the cylinder. Show that it oscillates like a simple pendulum, provided the can is smooth.

4. In the preceding problem, the height of the can is equal to its diameter, and the can weighs 5 lbs. The water weighs 31 lbs. Find the length of the equivalent simple pendulum.

5. A circular tube, smooth inside, plane vertical, is partly filled with water. The tube is held fast and the water is displaced, then released from rest. Show that it oscillates like a simple pendulum, and determine the length of the latter.

6. A bent tube in the form of an L is mounted so that it can slide freely on a smooth table. The vertical arm is filled with water, and the system is released from rest. How fast will it be moving where the vertical arm has just been emptied?

Assume the tube smooth inside; and also take the weight of the tube with its mount equal to the weight of the water.

7. The can of Question 4 is allowed to roll down a rough inclined plane, starting from rest. Find the acceleration of the centre of gravity.

CHAPTER VIII

IMPACT

1. Impact of Particles. Let two particles, of masses m_1 and m_2, be moving in the same straight line with velocities u_1 and u_2, and let them impinge on each other. To find their velocities after the impact.

Isolate the system consisting of the two particles. Then no external forces act, and so the momentum remains unchanged. Hence

Fig. 128

$$m_1 u_1' - m_1 u_1 + m_2 u_2' - m_2 u_2 = 0. \tag{1}$$

As yet, nothing has been said about the elasticity of the particles. The extreme cases are: perfect elasticity (like two billiard balls) and perfect inelasticity (like two balls of putty). In each case there is deformation of the bodies — for now we will no longer think of particles, but, say, of spheres, and the velocities u_1, u_2, etc. refer to their centres of gravity.

During the deformation the mutual pressures mount high, and even if other (ordinary) forces act, their effect is negligible, compared with the pressures in question. In the case of perfect inelasticity, there is no tendency toward a restitution of shape, and so, when the maximum deformation has been reached, the mutual pressures drop to nothing at all. At this point, the velocities of the two centres of gravity are the same,

$$u_1' = u_2',$$

and hence this common velocity, which we will denote by U, is given by the formula:

$$U = \frac{m_1 u_1 + m_2 u_2}{m_1 + m_2}. \tag{2}$$

Thus the problem is solved for perfect inelasticity. For partial elasticity, it is helpful to picture the impact as follows. The motion of the centre of gravity of each ball is given by the equation:

$$(3) \qquad m_1 \frac{d^2 x_1}{dt^2} = -R, \qquad m_2 \frac{d^2 x_2}{dt^2} = R.$$

For the first stage of the impact, *i.e.* up to the time of greatest deformation, $t = T$, we have, on integrating each side of each equation between the limits 0 and T:

$$(4) \qquad m_1 u_1 \Big|_{t=0}^{t=T} = -\int_0^T R\,dt, \qquad m_2 u_2 \Big|_{t=0}^{t=T} = \int_0^T R\,dt.$$

The integral is called an impulse,* and is denoted by P:

$$(5) \qquad P = \int_0^T R\,dt.$$

Hence

$$(6) \qquad \begin{cases} m_1 U - m_1 u_1 = -P \\ m_2 U - m_2 u_2 = P. \end{cases}$$

The second stage of the impact now begins, as the balls are kicked apart by their mutual pressures. On integrating the equations (3) between the limits T and T', we have:

$$(7) \qquad m_1 u_1 \Big|_{T}^{T'} = -\int_T^{T'} R'\,dt', \qquad m_2 u_2 \Big|_{T}^{T'} = \int_T^{T'} R'\,dt'.$$

Now it is easily intelligible physically if we assume that, in the case of partial elasticity, the value of R' stands in a constant ratio to the value of R at corresponding instants of time, or that

$$(8) \qquad R' = eR,$$

when

$$T - t = t' - T.$$

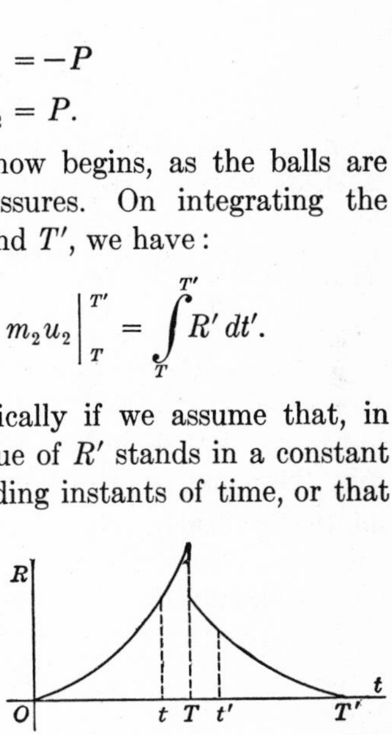

Here the physical constant e is called the *coefficient of restitution*. It lies between 0 and 1:

FIG. 129

$$(9) \qquad 0 < e < 1,$$

* Sometimes spoken of as an *impulsive force*; but this nomenclature is unfortunate, since P is not of the nature of a *force*, which is a push or a pull, but rather is expressed by a *change of momentum*. Moreover, the dimensions of impact are ML/T, not ML/T^2.

being 0 in the case of perfect inelasticity and 1 for perfect elasticity. Hence

$$(10) \qquad P' = \int_T^{T'} R'\,dt' = e\int_0^T R\,dt = eP.$$

Equations (7) thus yield the following:

$$(11) \qquad \begin{cases} m_1 u_1' - m_1 U = -eP \\ m_2 u_2' - m_2 U = eP. \end{cases}$$

The four equations, (6) and (11), contain the solution of the problem. Between them, U and P can be eliminated, and the resulting equations can then be solved for u_1', u_2'. The result is:

$$(12) \qquad \begin{cases} u_1' = \dfrac{(m_1 - em_2)\,u_1 + (1+e)\,m_2 u_2}{m_1 + m_2} \\[2ex] u_2' = \dfrac{(1+e)\,m_1 u_1 + (m_2 - em_1)\,u_2}{m_1 + m_2}. \end{cases}$$

The Case $m_2 = \infty$. If in Equations (12) we allow m_2 to increase without limit, we obtain the equations:

$$(13) \qquad \begin{cases} u_1' = -eu_1 + (1+e)\,u_2 \\ u_2' = u_2. \end{cases}$$

These equations do not prove that, when the mass m_2 is held fast, or is moving with unchanging velocity u_2, the velocity of the mass m_1 after the impact will be given by the first equation (13), but they suggest it. The proof is given by means of the first of the equations (6) and (11), resulting as they do respectively from the first of the equations (4) and (7), combined with (10); U having here the known value u_2.

If, in particular, $u_2 = 0$, we have:

$$(14) \qquad u_1' = -eu_1.$$

Perfect Elasticity, $e = 1$. Equation (14) becomes in this case $u_1' = -u_1$, and the ball recedes with the same velocity as that with which it impinged.

If the masses are equal, $m_1 = m_2$, Equations (12) become:

$$u_1' = u_2, \qquad u_2' = u_1,$$

and the balls interchange their velocities. This latter phenomenon can be illustrated suggestively by two equal ivory balls suspended side by side from strings of equal length, after the manner of two pendulums. If one ball hangs vertically at rest, and the other is released from an angle with the vertical, the second ball will be reduced to rest by the impact, and the first will rise to the same height on its side of the vertical as that from which the second ball was released. Thus the velocities will be successively interchanged at the lowest point of the circular arc.

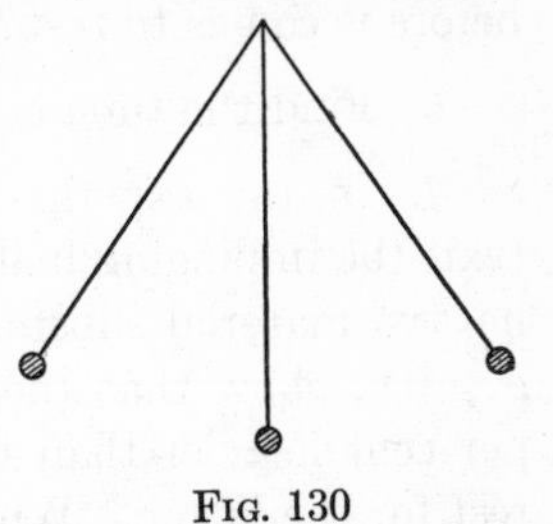

FIG. 130

Critique of the Hypothesis (8). In this hypothesis we have taken for granted an amount of detail in the phenomenon before us far in excess of what the physicist will admit as reasonable in viewing the actual situation, and he may easily be repelled by so dogmatic an assumption in a case that cannot be submitted to direct physical experiment and which, after all, is far less simple than we have led the reader to suppose, since the problem is essentially one in the elasticity of three-dimensional distributions of matter. The objection, however, is easily met. *We may take Equation* (10):

$$P' = eP,$$

as the physical postulate governing impact.

EXERCISES

1. A ball of 6 lbs. mass, moving at the rate of 10 m. an h. overtakes a ball of 4 lbs. mass moving at the rate of 5 m. an h. Determine their velocities after impact, assuming that the coefficient of restitution is $\frac{1}{2}$. *Ans.* 7 and 9.5 m. an h.

2. The same problem, when the balls are moving in opposite directions.

3. A perfectly elastic sphere impinges on a second perfectly elastic sphere of twice the mass. Find the velocity of each after the impact.

4. Newton found that the coefficient of restitution for glass is $\frac{15}{16}$. If a glass marble is dropped from a height of two feet on a glass slab, how high will it rise?

5. In the last question, what will be the height of the second rebound? What will be the total distance covered by the marble before it comes to rest? *

6. Find the time it takes the marble to come to rest.

7. In the experiment with the pendulums described in the text, the impinging ball will not be quite reduced to rest, because no two material substances are quite elastic. If, for given balls, $e = 0.9$, show that the ball which is at rest should be about 11 per cent heavier than the other one, in order to attain complete rest for the latter. What per cent larger should its radius be?

8. If, in the last question, the pendulum bobs are of glass, $e = \frac{15}{16}$, find the ratio of their diameters.

9. If two perfectly elastic balls impinge on each other with equal velocities, show that one of them will be brought to rest if it is three times as heavy as the other.

10. Determine the coefficient of restitution for a tennis ball by dropping it and comparing the height of the rebound with the height from which it was dropped.

11. Some pitchers used to deliver a slow ball to Babe Ruth, believing that he could not make a home run so easily as on a fast ball. Discuss the mechanics of the situation.

2. Continuation. Oblique Impact. Let two spheres impinge at an angle, and suppose them to be perfectly smooth. To determine the velocities after the impact.

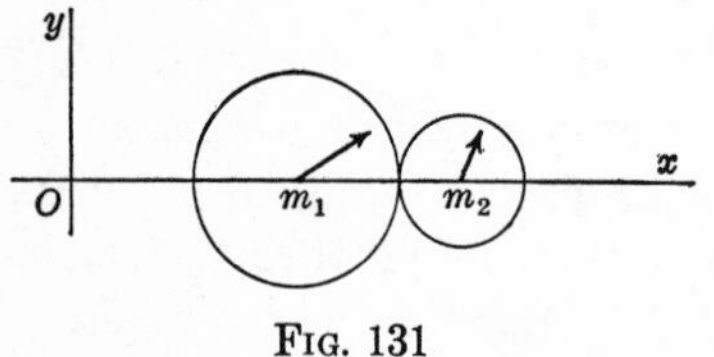

FIG. 131

Let the line of centres be taken as the axis of x. The deformation of each sphere is slight, and the force exerted by the other sphere, spread out as it is over a very small area and acting normally at each point of this area, will yield a resultant force, R, nearly parallel to the axis of x. For the first sphere we have:

* The physics of the second part of this problem (and of the next) is altogether phantastic. After a few rebounds we pass beyond the domain within which the physical hypothesis of the text applies, and the further motion becomes a purely mathematical fiction. It is amusing for those who have a sense of humor in science. But for the literal-minded person, be he physicist or mathematician, it is dangerous.

$$(15)\qquad m_1\frac{d^2x_1}{dt^2}=-X,\qquad m_1\frac{d^2y_1}{dt^2}=Y,$$

where

$$X=R\cos\epsilon,\qquad Y=R\sin\epsilon,$$

ϵ being numerically small and x_1, y_1 referring to the centre of gravity, and for the second sphere,

$$(16)\qquad m_2\frac{d^2x_2}{dt^2}=X,\qquad m_2\frac{d^2y_2}{dt^2}=-Y.$$

On integrating (15) we obtain:

$$(17)\qquad m_1\frac{dx_1}{dt}\Big|_{t=0}^{T}=-\int_0^T X\,dt,\qquad m_1\frac{dy_1}{dt}\Big|_{t=0}^{T}=\int_0^T Y\,dt.$$

And now we denote the first impulse by P, and lay down the *postulate* that the second impulse is 0:

$$(18)\qquad \int_0^T X\,dt=P,\qquad \int_0^T Y\,dt=0.$$

Thus the integrals of (15) and (16) lead to the equations:

$$(19)\qquad \begin{cases} m_1U-m_1u_1=-P\\ m_2U-m_2u_2=P\end{cases}\qquad \begin{cases} m_1V-m_1v_1=0\\ m_2V-m_2v_2=0,\end{cases}$$

which hold for the first epoch of the impact, the equations for the second epoch being, as in the corresponding case of § 1, the following:

$$(20)\qquad \begin{cases} m_1u_1'-m_1U=-eP\\ m_2u_2'-m_2U=eP\end{cases}\qquad \begin{cases} m_1v_1'-m_1V=0\\ m_2v_2'-m_2V=0.\end{cases}$$

For we assume as there the physical postulate:

$$(21)\qquad P'=eP.$$

The result at which we have arrived is seen to be the following. The component of the velocity of each sphere perpendicular to the line of centres has been unchanged by the impact,

$$(22)\qquad v_1'=v_1,\qquad v_2'=v_2.$$

The components of the velocity along the line of centres are changed precisely as in the case of direct impact, § 1, Equations (12):

$$
(23) \qquad \begin{cases} u_1' = \dfrac{(m_1 - em_2)\,u_1 + (1+e)\,m_2 u_2}{m_1 + m_2} \\[2ex] u_2' = \dfrac{(1+e)\,m_1 u_1 + (m_2 - em_1)\,u_2}{m_1 + m_2}. \end{cases}
$$

Kinetic Energy. When $e = 1$, *i.e.* when the spheres are perfectly elastic, the total kinetic energy is unchanged by the impact, for then

$$
(24) \qquad \frac{m_1 u_1'^2}{2} + \frac{m_2 u_2'^2}{2} = \frac{m_1 u_1^2}{2} + \frac{m_2 u_2^2}{2},
$$

as is shown by direct computation from (23), and the equations (22) hold in all cases, whether $e = 1$ or $e < 1$.

When $e = 0$, *i.e.* when the spheres are totally inelastic, an easy computation shows that the kinetic energy has been diminished.

The intermediate case, $0 < e < 1$, is treated in the same way. It follows from direct computation that the left-hand side of Equation (24) has the value:

$$
\frac{(m_1 u_1 + m_2 u_2)^2 + m_1 m_2 (u_1 - u_2)^2 e^2}{2\,(m_1 + m_2)},
$$

and this is at once shown to be less than the right-hand side. The terms arising from Equations (22) do not, of course, affect the result.

EXERCISES

1. A smooth ball travelling south-east strikes an equal ball travelling north-east with one-quarter the velocity, their line of centres at the time of impact being east and west. If $e = \frac{1}{2}$, find the velocities of the balls after impact.

2. A smooth ball strikes a horizontal pavement at an angle of 45°. Find the angle of rebound if the coefficient of restitution is $\frac{3}{4}$.

3. Show that the kinetic energy of the balls of Question 1 is diminished in the ratio of 245/272 by the impact.

4. The corresponding question for the ball of Question 2.

3. Rigid Bodies. Let a rigid body be acted on by a single impulse. By that is meant the postulates about to be laid down, suggested by the following physical picture. A force **F** acts at a point (x, y) for a short time, mounting high in intensity. Ordinary forces, if present, produce in this interval of time, $0 \leqq t \leqq T$, only slight results, and in the ultimate postulates do not appear, so they are not considered in the present picture.

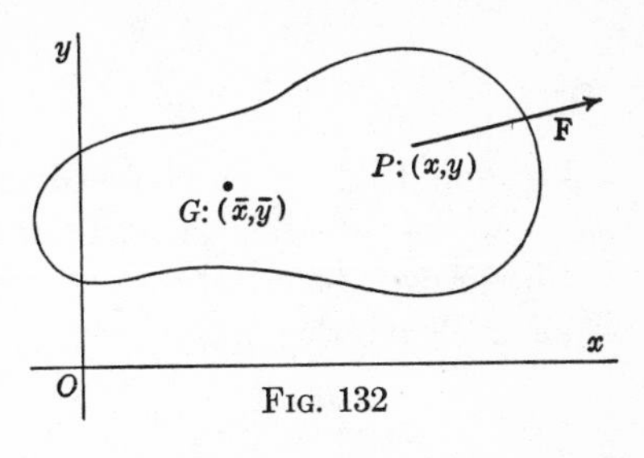

FIG. 132

The three equations which govern the motion are:

$$(1)\quad \begin{cases} M\dfrac{d^2\bar{x}}{dt^2} = X, \qquad M\dfrac{d^2\bar{y}}{dt^2} = Y, \\ I\dfrac{d^2\theta}{dt^2} = (x - \bar{x})\,Y - (y - \bar{y})\,X. \end{cases}$$

On integrating with respect to the time we find:

$$(2)\quad \begin{cases} M(\bar{u}' - \bar{u}) = \displaystyle\int_0^T X\,dt, \qquad M(\bar{v}' - \bar{v}) = \int_0^T Y\,dt \\ I(\Omega' - \Omega) = \displaystyle\int_0^T (x - \bar{x})\,Y\,dt - \int_0^T (y - \bar{y})\,X\,dt. \end{cases}$$

Concerning **F** we will assume that the vector changes continuously in magnitude and direction during the interval of time in question, and that the point of application, (x, y), also moves continuously, remaining near a fixed point (a, b) throughout the interval. Let

$$(3)\qquad x = a + \xi, \qquad y = b + \eta.$$

Then ξ, η are infinitesimal with T. Let

$$(4)\qquad P = \int_0^T X\,dt, \qquad Q = \int_0^T Y\,dt.$$

The last Equation (2) now becomes:

$$(5) \qquad I(\Omega' - \Omega) = (a - \bar{x})\,Q - (b - \bar{y})\,P + \int_0^T \xi\,Y\,dt - \int_0^T \eta\,X\,dt.$$

We should like to infer mathematically that from the hypothesis that the integrals (4) approach limits when T approaches 0, the integrals in the last line of (5) converge toward 0; for then we should have the equations:

$$(6) \qquad \begin{cases} M(\bar{u}' - \bar{u}) = P, \qquad M(\bar{v}' - \bar{v}) = Q, \\ I(\Omega' - \Omega) = (a - \bar{x})\,Q - (b - \bar{y})\,P, \end{cases}$$

P and Q here denoting the limiting values of the integrals (4).

This inference can in fact be drawn, provided the angle through which the vector $\mathbf{F}$ ranges is less than 180°. Equations (6) then hold, and, in particular, it follows, on eliminating P and Q between them, that

$$(7) \qquad I(\Omega' - \Omega) = M(a - \bar{x})(\bar{v}' - \bar{v}) - M(b - \bar{y})(\bar{u}' - \bar{u})$$

or

$$(8) \qquad k^2(\Omega' - \Omega) = (a - \bar{x})(\bar{v}' - \bar{v}) - (b - \bar{y})(\bar{u}' - \bar{u}).$$

An Example. A rod is rotating about one end, and it strikes an obstruction, which brings it suddenly to rest without any reaction on the support. What point of the rod comes into contact with the obstruction?

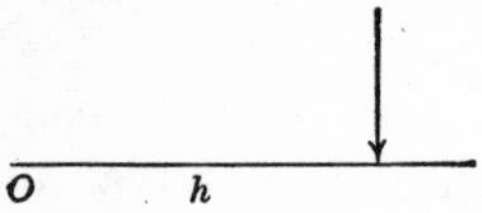

Fig. 133

Let the distance from the stationary end be h, and let l be the length of the rod. Let

$$\bar{v} = C; \qquad \text{then} \qquad \Omega = \frac{2}{l}C.$$

Since

$$\bar{u} = 0, \qquad \bar{u}' = 0, \qquad \bar{v}' = 0, \qquad \Omega' = 0, \qquad I = \frac{Ml^2}{12},$$

we have:

$$\frac{l^2}{12}\left(-\frac{2}{l}C\right) = \left(h - \frac{l}{2}\right)(-C).$$

Hence

$$h = \tfrac{2}{3}l.$$

The point is called the *centre of percussion*.

4. Proof of the Theorem. The proof is given by means of the Law of the Mean, which is as follows. Let $f(x)$, $\varphi(x)$ be two functions which are continuous in the closed interval $a \leqq x \leqq b$, and let $\varphi(x)$ not change sign there. Then

$$(9) \qquad \int_a^b f(x)\,\varphi(x)\,dx = f(x')\int_a^b \varphi(x)\,dx, \qquad a < x' < b.$$

In the present case the axes can be so chosen that

$$0 \leqq Y.$$

Hence

$$(10) \qquad \int_0^T \xi\, Y\,dt = \xi' \int_0^T Y\,dt,$$

where ξ' is the value of ξ at a suitable point, $t = t'$, in the interval $0 \leqq t \leqq T$. Now, by hypothesis, ξ and η approach 0 uniformly, *i.e.* the largest numerical value that either has in the interval $0 \leqq t \leqq T$ approaches 0; and furthermore, also by hypothesis, the integral on the right approaches a limit, Q. Hence the integral on the left approaches 0.

If the range of the angle of **F** does not exceed 90°, the axes of coordinates can be so chosen that neither X nor Y changes sign in the interval $0 \leqq t \leqq T$, and then it can be shown as above that both integrals in the second line of (5) approach 0.

In the more general case that **F** is contained merely within an angle less than π, the axes can be chosen in more ways than one so that Y will not change sign. If (x, y) refer to one such choice and (x', y') to a second, then

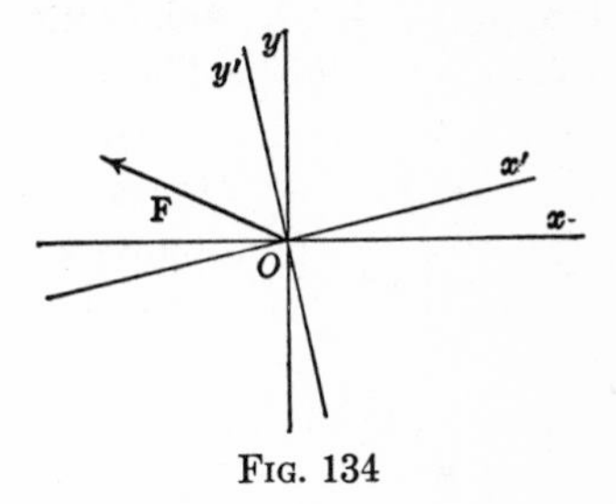

FIG. 134

$$(11) \qquad \begin{cases} x' = ax + by \\ y' = cx + dy \end{cases}$$

where

$$a = \cos\gamma, \qquad b = \sin\gamma,$$
$$c = -\sin\gamma, \qquad d = \cos\gamma.$$

The same transformation holds* with respect to the vector **F**:

* It is in such a case as the present one that the scientific importance of the proper definition of Cartesian coordinates, laid down in Analytic Geometry, is revealed. That definition begins with directed line segments on a line, proceeds to the theorem that the sum of the projections of two broken lines having the same

$$(12)\qquad \begin{cases} X' = aX + bY \\ Y' = cX + dY \end{cases}$$

and also with respect to (ξ, η):

$$(13)\qquad \begin{cases} \xi' = a\xi + b\eta \\ \eta' = c\xi + d\eta. \end{cases}$$

Hence

$$(14)\quad \int_0^T \xi'Y'\,dt = ac\int_0^T \xi X\,dt + bc\int_0^T \eta X\,dt + bd\int_0^T \eta Y\,dt + ad\int_0^T \xi Y\,dt.$$

The integral on the left, and the last two integrals on the right, approach the limit 0 with T, as has been shown above. We will show that this is true also of the other integrals, and hence in particular of the last integral in (5). To do this, write down Equation (14) for two choices of axes (x', y') subject to the above conditions and characterized by two values of γ: γ_1 and γ_2, where $\gamma_1 \neq 0$, $\gamma_2 \neq 0$, $\gamma_1 \neq \gamma_2$, and solve the resulting equations for the two integrals in question. The determinant of the equations,

$$\begin{vmatrix} a_1c_1 & b_1c_1 \\ a_2c_2 & b_2c_2 \end{vmatrix},$$

has the value

$$\sin\gamma_1 \sin\gamma_2 \sin(\gamma_2 - \gamma_1),$$

and so does not vanish. Thus the integrals for which we are solving are seen to be linear functions of integrals that are known to approach 0, and this completes the proof.

The Restriction. The theorem is not true when **F** is required merely to vary continuously with t in the interval $0 \leqq t \leqq T$, as the following example shows. Let

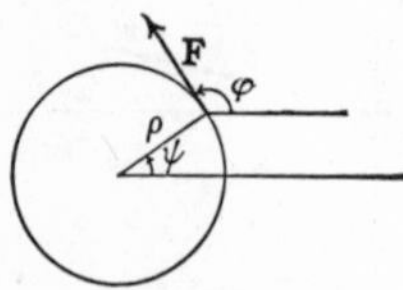

FIG. 135

$$X = F\cos\varphi, \qquad Y = F\sin\varphi;$$

$$\xi = \rho\cos\psi, \qquad \eta = \rho\sin\psi.$$

Then

$$\xi Y - \eta X = \rho F \sin(\varphi - \psi).$$

extremities, on an arbitrary line, is the same for both lines, and ends by declaring the coordinates of a point as the projections on the axes of the vector whose initial point is the origin and whose terminal point is the point in question. With that definition, Equations (12) and (13) are merely particular cases of Equations (11), since both sets of equations express the projections of a vector on the coordinate axes.

Let F and ρ be constants, and let

$$\varphi = \frac{2\pi t}{T} + \frac{\pi}{2}, \qquad \psi = \frac{2\pi t}{T}.$$

Then

$$P = \int_0^T X\,dt = -F\int_0^T \sin\frac{2\pi t}{T}\,dt = 0,$$

$$Q = \int_0^T Y\,dt = F\int_0^T \cos\frac{2\pi t}{T}\,dt = 0,$$

$$\int_0^T (\xi Y - \eta X)\,dt = \rho F T.$$

If, now, we set

$$\rho = T, \qquad F = \frac{1}{T^2},$$

the integrals (4), being always 0, each approach limits, and so the P and Q of Formulas (6) have each the value 0. But the third Equation (6) does not hold.

But may it not still be sufficient, in order to secure the vanishing of the limit of the integral

$$\int_0^T (\xi Y - \eta X)\,dt,$$

to demand that $Y \geqq 0$? That this is not enough, is shown by modifying the above example as follows. Let

$$X = F\cos\varphi, \qquad Y = 0.$$

The integral then has half the value it had before; hence, etc.

EXERCISES

1. A uniform rod at rest is struck a blow at one end, at right angles to the rod. About what point will it begin to rotate?

2. A packing box is sliding over an asphalt pavement, when it strikes the curbstone. Find the speed at which it begins to rotate.

3. If, in the preceding question, the pavement is icy, and if the box, before it reaches the curb, comes to a bare spot, $\mu = \frac{1}{4}$, what is the condition that it should not tip?

4. If the box tips, find whether it will slide, or rotate about a fixed line.

5. Show that greater braking power is available when the brakes are applied to the wheels of the forward truck of a railroad car.

6. If all four wheels of an automobile are locked, compare the pressure of the forward wheels on the ground with that of the rear wheels.

7. A lamina is rotating in its own plane about a point O, when it is suddenly brought to rest by an obstruction at a point P situated in the line OG produced. Show that OP is equal to the length of the equivalent simple pendulum, when the lamina is supported at O and allowed to oscillate under gravity in a vertical plane.

5. Tennis Ball, Returned with a Lawford. Consider a tennis ball, returned over the net with flat trajectory and rotation such that the lowest point of the ball is moving backward. The ground thus exerts a forward force, and we will assume that this state of affairs holds throughout the impact. We shall have, then, the following formulation of the problem:

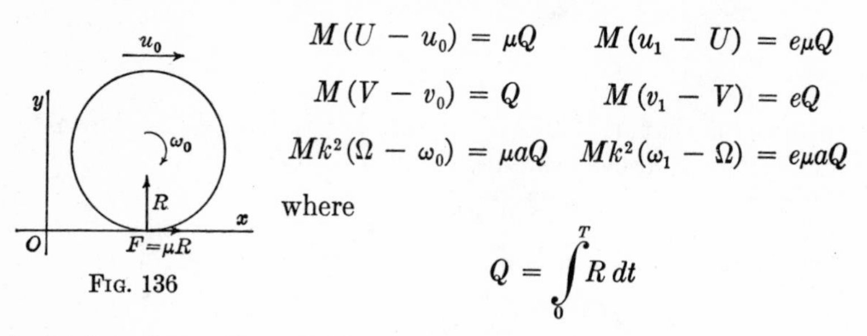

FIG. 136

$$M(U - u_0) = \mu Q \qquad M(u_1 - U) = e\mu Q$$

$$M(V - v_0) = Q \qquad M(v_1 - V) = eQ$$

$$Mk^2(\Omega - \omega_0) = \mu a Q \quad Mk^2(\omega_1 - \Omega) = e\mu a Q$$

where

$$Q = \int_0^T R\,dt$$

is the impulse. First of all,

$$V = 0,$$

for the point of greatest deformation is marked by the centre of gravity of the ball's ceasing to descend. Thus we have seven equations for the seven unknowns, u_1, v_1, ω_1, U, V, Ω, Q.

It is now easy to solve. Observe that

$$v_0 < 0, \qquad \omega_0 < 0, \qquad u_0 > 0.$$

We have, then:

$$Q = M(-v_0), \qquad v_1 = e(-v_0),$$

$$u_1 = u_0 + (1+e)\mu(-v_0).$$

The value of ω_1 is not interesting. What we do want to know is the slope of the trajectory at the end of the impact; *i.e.*

$$\frac{v_1}{u_1} = \frac{e(-v_0)}{u_0 + (1+e)\mu(-v_0)} = \frac{e\lambda}{1 + (1+e)\mu\lambda},$$

where $\lambda = (-v_0)/u_0$ is the numerical value of the slope before the impact.

As the ball has a drop due to the cut, λ will be considerably larger numerically than the slope in the part of the trajectory just preceding the last ten feet or so before touching the ground. It might conceivably have a value as great as $\frac{1}{5}$. The value of e is about 0.8. μ varies considerably and might be as high as $\frac{1}{4}$. Thus

$$\frac{v_1}{u_1} = .15,$$

as against $\lambda = .20$, or the steepness of the rebound is only three-fourths the steepness of the incident path.

Not only does the ball rise at a smaller angle, but the horizontal velocity is increased by nearly 10 per cent; for

$$u_1 = u_0[1 + (1+e)\mu\lambda] = 1.09\,u_0.$$

For this discussion to be correct it is essential that the ball maintain its spin throughout the whole impact. This explains the nature of the stroke. The racquet has a high upward velocity while the ball is on the guts.

The ball loses spin during the flight before the impact, due to the air resistance causing the drop, and this loss may easily be comparable with the loss during the impact. It would be interesting to take motion pictures of the ball, showing the trajectory just before and just after the impact.

EXERCISES

1. A billiard ball, rotating about a horizontal axis, falls on a partially elastic table. Find the direction of the rebound if $\mu = \frac{1}{5}$ and $e = .9$.

2. A rod, moving in a vertical plane, strikes a partially elastic smooth table. Determine the subsequent motion.

3. The preceding question, with the change that the table is rough, $\mu = \frac{1}{4}$.

4. Question 2 for a table that is wholly inelastic and infinitely rough.

5. A rigid lamina is oscillating in a vertical plane about a point O when it strikes an obstacle at P whose distance from O is equal to the length of the equivalent simple pendulum. Show that it will be brought to rest without any reaction on the axis.

For this reason P is called the *centre of percussion*.

6. A rigid lamina, at rest, is struck a blow at a point O. Find the point about which it will begin to rotate.

7. A solid cone, at rest, is struck a blow at the vertex in a direction at right angles to the axis. About what line will it begin to rotate?

CHAPTER IX

RELATIVE MOTION AND MOVING AXES

1. Relative Velocity. It is sometimes convenient to refer the motion of a system to moving axes. Let O be a point fixed in space. Let O' be a point moving in any manner, like the centre of gravity of a material body, or the centre of geometric symmetry of a body whose centre of gravity is not at O'; it is a point whose motion is known, or on which we wish particularly to focus our attention. Finally, let P be any point of the system whose motion we are studying. Then

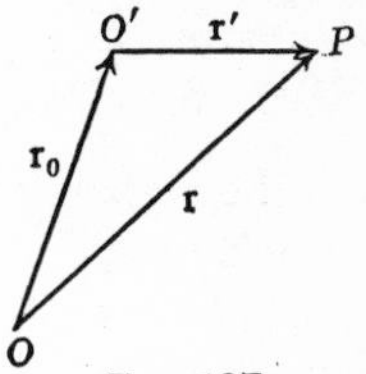

(1) $$\mathbf{r} = \mathbf{r}_0 + \mathbf{r}',$$

$$\frac{d\mathbf{r}}{dt} = \frac{d\mathbf{r}_0}{dt} + \frac{d\mathbf{r}'}{dt},$$

or

(2) $$\mathbf{v} = \mathbf{v}_0 + \mathbf{v}'.$$

FIG. 137

The choice of notation is here particularly important — boldface letters denote as usual vectors — because we have *two* analyses to emphasize, namely, *i*) the breaking up of the velocity $\mathbf{v}$ into the two velocities $\mathbf{v}_0$ and $\mathbf{v}'$; and *ii*) the breaking up of $\mathbf{v}'$ into the two velocities:

(3) $$\mathbf{v}' = \mathbf{v}_r + \mathbf{v}_e,$$

where $\mathbf{v}_r$, the *relative velocity*, and $\mathbf{v}_e$ the *vitesse d'entraînement* are presently to be defined. For this purpose we must first recall the results of an earlier study.

2. Linear Velocity in Terms of Angular Velocity. In Chapter V, § 8, we have studied the motion of a system referred to moving axes (ξ, η, ζ) with fixed origin O. Here,

(4) $$\mathbf{r} = \xi\alpha + \eta\beta + \zeta\gamma$$

and

(5) $$\frac{d\mathbf{r}}{dt} = \frac{d\xi}{dt}\alpha + \frac{d\eta}{dt}\beta + \frac{d\zeta}{dt}\gamma$$
$$+ \xi\dot{\alpha} + \eta\dot{\beta} + \zeta\dot{\gamma}.$$

This equation represents an analysis of the velocity

$$\mathbf{v} = \frac{d\mathbf{r}}{dt} \tag{6}$$

of the point P into two velocities, namely,

$$\mathbf{v} = \mathbf{v}_r + \mathbf{v}_e, \tag{7}$$

where

$$\mathbf{v}_r = \frac{d\xi}{dt}\alpha + \frac{d\eta}{dt}\beta + \frac{d\zeta}{dt}\gamma \tag{8}$$

is the *relative velocity* of P with respect to the moving axes; *i.e.* the absolute velocity which P would have if the (ξ, η, ζ)-axes were at rest and the point P moved relatively to them just as it does:

$$\xi = f(t), \qquad \eta = \varphi(t), \qquad \zeta = \psi(t). \tag{9}$$

Secondly,

$$\mathbf{v}_e = \xi\dot{\alpha} + \eta\dot{\beta} + \zeta\dot{\gamma} \tag{10}$$

is the *vitesse d'entraînement*, the *Schleppgeschwindigkeit*, the velocity with which that point Q fixed in the moving space and flashing through P at the one instant, t, is moving in space. Let (ω) be the vector angular velocity of the moving axes:

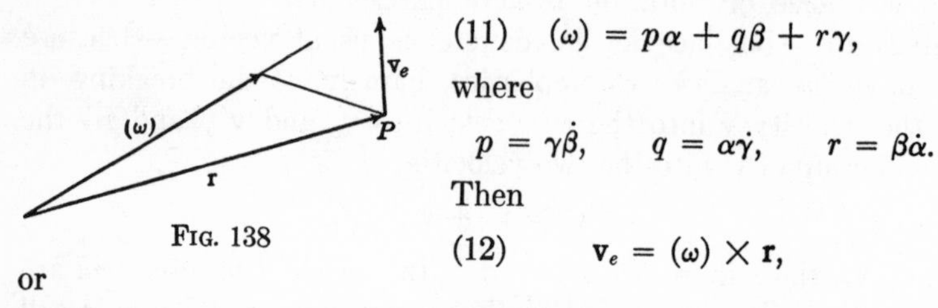

Fig. 138

$$(\omega) = p\alpha + q\beta + r\gamma, \tag{11}$$

where

$$p = \gamma\dot{\beta}, \qquad q = \alpha\dot{\gamma}, \qquad r = \beta\dot{\alpha}.$$

Then

$$\mathbf{v}_e = (\omega) \times \mathbf{r}, \tag{12}$$

or

$$\mathbf{v}_e = (\zeta q - \eta r)\,\alpha + (\xi r - \zeta p)\,\beta + (\eta p - \xi q)\,\gamma. \tag{13}$$

The final result is as follows: The components of $\mathbf{v}$, or $d\mathbf{r}/dt$, along the axes are

$$\left\{\begin{aligned} v_\xi &= \frac{d\xi}{dt} + \zeta q - \eta r \\ v_\eta &= \frac{d\eta}{dt} + \xi r - \zeta p \\ v_\zeta &= \frac{d\zeta}{dt} + \eta p - \xi q \end{aligned}\right. \tag{14}$$

I repeat: These are the formulas when the moving axes have their origin, O', fixed: $\mathbf{r}_0 = 0$, $\mathbf{v}_0 = 0$, $\mathbf{v} = \mathbf{v}'$.

3. Acceleration. Returning now to the point P of § 1 and Equations (1) and (2), we define its acceleration as the vector:

$$\mathfrak{a} = \frac{d\mathbf{v}}{dt} = \frac{d^2\mathbf{r}}{dt^2}. \tag{15}$$

Hence

$$\mathfrak{a} = \frac{d\mathbf{v}_0}{dt} + \frac{d\mathbf{v}'}{dt}, \tag{16}$$

or

$$\mathfrak{a} = \mathfrak{a}_0 + \mathfrak{a}'. \tag{17}$$

The first term on the right, $\mathfrak{a}_0$, requires no further comment. It is merely the acceleration in fixed space of the known point O'. The second term, $\mathfrak{a}'$, relates to the rotation and admits of a number of important evaluations.

First Evaluation. The first of these is as follows. Let $\mathfrak{a}'$ be denoted by $\mathbf{a}$. Then

$$\mathbf{a} = \frac{d\mathbf{v}'}{dt}. \tag{18}$$

We may identify the variable vector $\mathbf{v}'$ with the variable vector $\mathbf{r}$ of § 2, Formula (4); for, of course, $\mathbf{r}$ was *any* vector, moving according to any law we wish. Now, we have evaluated the right-hand side of (18) by means of Equations (14). Hence the components of the right-hand side of (18) are obtained by substituting in the right-hand side of (14) for ξ, η, ζ respectively v_ξ, v_η, v_ζ.

On the other hand, write

$$\mathbf{a} = a_\xi \alpha + a_\eta \beta + a_\zeta \gamma. \tag{19}$$

Thus we arrive at the final determination of $\mathbf{a}$ in terms of known functions:

$$\begin{cases} a_\xi = \dfrac{dv_\xi}{dt} + q v_\zeta - r v_\eta \\[2ex] a_\eta = \dfrac{dv_\eta}{dt} + r v_\xi - p v_\zeta \\[2ex] a_\zeta = \dfrac{dv_\zeta}{dt} + p v_\eta - q v_\xi \end{cases} \tag{20}$$

These are the formulas referred to as the First Evaluation.

Second Evaluation. The Theorem of Coriolis. Another form for the vector **a** can be obtained by differentiating (5), § 2:

$$\mathbf{v}' = \frac{d\xi}{dt}\alpha + \frac{d\eta}{dt}\beta + \frac{d\zeta}{dt}\gamma$$
$$+ \xi\dot{\alpha} + \eta\dot{\beta} + \zeta\dot{\gamma}.$$

Thus

$$\mathbf{a} = \frac{d^2\xi}{dt^2}\alpha + \frac{d^2\eta}{dt^2}\beta + \frac{d^2\zeta}{dt^2}\gamma \tag{21}$$
$$+ 2\left(\frac{d\xi}{dt}\frac{d\alpha}{dt} + \frac{d\eta}{dt}\frac{d\beta}{dt} + \frac{d\zeta}{dt}\frac{d\gamma}{dt}\right)$$
$$+ \quad \xi\ddot{\alpha} + \eta\ddot{\beta} \quad + \zeta\ddot{\gamma}.$$

This result is due to Coriolis.

The first and third lines admit immediate interpretations. For,

$$\mathbf{a}_r = \frac{d^2\xi}{dt^2}\alpha + \frac{d^2\eta}{dt^2}\beta + \frac{d^2\zeta}{dt^2}\gamma \tag{22}$$

is the *relative acceleration*, or the acceleration of P referred to the (ξ, η, ζ)-axes as fixed. Next,

$$\mathbf{a}_e = \xi\frac{d^2\alpha}{dt^2} + \eta\frac{d^2\beta}{dt^2} + \zeta\frac{d^2\gamma}{dt^2} \tag{23}$$

is the *acceleration d'entraînement*, or the *Schleppbeschleunigung*, the acceleration with which the point Q, fixed in the moving space and coinciding at the instant t with P, is being carried along in fixed space.

The vector (23):

$$\mathbf{a}_e = \xi\ddot{\alpha} + \eta\ddot{\beta} + \zeta\ddot{\gamma},$$

can be computed as follows. Since — as is geometrically, or kinematically, immediately obvious —

$$\dot{\alpha} = r\beta - q\gamma, \qquad \dot{\beta} = p\gamma - r\alpha, \qquad \dot{\gamma} = q\alpha - p\beta, \tag{24}$$

we have:

$$\ddot{\alpha} = \frac{dr}{dt}\beta - \frac{dq}{dt}\gamma \tag{25}$$
$$+ r\dot{\beta} - q\dot{\gamma}.$$

The last line has the value:

$$-\omega^2\alpha + p(\omega),$$

where

$$(\omega) = p\alpha + q\beta + r\gamma.$$

Hence

$$(26)\quad \mathbf{a}_e = \left(\zeta\frac{dq}{dt} - \eta\frac{dr}{dt}\right)\alpha + \left(\xi\frac{dr}{dt} - \zeta\frac{dp}{dt}\right)\beta + \left(\eta\frac{dp}{dt} - \xi\frac{dq}{dt}\right)\gamma$$

$$- \omega^2(\xi\alpha + \eta\beta + \zeta\gamma) + (p\xi + q\eta + r\zeta)(p\alpha + q\beta + r\gamma),$$

or

$$(27)\quad \mathbf{a}_e = (\omega') \times \mathbf{r} - \omega^2\mathbf{r} + ((\omega)\mathbf{r})(\omega),$$

where

$$(28)\quad (\omega') = \frac{dp}{dt}\alpha + \frac{dq}{dt}\beta + \frac{dr}{dt}\gamma.$$

This vector (ω') is the velocity relative to the fixed axes (ξ, η, ζ), with which the terminal point of (ω) is moving when the initial point is at O'; it is the *relative angular acceleration*, referred to the (ξ, η, ζ)-axes as fixed.

Finally, the vector

$$(29)\quad \mathbf{a}_i = \frac{d\xi}{dt}\frac{d\alpha}{dt} + \frac{d\eta}{dt}\frac{d\beta}{dt} + \frac{d\zeta}{dt}\frac{d\gamma}{dt},$$

can be expressed in the form:

$$(30)\quad \mathbf{a}_i = (\omega) \times \mathbf{v}_r,$$

or:

$$\mathbf{a}_i = \left(q\frac{d\zeta}{dt} - r\frac{d\eta}{dt}\right)\alpha + \left(r\frac{d\xi}{dt} - p\frac{d\zeta}{dt}\right)\beta + \left(p\frac{d\eta}{dt} - q\frac{d\xi}{dt}\right)\gamma.$$

For, on recurring to Formula (10) of § 2 and taking, as the arbitrary vector $\mathbf{r}$, the vector $\mathbf{v}_r$, which is given by (8), the right-hand side of (10) comes to coincide with the right-hand side of (29). With the aid of (12), this vector can be written in the form of the right-hand side of (30), and this completes the proof.

To sum up, then: — From (17),

$$(31)\quad \mathfrak{a} = \mathfrak{a}_0 + \mathfrak{a}',$$

where $\mathfrak{a}' = \mathbf{a}$, and $\mathbf{a}$ is given by (20). A second evaluation of $\mathbf{a}$ is given by (21),

$$(32)\quad \mathbf{a} = \mathbf{a}_r + 2\mathbf{a}_i + \mathbf{a}_e,$$

where $\mathbf{a}_r$ is given by (22) and $\mathbf{a}_e$ by (23); the latter, in a different form, by (26) or (27). Finally, $\mathbf{a}_i$ is given by (29) or (30).

4. The Dynamical Equations. From Newton's Second Law of Motion, written in the form:

$$(1) \qquad m\mathfrak{a} = \mathbf{F},$$

it follows that

$$(2) \qquad m\mathfrak{a}_0 + m\mathbf{a} = \mathbf{F},$$

where $\mathfrak{a}_0$ is the acceleration of the moving origin, O', and

$$\mathbf{a} = \mathbf{a}_r + 2\mathbf{a}_i + \mathbf{a}_e.$$

The vector $\mathbf{a}_r$ is the relative acceleration and is given by Formula (22), § 3. The vector $\mathbf{a}_e$ is the *acceleration d'entraînement* and is defined by (23); it is represented by (26) or (27). Finally, $\mathbf{a}_i$ is defined by (29) and is represented by (30).

If the motion of the moving axes is regarded as known, then $\mathfrak{a}_0$ is a known function of t, and (ω), *i.e.* p, q, r are known from (11), § 2. Equation (2) can now be written in the form:

$$(3) \qquad m\mathbf{a}_r = \mathbf{F} - m\mathfrak{a}_0 - 2m\mathbf{a}_i - m\mathbf{a}_e.$$

On substituting for $\mathbf{a}_i$ its value from (30) and for $\mathbf{a}_e$ its value from (26), a system of differential equations is found for determining ξ, η, ζ:

$$(4) \qquad \begin{cases} m\dfrac{d^2\xi}{dt^2} = \mathsf{X} + f\left(\dfrac{d\xi}{dt}, \dfrac{d\eta}{dt}, \dfrac{d\zeta}{dt}, \xi, \eta, \zeta\right) \\[2ex] m\dfrac{d^2\eta}{dt^2} = \mathsf{Y} + \varphi\left(\dfrac{d\xi}{dt}, \dfrac{d\eta}{dt}, \dfrac{d\zeta}{dt}, \xi, \eta, \zeta\right) \\[2ex] m\dfrac{d^2\zeta}{dt^2} = \mathsf{Z} + \psi\left(\dfrac{d\xi}{dt}, \dfrac{d\eta}{dt}, \dfrac{d\zeta}{dt}, \xi, \eta, \zeta\right) \end{cases}$$

where the functions f, φ, ψ can be written down explicitly from the above formulas.

More generally, Equation (1) can be thrown into the form required in a given problem by using a suitable form for $\mathfrak{a}$, $\mathfrak{a}_0$, $\mathbf{a}_r$, $\mathbf{a}_i$, $\mathbf{a}_e$ as pointed out at the end of § 3. Each one of these accelerations must be studied in the particular case. There is no single choice of sufficient importance to justify writing down the long formulas. But the student will do well to make his own syllabus, writing down the value of $\mathbf{a}_r$ and each form for $\mathbf{a}_e$, $\mathbf{a}_i$.

EXERCISE

Obtain the dynamical equations in explicit form from Lagrange's Equations, Chapter X. Observe that

$$T = \frac{m}{2}\Big\{(\dot{\xi} + \zeta q - \eta r + v_{0\alpha})^2 + (\dot{\eta} + \xi r - \zeta p + v_{0\beta})^2 + (\dot{\zeta} + \eta p - \xi q + v_{0\gamma})^2\Big\},$$

where

$$v_{0\alpha} = \mathbf{v}_0 \cdot \alpha = l_1 \frac{dx_0}{dt} + m_1 \frac{dy_0}{dt} + n_1 \frac{dz_0}{dt},$$

$$v_{0\beta} = \mathbf{v}_0 \cdot \beta = l_2 \frac{dx_0}{dt} + m_2 \frac{dy_0}{dt} + n_2 \frac{dz_0}{dt},$$

$$v_{0\gamma} = \mathbf{v}_0 \cdot \gamma = l_3 \frac{dx_0}{dt} + m_3 \frac{dy_0}{dt} + n_3 \frac{dz_0}{dt}.$$

5. The Centrifugal Field. Let space rotate with constant angular velocity about a fixed line through O, the (ξ, η, ζ)-axes being fixed in the moving space. Then the vector angular velocity (ω) is constant, and $\mathfrak{a}_0 = 0$. The vector $\mathbf{a}_e$, § 3, (27) reduces to:

$$\mathbf{a}_e = -\omega^2\mathbf{r} + ((\omega)\cdot\mathbf{r})(\omega), \tag{1}$$

and is easily interpreted. Kinematically, it is, of course, the centripetal acceleration; geometrically, it is a vector drawn from the point P toward the axis and of length $\omega^2\rho$, where ρ is the distance from P to the axis.

Newton's Law takes the form:

$$m\mathbf{a} = \mathbf{F},$$

where $\mathbf{a}$ is given by § 3, (32), and thus

$$\begin{aligned} \text{(2)}\quad \mathbf{a} = &\left\{\frac{d^2\xi}{dt^2} + 2q\frac{d\zeta}{dt} - 2r\frac{d\eta}{dt} - \omega^2\xi + p(p\xi + q\eta + r\zeta)\right\}\alpha \\ &+ \left\{\frac{d^2\eta}{dt^2} + 2r\frac{d\xi}{dt} - 2p\frac{d\zeta}{dt} - \omega^2\eta + q(p\xi + q\eta + r\zeta)\right\}\beta \\ &+ \left\{\frac{d^2\zeta}{dt^2} + 2p\frac{d\eta}{dt} - 2q\frac{d\xi}{dt} - \omega^2\zeta + r(p\xi + q\eta + r\zeta)\right\}\gamma \end{aligned}$$

Axis of ζ, the Axis of Rotation. In this case, $p = q = 0$, $r = \omega$, and the equations reduce to the following:

$$(3) \qquad \begin{cases} m\dfrac{d^2\xi}{dt^2} = \mathsf{X} + 2m\omega\dfrac{d\eta}{dt} + m\omega^2\xi \\[2ex] m\dfrac{d^2\eta}{dt^2} = \mathsf{Y} - 2m\omega\dfrac{d\xi}{dt} + m\omega^2\eta \\[2ex] m\dfrac{d^2\zeta}{dt^2} = \mathsf{Z} \end{cases}$$

Thus the motion along the axis of ζ is the same as it would be if space were not rotating. The projection of the path on the (ξ, η)-plane is the same as the path of a particle in fixed, or stationary, space, when acted on *i*) by the applied force **F**; *ii*) by a force $m\omega^2\rho$ directed away from O; and *iii*) by a force at right angles to the path, equal in magnitude to $2m\omega v$, and so oriented to the vector velocity **v** as the positive axis of ξ is, with respect to the positive axis of η.

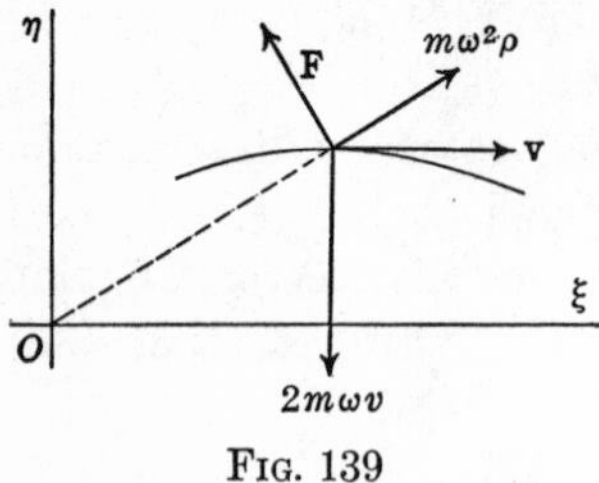

FIG. 139

This third force is known as the *Coriolis force.* In the case of the Centrifugal Oil Cup, and the corresponding revolving tennis court, Chapter III, § 23 it was enough, for problems in statics, to take into account the "centrifugal force," or the force *ii*) above. But for problems in motion, this is not sufficient. There is the Coriolis force *iii*) at right angles to the path, like the force an electro-magnetic field exerts on a moving charge of electricity.

6. Foucault Pendulum. Consider the motion of a pendulum when the rotation of the earth is taken into account. We may think, then, of the earth as rotating about a fixed axis through the poles, which we will take as the axis of z, the axes of x and y lying in the plane of the equator.

Let P be a point of the northern hemisphere, and let its distance from the axis be ρ. By the *vertical* through P is meant the line in which a plumb bob hangs at rest, or, more precisely, the normal to a level surface. Let ζ be taken along the vertical, directed upward; let ξ be tangent, as shown, to the meridian through the point of support of the pendulum; then η will be tangent to the parallel of latitude through the point of support, and directed west. Let λ be the latitude of P; *i.e.* the angle that ζ makes with the plane of the equator.

The earth rotates about its axis from west to east, and so the vector angular velocity, (ω), is directed downward. Thus

$$p = \omega \cos \lambda, \qquad q = 0,$$

$$r = -\omega \sin \lambda,$$

$$\omega = \frac{2\pi}{24 \cdot 60 \cdot 60} = .000727.$$

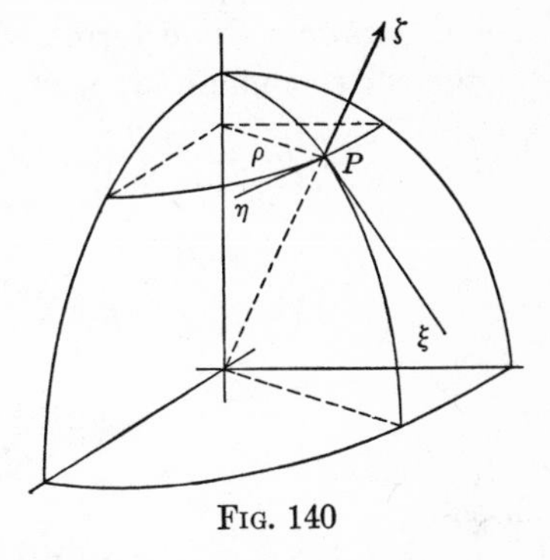

FIG. 140

We can now write down the differential equations that govern the motion. These are contained in the single vector equation of § 4:

$$m\mathfrak{a}_0 + m\mathbf{a} = \mathbf{F}.$$

Let P be the point of support, and let (ξ, η, ζ) be the coordinates of the pendulum; l, its length.

First, compute $\mathbf{F}$:

$$\mathbf{F} = \mathbf{G} + \mathbf{N},$$

where

$$\mathbf{G} = \frac{\partial U}{\partial \xi}\alpha + \frac{\partial U}{\partial \eta}\beta + \frac{\partial U}{\partial \zeta}\gamma$$

is the force due to gravity, or the attraction of the earth; and

$$\mathbf{N} = -\frac{\xi}{l}N\alpha - \frac{\eta}{l}N\beta - \frac{\zeta}{l}N\gamma$$

is the tension of the string.

Next, $\mathfrak{a}_0$ is the centripetal acceleration, or:

$$\mathfrak{a}_0 = -\omega^2 \rho(\alpha \sin \lambda + \gamma \cos \lambda).$$

Finally, $\mathbf{a}$ is given by the formulas (2) of § 5. For, although these were written down for the particular case $\mathfrak{a}_0 = 0$, they apply generally, where $\mathfrak{a}_0$ is arbitrary, provided the vector angular velocity of the moving space is constant.

Thus we can write down explicitly the three equations of motion. These we replace by approximate equations obtained as follows. Approximate, first, to the field of force by the gravity field,

$$U = -\, mg\zeta.$$

Next, suppress those terms which contain ω^2 as a factor, or are of the corresponding order of small quantities. Thus

$$m\left(\frac{d^2\xi}{dt^2} + 2\omega \sin\lambda \frac{d\eta}{dt}\right) = -\frac{\xi}{l} N$$

$$m\left(\frac{d^2\eta}{dt^2} - 2\omega \sin\lambda \frac{d\xi}{dt} - 2\omega \cos\lambda \frac{d\zeta}{dt}\right) = -\frac{\eta}{l} N$$

$$m\left(\frac{d^2\zeta}{dt^2} + 2\omega \cos\lambda \frac{d\eta}{dt}\right) = -\frac{\zeta}{l} N - mg$$

Finally,

$$\xi^2 + \eta^2 + \zeta^2 = l^2,$$

$$\zeta = -\, l\left(1 - \frac{\xi^2 + \eta^2}{l^2}\right)^{\frac{1}{2}} = -l + \frac{\xi^2 + \eta^2}{2l} + \text{terms of higher order.}$$

We introduce the further approximations which consist in suppressing the term in $d\zeta/dt$ in the second equation, and setting $N = mg$. The first two equations thus become:

$$\text{A)} \qquad \begin{cases} \dfrac{d^2\xi}{dt^2} = -\dfrac{g}{l}\,\xi - 2\omega \sin\lambda \dfrac{d\eta}{dt} \\[2ex] \dfrac{d^2\eta}{dt^2} = -\dfrac{g}{l}\,\eta + 2\omega \sin\lambda \dfrac{d\xi}{dt} \end{cases}$$

Discussion of the Equations. Multiply the first equation A) through by $d\xi/dt$, the second by $d\eta/dt$, and add. The resulting equation,

$$\frac{d\xi}{dt}\frac{d^2\xi}{dt^2} + \frac{d\eta}{dt}\frac{d^2\eta}{dt^2} = -\frac{g}{l}\left(\xi\frac{d\xi}{dt} + \eta\frac{d\eta}{dt}\right)$$

integrates into the equation of energy:

$$\frac{v^2}{2} = -\frac{g}{l}\frac{\xi^2 + \eta^2}{2} + k,$$

or, on introducing polar coordinates,

$$\text{(1)} \qquad \frac{dr^2}{dt^2} + r^2\frac{d\theta^2}{dt^2} = -\frac{g}{l}r^2 + h.$$

Next, multiply Equations A) by η and ξ respectively, and subtract:

$$\xi \frac{d^2\eta}{dt^2} - \eta \frac{d^2\xi}{dt^2} = 2\omega \sin \lambda \left(\xi \frac{d\xi}{dt} + \eta \frac{d\eta}{dt} \right).$$

This integrates into

$$\xi \frac{d\eta}{dt} - \eta \frac{d\xi}{dt} = \omega' r^2 + C,$$

or

$$(2) \qquad r^2 \frac{d\theta}{dt} = \omega' r^2 + C,$$

where $\omega' = \omega \sin \lambda$.

A Special Case. Let the pendulum be projected with a small velocity from the point of equilibrium. Then initially $r = 0$; hence $C = 0$ and

$$\frac{d\theta}{dt} = \omega'.$$

It follows, then, that

$$\theta = \omega' t.$$

This means that, if the motion be referred to moving axes, so chosen that ζ' coincides with ζ, but ξ' makes an angle $\omega' t$ with ξ, the pendulum will swing in the (ξ', ζ')-plane. It is now easy to determine r as a function of t from (1); r executes simple harmonic motion.

The General Case. Returning now to the general case, let the motion be referred to a moving plane through O (the point of equilibrium of the pendulum), perpendicular to the ζ-axis, and rotating with constant angular velocity ω' about O. Then

$$(3) \qquad \varphi = \theta - \omega' t$$

is the angular coordinate in the new plane. Equation (2) now becomes:

$$(4) \qquad r^2 \frac{d\varphi}{dt} = C,$$

and this is the equation of areas in its usual form.

Equation (1) goes over into:

$$\frac{dr^2}{dt^2} + r^2 \left(\frac{d\varphi}{dt} + \omega' \right)^2 = - \frac{g}{l} r^2 + h$$

or

$$(5) \qquad \frac{dr^2}{dt^2} + r^2 \frac{d\varphi^2}{dt^2} + 2\omega' C + r^2 \omega'^2 = - \frac{g}{l} r^2 + h.$$

On suppressing the term $r^2\omega'^2$ because of its smallness, we find:

$$(6)\qquad \frac{dr^2}{dt^2} + r^2\frac{d\varphi^2}{dt^2} = -\frac{g}{l}r^2 + h'.$$

But this is precisely the equation of energy corresponding to an attracting central force of intensity $\frac{mg}{l}r$. Hence the motion is elliptic with O at the centre; *i.e.* the pendulum, once released, describes a fixed ellipse in the moving plane. The axes of this ellipse rotate in the positive sense, *i.e.* the clockwise sense, as one looks down on the earth. But the pendulum describes the ellipse in either sense, according to the initial conditions, the degenerate case of the straight line lying between the description in positive sense and that in negative sense. In the Foucault experiment in the Pantheon the pendulum was slightly displaced from the position of equilibrium and released from rest relative to the earth. It then described the ellipse in the negative sense. For initially dr/dt was 0, so that it started from the extremity of an axis (obviously the major axis) and its initial motion relative to the moving plane was in the negative sense of rotation; *i.e.* counter clockwise. At the end of twenty-four hours, t has the value: $t = 24 \times 60 \times 60$, and hence

$$\theta = \omega' t = 2\pi \sin \lambda.$$

The result checks, for at the equator θ should be 0, and at the North Pole, 2π.

EXERCISE

Obtain the equations of motion A), directly from Lagrange's Equations, Chapter X.

CHAPTER X

LAGRANGE'S EQUATIONS AND VIRTUAL VELOCITIES

INTRODUCTION

In the preceding chapters, the treatment has been based on Newton's Second Law of Motion. Work and Energy have entered as derived concepts. It is true that certain general theorems have been established, whereby some of the forces of constraint have been eliminated, like the theorem relating to the motion of the centre of gravity, and the theorem of rotation of a rigid body. But in the last analysis, when there have been forces of constraint which have not annulled one another in pairs, the setting up of the problem has involved explicitly any unknown forces of constraint, as well as the known forces, and the former have then been eliminated analytically, anew in each new problem.

We turn now to methods whereby, in certain important cases, the forces of constraint can be eliminated once for all, so that they will not even enter in setting up the equations on whose solution the problem depends. Moreover, we introduce *intrinsic coordinates* and *intrinsic functions*. The *intrinsic coordinates* are a minimum number of independent variables whose values locate completely the system. They are often called *generalized coordinates*, and are denoted by $q_1, \cdots, q_m$. The intrinsic functions are the kinetic energy, the work function or its negative, the potential energy, and the Lagrangean function L. These we have called *intrinsic* because they do not depend on any special coordinate system, or on any special choice of the q's. Later, we shall consider intermediate cases in which the number of q's, though highly restricted, is not a minimum, and in which, moreover, the unknown forces, or constraints, have not been wholly eliminated.

1. The Problem. A material system may be determined in its position by one or more coordinates, $q_1, \cdots, q_n$,* and the

* We choose, in general, the letter m to denote the number of the q's. But we replace it by n in these early examples to avoid confusion with the m that refers to the mass of the particle.

time, t. For example, let a bead of mass m slide freely on a smooth circular wire, which rotates in a horizontal plane about one of its points, O, with constant angular velocity. The angle φ that the radius drawn from O makes with a fixed horizontal line is given explicitly,

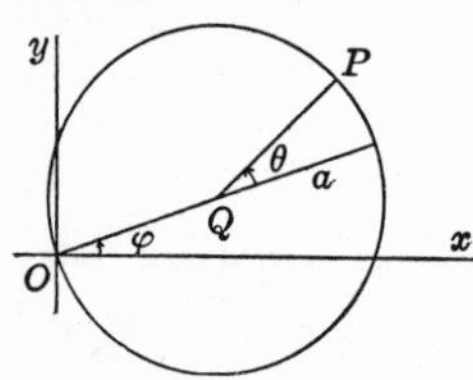

FIG. 141

$$\varphi = \omega t.$$

Let θ be the angle from OQ produced to the radius, QP, drawn to the bead. Then the position of m is fully determined by θ and t. Thus if we set $\theta = q$,

$$x = f(q, t), \qquad y = \psi(q, t).$$

The problem of determining the motion is that of finding q as a function of t.

More generally, let a smooth wire, carrying a bead, move according to any law, and let the bead be acted on by any forces. To determine the motion. We will treat this problem in detail presently.

As the second illustrative example, consider n masses, $m_1, \cdots, m_n$, fastened to a weightless inextensible flexible string, one end, O, of which is held fast, and let the system be slightly displaced from the position of equilibrium. To determine the oscillation.

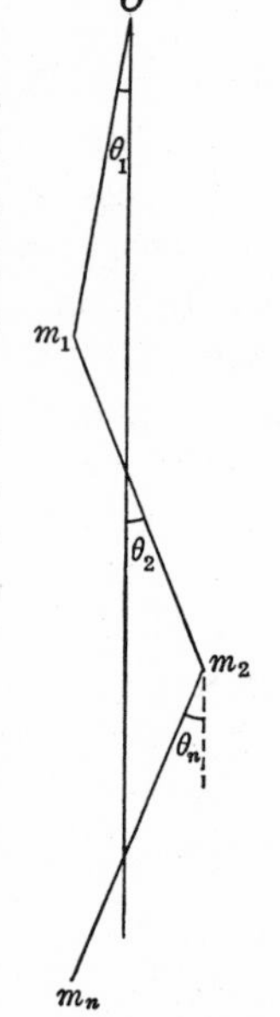

FIG. 142

Finally, we may think of a rigid body, acted on by any forces. If there are no constraints, it will require six coordinates, $q_1, q_2, \cdots, q_6$, to determine the position. These may be the three coordinates of one of the points of the body, as the centre of gravity, $\bar{x}$, $\bar{y}$, $\bar{z}$; and the three Eulerian angles, θ, φ, ψ, which determine the orientation of the body.

We may, however, also introduce constraints. If one point is fixed, there are three degrees of freedom, and so three coordinates, q_1, q_2, q_3, — for example, the Eulerian angles — are required. Or, again, the body might rotate about a fixed axis. Then $n = 1$, and $q_1 = q$, — a single coordinate — would be sufficient. Or, finally, the body might be free to slide along a fixed line and rotate about it. Here $n = 2$ and q_1, q_2 are the coordinates.

Each of the last two examples may be varied by causing the line to move in an altogether specified manner. Then, beside $q_1 = q$, or q_1 and q_2, the time, t, would enter explicitly.

In all such cases, the motion is determined by *Lagrange's Equations*, which, when there is a force function U, take the form:

$$\text{A)} \qquad \frac{d}{dt}\frac{\partial T}{\partial \dot{q}_r} - \frac{\partial T}{\partial q_r} = \frac{\partial U}{\partial q_r}, \qquad r = 1, \cdots, m,$$

where T denotes the kinetic energy, and $\dot{q}_r = dq_r/dt$. We turn now to the establishment of these equations, beginning with the simplest cases.

2. Lagrange's Equations in the Simplest Case. Let a bead slide on a smooth wire whose form as well as position varies with the time:

$$(1) \qquad x = f(q, t), \qquad y = \varphi(q, t), \qquad z = \psi(q, t),$$

where the functions $f(q, t)$, $\varphi(q, t)$, $\psi(q, t)$ are continuous together with whatever derivatives we need to use, and where

$$\frac{\partial x}{\partial q}, \qquad \frac{\partial y}{\partial q}, \qquad \frac{\partial z}{\partial q}$$

do not all vanish simultaneously. The motion is determined by the equations:

$$(2) \qquad \left\{ \begin{aligned} m\frac{d^2x}{dt^2} &= X + \mathsf{X} \\ m\frac{d^2y}{dt^2} &= Y + \mathsf{Y} \\ m\frac{d^2z}{dt^2} &= Z + \mathsf{Z}, \end{aligned} \right.$$

where X, Y, Z refer to the given, or *applied forces*, *i.e.* forces other than the reaction, $(\mathsf{X}, \mathsf{Y}, \mathsf{Z})$, of the wire, and

$$(3) \qquad \lambda\mathsf{X} + \mu\mathsf{Y} + \nu\mathsf{Z} = 0,$$

where (λ, μ, ν) are the direction components of the tangent to the wire — for the reaction of the wire is normal to the wire, though otherwise unknown.

Multiply these equations through by $\partial x/\partial q$, $\partial y/\partial q$, $\partial z/\partial q$ respectively and add:

$$m\left(\frac{d^2x}{dt^2}\frac{\partial x}{\partial q} + \frac{d^2y}{dt^2}\frac{\partial y}{\partial q} + \frac{\partial^2 z}{\partial t^2}\frac{\partial z}{\partial q}\right) = Q, \tag{4}$$

$$Q = X\frac{\partial x}{\partial q} + Y\frac{\partial y}{\partial q} + Z\frac{\partial z}{\partial q}, \tag{5}$$

the remaining terms, namely:

$$\mathsf{X}\frac{\partial x}{\partial q} + \mathsf{Y}\frac{\partial y}{\partial q} + \mathsf{Z}\frac{\partial z}{\partial q},$$

vanishing because of (3), since $\partial x/\partial q$, $\partial y/\partial q$, $\partial z/\partial q$ are the direction components of the tangent to the wire.

The left-hand side of Equation (4) can be transformed as follows. We write:

$$T = \frac{m}{2}(\dot{x}^2 + \dot{y}^2 + \dot{z}^2), \tag{6}$$

where the dot notation means a time derivative:

$$\dot{x} = \frac{dx}{dt}, \qquad \dot{q} = \frac{dq}{dt}, \qquad \text{etc.}$$

From (1)

$$\frac{dx}{dt} = \frac{\partial x}{\partial q}\frac{dq}{dt} + \frac{\partial x}{\partial t},$$

or

$$\dot{x} = \frac{\partial x}{\partial q}\dot{q} + \frac{\partial x}{\partial t}, \tag{7}$$

with similar expressions for $\dot{y}$ and $\dot{z}$. On substituting these values in (6), T becomes a function of q, $\dot{q}$, t. And now the left-hand side of Equation (4) turns out to be expressible in the form:

$$\frac{d}{dt}\frac{\partial T}{\partial \dot{q}} - \frac{\partial T}{\partial q}. \tag{8}$$

For, first, we have:

$$\frac{\partial T}{\partial \dot{q}} = m\left(\dot{x}\frac{\partial \dot{x}}{\partial \dot{q}} + \dot{y}\frac{\partial \dot{y}}{\partial \dot{q}} + \dot{z}\frac{\partial \dot{z}}{\partial \dot{q}}\right).$$

From (7) it follows that

$$\frac{\partial \dot{x}}{\partial \dot{q}} = \frac{\partial x}{\partial q}$$

with similar expressions for $\partial \dot{y}/\partial \dot{q}$ and $\partial \dot{z}/\partial \dot{q}$. Thus

$$\frac{\partial T}{\partial \dot{q}} = m\left(\dot{x}\frac{\partial x}{\partial q} + \dot{y}\frac{\partial y}{\partial q} + \dot{z}\frac{\partial z}{\partial q}\right).$$

Next, differentiate with respect to the time:

$$\frac{d}{dt}\frac{\partial T}{\partial \dot{q}} = m\left(\frac{d^2x}{dt^2}\frac{\partial x}{\partial q} + \frac{d^2y}{dt^2}\frac{\partial y}{\partial q} + \frac{d^2z}{dt^2}\frac{\partial z}{\partial q}\right) \tag{9}$$

$$+ m\left(\dot{x}\frac{d}{dt}\frac{\partial x}{\partial q} + \dot{y}\frac{d}{dt}\frac{\partial y}{\partial q} + \dot{z}\frac{d}{dt}\frac{\partial z}{\partial q}\right).$$

On the other hand,

$$\frac{\partial T}{\partial q} = m\left(\dot{x}\frac{\partial \dot{x}}{\partial q} + \dot{y}\frac{\partial \dot{y}}{\partial q} + \dot{z}\frac{\partial \dot{z}}{\partial q}\right). \tag{10}$$

Now,

$$\frac{\partial \dot{x}}{\partial q} = \frac{d}{dt}\frac{\partial x}{\partial q}. \tag{11}$$

For, from (7),

$$\frac{\partial \dot{x}}{\partial q} = \frac{\partial^2 x}{\partial q^2}\dot{q} + \frac{\partial^2 x}{\partial q\,\partial t};$$

and, of course,

$$\frac{d}{dt}\frac{\partial x}{\partial q} = \frac{\partial^2 x}{\partial q^2}\frac{dq}{dt} + \frac{\partial^2 x}{\partial t\,\partial q}.$$

Substituting the value given by (11) and the corresponding values for $\partial \dot{y}/\partial q$, $\partial \dot{z}/\partial q$ in the right-hand side of (10), we see that $\partial T/\partial q$ is equal to the last half of the right-hand side of (9), and thus the proof is complete: — the left-hand side of (4) has the value (8). We arrive, then, at the final result:

$$\frac{d}{dt}\frac{\partial T}{\partial \dot{q}} - \frac{\partial T}{\partial q} = Q. \tag{12}$$

This is precisely *Lagrange's Equation* for the present case.

The case that the applied forces have a *force function*, U, is of prime importance in practice. Here

$$X = \frac{\partial U}{\partial x}, \qquad Y = \frac{\partial U}{\partial y}, \qquad Z = \frac{\partial U}{\partial z},$$

and thus Q becomes:

$$Q = \frac{\partial U}{\partial x}\frac{\partial x}{\partial q} + \frac{\partial U}{\partial y}\frac{\partial y}{\partial q} + \frac{\partial U}{\partial z}\frac{\partial z}{\partial q}.$$

Lagrange's Equation now takes the form:

$$\frac{d}{dt}\frac{\partial T}{\partial \dot{q}} - \frac{\partial T}{\partial q} = \frac{\partial U}{\partial q}. \tag{13}$$

Example. Consider the problem stated at the opening of the paragraph. Here, the applied forces are absent, and so U = const. Furthermore,

$$\begin{cases} x = a\cos\omega t + a\cos(\theta + \omega t) \\ y = a\sin\omega t + a\sin(\theta + \omega t) \end{cases}$$

where $q = \theta$; and

$$\begin{cases} \dot{x} = -a\omega\sin\omega t - a(\dot{\theta} + \omega)\sin(\theta + \omega t) \\ \dot{y} = a\omega\cos\omega t + a(\dot{\theta} + \omega)\cos(\theta + \omega t) \end{cases}$$

$$T = \frac{ma^2}{2}\left((\dot{\theta} + \omega)^2 + 2\omega(\dot{\theta} + \omega)\cos\theta + \omega^2\right)$$

$$\frac{\partial T}{\partial \dot{\theta}} = ma^2(\dot{\theta} + \omega + \omega\cos\theta), \qquad \frac{\partial T}{\partial \theta} = -ma^2\omega(\dot{\theta} + \omega)\sin\theta,$$

$$\frac{d}{dt}\frac{\partial T}{\partial \dot{\theta}} - \frac{\partial T}{\partial \theta} = ma^2\left(\frac{d^2\theta}{dt^2} + \omega^2\sin\theta\right) = 0,$$

or

$$\frac{d^2\theta}{dt^2} = -\omega^2\sin\theta.$$

This last is the equation of Simple Pendulum Motion,

$$\frac{d^2\theta}{dt^2} = -\frac{g}{l}\sin\theta.$$

Thus the bead oscillates about the moving line OQ as a simple pendulum of length

$$l = \frac{g}{\omega^2}$$

would oscillate about the vertical.

EXERCISES

1. A bead slides on a smooth circular wire which is rotating with constant angular velocity about a fixed vertical diameter. Show that

$$\frac{d^2\theta}{dt^2} = \omega^2\sin\theta\cos\theta + \frac{g}{a}\sin\theta.$$

2. If the bead is released with no vertical velocity from a point on the level of the centre of the circle, show that it will not reach the lowest point if

$$\omega > \sqrt{\frac{2g}{a}}.$$

3. A bead slides on a smooth rod which is rotating about one end in a vertical plane with uniform angular velocity. Show that

$$\frac{d^2r}{dt^2} = \omega^2 r + g \sin \omega t.$$

4. Integrate the differential equation of the preceding question.

5. A bead slides on a smooth rod, one end of which is fixed, and the inclination of which does not change. Determine the motion, if the vertical plane through the rod rotates with constant angular velocity.

6. In the problem discussed in the text, determine the reaction, N, of the wire. *Ans.* $N = ma^2\left[\omega^2 \cos\theta + (\dot{\theta} + \omega)^2\right]$.

7. A smooth circular wire rotates with constant angular velocity about a vertical axis which lies in the plane of the circle. A bead slides on the wire. Determine the motion.

8. Show that if, in Question 1, the axis is a horizontal diameter, the motion is given by the equation:

$$\frac{d^2\varphi}{dt^2} - \left(\omega^2 \cos\varphi + \frac{g}{a} \cos \omega t\right) \sin\varphi = 0,$$

where φ is the angle which the radius drawn to the bead makes with the radius perpendicular to the axis.

9. A bead can slide on a smooth circular wire which is expanding, always remaining in a fixed plane. One point of the wire is fixed, and the centre describes a right line with constant velocity. Determine the motion of the bead. *Ans.* $\theta = \alpha + \frac{\beta}{t}$.

10. The same problem if the centre is at rest and the radius increases at an arbitrary rate.

3. Continuation. Particle on a Fixed or Moving Surface. Let a particle, of mass m, be constrained to move on a smooth surface, which can vary in size and shape,

(1) $\quad x = f(q_1, q_2, t), \qquad y = \varphi(q_1, q_2, t), \qquad z = \psi(q_1, q_2, t),$

where the functions on the right are continuous, together with whatever derivatives we wish to use, and the Jacobians

$$(2)\qquad \frac{\partial(y, z)}{\partial(q_1, q_2)}, \qquad \frac{\partial(z, x)}{\partial(q_1, q_2)}, \qquad \frac{\partial(x, y)}{\partial(q_1, q_2)}$$

do not vanish simultaneously. The motion is given as before by Equations (2) of § 2, where now, however, the reaction, due to the surface, is known in direction completely.

Multiply these equations through respectively by $\partial x/\partial q_1$, $\partial y/\partial q_1$, $\partial z/\partial q_1$, and add. On the right-hand side there remains only

$$(3)\qquad Q_1 = X\frac{\partial x}{\partial q_1} + Y\frac{\partial y}{\partial q_1} + Z\frac{\partial z}{\partial q_1},$$

since $\partial x/\partial q_1$, etc. are the direction components of a certain line in the surface, drawn from (x, y, z), and X, Y, Z are the components of a force normal to the surface. Hence

$$(4)\qquad m\left(\frac{d^2x}{dt^2}\frac{\partial x}{\partial q_1} + \frac{d^2y}{dt^2}\frac{\partial y}{\partial q_1} + \frac{d^2z}{dt^2}\frac{\partial z}{\partial q_1}\right) = Q_1.$$

The reduction of the left-hand side is similar to the reduction in the earlier case. It is, however, just as easy to carry this reduction through for a system with n degrees of freedom, and this is done in the following paragraph. Thus we see that

$$\frac{d}{dt}\frac{\partial T}{\partial \dot{q}_1} - \frac{\partial T}{\partial q_1} = Q_1;$$

and, similarly:

$$\frac{d}{dt}\frac{\partial T}{\partial \dot{q}_2} - \frac{\partial T}{\partial q_2} = Q_2.$$

These are *Lagrange's Equations* for the case of two variables; *i.e.* the case of (q_1, q_2). If, in particular, a force function, U, exists, then Lagrange's Equations take the form:

$$\text{A)}\qquad \frac{d}{dt}\frac{\partial T}{\partial \dot{q}_1} - \frac{\partial T}{\partial q_1} = \frac{\partial U}{\partial q_1}, \qquad \frac{d}{dt}\frac{\partial T}{\partial \dot{q}_2} - \frac{\partial T}{\partial q_2} = \frac{\partial U}{\partial q_2}.$$

Example. A particle is constrained to move, without friction, in a plane which is rotating with constant angular velocity about a horizontal axis. Determine the motion.

Let the axis of rotation be taken as the axis of x, and let r denote the distance of the particle from the axis. The coordinates of the particle are (x, y, z), where

$$y = r\cos\theta, \quad z = r\sin\theta, \quad \theta = \omega t,$$

and we take $q_1 = x$, $q_2 = r$. Then

$$\begin{cases} \dot{y} = \dot{r}\cos\omega t - r\omega\sin\omega t, \\ \dot{z} = \dot{r}\sin\omega t + r\omega\cos\omega t, \end{cases}$$

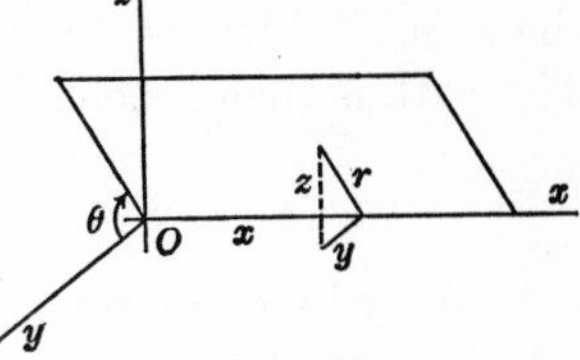

FIG. 143

$$T = \frac{m}{2}(\dot{x}^2 + \dot{y}^2 + \dot{z}^2) = \frac{m}{2}(\dot{x}^2 + \dot{r}^2 + \omega^2 r^2),$$

a result that may be read off directly, without the intervention of y, z. Furthermore,

$$U = -mgz = -mgr\sin\omega t.$$

Lagrange's Equations now become:

$$\frac{\partial T}{\partial \dot{x}} = m\dot{x}, \qquad \frac{\partial T}{\partial x} = 0, \qquad \frac{\partial U}{\partial x} = 0;$$

$$m\frac{d^2x}{dt^2} = 0,$$

a result immediately obvious. Next,

$$\frac{\partial T}{\partial \dot{r}} = m\dot{r}, \qquad \frac{\partial T}{\partial r} = m\omega^2 r, \qquad \frac{\partial U}{\partial r} = -\,mg\sin\omega t;$$

$$m\ddot{r} - m\omega^2 r = -\,mg\sin\omega t,$$

or

$$\frac{d^2r}{dt^2} - \omega^2 r = -g\sin\omega t.$$

A special solution of this equation is found, either by the method of the *variation of constants* or, more simply, by inspection, to be:

$$r = \frac{g\sin\omega t}{2\omega^2}.$$

Hence the general solution is

$$r = A\,e^{\omega t} + B\,e^{-\omega t} + \frac{g\sin\omega t}{2\omega^2}.$$

Since $\theta = \omega t$, the equation can be written in the form:

$$r = A\,e^{\theta} + B\,e^{-\theta} + \frac{g\sin\theta}{2\omega^2}.$$

A particular case of interest is that in which $A = B = 0$. This corresponds to the initial conditions of launching the particle from a point in the axis with a velocity whose projection along the axis is arbitrary, the projection normal to the axis being $g/2\omega$. The path is then a helix.

EXERCISES

1. A cylinder of revolution is rotating with constant angular velocity about a vertical axis, exterior to the cylinder, the axis of the cylinder being always vertical. A particle is projected along the inner surface, which is smooth. Determine the motion.

2. Use Lagrange's Equations to determine the motion of a particle in a plane, referred to polar coordinates:

$$m\left(\frac{d^2r}{dt^2} - r\frac{d\theta^2}{dt^2}\right) = R, \qquad \frac{m}{r}\frac{d}{dt}\left(r^2\frac{d\theta}{dt}\right) = \Theta.$$

4. The Spherical Pendulum. Consider the spherical pendulum; *i.e.* a particle moving under gravity and constrained to lie on a smooth sphere. Take as coordinates the colatitude, θ, and the longtitude, φ, the north pole being the point of unstable equilibrium. Then

$$T = \frac{ma^2}{2}(\dot\theta^2 + \dot\varphi^2 \sin^2\theta), \qquad U = -mgz = -mga\cos\theta;$$

$$\frac{\partial T}{\partial\dot\theta} = ma^2\dot\theta, \qquad \frac{\partial T}{\partial\theta} = ma^2\dot\varphi^2\sin\theta\cos\theta, \qquad \frac{\partial U}{\partial\theta} = mga\sin\theta,$$

and the first of Lagrange's Equations becomes:

$$ma^2\ddot\theta - ma^2\dot\varphi^2\sin\theta\cos\theta = mga\sin\theta,$$

or

$$\frac{d^2\theta}{dt^2} - \left(\frac{d\varphi}{dt}\right)^2\sin\theta\cos\theta = \frac{g}{a}\sin\theta. \tag{1}$$

Proceeding now to the second equation, we have:

$$\frac{\partial T}{\partial\dot\varphi} = ma^2\dot\varphi\sin^2\theta, \qquad \frac{\partial T}{\partial\varphi} = 0, \qquad \frac{\partial U}{\partial\varphi} = 0.$$

Hence

$$\frac{d}{dt}\left(\frac{d\varphi}{dt}\sin^2\theta\right) = 0.$$

This equation integrates into

$$\frac{d\varphi}{dt}\sin^2\theta = h, \tag{2}$$

where the constant h is of dimension -1 in the time, $[T^{-1}]$.

Combining Equations (1) and (2), we obtain:

$$\frac{d^2\theta}{dt^2} = \frac{h^2\cos\theta}{\sin^3\theta} + \frac{g}{a}\sin\theta. \tag{3}$$

This is the Equation of Spherical Pendulum Motion. In case the motion is to be studied for small oscillations near the lowest point of the sphere, it is well to replace θ by its supplement, $\theta' = \pi - \theta$. Equation (3) then becomes:

$$\frac{d^2\theta'}{dt^2} = \frac{h^2\cos\theta'}{\sin^3\theta'} - \frac{g}{a}\sin\theta'. \tag{3'}$$

This last equation reduces to the Equation of Simple Pendulum Motion when $h = 0$. Any differential equation of the form:

$$\frac{d^2\theta}{dt^2} = A\frac{\cos\theta}{\sin^3\theta} + B\sin\theta, \tag{4}$$

where $A > 0$ and $B > 0$ are arbitrary constants, can obviously be interpreted in terms of spherical pendulum motion.

A first integral of (3) can be obtained in the usual manner:

$$2\frac{d\theta}{dt}\frac{d^2\theta}{dt^2} = \frac{2h^2\cos\theta}{\sin^3\theta}\frac{d\theta}{dt} + \frac{2g}{a}\sin\theta\frac{d\theta}{dt}.$$

Integrating each side with respect to θ, we obtain:

$$\frac{d\theta^2}{dt^2} = -\frac{h^2}{\sin^2\theta} - \frac{2g}{a}\cos\theta + k. \tag{5}$$

For a further discussion of the problem, cf. Appell, *Mécanique rationelle*, vol. i, § 277.

EXERCISES

1. Give an approximate solution for small oscillations near the point of stable equilibrium, O, using Cartesian coordinates. Here,

$$T = \frac{m}{2}(\dot{x}^2 + \dot{y}^2), \qquad U = -\frac{mg}{2a}(x^2 + y^2),$$

and the approximate path is an ellipse with O as centre.

2. Treat the motion of a particle constrained to move on a smooth surface,

$$2z = \alpha^2 x^2 + \beta^2 y^2,$$

for small oscillations near the origin.

3. Show that a top which is not spinning moves like a spherical pendulum. More precisely, we mean the body of Chapter VI, § 18, when $\nu = 0$.

5. Geodesics. Let a particle be constrained to move on a smooth surface under no applied forces. The path is a geodesic.*

Let the surface be given by the equations:

$$x = f(u, v), \qquad y = \varphi(u, v), \qquad z = \psi(u, v),$$

where these functions are continuous together with their first derivatives, and not all the Jacobians

$$\frac{\partial(y, z)}{\partial(u, v)}, \qquad \frac{\partial(z, x)}{\partial(u, v)}, \qquad \frac{\partial(x, y)}{\partial(u, v)}$$

vanish. The element of arc is given by the formula:

$$ds^2 = E du^2 + 2F du dv + G dv^2,$$

where the coefficients are easily computed,

$$E = \frac{\partial x^2}{\partial u^2} + \frac{\partial y^2}{\partial u^2} + \frac{\partial z^2}{\partial u^2}, \quad \text{etc.}$$

The kinetic energy has the value:

$$T = \frac{m}{2}(E\dot{u}^2 + 2F\dot{u}\dot{v} + G\dot{v}^2).$$

Lagrange's Equations now become, since $U = 0$:

$$(6) \qquad \begin{cases} \dfrac{d}{dt}(E\dot{u} + F\dot{v}) - \tfrac{1}{2}(E_u \dot{u}^2 + 2F_u \dot{u}\dot{v} + G_u \dot{v}^2) = 0, \\[2ex] \dfrac{d}{dt}(F\dot{u} + G\dot{v}) - \tfrac{1}{2}(E_v \dot{u}^2 + 2F_v \dot{u}\dot{v} + G_v \dot{v}^2) = 0. \end{cases}$$

On the other hand, the geodesics, in their capacity of being the shortest lines on the surface, are given as extremals of the integral

$$L = \int_{\lambda_0}^{\lambda_1} \sqrt{E u'^2 + 2F u' v' + G v'^2}\, d\lambda$$

* By a *geodesic* is meant a line of minimum length on a surface — minimum, at least, if the points it connects are not too far apart; cf. *Advanced Calculus*, p. 411.

in the Calculus of Variations (cf. the Author's *Advanced Calculus*, p. 411) by the equations:

$$
(7)\quad \begin{cases} \dfrac{d}{d\lambda}\dfrac{Eu' + Fv'}{\sqrt{Eu'^2 + 2Fu'v' + Gv'^2}} - \dfrac{E_u u'^2 + 2F_u u'v' + G_u v'^2}{2\sqrt{Eu'^2 + 2Fu'v' + Gv'^2}} = 0 \\[2ex] \dfrac{d}{d\lambda}\dfrac{Fu' + Gv'}{\sqrt{Eu'^2 + 2Fu'v' + Gv'^2}} - \dfrac{E_v u'^2 + 2F_v u'v' + G_v v'^2}{2\sqrt{Eu'^2 + 2Fu'v' + Gv'^2}} = 0 \end{cases}
$$

The parameter λ can be replaced by any other parameter, μ:

$$\mu = f(\lambda),$$

provided that $f(\lambda)$ is continuous, together with its first derivative, and $f'(\lambda) \neq 0$. In particular, then, the choice $t = \mu$ is a possible one. But then, because of the equation of energy, $T = h$, or:

$$
(8)\quad \frac{m}{2}(Eu'^2 + 2Fu'v' + Gv'^2) = h, \qquad u' = \dot{u}, \qquad v' = \dot{v},
$$

it follows that Equations (7) reduce to Equations (6).

Since the velocity along the path is constant, the only force being normal to the path, t is proportional to s. In fact, (8) says that

$$\frac{m}{2}\frac{ds}{dt} = h.$$

Thus the transformation of the parameter from λ to t amounts in substance to a transformation to s; *i.e.* Equations (6) are virtually the intrinsic differential equations of the geodesic:

$$
(9)\quad \begin{cases} \dfrac{d}{ds}(Eu' + Fv') - \frac{1}{2}(E_u u'^2 + 2F_u u'v' + G_u v'^2) = 0 \\[2ex] \dfrac{d}{ds}(Fu' + Gv') - \frac{1}{2}(E_v u'^2 + 2F_v u'v' + G_v v'^2) = 0 \end{cases}
$$

$$u' = \frac{du}{ds}, \qquad v' = \frac{dv}{ds}.$$

EXERCISES

1. Obtain the geodesics on a cylinder of revolution. Observe that, when the cylinder is rolled out on a plane, the geodesics must go over into straight lines.

2. The same problem for a cone of revolution.

3. Show that the geodesics on an anchor ring, or torus, are given by the differential equations:

$$(10) \qquad \begin{cases} \dfrac{d^2\theta}{dt^2} + \dfrac{h^2 \sin\theta}{a\,(b + a\cos\theta)^3} = 0 \\[2ex] (b + a\cos\theta)^2 \dfrac{d\varphi}{dt} = h, \end{cases}$$

where a and $b > a$ are the constants of the anchor ring, and h is a constant of integration.

4. Show that if, in the preceding question, initially $\theta = \pi/2$, $d\theta/dt = \alpha$, $d\varphi/dt = \beta$, then

$$(11) \qquad \begin{cases} \dfrac{d\theta^2}{dt^2} = \dfrac{b^4\beta^2}{a^2(b + a\cos\theta)^2} - \dfrac{b^2\beta^2}{a^2} + \alpha^2, \\[2ex] \dfrac{d\varphi}{dt} = \dfrac{b^2\beta}{(b + a\cos\theta)^2}. \end{cases}$$

6. Lemma. We have seen in § 2 that, in the case of a single q,

$$(1) \qquad \frac{d}{dt}\frac{\partial T}{\partial \dot q} - \frac{\partial T}{\partial q} = m\left(\frac{d^2x}{dt^2}\frac{\partial x}{\partial q} + \frac{d^2y}{dt^2}\frac{\partial y}{\partial q} + \frac{d^2z}{dt^2}\frac{\partial z}{\partial q}\right).$$

It is important to recognize this equation as a purely analytic identity, irrespective of any physical meaning to be attached to T. It says that, if

$$T = \frac{m}{2}(\dot x^2 + \dot y^2 + \dot z^2)$$

and

$$x = f(q, t), \qquad y = \varphi(q, t), \qquad z = \psi(q, t),$$

where these functions are any functions subject merely to the ordinary requirements of continuity, then Equation (1) is true.

We turn now to the general case of n particles, m_i with the coordinates (x_i, y_i, z_i), $i = 1, 2, \cdots, n$. Let the position of the system be determined by m parameters, or *generalized coordinates*, $q_1, \cdots, q_m$, and the time, t:

$$\text{A)} \qquad \begin{cases} x_i = f_i(q_1, \cdots, q_m, t) \\ y_i = \varphi_i(q_1, \cdots, q_m, t) \\ z_i = \psi_i(q_1, \cdots, q_m, t) \end{cases}$$

where the functions f_i, φ_i, ψ_i, are continuous together with their partial derivatives of the first two orders, and where the rank of the matrix a) is m:

a)
$$\left\| \begin{matrix} \dfrac{\partial x_1}{\partial q_1} & \cdots\cdots & \dfrac{\partial x_1}{\partial q_m} \\ \cdots & \cdots & \cdots \\ \dfrac{\partial z_n}{\partial q_1} & \cdots\cdots & \dfrac{\partial z_n}{\partial q_m} \end{matrix} \right\|$$

The kinetic energy,

$$T = \sum_{i=1}^{n} \frac{m_i}{2} (\dot{x}_i^2 + \dot{y}_i^2 + \dot{z}_i^2), \tag{2}$$

can be expressed in terms of $q_1, \cdots, q_m, \dot{q}_1, \cdots, \dot{q}_m, t$, for

$$\dot{x}_i = \sum_{r=1}^{m} \frac{\partial x_i}{\partial q_r} \dot{q}_r + \frac{\partial x_i}{\partial t}, \qquad i = 1, \cdots, n, \tag{3}$$

with similar formulas for $\dot{y}_i$, $\dot{z}_i$.

Our aim is to establish the general fundamental formula corresponding to (1):

$$\text{I.} \quad \frac{d}{dt}\frac{\partial T}{\partial \dot{q}_r} - \frac{\partial T}{\partial q_r} = \sum_{i=1}^{n} m_i \left(\frac{d^2 x_i}{dt^2} \frac{\partial x_i}{\partial q_r} + \frac{d^2 y_i}{dt^2} \frac{\partial y_i}{\partial q_r} + \frac{d^2 z_i}{dt^2} \frac{\partial z_i}{\partial q_r} \right), \qquad r = 1, \cdots, m.$$

The independent variables in the partial differentiations are $(q_1, \cdots, q_m, \dot{q}_1, \cdots, \dot{q}_m, t)$, and $\dot{x}_i$, $\dot{y}_i$, $\dot{z}_i$ are given by (3). We have:

$$\frac{\partial T}{\partial \dot{q}_r} = \sum_i m_i \left(\dot{x}_i \frac{\partial \dot{x}_i}{\partial \dot{q}_r} + \dot{y}_i \frac{\partial \dot{y}_i}{\partial \dot{q}_r} + \dot{z}_i \frac{\partial \dot{z}_i}{\partial \dot{q}_r} \right).$$

From (3):

$$\frac{\partial \dot{x}_i}{\partial \dot{q}_r} = \frac{\partial x_i}{\partial q_r}, \qquad \frac{\partial \dot{y}_i}{\partial \dot{q}_r} = \text{etc.}$$

Hence

$$\frac{\partial T}{\partial \dot{q}_r} = \sum_i m_i \left(\dot{x}_i \frac{\partial x_i}{\partial q_r} + \dot{y}_i \frac{\partial y_i}{\partial q_r} + \dot{z}_i \frac{\partial z_i}{\partial q_r} \right).$$

Differentiate with respect to t along a given curve:

$$\frac{d}{dt}\frac{\partial T}{\partial \dot{q}_r} = \sum_i m_i \left(\frac{d^2 x_i}{dt^2} \frac{\partial x_i}{\partial q_r} + \frac{d^2 y_i}{dt^2} \frac{\partial y_i}{\partial q_r} + \frac{d^2 z_i}{dt^2} \frac{\partial z_i}{\partial q_r} \right) \tag{4}$$

$$+ \sum_i m_i \left(\dot{x}_i \frac{d}{dt}\frac{\partial x_i}{\partial q_r} + \dot{y}_i \frac{d}{dt}\frac{\partial y_i}{\partial q_r} + \dot{z}_i \frac{d}{dt}\frac{\partial z_i}{\partial q_r} \right).$$

On the other hand,

$$(5)\qquad \frac{\partial T}{\partial q_r} = \sum_i m_i\left(\dot{x}_i\frac{\partial \dot{x}_i}{\partial q_r} + \dot{y}_i\frac{\partial \dot{y}_i}{\partial q_r} + \dot{z}_i\frac{\partial \dot{z}_i}{\partial q_r}\right).$$

Now,

$$\frac{\partial \dot{x}_i}{\partial q_r} = \frac{d}{dt}\frac{\partial x_i}{\partial q_r}.$$

For, from (3),

$$\frac{\partial \dot{x}_i}{\partial q_r} = \sum_s \frac{\partial^2 x_i}{\partial q_r \partial q_s}\dot{q}_s + \frac{\partial^2 x_i}{\partial q_r \partial t},$$

while

$$\frac{d}{dt}\frac{\partial x_i}{\partial q_r} = \sum_s \frac{\partial^2 x_i}{\partial q_s \partial q_r}\dot{q}_s + \frac{\partial^2 x_i}{\partial t\, \partial q_r}.$$

Similar relations hold for $\partial\dot{y}_i/\partial q_r$ and $\partial\dot{z}_i/\partial q_r$. On substituting these values in (5), it is seen that the last sum in (4) has precisely the value $\partial T/\partial q_r$, and thus the relation I. is established.

7. Lagrange's Equations in the General Case. Let a system of particles m_i with the coordinates (x_i, y_i, z_i) be acted on by any forces whatever, X_i, Y_i, Z_i. By Newton's Second Law of Motion

$$(1)\qquad \begin{cases} m_i\dfrac{d^2x_i}{dt^2} = X_i \\[2ex] m_i\dfrac{d^2y_i}{dt^2} = Y_i \\[2ex] m_i\dfrac{d^2z_i}{dt^2} = Z_i \end{cases}$$

Let the position of the system be determined, as in § 6, by m coordinates $q_1, \cdots, q_m$ and t:

$$\text{A)}\qquad \begin{cases} x_i = f_i(q_1, \cdots, q_m, t) \\ y_i = \varphi_i(q_1, \cdots, q_m, t) \\ z_i = \psi_i(q_1, \cdots, q_m, t) \end{cases}$$

where the functions f_i, φ_i, ψ_i are continuous together with their derivatives of the first two orders, the matrix

$$\text{a)}\qquad \left\|\begin{matrix} \dfrac{\partial x_1}{\partial q_1} & \cdots\cdots & \dfrac{\partial x_1}{\partial q_m} \\ \cdots & \cdots\cdots & \cdots \\ \dfrac{\partial z_n}{\partial q_1} & \cdots\cdots & \dfrac{\partial z_n}{\partial q_m} \end{matrix}\right\|$$

being of rank m. Multiply the first of Equations (1) by $\partial x_i/\partial q_r$, the second by $\partial y_i/\partial q_r$, the third by $\partial z_i/\partial q_r$, and add:

$$\sum_{i=1}^{n} m_i\left(\frac{d^2x_i}{dt^2}\frac{\partial x_i}{\partial q_r} + \frac{d^2y_i}{dt^2}\frac{\partial y_i}{\partial q_r} + \frac{d^2z_i}{dt^2}\frac{\partial z_i}{\partial q_r}\right) \tag{2}$$

$$= \sum_{i=1}^{n}\left(X_i\frac{\partial x_i}{\partial q_r} + Y_i\frac{\partial y_i}{\partial q_r} + Z_i\frac{\partial z_i}{\partial q_r}\right),$$

$$r = 1, \cdots, m.$$

The left-hand side has the value expressed by the Fundamental Equation I. of § 6. Let the right-hand side be denoted by Q_r:

$$Q_r = \sum_{i=1}^{n}\left(X_i\frac{\partial x_i}{\partial q_r} + Y_i\frac{\partial y_i}{\partial q_r} + Z_i\frac{\partial z_i}{\partial q_r}\right). \tag{3}$$

It thus appears that

$$\text{I.} \qquad \frac{d}{dt}\frac{\partial T}{\partial \dot{q}_r} - \frac{\partial T}{\partial q_r} = Q_r, \qquad r = 1, \cdots, m.$$

These are known as *Lagrange's Equations.*

We have deduced Lagrange's Equations from Newton's Second Law of Motion. They include Newton's Law as a particular case. For if we set

$$x_i = q_{3i}, \qquad y_i = q_{3i+1}, \qquad z_i = q_{3i+2}, \qquad i = 1, \cdots, n,$$

then T becomes:

$$T = \sum_{i=1}^{n}\frac{m_i}{2}\left(\dot{q}_{3i}^2 + \dot{q}_{3i+1}^2 + \dot{q}_{3i+2}^2\right),$$

and Q_{3i}, Q_{3i+1}, Q_{3i+2} are now the components of the force which acts on m_i. Thus Newton's Equations result at once.

8. Discussion of the Equations. Holonomic and Non-Holonomic Systems. We have before us the most general case. No restrictions have been made on the forces. These may, then, comprise dissipative forces, like those of friction or air resistance.

On the other hand, there may be one or more equations of the form:

$$F(q_1, \cdots, q_m, t) = 0, \tag{4}$$

where the function F does not depend on the initial conditions.

Moreover it may happen, whether there are relations of the form (4) or not, that the q_r's and their time derivatives are bound together by one or more equations:

$$(5) \qquad \Phi(q_1, \cdots, q_m, \dot{q}_1, \cdots, \dot{q}_m, t) = 0.$$

An airplane, rising at a given angle, would be an example.

The case which is most important in practice, is that in which Φ is linear in the $\dot{q}_r$:

$$(6) \qquad A_1\dot{q}_1 + \cdots + A_m\dot{q}_m + A = 0,$$

where the A's are functions of $(q_1, \cdots, q_m, t)$, independent of the initial conditions.

It may happen that a relation of the form (6) is equivalent to one of the form (4). Thus if

$$(7) \qquad \frac{ds}{dt} - a\frac{d\theta}{dt} = 0,$$

this relation is equivalent to the equation:

$$(8) \qquad s - a\theta = c,$$

which is essentially of the form (4).

If no relations (5) or (6) are present; or if such relations (5) or (6) as may have entered in the formulation of the problem are *all* capable of being replaced by equations of the form (4), the system is said to be *holonomic*. Examples of *non-holonomic* systems are the Cart Wheels of § 24 infra, and the Billiard Ball on the rough table, rolling and pivoting without slipping, p. 240; also the coin on the rough table, and the bicycle.* But when the Billiard Ball slips, p. 237, the system is holonomic, for the unknown reaction of the table can be computed explicitly, as the reader can easily verify, in terms of the velocity of the point of contact, and thus its components are expressed in terms of the time derivatives of the generalized coordinates.

We are still leaving in abeyance the question of whether Lagrange's Equations admit a unique solution. Our conditions are necessary for a solution of the mechanical problem, but not always sufficient. The study of sufficient conditions will be taken up in § 17 and in Appendix D.

* Appell, *Mécanique rationelle*, Vol. II, Chaps. XXI, XXII.

9. Continuation. The Forces. The question of holonomic or non-holonomic has to do with the left-hand side of Lagrange's Equations, *i.e.* with conditions on the q_r, $\dot{q}_r$, t which do not involve the forces or contain the constants of the initial conditions.

The forces appear on the right, and it is to these that we now turn our attention. It may happen that the total force X_i, Y_i, Z_i can be decomposed into two forces:

$$(9) \qquad X_i = X_i' + X_i^*, \qquad Y_i = Y_i' + Y_i^*, \qquad Z_i = Z_i' + Z_i^*$$

in such a manner that the X_i', Y_i', Z_i' will be essentially simpler than the X_i, Y_i, Z_i, and that the X_i^*, Y_i^*, Z_i^* disappear altogether from Lagrange's Equations. For example, the X_i', Y_i', Z_i' may be expressible in terms of a force function:

$$(10) \qquad X_i' = \frac{\partial U}{\partial x_i}, \qquad Y_i' = \frac{\partial U}{\partial y_i}, \qquad Z_i' = \frac{\partial U}{\partial z_i},$$

where U is known explicitly in terms of x_i, y_i, z_i, t.

As regards the disappearance of the X_i^*, Y_i^*, Z_i^* the problems discussed in §§ 1–6 have afforded ample illustration. These were the so-called *forces of constraint*, and they did not appear in Lagrange's Equations.

Returning now to the general case, we observe that it may happen that the X_i^*, Y_i^*, Z_i^* fulfil the condition:

$$(11) \qquad \sum_{i=1}^{n} \left(X_i^* \frac{\partial x_i}{\partial q_r} + Y_i^* \frac{\partial y_i}{\partial q_r} + Z_i^* \frac{\partial z_i}{\partial q_r} \right) = 0,$$

$r = 1, \cdots, m$. When this is true, the Q_r on the right of Lagrange's Equations take on the simpler form:

$$(12) \qquad Q_r = \sum_{i=1}^{n} \left(X_i' \frac{\partial x_i}{\partial q_r} + Y_i' \frac{\partial y_i}{\partial q_r} + Z_i' \frac{\partial z_i}{\partial q_r} \right).$$

This case is important in practice because it enables us to get rid of some or all of the unknown forces of the problem arising from *constraints*; cf. § 15. But even when all of the latter forces cannot be eliminated in this way, their number can be reduced to a minimum; and then the method of multipliers set forth in the next paragraph leads to the final elimination.

10. Conclusion. Lagrange's Multipliers. Consider a system, the motion of which is given by Lagrange's Equations:

$$\frac{d}{dt}\frac{\partial T}{\partial \dot{q}_r} - \frac{\partial T}{\partial q_r} = Q_r, \qquad r = 1, \cdots, m. \tag{13}$$

It can happen that the Q_r's can be split in two:

$$Q_r = Q_r' + Q_r^*, \qquad r = 1, \cdots, m, \tag{14}$$

in such a manner that the Q_r' are essentially simpler than the Q_r — known functions, for example — whereas the Q_r^* have the property that

$$Q_1^*\pi_1 + \cdots + Q_m^*\pi_m = 0, \tag{15}$$

where $\pi_1, \cdots, \pi_m$ are any m numbers which satisfy the equations:

$$a_{1s}\pi_1 + \cdots + a_{ms}\pi_m = 0, \qquad s = 1, \cdots, \mu < m. \tag{16}$$

Let the rank of the matrix

$$\left\| \begin{matrix} a_{11} & \cdots\cdots\cdots & a_{m1} \\ \cdots & \cdots\cdots\cdots & \cdots \\ a_{1\mu} & \cdots\cdots\cdots & a_{m\mu} \end{matrix} \right\| \tag{17}$$

be μ.

From Equations (13) it follows that

$$\sum_{r=1}^{m}\left(\frac{d}{dt}\frac{\partial T}{\partial \dot{q}_r} - \frac{\partial T}{\partial q_r} - Q_r\right)\pi_r = 0,$$

no matter what numbers the π_r may be. If, in particular, the Q_r and the π_r are subject to the condition expressed in (14), (15), and (16), then

$$\sum_{r=1}^{m}\left(\frac{d}{dt}\frac{\partial T}{\partial \dot{q}_r} - \frac{\partial T}{\partial q_r} - Q_r'\right)\pi_r = 0. \tag{18}$$

Multiply the μ equations (16) respectively by arbitrary numbers, $\lambda_1, \cdots, \lambda_\mu$, and subtract from (18):

$$\sum_{r=1}^{m}\left(\frac{d}{dt}\frac{\partial T}{\partial \dot{q}_r} - \frac{\partial T}{\partial q_r} - Q_r' - \sum_{s=1}^{\mu}\lambda_s a_{rs}\right)\pi_r = 0. \tag{19}$$

Of the m numbers $\pi_1, \cdots, \pi_m$ it is possible to choose some set of $m - \mu$ arbitrarily, and then the rest are determined by (16). For definiteness, suppose

$$(20)\qquad \begin{vmatrix} a_{11} & \cdots & a_{\mu 1} \\ \cdots & \cdots & \cdots \\ a_{1\mu} & \cdots & a_{\mu\mu} \end{vmatrix} \neq 0.$$

Then $\pi_{\mu+1}, \cdots, \pi_m$ are arbitrary, and $\pi_1, \cdots, \pi_\mu$ are determined.

Now let the $\lambda_1, \cdots, \lambda_\mu$ be so chosen that

$$(21)\qquad \frac{d}{dt}\frac{\partial T}{\partial \dot{q}_r} - \frac{\partial T}{\partial q_r} - Q_r' - \sum_{s=1}^{\mu} \lambda_s a_{rs} = 0, \qquad r = 1, \cdots, \mu.$$

This is possible because the determinant of these μ linear equations in $\lambda_1, \cdots, \lambda_\mu$ is not zero. For these values of λ Equations (19) reduce to the following $m - \mu$ equations:

$$(22)\qquad \sum_{r=\mu+1}^{m} \left(\frac{d}{dt}\frac{\partial T}{\partial \dot{q}_r} - \frac{\partial T}{\partial q_r} - Q_r' - \sum_{s=1}^{\mu} \lambda_s a_{rs}\right) \pi_r = 0.$$

The determination of the λ_s's by (21) is independent of any choice of the π_r's. The numbers $\pi_{\mu+1}, \cdots, \pi_m$ are wholly arbitrary. Hence each coefficient in (22) must vanish. We have thus established the following

THEOREM. *When Q_r can be written in the form:*

$$(23)\qquad Q_r = Q_r' + Q_r^*, \qquad r = 1, \cdots, m,$$

where

$$(24)\qquad Q_1^* \pi_1 + \cdots + Q_m^* \pi_m = 0,$$

provided

$$(25)\qquad a_{1s}\pi_1 + \cdots + a_{ms}\pi_m = 0, \qquad s = 1, \cdots, \mu < m,$$

the rank of the matrix

$$(26)\qquad \begin{Vmatrix} a_{11} & \cdots & a_{m1} \\ \cdots & \cdots & \cdots \\ a_{1\mu} & \cdots & a_{m\mu} \end{Vmatrix}$$

being μ, then it is possible to find μ numbers $\lambda_1, \cdots, \lambda_\mu$ such that

$$(27)\qquad \frac{d}{dt}\frac{\partial T}{\partial \dot{q}_r} - \frac{\partial T}{\partial q_r} = Q_r' + \sum_{s=1}^{\mu} \lambda_s a_{rs}, \qquad r = 1, \cdots, m.$$

These numbers are determined by μ of the Equations (27), and the values thus obtained are then substituted in the remaining $m - \mu$ equations.

Applications of this theorem occur in practice in a variety of problems in which the q_r, $\dot{q}_r$, and t are connected by relations of the form (6), § 7:

$$a_{1s}\dot{q}_1 + \cdots + a_{ms}\dot{q}_m + a_s = 0, \qquad s = 1, \cdots, \mu. \tag{28}$$

where the a_{rs}, a_s depend on the q_r and t, but not on the initial conditions. As a matter of fact, in a number of such cases the coefficients a_{rs} in (28) do lead to a system of equations (16) which control a set of numbers $\pi_1, \cdots, \pi_m$ for which an analysis (14) of Q_r with the resulting relation (15) is possible.

In other problems, however, the a_{rs} of equations (16) have nothing to do with any such equations as (28), if indeed the latter exist, but may even themselves depend on $\dot{q}_r$, as well as q_r and t. For a complete discussion cf. Appendix D.

We turn now to a direct determination of the Q_r from purely mechanical considerations.

11. Virtual Velocities and Virtual Work. Let a system of particles m_i with the coordinates (x_i, y_i, z_i) be given $(i = 1, \cdots, n)$. Let δx_i, δy_i, δz_i be any $3n$ numbers, and let m_i be carried to the point $(x_i + \delta x_i, y_i + \delta y_i, z_i + \delta z_i)$. Then the system is said to experience a *virtual displacement* $(\delta x_i, \delta y_i, \delta z_i)$, the word "virtual" expressing the fact that the actual system may not be capable of such a displacement, even approximately. Thus a particle constrained to move on a curve or a surface would in general be taken off its constraint, and not lie even in the tangent line or plane.

If forces (X_i, Y_i, Z_i) act on the system, the quantity

$$W_\delta = \sum_{i=1}^{n} (X_i\,\delta x_i + Y_i\,\delta y_i + Z_i\,\delta z_i) \tag{1}$$

is defined as the *virtual work* due to the virtual displacement.

It is convenient in many applications to restrict the virtual displacements admitted to consideration by linear homogeneous equations between the δx_i, δy_i, δz_i. Consider, in particular, a system of particles whose coordinates are given by Equations A), § 7:

$$\text{A)} \qquad \begin{cases} x_i = f_i(q_1, \cdots, q_m, t) \\ y_i = \varphi_i(q_1, \cdots, q_m, t) \\ z_i = \psi_i(q_1, \cdots, q_m, t) \end{cases}$$

where the rank of the matrix

$$\text{a)} \qquad \left\| \begin{matrix} \dfrac{\partial x_1}{\partial q_1} & \cdots & \dfrac{\partial x_1}{\partial q_m} \\ \cdots & \cdots & \cdots \\ \dfrac{\partial z_n}{\partial q_1} & \cdots & \dfrac{\partial z_n}{\partial q_m} \end{matrix} \right\|$$

is m. Let

$$(2) \qquad \begin{cases} \delta x_i = \dfrac{\partial x_i}{\partial q_1} \delta q_1 + \cdots + \dfrac{\partial x_i}{\partial q_m} \delta q_m \\ \delta y_i = \dfrac{\partial y_i}{\partial q_1} \delta q_1 + \cdots + \dfrac{\partial y_i}{\partial q_m} \delta q_m \\ \delta z_i = \dfrac{\partial z_i}{\partial q_1} \delta q_1 + \cdots + \dfrac{\partial z_i}{\partial q_m} \delta q_m \end{cases}$$

where $\delta q_1, \cdots, \delta q_m$ are m arbitrary quantities. If $m < 3n$, the $\delta x_i, \delta y_i, \delta z_i$ are subject to one or more linear homogeneous equations. Thus only a limited number of them can now be chosen arbitrarily, the rest being then determined.

Consider the actual displacement $(\Delta x_i, \Delta y_i, \Delta z_i)$ which the system experiences in time Δt as it describes its natural path. It is:

$$(3) \qquad \begin{cases} \Delta x_i = f_i(q_1 + \Delta q_1, \cdots, q_m + \Delta q_m, t + \Delta t) - f_i(q_1, \cdots, q_m, t) \\ \Delta y_i = \varphi_i(q_1 + \Delta q_1, \cdots, q_m + \Delta q_m, t + \Delta t) - \varphi_i(q_1, \cdots, q_m, t) \\ \Delta z_i = \psi_i(q_1 + \Delta q_1, \cdots, q_m + \Delta q_m, t + \Delta t) - \psi_i(q_1, \cdots, q_m, t) \end{cases}$$

Since the δq_r are arbitrary, it is possible to choose them equal to the Δq_r, or $\delta q_r = \Delta q_r$. It does not follow, however, that the corresponding $\delta x_i, \delta y_i, \delta z_i$ will differ from $\Delta x_i, \Delta y_i, \Delta z_i$ by infinitesimals of higher order with respect to Δt. This will, in fact, be the case if the functions f_i, φ_i, ψ_i do not contain the time, *i.e.* if $\partial f_i/\partial t = 0$, etc. But otherwise in general not.

The virtual work has the value:

$$(4) \qquad W_\delta = \sum_{r=1}^{m} \sum_{i=1}^{n} \left(X_i \frac{\partial x_i}{\partial q_r} + Y_i \frac{\partial y_i}{\partial q_r} + Z_i \frac{\partial z_i}{\partial q_r} \right) \delta q_r,$$

where all m of the numbers δq_r are arbitrary. If, in particular, the forces can be broken up as in (9) § 9:

$$(5) \qquad X_i = X_i' + X_i^*, \qquad Y_i = Y_i' + Y_i^*, \qquad Z_i = Z_i' + Z_i^*,$$

so that (11) holds:

$$(6) \qquad \sum_{i=1}^{n} \left(X_i^* \frac{\partial x_i}{\partial q_r} + Y_i^* \frac{\partial y_i}{\partial q_r} + Z_i^* \frac{\partial z_i}{\partial q_r} \right) = 0,$$

$r = 1, \cdots, m$, then (4) takes on the simpler form:

$$(7) \qquad W_\delta = \sum_{r=1}^{m} \sum_{i=1}^{n} \left(X_i' \frac{\partial x_i}{\partial q_r} + Y_i' \frac{\partial y_i}{\partial q_r} + Z_i' \frac{\partial z_i}{\partial q_r} \right) \delta q_r.$$

In either case,

$$(8) \qquad W_\delta = Q_1 \delta q_1 + \cdots + Q_m \, \delta q_m.$$

Consider the actual displacement $(\Delta x_i, \Delta y_i, \Delta z_i)$ of the system in time Δt as it describes its natural path. If f_i, φ_i, ψ_i do not contain t, the virtual work W_δ will differ from the actual work, ΔW, by an infinitesimal of higher order than Δt; otherwise, this will not in general be the case.

12. Computation of Q_r. In Equation (8), § 11 the δq_r are m arbitrary numbers. We may, then, set $\delta q_k = 0$, $k \neq r$; $\delta q_r \neq 0$, and compute the corresponding value of W_δ. We shall then have:

$$(9) \qquad Q_r = \frac{W_\delta}{\delta q_r}.$$

Consider, for example, a particle that is constrained to lie on a moving surface. Its coordinates are subject to the conditions:

$$(10) \qquad x = f(q_1, q_2, t), \qquad y = \varphi(q_1, q_2, t), \qquad z = \psi(q_1, q_2, t).$$

A virtual displacement means that we fix our attention on an arbitrary instant of time, t, and consider the surface represented by (10) for this value of t. Next, consider a point (x, y, z) of this surface. Then a virtual displacement $(\delta x, \delta y, \delta z)$ of this point means an arbitrary displacement in the tangent plane to the surface at the point in question. In particular, if we set $\delta q_2 = 0$ and take $\delta q_1 \neq 0$, then the virtual displacement takes place along the tangent to that curve in the surface whose coordinates are represented by (10) when q_2 and t are held fast.

Now, the natural path of the particle under the forces that act does not in general lie in the surface just considered, nor is it tangent to the surface. If the surface is smooth, the reaction on the particle will be normal to the surface, and so the virtual work W_δ due to the reaction will be 0. But the actual work done by the reaction in Δt seconds along the natural path will in general be an infinitesimal of the same order as Δt.

It is now easy to see how to compute Q_1 and Q_2 in case the surface is smooth. The virtual work of the reaction of the surface is nil, and so we need consider only the other forces. The virtual displacement takes place in the tangent plane to the surface, and we can compute directly the virtual work corresponding to the successive virtual displacements given by $\delta q_1 \neq 0,\ \delta q_2 = 0$ and $\delta q_1 = 0,\ \delta q_2 \neq 0$; cf. further § 16 infra.

13. Virtual Velocities, an Aid in the Choice of the π_r. In the general theorem of § 10 there was no indication as to how the π_r may be chosen. In certain cases which arise in practice, the motion being subject to Lagrange's Equations:

$$\text{(1)} \qquad \frac{d}{dt}\frac{\partial T}{\partial \dot{q}_r} - \frac{\partial T}{\partial q_r} = Q_r, \qquad r = 1, \cdots, m,$$

it happens that there are geometric or kinematical relations between the q_r's of the form (28), § 10:

$$\text{(2)} \qquad a_{1s}\dot{q}_1 + \cdots + a_{ms}\dot{q}_m + a_s = 0, \qquad s = 1, \cdots, \mu < m,$$

where the a_{rs}, a_s are known functions of $q_1, \cdots, q_m, t$,* which do not depend on the initial conditions, and where the rank of the matrix

$$\text{(3)} \qquad \left\| \begin{matrix} a_{11} & \cdots & a_{1\mu} \\ \cdots & \cdots & \cdots \\ a_{m1} & \cdots & a_{m\mu} \end{matrix} \right\|$$

is μ.

If, now, the possible virtual displacements corresponding to an arbitrary choice of $\delta q_1, \cdots, \delta q_m$ are so restricted that

$$\text{(4)} \qquad a_{1s}\,\delta q_1 + \cdots + a_{ms}\,\delta q_m = 0, \qquad s = 1, \cdots, \mu,$$

it turns out that the virtual work of certain forces ("forces of constraint") will vanish. Hence by identifying these "forces of constraint" with the Q_r^* and setting $\delta q_r = \pi_r$, the hypotheses of that theorem are fulfilled.

* It may happen that some or all of these equations may be integrated in the form: $F(q_1, \cdots, q_m, t) = 0$, where F does not depend on the initial conditions; but it is not important to distinguish this case.

Example. Consider the disc of Chapter VI, § 24 as free to roll without slipping on a rough horizontal plane which is moving in its own plane according to any given law; for example, rotating about a fixed point with constant velocity. The force which the plane exerts on the disc at the point of contact will do work on the disc. But the virtual work of this force, when the virtual displacement is restricted as above, is nil. Thus we are led to a suitable set of multipliers π_r, namely, the δq_r thus restricted.

14. On the Number m of the q_r. For a system of particles the q_r's, as has already been pointed out, can always be identified with the coordinates:

$$x_i = q_{3i}, \qquad y_i = q_{3i+1}, \qquad z_i = q_{3i+2}, \qquad i = 1, \cdots, n.$$

Here, $m = 3n$, and Lagrange's Equations become identical with Newton's Equations.

In theory, then, there is no difference between the two systems. In practice, Lagrange's Equations provide in many cases an elimination of forces in which we are not interested.

Consider a ladder sliding down a wall. If the wall and floor are smooth, we may take $m = 1$, $q = \theta$, and all the forces in which we are not interested will be eliminated. Here,

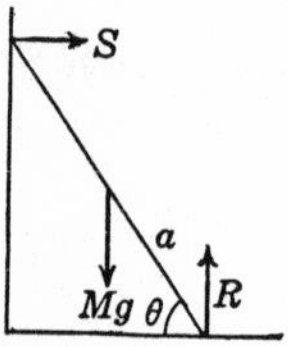

(1) $$T = \tfrac{1}{2}[M(\dot{\bar{x}}^2 + \dot{\bar{y}}^2) + Mk^2\dot{\theta}^2],$$

(2) $$\bar{x} = a\cos\theta, \qquad \bar{y} = a\sin\theta.$$

Hence

(3) $$T = \frac{M}{2}(a^2 + k^2)\,\dot{\theta}^2.$$

Fig. 144

Lagrange's Equation becomes:

(4) $$\frac{d}{dt}\frac{\partial T}{\partial \dot{\theta}} - \frac{\partial T}{\partial \theta} = \frac{\partial U}{\partial \theta},$$

where

$$U = -\,Mg\,a\sin\theta.$$

Thus we find as the equation governing the motion:

$$M(a^2 + k^2)\frac{d^2\theta}{dt^2} = -\,Mga\cos\theta$$

or

(5) $$\frac{d^2\theta}{dt^2} = -\,\frac{ag}{a^2 + k^2}\cos\theta.$$

In this example, the maximum of elimination has been attained at one blow. If we think of the ladder as made up of a huge number of particles connected by weightless rods, the forces in the rods have been eliminated, and also the forces exerted by the floor and the wall.

Suppose, however, that the floor and the wall are rough. We can still write down a single Lagrangean equation,

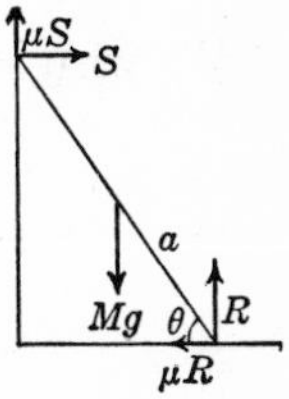

$$\frac{d}{dt}\frac{\partial T}{\partial \dot{\theta}} - \frac{\partial T}{\partial \theta} = Q, \tag{6}$$

the left-hand side being as before. But now

FIG. 145

$$Q = \frac{W_\delta}{\delta q} = -\,Mga\cos\theta + 2a\cos\theta\,\mu S + 2a\sin\theta\,\mu R. \tag{7}$$

We have not equations enough to solve the problem.

The difficulty can be met by taking $m = 3$ and setting

$$q_1 = \bar{x}, \qquad q_2 = \bar{y}, \qquad q_3 = \theta. \tag{8}$$

T is given by (1). And now

$$\begin{cases} Q_1 = S - \mu R, \qquad Q_2 = R + \mu S - Mg, \\ Q_3 = a(\sin\theta + \mu\cos\theta)\,S + a(\mu\sin\theta - \cos\theta)\,R. \end{cases} \tag{9}$$

Lagrange's three equations become:

$$\begin{cases} M\dfrac{d^2\bar{x}}{dt^2} = S - \mu R \\[2ex] M\dfrac{d^2\bar{y}}{dt^2} = R + \mu S - Mg \\[2ex] Mk^2\dfrac{d^2\theta}{dt^2} = a(\sin\theta + \mu\cos\theta)\,S + a(\mu\sin\theta - \cos\theta)\,R. \end{cases} \tag{10}$$

The first two of these are the equations of motion of the centre of gravity; the third, the equation of moments about the centre of gravity.

What Lagrange's method here has done, is first to eliminate the internal forces between the particles, just as we did in Chapter IV, § 1, when we proved the theorem about the motion of the centre of gravity; and similarly, when we proved the theorem of moments, Chapter IV, § 3 and § 9.

Let

$$Q_r = Q'_r + Q^*_r,$$

where

$$Q^*_1 = S - \mu R, \qquad Q^*_2 = R + \mu S,$$

$$Q^*_3 = a(\sin\theta + \mu\cos\theta)\,S + a(\mu\sin\theta - \cos\theta)\,R.$$

We wish to find three multipliers, π_1, π_2, π_3, such that

$$(11) \qquad Q^*_1\pi_1 + Q^*_2\pi_2 + Q^*_3\pi_3 = 0.$$

This can easily be done algebraically, with the result that

$$(12) \qquad \begin{cases} \pi_1 + \mu\pi_2 + a(\sin\theta + \mu\cos\theta)\,\pi_3 = 0 \\ -\mu\pi_1 + \pi_2 + a(\mu\sin\theta - \cos\theta)\,\pi_3 = 0 \end{cases}$$

a solution of these equations being:

$$\pi_1 = -\sin(\theta + 2\lambda), \qquad \pi_2 = \cos(\theta + 2\lambda), \qquad \pi_3 = 1/a.$$

But a mechanical derivation is easy, too. Consider the resultant of the forces R and μR. Draw a perpendicular to it through the lower end of the rod, and displace this end along this line. Do the same thing at the upper end of the rod, and displace the upper end along this line. The result is, that

$$(13) \qquad \begin{cases} \delta q_1 = -a\sin(\theta + 2\lambda)\,\delta q_3 \\ \delta q_2 = \quad a\cos(\theta + 2\lambda)\,\delta q_3 \end{cases}$$

Corresponding to such a displacement the virtual work of the "constraints" must vanish. And now it is merely a question of trigonometry to show that our expectation is fulfilled:

$$(14) \qquad Q^*_1\delta q_1 + Q^*_2\delta q_2 + Q^*_3\delta q_3 = 0.$$

Equation (11) corresponds to Equation (15) of § 10, and Equations (12) are the Equations (16) of that paragraph. But Equations (13), though corresponding to Equations (4), § 13, do not have their origin in Equations (2), § 13. The latter would arise from differentiating (2).

Returning now to Equations (10), we see that the unknown reactions R and S are eliminated by (14), where $\delta q_1, \delta q_2, \delta q_3$ satisfy (13). Hence

$$(15) \qquad M\ddot{\bar{x}}\,\delta q_1 + (M\ddot{\bar{y}} + Mg)\,\delta q_2 + Mk^2\ddot{\theta}\,\delta q_3 = 0,$$

or

$$(16) \qquad -M\ddot{\bar{x}}\,a\sin(\theta + 2\lambda) + (M\ddot{\bar{y}} + Mg)\,a\cos(\theta + 2\lambda) + Mk^2\,\ddot{\theta} = 0.$$

This equation, combined with Equations (2), leads at once to the solution of the problem:

$$(17)\quad (k^2 + a^2 \cos 2\lambda)\frac{d^2\theta}{dt^2} + a^2 \sin 2\lambda \frac{d\theta^2}{dt^2} + ag \cos(\theta + 2\lambda) = 0.$$

15. Forces of Constraint. A definition of "forces of constraint" from the point of view of physics, which shall be both accurate and comprehensive, has, so far as the author knows, never been given. They would be included in such forces as the X_i^*, Y_i^*, Z_i^* of § 9, which disappear from the Q_r; l.c., Equation (11). And still again, the Q_r^* of § 10 arise from unknown forces and are eliminated by the method of multipliers.

Perhaps these two cases are comprehensive in Rational Mechanics. Are there problems in this science not included here? If not, the asterisk forces could be defined as the *forces of constraint*; cf. Appendix D.

16. Euler's Equations, Deduced from Lagrange's Equations. When a rigid body rotates about a fixed point, the kinetic energy is

$$(1)\qquad T = \tfrac{1}{2}(Ap^2 + Bq^2 + Cr^2).$$

Let p, q, r be expressed in terms of Euler's angles, Chapter VI, § 15:

$$(2)\qquad \begin{cases} p = \dot\theta \sin\varphi - \dot\psi \sin\theta \cos\varphi \\ q = \dot\theta \cos\varphi + \dot\psi \sin\theta \sin\varphi \\ r = \dot\varphi + \dot\psi \cos\theta \end{cases}$$

The second of Lagrange's Equations is readily computed:

$$\frac{\partial T}{\partial \dot\varphi} = Ap\frac{\partial p}{\partial \dot\varphi} + Bq\frac{\partial q}{\partial \dot\varphi} + Cr\frac{\partial r}{\partial \dot\varphi} = Cr;$$

$$\frac{\partial T}{\partial \varphi} = Ap\frac{\partial p}{\partial \varphi} + Bq\frac{\partial q}{\partial \varphi} + Cr\frac{\partial r}{\partial \varphi},$$

$$\frac{\partial p}{\partial \varphi} = \dot\theta \cos\varphi + \dot\psi \sin\theta \sin\varphi = q,$$

$$\frac{\partial q}{\partial \varphi} = -\dot\theta \sin\varphi + \dot\psi \sin\theta \cos\varphi = -p,$$

Hence

$$C\frac{dr}{dt} - (A - B)pq = \Phi.$$

To compute Φ, observe that, no matter what forces may act, they can be replaced by a force at O and a couple. The latter can be realized by means of three forces:

i) a force $L\gamma$ acting at the point* $\mathbf{r} = \beta$;

ii) " $M\alpha$ " " $\mathbf{r} = \gamma$;

iii) " $N\beta$ " " $\mathbf{r} = \alpha$.

A virtual displacement $\delta\theta = 0$, $\delta\varphi \neq 0$, $\delta\psi = 0$ gives as the virtual work

$$N\,\delta\varphi$$

and this is equal to $\Phi\,\delta\varphi$. Hence $\Phi = N$ and we have:

$$C\frac{dr}{dt} - (A - B)\,pq = N.$$

This is the third of Euler's Dynamical Equations. The other two follow from this one by symmetry, and are obtained by advancing the letters cyclically.

EXERCISE

Obtain the six equations of motion of a rigid body by means of Lagrange's Equations.

17. Solution of Lagrange's Equations. We have seen in § 10 that if the Q_r satisfy the conditions of the Theorem of that paragraph, then

$$(1)\qquad \frac{d}{dt}\frac{\partial T}{\partial \dot{q}_r} - \frac{\partial T}{\partial q_r} = Q'_r + \sum_{s=1}^{\mu} a_{rs}\lambda_s, \qquad r = 1, \cdots, m,$$

where the matrix:

$$(2)\qquad \left\| \begin{matrix} a_{11} & \cdots & a_{m1} \\ \cdots & \cdots & \cdots \\ a_{1\mu} & \cdots & a_{m\mu} \end{matrix} \right\|$$

is of rank $\mu < m$. In the cases which arise most frequently in practice, there are μ equations of the form:

* By the "point $\mathbf{r}$" is meant the terminal point of the vector $\mathbf{r}$ when the initial point is at O.

(3) $$a_{1s}\dot{q}_1 + \cdots + a_{ms}\dot{q}_m = a_s, \qquad s = 1, \cdots, \mu,$$

where the a's are functions of $q_1, \cdots, q_m, t$.

The kinetic energy T is a positive definite quadratic form in the $\dot{q}_1, \cdots, \dot{q}_m$, but not necessarily homogeneous:

$$T = T_2 + T_1 + T_0,$$

where

$$T_2 = \sum_{i,j} A_{ij}\dot{q}_i\dot{q}_j, \qquad A_{ij} = A_{ji},$$

and T_1, T_0 are homogeneous of degree 1 or 0 in the $\dot{q}_r$, or vanish identically. The coefficients are functions of $q_1, \cdots, q_m, t$. The form T_2, or

(4) $$\sum_{i,j} A_{ij}\xi_i\xi_j,$$

is a positive definite homogeneous quadratic form. For T can be written in the form:

$$T = \sum_{i,j} A_{ij}(\dot{q}_i - \mu_i)(\dot{q}_j - \mu_j) + C,$$

where $\mu_1, \cdots, \mu_m$, C are functions of $q_1, \cdots, q_m, t$.

Finally, let Q'_r be a known function of $q_r, \dot{q}_r, t$.

THEOREM. *The $m + \mu$ Equations:*

$$\frac{d}{dt}\frac{\partial T}{\partial \dot{q}_r} - \frac{\partial T}{\partial q_r} = Q'_r + \sum_{s=1}^{\mu} a_{rs}\lambda_s, \qquad r = 1, \cdots, m;$$

$$a_{1s}\dot{q}_1 + \cdots + a_{ms}\dot{q}_m = a_s, \qquad s = 1, \cdots, \mu,$$

determine uniquely the $m + \mu$ functions $q_1, \cdots, q_m, \lambda_1, \cdots, \lambda_\mu$.

Proof. The first m of these equations have the form:

(5) $$A_{1r}\ddot{q}_1 + \cdots + A_{mr}\ddot{q}_m - a_{r1}\lambda_1 - \cdots - a_{r\mu}\lambda_\mu = B_r,$$

$r = 1, \cdots, m$, where B_r is a function of the $q_r, \dot{q}_r$, and t. The remaining μ equations give, on differentiating:

(6) $$a_{1s}\ddot{q}_1 + \cdots + a_{ms}\ddot{q}_m = C_s, \qquad s = 1, \cdots, \mu,$$

where C_s is likewise a function of the $q_r, \dot{q}_r$, and t. Thus we have $m + \mu$ linear equations in the $m + \mu$ unknowns: $\ddot{q}_1, \cdots, \ddot{q}_m, \lambda_1, \cdots, \lambda_\mu$. Their determinant:

$$
(7) \qquad D = \pm \begin{vmatrix} A_{11} & \cdots & A_{m1} & a_{11} & \cdots & a_{1\mu} \\ \cdots & \cdots & \cdots & \cdots & \cdots & \cdots \\ A_{1m} & \cdots & A_{mm} & a_{m1} & \cdots & a_{m\mu} \\ a_{11} & \cdots & a_{m1} & 0 & \cdots & 0 \\ \cdots & \cdots & \cdots & \cdots & \cdots & \cdots \\ a_{1\mu} & \cdots & a_{m\mu} & 0 & \cdots & 0 \end{vmatrix}
$$

does not vanish. For otherwise the $m + \mu$ linear homogeneous equations:

$$
(8) \qquad \begin{cases} A_{11}\xi_1 + \cdots + A_{m1}\xi_m + a_{11}\eta_1 + \cdots + a_{1\mu}\eta_\mu = 0 \\ \cdots\cdots\cdots\cdots\cdots\cdots\cdots\cdots \\ A_{1m}\xi_1 + \cdots + A_{mm}\xi_m + a_{m1}\eta_1 + \cdots + a_{m\mu}\eta_\mu = 0 \\ a_{11}\xi_1 + \cdots + a_{m1}\xi_m = 0 \\ \cdots\cdots\cdots\cdots\cdots\cdots\cdots\cdots \\ a_{1\mu}\xi_1 + \cdots + a_{m\mu}\xi_m = 0 \end{cases}
$$

would admit a solution $(\xi_1, \cdots, \xi_m, \eta_1, \cdots, \eta_\mu)$ not the identity. Moreover, not all the $\xi_1, \cdots, \xi_m$ in this solution could vanish; for then we should have:

$$
\begin{cases} a_{11}\eta_1 + \cdots + a_{1\mu}\eta_\mu = 0 \\ \cdots\cdots\cdots\cdots\cdots\cdots \\ a_{m1}\eta_1 + \cdots + a_{m\mu}\eta_\mu = 0. \end{cases}
$$

But the rank of the matrix of these equations, namely the matrix (2), is μ. Hence all the $\eta_1, \cdots, \eta_\mu$ of the solution vanish — a contradiction.

Next, multiply the r-th equation (8) by ξ_r, $r = 1, \cdots, m$, and add. The terms in $\eta_1, \cdots, \eta_\mu$ drop out because of the last μ of the equations (8), and so there results the equation:

$$
\sum_{i,j} A_{ij}\xi_i\xi_j = 0,
$$

where not all the $\xi_1, \cdots, \xi_m$ are 0. This is impossible, since (4) is a positive definite quadratic form.

Equations (5) and (6) admit, therefore, a solution:

$$
(9) \qquad \begin{cases} \ddot{q}_r = F_r(q_1, \cdots, q_m, \dot{q}_1, \cdots, \dot{q}_m, t), & r = 1, \cdots, m; \\ \lambda_s = \Phi_s(q_1, \cdots, q_m, \dot{q}_1, \cdots, \dot{q}_m, t) & s = 1, \cdots, \mu. \end{cases}
$$

Assuming for definiteness that the determinant

$$\begin{vmatrix} a_{11} & \cdots & a_{\mu 1} \\ \cdots & \cdots & \cdots \\ a_{1\mu} & \cdots & a_{\mu\mu} \end{vmatrix} \neq 0,$$

we see that the first m of the Equations (9) admit the integral given by (3), or:

$$(10) \qquad \dot{q}_s = f_s(q_1, \cdots, q_m, \dot{q}_{\mu+1}, \cdots, \dot{q}_m, t), \qquad s = 1, \cdots, \mu,$$

where, in particular, f_s is linear in $\dot{q}_{\mu+1}, \cdots, \dot{q}_m$. This is a particular integral which is independent of the initial conditions of the mechanical problem. Let

$$(11) \qquad \dot{q}_\alpha = \kappa_\alpha, \qquad \alpha = \mu + 1, \cdots, m.$$

The system of m differential equations (9) is now seen to be equivalent, under the restrictions of the dynamical problem, to the system:

$$(12) \quad \begin{cases} \dfrac{dq_s}{dt} = f_s(q_1, \cdots, q_m, \kappa_{\mu+1}, \cdots, \kappa_m, t), & s = 1, \cdots, \mu; \\ \dfrac{dq_\alpha}{dt} = \kappa_\alpha, & \alpha = \mu + 1, \cdots, m; \\ \dfrac{d\kappa_\alpha}{dt} = g_\alpha(q_1, \cdots, q_m, \kappa_{\mu+1}, \cdots, \kappa_m, t), & \alpha = \mu + 1, \cdots, m. \end{cases}$$

Here is a system of $2m - \mu$ differential equations of the first order for determining the $2m - \mu$ unknown functions q_r, κ_α. Their solution yields the m desired functions, $q_1, \cdots, q_m$. The λ_s are now uniquely determined as functions of t and the initial conditions. As regards the freedom of the initial conditions, on which of course the determination of the constants of integration depends, the initial values q_r^0, $\dot{q}_r^0$ of q_r, $\dot{q}_r$ are restricted by the equation

$$a_{1s}\dot{q}_1 ++ \cdots + a_{ms}\dot{q}_m = a_s.$$

It is important here, as in so many problems of the kind discussed in this chapter, to distinguish between constants that are connected with the choice of coordinates and constants that arise from the initial conditions of the mechanical problem. Thus in the problem of Chap. IV, § 13, p. 141, Fig. 84, s might equally well have been measured from a different level, and then the relation would have been:

$$s = a\theta + c.$$

18. Equilibrium. Let a dynamical system be given, with n particles m_i, the motion being subject to Newton's Law:

$$(1) \qquad m_i\ddot{x}_i = X_i, \quad m_i\ddot{y}_i = Y_i, \quad m_i\ddot{z}_i = Z_i, \qquad i = 1, \cdots, n.$$

The forces are said to be *in equilibrium* if

$$(2) \qquad X_i = 0, \quad Y_i = 0, \quad Z_i = 0, \qquad i = 1, \cdots, n.$$

A necessary and sufficient condition for equilibrium is, that

$$(3) \qquad \ddot{x}_i = 0, \quad \ddot{y}_i = 0, \quad \ddot{z}_i = 0.$$

We are not interested in the general case, which, in accord with the definition just given, relates to a single instant of time, the forces not in general being in equilibrium at any other instant. We have concern rather with a permanent state of rest of a system capable of certain motions which are subject to geometric conditions. We are thinking primarily of such problems in the statics of particles and rigid bodies as were studied in Chapters I and II; but also of more general problems, like the following: — A uniform circular disc has a particle attached to its rim. The disc rests on a smooth ellipsoid and a rough table which contains two axes of the ellipsoid. Find the positions of equilibrium.

More precisely, the system shall be capable of assuming the positions defined by the equations:

$$(4) \qquad \begin{cases} x_i = f_i(q_1, \cdots, q_m) \\ y_i = \varphi_i(q_1, \cdots, q_m) \\ z_i = \psi_i(q_1, \cdots, q_m) \end{cases}$$

where the functions f_i, φ_i, ψ_i do not depend on t, and where the rank of the matrix

$$(5) \qquad \left\| \begin{matrix} \dfrac{\partial x_1}{\partial q_1} & \cdots & \dfrac{\partial x_1}{\partial q_m} \\ \cdots & \cdots & \cdots \\ \dfrac{\partial z_n}{\partial q_1} & \cdots & \dfrac{\partial z_n}{\partial q_m} \end{matrix} \right\|$$

is m.

Observe that this last requirement does not imply that m has the least value for which the x_i, y_i, z_i can be represented by equations of the form (4), satisfying the above requirements. It is still possible that the $q_1, \cdots, q_m$ may be connected by relations of the form:

$$F_j(q_1, \cdots, q_m) = 0, \qquad j = 1, \cdots, \rho < m. \tag{6}$$

On the other hand it does imply that if the $\dot{x}_i, \dot{y}_i, \dot{z}_i$, all vanish, then this is true of the $\dot{q}_r$, and conversely; and if, furthermore, both the $\dot{x}_i, \dot{y}_i, \dot{z}_i$, and the $\ddot{x}_i, \ddot{y}_i, \ddot{z}_i$ all vanish, then this is true of the $\dot{q}_r$ and the $\ddot{q}_r$, and conversely.

The motion of the system is, first of all, subject to the equations:

$$[T]_r = Q_r, \qquad r = 1, \cdots, \mu < m, \tag{7}$$

where by definition

$$[T]_r \equiv \frac{d}{dt}\frac{\partial T}{\partial \dot{q}_r} - \frac{\partial T}{\partial q_r}.$$

To these may be added further equations:

$$a_{1s}\dot{q}_1 + \cdots + a_{ms}\dot{q}_m = 0, \qquad s = 1, \cdots, \mu < m, \tag{8}$$

where the rank of the matrix

$$\left\| \begin{matrix} a_{11} & \cdots & a_{m1} \\ \cdots & \cdots & \cdots \\ a_{1\mu} & \cdots & a_{m\mu} \end{matrix} \right\| \tag{9}$$

is μ. It is possible that some or all of these equations can be expressed in the form (6), but this is unimportant.

A first necessary and sufficient condition for equilibrium is that

$$Q_r = 0, \qquad r = 1, \cdots, m. \tag{10}$$

For, a necessary and sufficient condition for the vanishing of the left-hand side of (7) for $\dot{q}_1 = 0, \cdots, \dot{q}_m = 0$ is that $\ddot{q}_1 = 0, \cdots, \ddot{q}_m = 0$.

If there are relations of the form (8), it may happen that the Q_r can be split up as follows:

$$Q_r = Q'_r + Q^*_r, \tag{11}$$

where

$$Q^*_1\delta q_1 + \cdots + Q^*_m\delta q_m = 0, \tag{12}$$

provided the δq_r are so chosen that

$$a_{1s}\delta q_1 + \cdots + a_{ms}\delta q_m = 0, \qquad s = 1, \cdots, \mu. \tag{13}$$

Under these circumstances a necessary and sufficient condition for equilibrium is that

(14) $$Q'_1 \delta q_1 + \cdots + Q'_m \delta q_m = 0,$$

provided the δq_r satisfy (13).

That the condition is necessary appears from the fact that (10) is true, and hence

(15) $$(Q'_1 + Q^*_1)\, \delta q_1 + \cdots + (Q'_m + Q^*_m)\, \delta q_m = 0$$

for all values of the δq_r. If the δq_r satisfy (13), it follows that (12) is true, and (14) now follows.

Suppose conversely that (12) and (14) hold when the δq_r are subject to (13). Then the system is in equilibrium. Suppose the statement false. From (12) and (14) it follows that (15) holds, provided (13) is true, and hence from (7) it follows that

(16) $$\sum_{r=1}^{m} [T]_r\, \delta q_r = 0,$$

provided (13) holds. Let

(17) $$\ddot{q}_r = c_r, \qquad r = 1, \cdots, m,$$

initially. Then not all the c_r are 0. Now,

(18) $$\delta q_r = c_r, \qquad r = 1, \cdots, m,$$

is a system of values satisfying (13). For, on differentiating (8) with respect to t and then setting $t = t_0$, $\dot{q}_r = 0$, these relations follow, namely:

$$a_{1s} c_1 + \cdots + a_{ms} c_m = 0, \qquad s = 1, \cdots, \mu.$$

Consequently (16) holds for these values of c_r. Now,

(19) $$T = \sum_{\alpha,\beta} A_{\alpha\beta}\, \dot{q}_\alpha \dot{q}_\beta.$$

Hence

$$\frac{\partial T}{\partial \dot{q}_r} = A_{1r}\, \dot{q}_1 + \cdots + A_{mr}\, \dot{q}_m,$$

$$\frac{d}{dt}\frac{\partial T}{\partial \dot{q}_r} = A_{1r}\, \ddot{q}_1 + \cdots + A_{mr}\, \ddot{q}_m + \text{terms in } (\dot{q}_1, \cdots, \dot{q}_m),$$

and so initially

$$\sum_{r=1}^{m} [T]_r\, \delta q_r = \sum_{r=1}^{m} (A_{1r}\, c_1 + \cdots + A_{mr}\, c_m)\, c_r = \sum_{\alpha,\beta} A_{\alpha\beta}\, c_\alpha c_\beta.$$

But here is a contradiction, since (19), being a positive definite quadratic form, can vanish only when all the arguments are 0. We can state the result as a

THEOREM. *A necessary and sufficient condition that a dynamical system, the motion of which is governed by the equations:*

$$(20)\qquad \begin{cases} \dfrac{d}{dt}\dfrac{\partial T}{\partial \dot{q}_r} - \dfrac{\partial T}{\partial q_r} = Q_r, & r = 1, \cdots, m; \\ a_{1s}\dot{q}_1 + \cdots + a_{ms}\dot{q}_m = 0, & s = 1, \cdots, \mu < m, \end{cases}$$

where the rank of the matrix of these last μ *equations is* μ, *be in equilibrium and at rest, is that* $\dot{q}_r = 0$, $r = 1, \cdots, m$, *and that*

$$(21)\qquad Q_1\delta q_1 + \cdots + Q_m\delta q_m = 0,$$

where $\delta q_1, \cdots, \delta q_m$ *are subject to the condition:*

$$(22)\qquad a_{1s}\delta q_1 + \cdots + a_{ms}\delta q_m = 0, \qquad s = 1, \cdots, \mu.$$

Here T *is a homogeneous positive definite quadratic form in* $\dot{q}_1, \cdots, \dot{q}_m$.

If, in particular,

$$(23)\qquad Q_r = Q'_r + Q^*_r, \qquad r = 1, \cdots, m,$$

and if it is known that

$$(24)\qquad Q^*_1\delta q_1 + \cdots + Q^*_m\delta q_m = 0,$$

where $\delta q_1, \cdots, \delta q_m$ *are subject to* (22), *then* (21) *can be replaced by*

$$(25)\qquad Q'_1\delta q_1 + \cdots + Q'_m\delta q_m = 0.$$

19. Small Oscillations. Two equal masses are knotted to a string, one end of which is made fast to a peg at O. Determine the motion in the case of small vibrations. Here,

$$T = \frac{ma^2}{2}\Big(2\dot{\theta}^2 + \dot{\varphi}^2 + 2\dot{\theta}\dot{\varphi}\cos(\varphi - \theta)\Big),$$

$$U = mga(2\cos\theta + \cos\varphi).$$

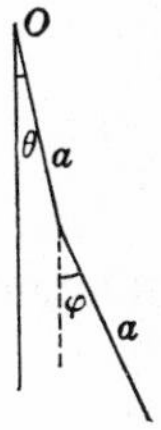

Since θ, $\dot{\theta}$, φ, $\dot{\varphi}$ are small, these functions can be replaced by the approximations:

FIG. 146

$$(1)\qquad \begin{cases} T = \dfrac{ma^2}{2}(2\dot{\theta}^2 + 2\dot{\theta}\dot{\varphi} + \dot{\varphi}^2), \\ U = -\,mga\left(\theta^2 + \dfrac{\varphi^2}{2}\right) + \text{const.} \end{cases}$$

Equations (1) are typical for an important class of problems in small oscillations of a system about a position of stable equilibrium, the applied forces being derived from a force function, U. Let T and U both be independent of t; let $q_r = 0$ for $r = 1, \cdots, m$, be the position of equilibrium, and let T, U be replaced by their approximate values when q_r, $\dot{q}_r$ are all small. Then

$$(2) \qquad T = \sum_{r,s} a_{rs}\,\dot{q}_r\dot{q}_s, \qquad U = -\sum_{r,s} b_{rs}\,q_r q_s,$$

$r, s = 1, \cdots, m$, where the coefficients $a_{rs} = a_{sr}$, $b_{rs} = b_{sr}$ are constants and each of the quadratic forms is definite.

Lagrange's Equations now take the form:

$$(3) \qquad a_{r1}\ddot{q}_1 + \cdots + a_{rm}\ddot{q}_m = -(b_{r1}q_1 + \cdots + b_{rm}q_m),$$
$$r = 1, \cdots, m.$$

To integrate these equations, it is convenient to introduce new variables as follows. It is a theorem of algebra * that by means of a suitable linear transformation with constant coefficients:

$$(4) \qquad q_r = \mu_{r1}q_1' + \cdots + \mu_{rm}q_m', \qquad r = 1, \cdots, m,$$

the quadratic forms (2) can each be reduced to a sum of squares:

$$(5) \qquad \begin{cases} T = \dot{q}_1'^2 + \cdots + \dot{q}_m'^2, \\ U = -n_1^2 q_1'^2 - \cdots - n_m^2 q_m'^2. \end{cases}$$

Lagrange's Equations now become:

$$(6) \qquad \frac{d^2 q_r'}{dt^2} = -n_r^2 q_r', \qquad r = 1, \cdots, m.$$

Their integrals take the form:

$$(7) \qquad q_r' = C_r \cos(n_r t + \gamma_r), \qquad r = 1, \cdots, m.$$

Returning to the original variables q_r, we find as the general solution of Equations (3) the following:

$$(8) \quad q_r = C_1\mu_{r1}\cos(n_1 t + \gamma_1) + \cdots + C_m\mu_{rm}\cos(n_m t + \gamma_m),$$
$$r = 1, \cdots, m,$$

where the C_r, γ_r are the $2m$ constants of integration.

* Bôcher, *Higher Algebra*, Chap. 13.

In this result, complete as it is in theory, there appear, however, the coefficients μ_{rs} of the linear transformation (4). These can be determined by the following consideration.

Let $C_r = 0$ in (8) when $r \neq s$; and let $C_s = 1$. Thus we have a special solution:

$$q_r^{(s)} = \lambda_r \cos (nt + \gamma), \tag{9}$$

where $\lambda_r = \mu_{rs}$ and $n = n_s$, $\gamma = \gamma_s$. Substitute $q_r^{(s)}$ in (3):

$$(b_{r1} - n^2 a_{r1}) \lambda_1 + \cdots + (b_{rm} - n^2 a_{rm}) \lambda_m = 0, \tag{10}$$

$$r = 1, \cdots, m.$$

A necessary condition that (9) be a solution is, that the m linear Equations (10) admit a solution in which the λ_r's are not all 0. Hence the determinant of these equations must vanish:

$$\begin{vmatrix} b_{11} - n^2 a_{11} & \cdots & b_{1m} - n^2 a_{1m} \\ \cdots & \cdots & \cdots \\ b_{m1} - n^2 a_{m1} & \cdots & b_{mm} - n^2 a_{mm} \end{vmatrix} = 0. \tag{11}$$

If the n_r^2 are all distinct, they form precisely the m roots of this equation in n^2. Moreover, each n_r^2 leads to a unique determination of the ratios of the λ's through the m Equations (10), and our problem is solved.

It may happen that k of the roots n^2 of (11) coincide. In that case, k of the λ's can be chosen arbitrarily, and so we still have k linearly independent solutions (9) corresponding to such a root n^2. More precisely, let n_1^2 be a multiple root of order k_1; n_2^2, a multiple root of order k_2; etc. Let n^2 be set $= n_1^2$ in Equations (10). Then, of the unknown $\lambda_1, \cdots, \lambda_m$, it is possible to choose a certain set of k_1 arbitrarily, and then the rest will be uniquely determined. Let all but one of these k_1 λ's be set $= 1$. Thus we get k_1 sets of $(\lambda_1, \cdots, \lambda_m)$, and each set gives a solution of Equations (3). Moreover these solutions are obviously linearly independent. Proceeding to n_2^2 we determine in the same manner k_2 further sets of $(\lambda_1, \cdots, \lambda_m)$, each set giving a solution of (3); these solutions are likewise linearly independent of one another and also of the earlier solutions. And so on, to the end. Thus in all cases the roots of (11) lead to m linearly independent solutions (9).

The variables q_r' are known as the *normal coordinates* of the problem. Each is uniquely determined, save as to a factor of proportionality, when the roots of (11) are distinct. But in the case of equal roots, an infinite number of different choices are possible.

EXERCISE

Carry through the example given at the beginning of the paragraph.

Ans. Two sets of linearly independent solutions are the following:

$$\begin{cases} \theta_1 = \cos(n_1 t + \gamma_1), \\ \varphi_1 = \sqrt{2}\cos(n_1 t + \gamma_1), \end{cases} \qquad \begin{cases} \theta_2 = \cos(n_2 t + \gamma_2); \\ \varphi_2 = -\sqrt{2}\cos(n_2 t + \gamma_2), \end{cases}$$

where

$$n_1^2 = (2 - \sqrt{2})\frac{g}{a}, \qquad n_2^2 = (2 + \sqrt{2})\frac{g}{a}.$$

The general solution is:

$$\begin{cases} \theta = C_1 \cos(n_1 t + \gamma_1) + C_2 \cos(n_2 t + \gamma_2), \\ \varphi = C_1\sqrt{2}\cos(n_1 t + \gamma_1) - C_2\sqrt{2}\cos(n_2 t + \gamma_2). \end{cases}$$

EXERCISES ON CHAPTER X

1. A smooth wedge rests on a table. A block is placed on the wedge, and the system is released from rest. Determine the motion.

2. Two billiard balls are placed one on top of the other, on a rough table, and released from rest, slightly displaced from the position of equilibrium. Determine the motion.

3. A uniform rod is pivoted at one end, and is acted on by gravity. Will it move like a spherical pendulum?

4. A uniform rod of length $2a$ and mass $3m$ can turn freely about its mid-point. A mass m is attached to one end of the rod. If the rod is rotating about a vertical axis with an angular velocity of $\sqrt{2ng/a}$, and so released, show that the heavy end will dip till the rod makes an angle of $\cos^{-1}(\sqrt{n^2 + 1} - n)$ with the vertical, and then rise again to the horizontal.

5. Obtain the equations of the top from Lagrange's Equations.

6. Determine the motion of a top whose peg, considered as a point, slides on a smooth horizontal plane.

7. The same question when the size of the peg is taken into account.

8. The ladder of p. 322, the initial position being oblique to the line of intersection of the wall and the floor.*

9. Two equal rods are hinged at their ends and project over a smooth horizontal plane. Determine the motion.

SUGGESTION. Take as coordinates (1) the x, y of the centre of gravity; (2) the inclination θ of the line through the centre of gravity and the hinge; (3) the angle α between this line and either of the rods.

Two of Lagrange's Equations control the motion of the centre of gravity. A third expresses the fact that the total moment of momentum with respect to the centre of gravity, is constant. And fourthly there is the equation of energy.†

10. A rough table is rotating about a vertical axis. Study the motion of a billiard ball on the table, assuming that there is no slipping.

11. The same problem with slipping.

12. Work the problem of § 19, p. 333, when the particles are not required to move in a vertical plane.

13. Two equal uniform rods are hinged at one of their ends, and the other end of one rod is pivoted. Find the motion for small oscillations in a vertical plane.

14. The same problem when the rods are not restricted to lying in a plane.

15. A uniform rod is supported by two strings of equal length, attached to its ends, their other ends being made fast at two points on the same level, whose distance apart is equal to the length of the rod. A smooth vertical wire passes through a small hole at the middle of the rod and bisects the line joining the fixed points. Determine the motion.

16. If in the preceding question the wire is absent, study the small oscillations of the rod about the position of equilibrium.

17. A bead can slide on a circular wire, no external forces acting. Determine the motion in two and in three dimensions. Begin with the case of no friction.

* Routh, *Elementary Rigid Dynamics*, p. 329.

† Appell, *Mécanique rationelle*, vol. ii, chap. 24, § 446. Many other problems of the present kind are found in this chapter.

CHAPTER XI

HAMILTON'S CANONICAL EQUATIONS

1. The Problem. The problem of this chapter is the deduction of Hamilton's Canonical Equations:

$$\frac{dq_r}{dt} = \frac{\partial H}{\partial p_r}, \quad \frac{dp_r}{dt} = -\frac{\partial H}{\partial q_r}, \qquad r = 1, \cdots, m,$$

from Lagrange's Equations:

$$\frac{d}{dt}\frac{\partial L}{\partial \dot{q}_r} - \frac{\partial L}{\partial q_r} = 0, \qquad r = 1, \cdots, m.$$

The transition is purely analytical, involving no physical concepts whatever, and for that reason it is well to set the theorem and proof apart in a separate chapter.

The problem can be stated as follows. We start out with a *Lagrangean System*. Such a system is defined as a material system which can be located by means of m generalized coordinates $q_1, \cdots, q_m$ and whose motion is determined by Lagrange's Equations. If we set $\dot{q}_r = \kappa_r$, these go over into the $2m$ equations:

$$\text{A)} \qquad \begin{cases} \dfrac{dq_r}{dt} = \kappa_r, \\[2ex] \dfrac{d}{dt}\dfrac{\partial L}{\partial \kappa_r} - \dfrac{\partial L}{\partial q_r} = 0, \end{cases}$$

$r = 1, \cdots, m$, where

$$(1) \qquad L = L(q_1, \cdots, q_m, \kappa_1, \cdots, \kappa_m, t) = L(q_r, \kappa_r, t)$$

is a function of the $2m + 1$ independent variables q_r, κ_r, t.

The function L is called the *Lagrangean Function.* In case there is a work function,

$$U = U(q_1, \cdots, q_m, t),$$

L is given by the equation:

$$(2) \qquad L = T + U,$$

where $\kappa_r = \dot{q}_r$ and

$$T = T(q_r, \dot{q}_r, t)$$

is the kinetic energy. In any case, the Hessian Determinant, the Jacobian:

$$\frac{\partial(L_1, \cdots, L_m)}{\partial(\kappa_1, \cdots, \kappa_m)}, \tag{3}$$

where

$$L_r = \frac{\partial L}{\partial \kappa_r}$$

shall not vanish.

The $(2m + 1)$-dimensional space S_{2m+1} of the variables (q_r, κ_r, t) shall be transformed on the $(2m + 1)$-dimensional space R_{2m+1} of the variables (q_r, p_r, t) by means of the transformation:

$$p_r = \frac{\partial L}{\partial \kappa_r}, \qquad r = 1, \cdots, m, \tag{4}$$

the q_r going over individually into themselves. The system of $2m$ differential equations of the first order in the $2m$ dependent variables q_r, κ_r, namely, Equations A), thereby goes over into a system of $2m$ like equations in the $2m$ dependent variables q_r, p_r. *These last equations are the following:*

$$\frac{dq_r}{dt} = \frac{\partial H}{\partial p_r}, \quad \frac{dp_r}{dt} = -\frac{\partial H}{\partial q_r}, \qquad r = 1, \cdots, m, \tag{5}$$

where $H = H(q_r, p_r, t)$ *is defined by the equation:*

$$H = \sum_{r=1}^{m} p_r \kappa_r - L, \tag{6}$$

the κ_r *being functions of* (q_r, p_r, t) *defined by* (4).

This is the theorem which is the subject of this chapter and to the proof of which we now turn. The converse is true under suitable restrictions.

2. A General Theorem. Let $F(x_1, \cdots, x_n)$ be any function, continuous together with its derivatives of the first two orders, and such that its Hessian Determinant, the Jacobian:

$$\frac{\partial(F_1, \cdots, F_n)}{\partial(x_1, \cdots, x_n)} \neq 0. \tag{1}$$

Let a transformation, T, be defined by the equations:

$$T: \qquad y_r = \frac{\partial F}{\partial x_r}, \qquad r = 1, \cdots, n.$$

Let $G(y_1, \cdots, y_n)$ be defined by the relation:

$$(2) \qquad G(y_1, \cdots, y_n) = \sum_{r=1}^{n} x_r y_r - F(x_1, \cdots, x_m),$$

where $x_1, \cdots, x_n$ are the functions of $y_1, \cdots, y_n$ defined by T^{-1}. Then

$$(3) \qquad x_r = \frac{\partial G}{\partial y_r}, \qquad r = 1, \cdots, n.$$

For, differentiate (2), regarded as an identity in the independent variables $y_1, \cdots, y_n$:

$$\frac{\partial G}{\partial y_r} = x_r + \sum_{s=1}^{n} y_s \frac{\partial x_s}{\partial y_r} - \sum_{s=1}^{n} \frac{\partial F}{\partial x_s} \frac{\partial x_s}{\partial y_r}.$$

The right-hand side, by the equations defining T, reduces to x_r, and the proof is complete. Furthermore,

$$(4) \qquad \frac{\partial (G_1, \cdots, G_n)}{\partial (y_1, \cdots, y_n)} \neq 0.$$

For, on performing first the transformation T, then the transformation T^{-1}, the result is the identical transformation. Hence

$$\frac{\partial (y_1, \cdots, y_n)}{\partial (x_1, \cdots, x_n)} \cdot \frac{\partial (x_1, \cdots, x_n)}{\partial (y_1, \cdots, y_n)} = 1,$$

or:

$$\frac{\partial (F_1, \cdots, F_n)}{\partial (x_1, \cdots, x_n)} \cdot \frac{\partial (G_1, \cdots, G_n)}{\partial (y_1, \cdots, y_n)} = 1.$$

In particular, then,

$$(5) \qquad \frac{\partial (G_1, \cdots, G_n)}{\partial (y_1, \cdots, y_n)} = \left[\frac{\partial (F_1, \cdots, F_n)}{\partial (x_1, \cdots, x_n)}\right]^{-1}.$$

These results may be stated in the following theorem.

THEOREM I. *Let $F(x_1, \cdots, x_n)$ be a function satisfying the condition:*

$$\frac{\partial (F_1, \cdots, F_n)}{\partial (x_1, \cdots, x_n)} \neq 0.$$

Perform the transformation:

$$T: \qquad y_r = \frac{\partial F}{\partial x_r}, \qquad r = 1, \cdots, n.$$

Let $G(y_1, \cdots, y_n)$ *be defined by the equation:*

$$(6) \qquad G(y_1, \cdots, y_n) = \sum_{r=1}^{n} x_r y_r - F(x_1, \cdots, x_n),$$

where x_r *is determined as a function of* $(y_1, \cdots, y_n)$ *by the inverse,* T^{-1}, *of* T. *Then the inverse of* T *is represented as follows:*

$$T^{-1}: \qquad x_r = \frac{\partial G}{\partial y_r}, \qquad r = 1, \cdots, n.$$

Moreover,

$$\frac{\partial(G_1, \cdots, G_n)}{\partial(y_1, \cdots, y_n)} \neq 0.$$

In particular,

$$\frac{\partial(G_1, \cdots, G_n)}{\partial(y_1, \cdots, y_n)} = \left[\frac{\partial(F_1, \cdots, F_n)}{\partial(x_1, \cdots, x_n)}\right]^{-1}.$$

The identical relation can be written in the symmetric form:

$$(7) \qquad F(x_1, \cdots, x_n) + G(y_1, \cdots, y_n) = \sum_{r=1}^{n} x_r y_r,$$

where

$$y_r = \frac{\partial F}{\partial x_r}, \qquad x_r = \frac{\partial G}{\partial y_r}, \qquad r = 1, \cdots, n,$$

and the Hessian Determinants of F, G *are* $\neq 0$.

We now proceed to a second theorem, which is of importance in the applications of the results of this chapter in mechanics.

THEOREM II. *If* F, G *are defined as before, and if each depends on a parameter,* ξ, *the relation*

$$(8) \qquad F(\xi;\, x_1, \cdots, x_n) + G(\xi;\, y_1, \cdots, y_n) = \sum_{r=1}^{n} x_r y_r$$

being an identity, because of T *or* T^{-1}, *either in the* $n + 1$ *arguments* $(\xi;\, x_1, \cdots, x_n)$ *or in the* $n + 1$ *arguments* $(\xi;\, y_1, \cdots, y_n)$, *then*

$$(9) \qquad \frac{\partial F}{\partial \xi} + \frac{\partial G}{\partial \xi} = 0.$$

Let $(\xi;\, x_1, \cdots, x_n)$ be the independent variables in (8). Then

$$\frac{\partial F}{\partial \xi} + \frac{\partial G}{\partial \xi} + \sum_{r=1}^{n} \frac{\partial G}{\partial y_r}\frac{\partial y_r}{\partial \xi} = \sum_{r=1}^{n} x_r \frac{\partial y_r}{\partial \xi}.$$

But

$$x_r = \frac{\partial G}{\partial y_r},$$

and the proof is complete.

3. Proof of Hamilton's Equations. We start out with the Lagrangean Function $L(q_r, \kappa_r, t)$, which fulfils the condition:

$$(1) \qquad \frac{\partial(L_1, \cdots, L_m)}{\partial(\kappa_1, \cdots, \kappa_m)} \neq 0,$$

and make the transformation:

$$(2) \qquad p_r = \frac{\partial L}{\partial \kappa_r}, \qquad r = 1, \cdots, m.$$

The Hamiltonian Function $H(q_r, p_r, t)$ is then defined by the equation:

$$(3) \qquad L + H = \sum_r p_r \kappa_r.$$

If, then, we set $x_r = \kappa_r$, $y_r = p_r$, and regard the q_r and t as parameters, all the conditions of the theorems of § 2 will be met. It follows, then, that the inverse of (2) is given by the equation:

$$(4) \qquad \kappa_r = \frac{\partial H}{\partial p_r};$$

and furthermore that

$$(5) \qquad \frac{\partial L}{\partial q_r} + \frac{\partial H}{\partial q_r} = 0, \qquad r = 1, \cdots, m.$$

It is also true that

$$(6) \qquad \frac{\partial L}{\partial t} + \frac{\partial H}{\partial t} = 0,$$

although this relation is not important for our present purposes.

Turn now to Lagrange's Equations:

$$(7) \qquad \begin{cases} \dfrac{dq_r}{dt} = \kappa_r \\[2ex] \dfrac{d}{dt}\dfrac{\partial L}{\partial \kappa_r} = \dfrac{\partial L}{\partial q_r}. \end{cases}$$

The first of these, combined with (4), gives:

$$(8) \qquad \frac{dq_r}{dt} = \frac{\partial H}{\partial p_r}, \qquad r = 1, \cdots, m.$$

From the second, combined with (2) and (5), we infer that

$$(9) \qquad \frac{dp_r}{dt} = -\frac{\partial H}{\partial q_r}, \qquad r = 1, \cdots, m.$$

But these are precisely the Hamiltonian Equations (5) of § 1:

$$\frac{dq_r}{dt} = \frac{\partial H}{\partial p_r}, \quad \frac{dp_r}{dt} = -\frac{\partial H}{\partial q_r}, \qquad r = 1, \cdots, m, \tag{10}$$

which we set out to establish.

The mathematical converse is simple. Given Equations (10) with the condition

$$\frac{\partial(H_1, \cdots, H_m)}{\partial(p_1, \cdots, p_m)} \neq 0,$$

Equations (4) define a transformation, and then L is defined by (3). Then (2) and (5) follow from the theorems of § 2. And now the first of Equations (10), combined with (4), gives

$$\frac{dq_r}{dt} = \kappa_r, \qquad r = 1, \cdots, m.$$

The second equation (10), combined with (2) and (5), leads to the equation:

$$\frac{d}{dt}\frac{\partial L}{\partial \kappa_r} = \frac{\partial L}{\partial q_r}, \qquad r = 1, \cdots, m.$$

Thus we arrive at Lagrange's Equations (7).

We see, then, that a knowledge of the function H is sufficient for a complete mathematical formulation of the motion. But what can we say of the physical meaning of H in the general case? There is an important restricted class of cases in which the definition is simple. Suppose there is a force function U depending on q_r, t alone, and furthermore that the kinetic energy T is a positive definite homogeneous quadratic form in the $\dot{q}_1, \cdots, \dot{q}_m$. Let L be defined by the equation:

$$L = T + U. \tag{11}$$

Let $\kappa_r = \dot{q}_r$. Then

$$\frac{\partial L}{\partial \kappa_r} = \frac{\partial T}{\partial \kappa_r}.$$

The transformation:

$$p_r = \frac{\partial L}{\partial \kappa_r}$$

now becomes:

$$p_r = \frac{\partial T}{\partial \kappa_r}.$$

Thus

$$\sum_{r=1}^{m} p_r \kappa_r = \sum_r \kappa_r \frac{\partial T}{\partial \kappa_r} = 2T.$$

Hence

$$H = \sum_r p_r \kappa_r - L = T - U. \tag{12}$$

Still more specially, if neither T nor U depends on t, then H becomes the *total energy* of the system. That H is here constant along any given path appears as follows. We have in the general case the relation:

$$\frac{dH}{dt} = \frac{\partial H}{\partial t}, \tag{13}$$

as is seen at once by differentiating:

$$\frac{dH}{dt} = \sum_r \frac{\partial H}{\partial q_r} \frac{dq_r}{dt} + \sum_r \frac{\partial H}{\partial p_r} \frac{dp_r}{dt} + \frac{\partial H}{\partial t},$$

and then making use of Hamilton's Equations. But in the general case $\partial H/\partial t \neq 0$, and so H is not constant along an arbitrary path. If, however, H does not contain t, then $\partial H/\partial t = 0$ and since now $dH/dt = 0$, we have:

$$H = h. \tag{14}$$

CHAPTER XII

D'ALEMBERT'S PRINCIPLE

1. The Problem. The general problem of Rational Mechanics, so far as it relates to a system of particles, can be formulated:

i) in terms of the $3n$ equations given by Newton's Second Law of Motion:

$$\text{A)} \qquad m_i\ddot{x}_i = X_i, \quad m_i\ddot{y}_i = Y_i, \quad m_i\ddot{z}_i = Z_i, \qquad i = 1, \cdots, n;$$

ii) in terms of further conditions equivalent to $3n$ relations between the $6n + 1$ variables $(x_i, y_i, z_i, X_i, Y_i, Z_i, t)$. A postulational treatment of these conditions will be found below in Appendix D.

Two extreme cases may be mentioned at the outset. First, each variable X_i, Y_i, Z_i, may be given as an explicit function of the x_i, y_i, z_i and their first derivatives with respect to the time, and t:

$$\begin{cases} X_j = \Phi_j(x_i, y_i, z_i, \dot{x}_i, \dot{y}_i, \dot{z}_i, t) \\ Y_j = \Psi_j(x_i, y_i, z_i, \dot{x}_i, \dot{y}_i, \dot{z}_i, t) \\ Z_j = \Omega_j(x_i, y_i, z_i, \dot{x}_i, \dot{y}_i, \dot{z}_i, t) \end{cases}$$

Thus A) reduces to a system of simultaneous differential equations for determining x_i, y_i, z_i as functions of the time, and with the solution of this problem the determination of the X_i, Y_i, Z_i is given by substitution.

Secondly, at the other extreme, the path of each particle and the velocity of the particle in its path may be given. Thus x_i, y_i, z_i become known functions of t, and again the X_i, Y_i, Z_i are found by substitution.

Between these two extremes there is a class of problems in which *constraints* occur which can be eliminated by a general principle due to d'Alembert. We shall not attempt to give a general definition of "constraints," for no such definition exists; but we can formulate a requirement which embraces the ordinary cases that arise in practice. Let $\delta x_i, \delta y_i, \delta z_i$ be any $3n$ quantities whatsoever. Then it is seen at once from A) that the following equation is true:

$$(1) \quad \sum_{i=1}^{n} (m_i\ddot{x}_i - X_i)\,\delta x_i + (m_i\ddot{y}_i - Y_i)\,\delta y_i + (m_i\ddot{z}_i - Z_i)\,\delta z_i = 0.$$

This equation is sometimes referred to as the *General Equation of Dynamics.*

Now it may happen that the force X_i, Y_i, Z_i can be broken up into two forces:

$$(2) \quad X_i = X_i' + X_i^*, \qquad Y_i = Y_i' + Y_i^*, \qquad Z_i = Z_i' + Z_i^*,$$

where the two new forces, namely, the X_i', Y_i', Z_i' and the X_i^*, Y_i^*, Z_i^*, are simpler than the old for the following reasons.

i) The X_i', Y_i', Z_i' are either explicit functions of the $x_i, y_i, z_i, \dot{x}_i, \dot{y}_i, \dot{z}_i, t$ or they involve in addition a restricted set of unknown functions arising from forces which are not given as functions of these $6n + 1$ variables.

ii) The X_i^*, Y_i^*, Z_i^* have the property that

$$(3) \qquad \sum_{i=1}^{n} X_i^*\,\delta x_i + Y_i^*\,\delta y_i + Z_i^*\,\delta z_i = 0$$

for all values of the $\delta x_i, \delta y_i, \delta z_i$ which satisfy the μ equations:

$$(4) \qquad \sum_{i=1}^{n} A_{i\alpha}\,\delta x_i + B_{i\alpha}\,\delta y_i + C_{i\alpha}\,\delta z_i = 0, \qquad \alpha = 1, \cdots, \mu,$$

where the coefficients are given functions of the $6n + 1$ variables $x_i, y_i, z_i, \dot{x}_i, \dot{y}_i, \dot{z}_i, t$, and the rank of the matrix:

$$(5) \quad \left\| \begin{matrix} A_{11} & \cdots & A_{n1} & B_{11} & \cdots & B_{n1} & C_{11} & \cdots & C_{n1} \\ \cdots & \cdots & \cdots & \cdots & \cdots & \cdots & \cdots & \cdots & \cdots \\ A_{1\mu} & \cdots & A_{n\mu} & B_{1\mu} & \cdots & B_{n\mu} & C_{1\mu} & \cdots & C_{n\mu} \end{matrix} \right\|$$

is μ; and conversely.

By means of Equations (4) the $3n$ quantities X_i^*, Y_i^*, Z_i^* can be expressed in terms of μ unknowns as follows. Multiply the α-th Equation (4) by an arbitrary number λ_α, and subtract the new equation from (3). Thus

$$(6) \quad \sum_{i=1}^{n} \left(X_i^* - \sum_{\alpha=1}^{\mu} A_{i\alpha}\lambda_\alpha\right) \delta x_i + \left(Y_i^* - \sum_{\alpha=1}^{\mu} B_{i\alpha}\lambda_\alpha\right) \delta y_i + \left(Z_i^* - \sum_{\alpha=1}^{\mu} C_{i\alpha}\lambda_\alpha\right) \delta z_i = 0.$$

Now, in Equations (4), a certain set of $3n - \mu$ of the $\delta x_i, \delta y_i, \delta z_i$ can be chosen at pleasure and then the remaining μ of these quantities will be uniquely determined, for at least one μ-rowed determinant from the matrix (5) does not vanish. It follows, then, that $\lambda_1, \cdots, \lambda_\mu$ can be determined uniquely from a suitable set of μ equations chosen from the $3n$ equations:

$$(7)\quad X_i^* = \sum_{\alpha=1}^{\mu} A_{i\alpha}\lambda_\alpha, \qquad Y_i^* = \sum_{\alpha=1}^{\mu} B_{i\alpha}\lambda_\alpha, \qquad Z_i^* = \sum_{\alpha=1}^{\mu} C_{i\alpha}\lambda_\alpha.$$

And now the remaining $3n - \mu$ Equations (7) will be satisfied by these values of the λ's. For Equation (6) has become an equation in which only those terms appear for which $\delta x_i, \delta y_i, \delta z_i$ are arbitrary, and hence their coefficients must each vanish.

Equations A) can now be written in the form:

$$(8)\qquad m_i\ddot{x}_i = X_i' + \sum_{\alpha=1}^{\mu} A_{i\alpha}\lambda_\alpha, \qquad m_i\ddot{y}_i = Y_i' + \sum_{\alpha=1}^{\mu} B_{i\alpha}\lambda_\alpha,$$

$$m\ddot{z}_i = Z_i' + \sum_{\alpha=1}^{\mu} C_{i\alpha}\lambda_\alpha,$$

where the $\lambda_1, \cdots, \lambda_\mu$ have come to us as linear combinations of μ suitably chosen X_i^*, Y_i^*, Z_i^*. On the other hand, they appear in Equations (8) merely as μ unknown functions, which can be determined from μ of these equations and then eliminated from the remainder.

Virtual Work. To put first things first was never more important than in the statement of d'Alembert's Principle. The $3n$ quantities $\delta x_i, \delta y_i, \delta z_i$ are to begin with $3n$ arbitrary numbers, and we then proceed to restrict them by the equations (4). Nevertheless, whatever values they may have, they determine by definition a virtual displacement of the system of points (x_i, y_i, z_i), and the quantity:

$$W_\delta = \sum_{i=1}^{n} X_i\,\delta x_i + Y_i\,\delta y_i + Z_i\,\delta z_i$$

is by definition the *virtual work* corresponding to this virtual displacement. Thus Equation (3) says that the force X_i^*, Y_i^*, Z_i^* is such that it does no virtual work when the virtual displacement is subject to the conditions (4).

D'ALEMBERT'S PRINCIPLE FOR A SYSTEM OF PARTICLES. *Given a system of particles, the motion of which is determined in part by Equations A). The discovery of an analysis of* X_i, Y_i, Z_i *by* (2) *and of the most general virtual displacement* δx_i, δy_i, δz_i, *whereby the virtual work of the force* X_i^*, Y_i^*, Z_i^* *vanishes, this virtual displacement being expressed by* (4); *finally the elimination of* δx_i, δy_i, δz_i, *and* X_i^*, Y_i^*, Z_i^*, *as above set forth, whereby* $3n - \mu$ *equations free from these unknowns result;— this is the spirit and content of d'Alembert's Principle.*

This enunciation of the Principle does not represent its historic origin, but rather its interpretation in the science today; cf. Appendix D.

2. Lagrange's Equations for a System of Particles, Deduced from d'Alembert's Principle. Let the coordinates x_i, y_i, z_i of the system of particles considered in § 1 be expressible in terms of m parameters and the time:

$$(1) \qquad \begin{cases} x_i = f_i(q_1, \cdots, q_m, t) \\ y_i = \varphi_i(q_1, \cdots, q_m, t) \\ z_i = \psi_i(q_1, \cdots, q_m, t) \end{cases}$$

where $(q_1, \cdots, q_m)$ is an arbitrary point of a certain region of the $(q_1, \cdots, q_m)$-space and the rank of the matrix is m. Let

$$(2) \qquad \delta x_i = \sum_{r=1}^{m} \frac{\partial f_i}{\partial q_r}\delta q_r, \qquad \delta y_i = \sum_{r=1}^{m} \frac{\partial \varphi_i}{\partial q_r}\delta q_r, \qquad \delta z_i = \sum_{r=1}^{m} \frac{\partial \psi_i}{\partial q_r}\delta q_r.$$

Then by the purely mathematical process of differentiation and substitution Equation (1) of § 1 yields:

$$(3) \qquad \sum_{r=1}^{m}\left(\frac{d}{dt}\frac{\partial T}{\partial \dot{q}_r} - \frac{\partial T}{\partial q_r} - Q_r\right)\delta q_r = 0,$$

where

$$T = \sum_{i=1}^{n} \frac{m_i}{2}\left(\dot{x}_i^2 + \dot{y}_i^2 + \dot{z}_i^2\right)$$

and

$$(4) \qquad Q_r = \sum_{i=1}^{n}\left(X_i\frac{\partial x_i}{\partial q_r} + Y_i\frac{\partial y_i}{\partial q_r} + Z_i\frac{\partial z_i}{\partial q_r}\right).$$

Now, the m quantities $\delta q_1, \cdots, \delta q_m$ are wholly arbitrary. Hence the coefficient of each term in (3) must vanish, and so we arrive at Lagrange's Equations in their most general form for a system of particles:

$$\frac{d}{dt}\frac{\partial T}{\partial \dot{q}_r} - \frac{\partial T}{\partial q_r} = Q_r, \qquad r = 1, \cdots, m. \tag{5}$$

Here, no restriction whatever is placed on the forces X_i, Y_i, Z_i, nor is the number of q_r's required to be a minimum.

In an important class of cases which arise in practice,

$$X_i = X'_i + X^*_i, \qquad Y_i = Y'_i + Y^*_i, \qquad Z_i = Z'_i + Z^*_i, \tag{6}$$

where

$$\sum_{i=1}^{n}\left(X^*_i\frac{\partial x_i}{\partial q_r} + Y^*_i\frac{\partial y_i}{\partial q_r} + Z^*_i\frac{\partial z_i}{\partial q_r}\right) = 0. \tag{7}$$

Hence

$$Q_r = \sum_{i=1}^{n}\left(X'_i\frac{\partial x_i}{\partial q_r} + Y'_i\frac{\partial y_i}{\partial q_r} + Z'_i\frac{\partial z_i}{\partial q_r}\right), \tag{8}$$

$$r = 1, \cdots, m.$$

Equations (5) now become Lagrange's Equations for this restricted case. The cases of constraints that do no virtual work are here included. Cf. further Appendix D.

3. The Six Equations for a System of Particles, Deduced from d'Alembert's Principle. Let the system of particles of § 1 be subject to internal forces such that the action and reaction between any two particles are equal and opposite and in the line through the particles:

$$X^*_{ij} + X^*_{ji} = 0, \qquad Y^*_{ij} + Y^*_{ji} = 0, \qquad Z^*_{ij} + Z^*_{ji} = 0;$$

$$\frac{X^*_{ij}}{x_j - x_i} = \frac{Y^*_{ij}}{y_j - y_i} = \frac{Z^*_{ij}}{z_j - z_i}.$$

And let any other forces X'_i, Y'_i, Z'_i act. Let

$$\begin{cases} \delta x_i = a + \beta z_i - \gamma y_i \\ \delta y_i = b + \gamma x_i - \alpha z_i \\ \delta z_i = c + \alpha y_i - \beta x_i \end{cases}$$

where a, b, c, α, β, γ are six arbitrary quantities. Since the internal forces destroy one another in pairs, and likewise, their moments, Equation (1) goes over into the following:

$$0 = a\sum_{i=1}^{n}(m_i\ddot{x}_i - X_i') + b\sum_{i=1}^{n}(m_i\ddot{y}_i - Y_i') + c\sum_{i=1}^{n}(m_i\ddot{z}_i - Z_i')$$
$$+ \alpha\sum_{i=1}^{n}\Big(m_i(y_i\ddot{z}_i - z_i\ddot{y}_i) - (y_iZ_i' - z_iY_i')\Big)$$
$$+ \beta\sum_{i=1}^{n}\Big(m_i(z_i\ddot{x}_i - x_i\ddot{z}_i) - (z_iX_i' - x_iZ_i')\Big)$$
$$+ \gamma\sum_{i=1}^{n}\Big(m_i(x_i\ddot{y}_i - y_i\ddot{x}_i) - (x_iY_i' - y_iX_i')\Big)$$

Now, set any five of the six quantities $a, b, c, \alpha, \beta, \gamma$ equal to 0, and the sixth equal to 1. Thus the six equations of motion, from which the internal reactions have been eliminated, emerge:

$$(1)\quad \left\{\begin{array}{l} \sum_{i=1}^{n} m_i\ddot{x}_i = \sum_{i=1}^{n} X_i', \quad \sum_{i=1}^{n} m_i\ddot{y}_i = \sum_{i=1}^{n} Y_i', \quad \sum_{i=1}^{n} m_i\ddot{z}_i = \sum_{i=1}^{n} Z_i'; \\ \sum_{i=1}^{n} m_i(y_i\ddot{z}_i - z_i\ddot{y}_i) = \sum_{i=1}^{n}(y_iZ_i' - z_iY_i'), \qquad \text{etc.} \end{array}\right.$$

In vector form these equations appear as the Equation of Linear Momentum:

$$\frac{d\rho}{dt} = \mathbf{F}$$

and as the Equation of Moment of Momentum:

$$\frac{d\sigma}{dt} = \mathbf{M}$$

We have used d'Alembert's Principle to deduce a set of *necessary* conditions. These are not in general *sufficient*, because the foregoing choice of δx_i, δy_i, δz_i is not in general the most general one.

4. Lagrange's Equations in the General Case, and d'Alembert's Principle. Consider an arbitrary system of masses, to which Lagrange's Equations, on the basis of suitable postulates, apply:

$$(1)\qquad \frac{d}{dt}\frac{\partial T}{\partial \dot{q}_r} - \frac{\partial T}{\partial q_r} = Q_r, \qquad r = 1, \cdots, m.$$

If we set by way of abbreviation:

$$(2)\qquad \frac{d}{dt}\frac{\partial T}{\partial \dot{q}_r} - \frac{\partial T}{\partial q_r} = [T]_r$$

then

$$\sum_{r=1}^{m}\Big([T]_r - Q_r\Big)\,\delta q_r = 0, \tag{3}$$

where $\delta q_1, \cdots, \delta q_m$ are any m quantities whatever. It may happen that Q_r can be expressed in the form:

$$Q_r = Q'_r + Q^*_r,$$

where Q'_r is for some reason simpler than Q_r and where, moreover,

$$Q^*_1\,\delta q_1 + \cdots + Q^*_m\,\delta q_m = 0, \tag{4}$$

provided

$$a_{\alpha 1}\,\delta q_1 + \cdots + a_{\alpha m}\,\delta q_m = 0, \qquad \alpha = 1, \cdots, \mu, \tag{5}$$

the rank of the matrix:

$$\left\|\begin{matrix} a_{11} & \cdots\cdots & a_{1m} \\ \cdots & \cdots\cdots & \cdots \\ a_{\mu 1} & \cdots\cdots & a_{\mu m} \end{matrix}\right\|$$

being μ. By reasoning precisely similar to that used in § 1, it is seen that the Q^*_r can be represented in the form:

$$Q^*_r = \sum_{\alpha=1}^{\mu} a_{\alpha r}\lambda_\alpha, \qquad r = 1, \cdots, m,$$

where the λ_α can be interpreted physically as certain linear combinations of a suitable set of μ of the quantities Q^*_r.

Moreover, Lagrange's Equations take on the form:

$$\frac{d}{dt}\frac{\partial T}{\partial \dot{q}_r} - \frac{\partial T}{\partial q_r} = Q'_r + \sum_{\alpha=1}^{\mu} a_{\alpha r}\lambda_\alpha, \qquad r = 1, \cdots, m,$$

where now the λ_α are thought of as unknown functions, which can be determined by μ of these equations and then eliminated from the remaining $m - \mu$ equations.

Virtual Work. In all cases the expression

$$W_\delta = Q_1\,\delta q_1 + \cdots + Q_m\,\delta q_m$$

can be interpreted as the *virtual work* done on the system by the forces which correspond to the Q_r. In particular, then, the condition (4) means that the virtual work of the forces which lead to the Q^*_r is nil, provided that the virtual displacement corresponds to the condition expressed by Equations (5).

5. Application: Euler's Dynamical Equations. Consider a rigid body, one point, O, of which is fixed, and which is acted on by any forces. Its position may be described geometrically in terms of Euler's Angles, Chapter VI, § 15:

$$(1) \qquad q_1 = \theta, \qquad q_2 = \psi, \qquad q_3 = \varphi.$$

Its kinetic energy is, by Chapter VII, § 6:

$$(2) \qquad T = \tfrac{1}{2}(Ap^2 + Bq^2 + Cr^2),$$

where

$$(3) \qquad \begin{cases} p = -\dot{\psi} \sin\theta \cos\varphi + \dot{\theta} \sin\varphi \\ q = \quad \dot{\psi} \sin\theta \sin\varphi + \dot{\theta} \cos\varphi \\ r = \quad \dot{\psi} \cos\theta + \dot{\varphi} \end{cases}$$

By d'Alembert's Principle, § 4:

$$(4) \qquad \sum_{r=1}^{3} \Big([T]_r - Q_r\Big)\, \delta q_r = 0,$$

where all three δq_r are arbitrary. Let $\delta q_1 = 0$, $\delta q_2 = 0$, $\delta q_3 \neq 0$. Compute

$$[T]_3 = \frac{d}{dt}\frac{\partial T}{\partial \dot{\varphi}} - \frac{\partial T}{\partial \varphi}.$$

The value is seen at once to be:

$$C\frac{dr}{dt} - (A - B)\, pq.$$

On the other hand, Q_3 can be computed as follows. Denote the vector moment of the applied forces, referred to O, as

$$\mathbf{M} = L\alpha + M\beta + N\gamma.$$

Then the virtual work corresponding to the virtual displacement $(\delta q_1, \delta q_2, \delta q_3) = (0, 0, \delta q_3)$ is seen to be:

$$Q_3\, \delta q_3 = N\, \delta q_3.$$

Hence $Q_3 = N$, and we find:

$$C\frac{dr}{dt} - (A - B)\, pq = N.$$

Thus one of Euler's Dynamical Equations is obtained, and the other two follow by symmetry, through advancing the letters cyclically.

The reader will say: "But this is precisely the same solution as that given earlier by Lagrange's Equations, Chap. X, § 16." True, so far as the analytic details of the solution go; and this is usually the case with applications of d'Alembert's Principle. It is the approach to the problem through the General Equation of Dynamics, § 1, which here yields (4), and the concept and use of virtual work. that brings the treatment under d'Alembert's Principle.

6. Examples. Consider the problem of the ladder sliding down a smooth wall; cf. Fig. 88, p. 147. Let us regard this problem as the motion of a lamina, moving in its own plane. As the generalized coordinates of the lamina we may take the coordinates of the centre of gravity:

$$q_1 = \bar{x}, \qquad q_2 = \bar{y}, \qquad q_3 = \theta.$$

Then

$$T = \tfrac{1}{2}M(\dot{\bar{x}}^2 + \dot{\bar{y}}^2) + \tfrac{1}{2}Mk^2\dot{\theta}^2, \tag{1}$$

where k is the radius of gyration about the centre of gravity. By d'Alembert's Principle,

$$\sum_{r=1}^{3}\Big([T]_r - Q_r\Big)\,\delta q_r = 0. \tag{2}$$

In the present case,

$$Q_1\delta q_1 = S\,\delta\bar{x}, \qquad Q_2\delta q_2 = (R - Mg)\,\delta\bar{y},$$

$$Q_3\,\delta q_3 = a(S\sin\theta - R\cos\theta)\,\delta\theta,$$

and thus Q_r is determined. Let

$$Q_r = Q_r' + Q_r^*,$$

where

$$Q_1^* = S, \qquad Q_2^* = R, \qquad Q_3^* = a(S\sin\theta - R\cos\theta).$$

Now, $\bar{x}$, $\bar{y}$, θ are connected by the relations:

$$\bar{x} = a\cos\theta, \qquad \bar{y} = a\sin\theta. \tag{3}$$

If, then, we subject $\delta\bar{x}$, $\delta\bar{y}$, $\delta\theta$ to the corresponding relations:

$$\delta\bar{x} = -\,a\sin\theta\,\delta\theta, \qquad \delta\bar{y} = a\cos\theta\,\delta\theta,$$

we see that

$$Q_1^*\delta q_1 + Q_2^*\delta q_2 + Q_3^*\delta q_3 = 0. \tag{4}$$

Thus the virtual work of the forces Q_r^*, corresponding to such a displacement, is seen to be nil, and so Equation (2) is replaced by the simpler equation:

$$(5) \qquad \sum_{r=1}^{3}\Big([T]_r - Q_r'\Big)\,\delta q_r = 0.$$

Hence

$$M\frac{d^2\bar{x}}{dt^2}(-\,a\sin\theta)\,\delta\theta + \Big(M\frac{d^2\bar{y}}{dt^2} + Mg\Big)\,a\cos\theta\,\delta\theta + Mk^2\frac{d^2\theta}{dt^2}\,\delta\theta = 0.$$

On replacing these second derivatives of $\bar{x}$ and $\bar{y}$ by their values from (3) a differential equation in the single dependent variable θ is obtained:

$$(6) \qquad \frac{d^2\theta}{dt^2} = \frac{ag}{a^2+k^2}\cos\theta,$$

and this determines the motion.

Rough Wall. Suppose, however, the wall is rough; Fig. 145, p. 323. Equations (1) and (2) still hold. But now

$$Q_1\,\delta q_1 = (S - \mu R)\,\delta\bar{x}, \qquad Q_2\,\delta q_2 = (R + \mu S - Mg)\,\delta\bar{y},$$

$$Q_3\,\delta q_3 = a\,[S\,(\sin\theta + \mu\cos\theta) + R\,(\mu\sin\theta - \cos\theta)]\,\delta\theta.$$

Let

$$Q_r = Q_r' + Q_r^*,$$

where

$$Q_1^* = S - \mu R, \qquad Q_2^* = R + \mu S,$$

$$Q_3^* = aS\,(\sin\theta + \mu\cos\theta) + aR\,(\mu\sin\theta - \cos\theta).$$

The values of δq_1, δq_2, δq_3 which make the virtual work of the force Q_r^* vanish:

$$Q_1^*\,\delta q_1 + Q_2^*\,\delta q_2 + Q_3^*\,\delta q_3 = 0,$$

are found by making the coefficients of R and S zero in this last equation:

$$-\mu\,\delta q_1 + \;\delta q_2 + a\,(\mu\sin\theta - \cos\theta)\,\delta q_3 = 0$$

$$\delta q_1 + \mu\,\delta q_2 + a\,(\sin\theta + \mu\cos\theta)\,\delta q_3 = 0$$

Hence

$$(7) \qquad \begin{cases} (1+\mu^2)\,\delta q_1 = a\,[-\,(1-\mu^2)\sin\theta - 2\mu\cos\theta]\,\delta q_3 \\ (1+\mu^2)\,\delta q_2 = a\,[-\,2\mu\sin\theta + (1-\mu^2)\cos\theta]\,\delta q_3 \end{cases}$$

Equation (5) now becomes:

$$M\frac{d^2\bar{x}}{dt^2}\,\delta q_1 + \left(M\frac{d^2\bar{y}}{dt^2} + Mg\right)\delta q_2 + Mk^2\frac{d^2\theta}{dt^2}\,\delta q_3 = 0,$$

and it remains only to substitute the values of δq_1, δq_2 from (7), and the values of $\bar{x}$, $\bar{y}$ from (3), and reduce. The result is:

$$\left((1-\mu^2)\,a^2 + (1+\mu^2)\,k^2\right)\frac{d^2\theta}{dt^2} + 2\mu a^2\left(\frac{d\theta}{dt}\right)^2 =$$

$$ag\,[2\mu\sin\theta - (1-\mu^2)\cos\theta].$$

The virtual displacement $(\delta q_1, \delta q_2, \delta q_3)$ which here led to the elimination of the unknown reactions R, S was not one which in any wise conformed to the "constraints" in the sense of the floor and the wall. If we replace R and μR by their resultant and draw a line L_1 through the bottom of the ladder perpendicular to it, and then do the same thing at the top of the ladder, thus obtaining a line L_2, the above virtual displacement corresponds to an actual displacement in which the bottom of the ladder is moved along L_1 and the top along L_2.

CHAPTER XIII

HAMILTON'S PRINCIPLE AND THE PRINCIPLE OF LEAST ACTION

1. Definition of δ. A new and independent foundation for Mechanics is given by Hamilton's Principle and certain other Principles of like nature. An integral, for which Hamilton's Integral:

$$\int_{t_0}^{t_1} (T + U)\, dt,$$

is typical, is extended along the natural path of the system, and then its value is considered for a neighboring, or varied, path. The Principle asserts that the integral is a minimum for the natural path, or at least that the integral is stationary for this path, *i.e.* that its variation vanishes:

$$\delta \int_{t_0}^{t_1} (T + U)\, dt = 0.$$

It is to the treatment of this subject that we now turn. Obviously we must begin by defining what is meant by a varied path and by a variation δ.

Let $F(x_1, \cdots, x_n, x_1', \cdots, x_n', u)$ be a function of the $2n + 1$ variables indicated. Here, $(x_1, \cdots, x_n)$ shall lie in a certain region R of the n-dimensional space of the variables $(x_1, \cdots, x_n)$; the variables $x_1', \cdots, x_n'$ shall be wholly unrestricted; and u shall lie in the interval: $a \leqq u \leqq b$. The function F shall be continuous, together with its partial derivatives of the first and second orders.* Let

* As regards assumptions of continuity, we lay down once and for all the requirement that whatever arbitrary functions are introduced shall be continuous, together with whatever derivatives we may wish to use, unless the contrary is stated.

For an introductory treatment of the Calculus of Variations cf. the author's *Advanced Calculus*, Chap. XVII.

$$C: \qquad x_i = x_i(u), \qquad a \leqq u \leqq b, \qquad i = 1, \cdots, n,$$

be a path lying in R. Let

$$x_i' = \frac{dx_i(u)}{du}.$$

Thus a path Γ in the $(2n + 1)$-dimensional space of the arguments of F is determined.

By a *varied path*, Γ', is meant the following. Let C' be a curve in R defined by the equations:

$$C': \qquad x_i = x_i(u, \epsilon), \qquad a \leqq u \leqq b, \qquad i = 1, \cdots, n,$$

where

$$x_i(u, 0) = x_i(u),$$

and ϵ is considered only in a region for which $|\epsilon|$ is small. Denote partial differentiation with respect to u by d. Let

$$x_i' = \frac{dx_i(u, \epsilon)}{du} = x_i'(u, \epsilon), \qquad i = 1, \cdots, n,$$

be chosen as the values of the $x_1', \cdots, x_n'$. The curve Γ' in the $(2n + 1)$-dimensional space is what is meant by a *varied curve.*

The *variation of* x_i, or δx_i, is defined by the equation:

$$\delta x_i = \left(\frac{\partial x_i(u, \epsilon)}{\partial \epsilon}\right)_{\epsilon=0}. \tag{1}$$

Since $x_i(u, \epsilon)$ is any function that conforms merely to the general requirements of continuity, we see that

$$\delta x_i = \eta_i(u), \qquad i = 1, \cdots, n,$$

is a wholly arbitrary function, restricted only by the above requirements of continuity.

The *variation of* x_i', or $\delta x_i'$ is not, however, arbitrary, but is defined by the equation:

$$\delta x_i' = \left(\frac{\partial^2 x_i(u, \epsilon)}{\partial u \, \partial \epsilon}\right)_{\epsilon=0}. \tag{2}$$

Thus

$$\delta x_i' = \frac{d\eta_i(u)}{du} = \eta_i'.$$

Hence

$$\frac{d}{du}\delta x_i = \delta \frac{dx_i}{du}. \tag{3}$$

It is now natural to lay down the further *definition*:

$$\delta \, dx_i = d \, \delta x_i. \tag{4}$$

Definition of δF. By the *variation of* $F(x_i, x'_i, u)$ is meant:

(5) $$\delta F = \left(\frac{\partial F}{\partial \epsilon}\right)_{\epsilon=0},$$

where x_i and x'_i on the right-hand side are set equal to $x_i(u, \epsilon)$ and $x'_i(u, \epsilon)$. Hence

(6) $$\delta F = \sum_{i=1}^{n} \left(\frac{\partial F}{\partial x_i}\,\delta x_i + \frac{\partial F}{\partial x'_i}\,\delta x'_i\right).$$

It is obvious that

$$\delta(F + \Phi) = \delta F + \delta \Phi;$$

$$\delta(F\Phi) = \Phi\,\delta F + F\,\delta\Phi;$$

$$\delta\frac{F}{\Phi} = \frac{\Phi\,\delta F - F\,\delta\Phi}{\Phi^2},$$

and also that

$$\delta\Omega(y_1, \cdots, y_m, u) = \sum_{k=1}^{m} \frac{\partial \Omega}{\partial y_k}\,\delta y_k,$$

where

$$y_k = \Psi_k(x_1, \cdots, x_n, x'_1, \cdots, x'_n, u), \qquad k = 1, \cdots, m.$$

Finally, the *definition*:

(7) $$\delta\,dF = d\,\delta F,$$

corresponding to the theorem:

(8) $$\delta\frac{dF}{du} = \frac{d\,\delta F}{du}.$$

And similarly,

(9) $$\int_a^b F\,\delta d\Phi = \int_a^b F\,d\,\delta\Phi = \int_a^b F\,\frac{d\,\delta\Phi}{du}\,du.$$

The dependent variables $x_1(u), \cdots, x_n(u)$ play a rôle in the foregoing definitions analogous to that of the independent variables in partial differentiation. But the analogy holds only up to a certain point, and to assume it beyond theorems like the above formulas which we can prove, has led to confusion and error in physics.

Variation of an Integral. Consider the integral:

$$J = \int_a^b F(x_1, \cdots, x_n, x'_1, \cdots, x'_n, u)\,du,$$

taken along the path Γ. By δJ is meant the following: — Extend the integral along the path Γ'. Thus a function $J(\epsilon)$ is defined. and now, by *definition*:

(10) $$\delta J = \left(\frac{dJ}{d\epsilon}\right)_{\epsilon=0} = J'(0).$$

It follows at once as a theorem that

(11) $$\delta \int_a^b F\, du = \int_a^b \sum_{i=1}^n \left(\frac{\partial F}{\partial x_i}\,\delta x_i + \frac{\partial F}{\partial x_i'}\,\delta x_i'\right) du.$$

The integral is said to be *stationary* for a particular path Γ if

$$\delta \int_a^b F\, du = 0,$$

no matter what functions $\delta x_i = \eta_i(u)$ may be. The condition is readily obtained in case δx_i is restricted to vanish for $u = a$ and for $u = b$:

(12) $$\delta x_i \,|_{u=a} = \eta_i(a) = 0; \qquad \delta x_i \,|_{u=b} = \eta_i(b) = 0.$$

For:

$$\frac{d}{du}\left(\frac{\partial F}{\partial x_i'}\,\delta x_i\right) = \frac{\partial F}{\partial x_i'}\,\delta x_i' + \frac{d}{du}\frac{\partial F}{\partial x_i'}\,\delta x_i.$$

Hence

(13) $$\delta \int_a^b F\, du = \int_a^b \sum_{i=1}^n \left(\frac{\partial F}{\partial x_i} - \frac{d}{du}\frac{\partial F}{\partial x_i'}\right) \delta x_i.$$

If, now, the integral in question is to vanish for an arbitrary choice of δx_i, it is easily seen that each parenthesis in the integrand of the last integral must vanish, or:

(14) $$\frac{\partial F}{\partial x_i} - \frac{d}{du}\frac{\partial F}{\partial x_i'} = 0, \qquad i = 1, \cdots, n.$$

These are known as *Euler's Equations*.

It is clear that

(15) $$\delta \int_a^b F\, du = \int_a^b \delta F\, du.$$

The limits of integration may be varied, too. Let

$$a' = \varphi(a, \epsilon), \qquad b' = \psi(b, \epsilon),$$

where $\varphi(a, 0) = a$, $\psi(b, 0) = b$. Let

$$J(\epsilon) = \int_{a'}^{b'} F[x_i(u, \epsilon), x_i'(u, \epsilon), u]\, du.$$

Then the *variation* of the integral is defined as before, by (10). It follows that

$$\delta \int_a^b F\, du = \int_a^b \delta F\, du + F(B_i, B_i', b)\, \delta b - F(A_i, A_i', a)\, \delta a, \tag{16}$$

where

$$A_i = x_i(a), \qquad A_i' = x_i'(a), \qquad B_i = x_i(b), \qquad B_i' = x_i'(b).$$

EXERCISE

Since

$$\frac{\partial^2 \Phi(u, \epsilon)}{\partial \epsilon\, \partial u} = \frac{\partial^2 \Phi(u, \epsilon)}{\partial u\, \partial \epsilon},$$

it follows (under the ordinary hypotheses of continuity), on letting ϵ approach 0, that the right-hand side approaches

$$\frac{d}{du}\, \delta \Phi.$$

The left-hand side approaches $\delta \Phi'$. Hence

$$\delta \frac{d\Phi}{du} = \frac{d}{du}\, \delta \Phi.$$

Thus Equation (7) is obtained as a theorem, and no new definition is necessary. Explain the error.

2. The Integral of Rational Mechanics. All that has gone before merely leads up to the definition of the variation of the following integral:

$$\int_{t_0}^{t_1} F(x_1, \cdots, x_n, \dot{x}_1, \cdots, \dot{x}_n, t)\, dt, \tag{1}$$

where the limits of integration may be constant or variable. The answer would seem to be simple, since t is the variable of integration and hence the independent variable in each of the functions

$$x_i = x_i(t).$$

But the symbol written down as the integral (1) is taken in Physics to mean something totally different. Let

$$(2) \qquad t = t(u), \qquad 0 \leqq u \leqq 1; \qquad t(0) = t_0, \qquad t(1) = t_1,$$

be any function of u such that, in the closed interval (0, 1),

$$0 < \frac{dt}{du} = t'.$$

Then the symbol (1) is taken to mean the integral:

$$(3) \qquad J = \int_0^1 F\left(x_1, \cdots, x_n, \frac{x_1'}{t'}, \cdots, \frac{x_n'}{t'}\right) t'\, du,$$

where $x_i' = dx_i/du$, and the "variation of the integral (1)" is understood to be the variation of this last integral. Thus

$$(4) \qquad \delta J = \int_0^1 \delta(Ft')\, du,$$

where u is the variable of integration, and the independent variable in each of the functions $x_i = x_i(u),\ t = t(u)$.

This last variation, (4): δJ, comes under the earlier definition of the variation of an integral. In particular, Equation (4) may be written in the form:

$$\delta J = \int_0^1 \delta F t'\, du + \int_0^1 F\, \delta t'\, du.$$

In each of these integrals the variable of integration may be changed back from u to t, and thus

$$(5) \qquad \delta J = \int_{t_0}^{t_1} \delta F\, dt + \int_{t_0}^{t_1} F\, d\,\delta t.$$

The variation of t, namely δt, is an arbitrary function of u:

$$\delta t = \tau(u),$$

and

$$\frac{d\,\delta t}{dt} = \frac{\tau'(u)}{t'(u)},$$

where u is the inverse function defined by (2). Moreover, by δF in (5) is meant the following:

$$\text{(6)}\quad \delta F = \delta F\left(x_i, \frac{x_i'}{t'}, t\right) = \sum_{i=1}^{n} \frac{\partial F}{\partial x_i}\,\delta x_i + \sum_{i=1}^{n} \frac{\partial F}{\partial \dot{x}_i}\,\delta\left(\frac{x_i'}{t'}\right) + \frac{\partial F}{\partial t}\,\delta t,$$

where

$$\delta\left(\frac{x_i'}{t'}\right) = \frac{t'\,\delta x_i' - x_i'\,\delta t'}{t'^2}, \qquad \delta x_i' = \frac{d}{du}\,\delta x_i, \qquad \delta t' = \frac{d}{du}\,\delta t.$$

We see, then, that

$$\text{(7)}\qquad \delta\dot{x}_i = \delta\left(\frac{x_i'}{t'}\right) = \frac{d\,\delta x_i}{dt} - \dot{x}_i\frac{d\,\delta t}{dt}.$$

Now, when t is the independent variable,

$$\text{(8)}\qquad \delta\dot{x}_i = \frac{d\,\delta x_i}{dt}.$$

The two formulas, (7) and (8), show that $\delta\dot{x}_i$ is not invariant of the independent variable. Why should it be? Similarly, the variation of an integral is not invariant of the variable of integration. Much of the confusion in the literature arises from losing sight of this fact. The "variation of the independent variable" is supposed to cover this case. It does so when and only when it becomes identical in substance with the above analysis.

3. Application to the Integral of Kinetic Energy. By definition, the kinetic energy

$$T = \tfrac{1}{2}\sum_{i=1}^{n} m_i(\dot{x}_i^2 + \dot{y}_i^2 + \dot{z}_i^2).$$

Hence

$$\text{(1)}\qquad \delta T = \sum_{i=1}^{n} m_i(\dot{x}_i\,\delta\dot{x}_i + \dot{y}_i\,\delta\dot{y}_i + \dot{z}_i\,\delta\dot{z}_i),$$

no matter what the independent variable and the dependent functions may be. If the former is u, then

$$(2)\qquad \delta T = \sum_{i=1}^{n} m_i\left[\dot{x}_i\,\delta\left(\frac{x_i'}{t'}\right) + \dot{y}_i\,\delta\left(\frac{y_i'}{t'}\right) + \dot{z}_i\,\delta\left(\frac{z_i'}{t'}\right)\right].$$

By Formula (7) of § 2 and the corresponding formulas involving $\dot{y}_i$, $\dot{z}_i$, this equation becomes:

$$(3)\qquad \delta T = \sum_{i=1}^{n} m_i\left(\dot{x}_i\frac{d\,\delta x_i}{dt} + \dot{y}_i\frac{d\,\delta y_i}{dt} + \dot{z}_i\frac{d\,\delta z_i}{dt}\right) - 2T\frac{d\,\delta t}{dt}.$$

Variation of the Integral:

$$(4)\qquad J = \int_{t_0}^{t_1} T\,dt.$$

Let the natural path in space be represented parametrically by the equations:

$$x_i = x_i(u), \qquad 0 \leqq u \leqq 1,$$

and let

$$t = t(u), \qquad 0 < \frac{dt}{du}.$$

The *variation* of this integral has, by § 2, (5) the value:

$$(5)\qquad \delta\int_{t_0}^{t_1} T\,dt = \int_{t_0}^{t_1} \delta T\,dt + \int_{t_0}^{t_1} T\,d\,\delta t.$$

By the aid of (3),

$$(6)\quad \int_{t_0}^{t_1}\delta T\,dt = \int_{t_0}^{t_1}\sum_{i=1}^{n} m_i\left(\dot{x}_i\frac{d\,\delta x_i}{dt} + \dot{y}_i\frac{d\,\delta y_i}{dt} + \dot{z}_i\frac{d\,\delta z_i}{dt}\right)dt - 2\int_{t_0}^{t_1} T\,d\,\delta t.$$

The first term on the right can be transformed by integration by parts, the integrand obviously having the value:

$$-\sum_{i=1}^{n} m_i(\ddot{x}_i\,\delta x_i + \ddot{y}_i\,\delta y_i + \ddot{z}_i\,\delta z_i) + \frac{d}{dt}\sum_{i=1}^{n} m_i(\dot{x}_i\,\delta x_i + \dot{y}_i\,\delta y_i + \dot{z}_i\,\delta z_i).$$

Hence

$$(7)\qquad \int_{t_0}^{t_1}\delta T\,dt = -\int_{t_0}^{t_1}\sum_{i=1}^{n} m_i(\ddot{x}_i\,\delta x_i + \ddot{y}_i\,\delta y_i + \ddot{z}_i\,\delta z_i)\,dt$$
$$+ \sum_{i=1}^{n} m_i(\dot{x}_i\,\delta x_i + \dot{y}_i\,\delta y_i + \dot{z}_i\,\delta z_i)\Big|_{t_0}^{t_1} - 2\int_{t_0}^{t_1} T\,d\,\delta t.$$

Finally, then:

$$(8)\qquad \delta\int_{t_0}^{t_1} T\,dt = -\int_{t_0}^{t_1}\sum_{i=1}^{n} m_i(\ddot{x}_i\,\delta x_i + \ddot{y}_i\,\delta y_i + \ddot{z}_i\,\delta z_i)\,dt$$

$$+\sum_{i=1}^{n} m_i(\dot{x}_i\,\delta x_i + \dot{y}_i\,\delta y_i + \dot{z}_i\,\delta z_i)\Big|_{t_0}^{t_1} - \int_{t_0}^{t_1} T\,d\delta t.$$

The variations δx_i, δy_i, δz_i, δt are $3n+1$ arbitrary functions, subject merely to the ordinary conditions of continuity. If, in particular, we impose on δx_i, δy_i, δz_i the condition that they vanish at the extremities of the interval of integration, *i.e.* for $t = t_0,\ t_1$, then

$$(9)\quad \int_{t_0}^{t_1} \delta T\,dt = -\int_{t_0}^{t_1}\sum_{i=1}^{n} m_i(\ddot{x}_i\,\delta x_i + \ddot{y}_i\,\delta y_i + \ddot{z}_i\,\delta z_i)\,dt - 2\int_{t_0}^{t_1} T\,d\delta t$$

and

$$(10)\quad \delta\int_{t_0}^{t_1} T\,dt = -\int_{t_0}^{t_1}\sum_{i=1}^{n} m_i(\ddot{x}_i\,\delta x_i + \ddot{y}_i\,\delta y_i + \ddot{z}_i\,\delta z_i) - \int_{t_0}^{t_1} T\,d\delta t.$$

4. Virtual Work. By the *virtual work* of the forces X_i, Y_i, Z_i, considered along the natural path:

$$(1)\quad x_i = x_i(u),\qquad y_i = y_i(u),\qquad z_i = z_i(u),\qquad u_0 \leqq u \leqq u_1,$$

is meant the quantity:

$$(2)\qquad W_\delta = \sum_{i=1}^{n} X_i\,\delta x_i + Y_i\,\delta y_i + Z_i\,\delta z_i,$$

where δx_i, δy_i, δz_i are $3n$ arbitrary functions of u, subject merely to the ordinary conditions of continuity.

This quantity is often denoted by δW; but it is not, in general, the variation of any function, and so it is better to avoid this confusion, writing δW only when W_δ is the variation of a function.

5. The Fundamental Equation. Combining Equation (9) of § 3 with Equation (2) of § 4 we have:

$$\int_{t_0}^{t_1}\left(\delta T + 2T\frac{d\,\delta t}{dt} + W_\delta\right)dt = \tag{1}$$

$$-\int_{t_0}^{t_1}\sum_{i=1}^{n}\Big((m_i\ddot{x}_i - X_i)\,\delta x_i + (m_i\ddot{y}_i - Y_i)\,\delta y_i + (m_i\ddot{z}_i - Z_i)\,\delta z_i\Big)\,dt.$$

Here the $3n + 1$ variations δx_i, δy_i, δz_i, δt are arbitrary except that the first $3n$ of these vanish for $t = t_0, t_1$; and the n forces X_i, Y_i, Z_i are any forces whatever. If these are the total forces acting on the particles, the right-hand side of (1) will vanish since each parenthesis vanishes by Newton's Law, and we shall have:

$$\int_{t_0}^{t_1}\left(\delta T + 2T\frac{d\,\delta t}{dt} + W_\delta\right)dt = 0. \tag{2}$$

Let the total force be broken up into two forces:

$$X_i = X_i' + X_i^*, \qquad Y_i = Y_i' + Y_i^*, \qquad Z_i = Z_i' + Z_i^*. \tag{3}$$

Then

$$W_\delta = W_{\delta'} + W_{\delta*}. \tag{4}$$

Suppose that $W_{\delta*}$ vanishes:

$$W_{\delta*} = \sum_{i=1}^{n} X_i^*\,\delta x_i + Y_i^*\,\delta y_i + Z_i^*\,\delta z_i = 0, \tag{5}$$

when the variations δx_i, δy_i, δz_i are chosen subject to certain conditions. Then Equation (2) takes the form:

$$\int_{t_0}^{t_1}\left(\delta T + 2T\frac{d\,\delta t}{dt} + W_{\delta'}\right)dt = 0, \tag{6}$$

where now δx_i, δy_i, δz_i satisfy these conditions, δt being still wholly arbitrary, and

$$W_{\delta'} = \sum_{i=1}^{n} X_i'\,\delta x_i + Y_i'\,\delta y_i + Z_i'\,\delta z_i. \tag{7}$$

These conditions usually take the form:

$$\sum_{i=1}^{n} A_{i\alpha}\,\delta x_i + B_{i\alpha}\,\delta y_i + C_{i\alpha}\,\delta z_i = 0, \qquad \alpha = 1, \cdots, \mu, \tag{8}$$

where $A_{i\alpha}$, $B_{i\alpha}$, $C_{i\alpha}$ are functions of x_i, y_i, z_i, $\dot{x}_i$, $\dot{y}_i$, $\dot{z}_i$, t, and the rank of the matrix:

$$(9) \qquad \left\| \begin{matrix} A_{11} & \cdots & C_{n1} \\ \cdots & \cdots & \cdots \\ A_{1\mu} & \cdots & C_{n\mu} \end{matrix} \right\|$$

is μ. This case includes both the holonomic and the non-holonomic cases. But it must be observed that δT is in general no longer, or not yet, the variation of a function.*

Generalized Coordinates. Suppose that the coordinates x_i, y_i, z_i of each mass m_i can be expressed in terms of m parameters $q_1, \cdots, q_m$ and the time:

$$(10) \qquad \begin{cases} x_i = f_i(q_1, \cdots, q_m, t) \\ y_i = \varphi_i(q_1, \cdots, q_m, t) \\ z_i = \psi_i(q_1, \cdots, q_m, t) \end{cases}$$

where the rank of the matrix

$$\left\| \begin{matrix} \dfrac{\partial f_1}{\partial q_1} & \cdots & \dfrac{\partial f_1}{\partial q_m} \\ \cdots & \cdots & \cdots \\ \dfrac{\partial \psi_n}{\partial q_1} & \cdots & \dfrac{\partial \psi_n}{\partial q_m} \end{matrix} \right\|$$

is m. Suppose further that Equation (5) is satisfied when

$$(11) \qquad \begin{cases} \delta x_i = \dfrac{\partial f_i}{\partial q_1}\delta q_1 + \cdots + \dfrac{\partial f_i}{\partial q_m}\delta q_m \\ \delta y_i = \dfrac{\partial \varphi_i}{\partial q_1}\delta q_1 + \cdots + \dfrac{\partial \varphi_i}{\partial q_m}\delta q_m \\ \delta z_i = \dfrac{\partial \psi_i}{\partial q_1}\delta q_1 + \cdots + \dfrac{\partial \psi_i}{\partial q_m}\delta q_m \end{cases}$$

the $\delta q_1, \cdots, \delta q_m$ being arbitrary. Let

$$(12) \qquad Q_r = \sum_{i=1}^{n}\left(X_i' \frac{\partial x_i}{\partial q_r} + Y_i' \frac{\partial y_i}{\partial q_r} + Z_i' \frac{\partial z_i}{\partial q_r}\right), \qquad r = 1, \cdots, m.$$

* The definition of the variation of a function, it will be recalled, is based on the dependence of the latter on certain arbitrary functions, whose variations may also be taken as arbitrary. These arbitrary functions are analogous, let us repeat, to the independent variables in the case of partial differentiation. And so further assumptions (*i.e.* postulates or definitions) are needed before δT can again mean a variation.

Then

(13) $$W_{\delta'} = Q_1 \delta q_1 + \cdots + Q_m \delta q_m.$$

On the other hand,

(14) $$T = T(q_1, \cdots, q_m, \dot{q}_1, \cdots, \dot{q}_m, t)$$
$$= T\left(q_1, \cdots, q_m, \frac{q_1'}{t'}, \cdots, \frac{q_m'}{t'}, t\right),$$

and δT as given by § 3, (2), becomes the variation of this latter function, where $q_1(u), \cdots, q_m(u), t(u)$ are the independent functions. Equation (6) of the present paragraph now takes the form:

I. $$\int_{t_0}^{t_1} \left(\delta T + 2T \frac{d\,\delta t}{dt} + W_\delta\right) dt = 0,$$

where δT means what it says — the *variation* of T — and where W_δ is the $W_{\delta'}$ of (13).

We will denote this equation as the *Fundamental Equation.* It embraces Equation (2) above, for the x_i, y_i, z_i can always be taken as $m = 3n$ generalized coordinates.

This equation is sometimes written in the form:

X) $$\int_{t_0}^{t_1} (\delta T + \delta W)\, dt + 2T\, \delta dt = 0,$$

where

(15) $$\delta W = Q_1 \delta q_1 + \cdots + Q_m\, \delta q_m.$$

Let us see just what this means. First of all, the equation is true under the hypotheses which led to Equation I. These were, that the path is the natural path of the system, given by the equations:

(16) $$q_r = q_r(u), \qquad t = t(u),$$

where $t(0) = t_0$, $t(1) = t_1$, and the variations

$$\delta q_r = \eta_r(u), \qquad \delta t = \eta_0(u),$$

are arbitrary functions subject merely to the conditions:

$$\eta_r(0) = 0, \quad r = 0, 1, \cdots, m; \qquad \eta_r(1) = 0, \quad r = 1, \cdots, m,$$

and possibly to a further restriction:

$$|\eta_r(u)| < h, \qquad |\eta_r'(u)| < h, \qquad r = 0, 1, \cdots, m,$$

where h is a definite positive constant.

Furthermore, in Equation X), $\delta W = W_\delta$ is not in general the variation of any function of $q_1, \cdots, q_m, t$. The Q_r have definite values at each point of the natural path, and so are definite functions of u; but they do not in general have any meaning at a point (q_r, t) not on the natural path, nor does δW.

Finally,

$$\delta T = \sum_{r=1}^{m} \frac{\partial T}{\partial \dot{q}_r} \delta \dot{q}_r + \sum_{r=1}^{m} \frac{\partial T}{\partial q_r} \delta q_r + \frac{\partial T}{\partial t} \delta t, \tag{17}$$

where

$$\delta \dot{q}_r = \delta \frac{q_r'}{t'} = \frac{t' \, \delta q_r' - q_r' \, \delta t'}{t'^2}$$

$$= \left(\frac{dt}{du}\frac{d\eta_r}{du} - \frac{dq_r}{du}\frac{d\eta_0}{du}\right) \Big/ \left(\frac{dt}{du}\right)^2.$$

The last term in the integral has the value:

$$\int_{t_0}^{t_1} 2T \frac{d\,\delta t}{dt} dt = \int_0^1 2T \, \eta_0'(u) \, du.$$

And now the meaning of Equation X) is this: — If q_r and t are set equal to the functions (16) which define the natural path and if δq_r, δt are chosen arbitrarily, subject merely to the general conditions above imposed, Equation X) will be fulfilled. Thus Equation X) expresses a necessary condition for the motion of the system — and this in all cases, be they holonomic or non-holonomic.

Since Equation X) is true for all variations δq_r, δt, it still represents a necessary condition when these functions are subject to any special restrictions we may choose to impose on them. For example, it may happen that the Q_r can be broken up into two functions:

$$Q_r = Q_r' + Q_r^*, \qquad r = 1, \cdots, m,$$

such that

$$Q_1^* \delta q_1 + \cdots + Q_m^* \delta q_m = 0,$$

provided that

$$a_{\alpha 1} \delta q_1 + \cdots + a_{\alpha m} \delta q_m = 0, \qquad \alpha = 1, \cdots, \mu,$$

where $a_{\alpha r} = a_{\alpha r}(q_1, \cdots, q_m, \dot{q}_1, \cdots, \dot{q}_m, t)$, and the rank of the matrix:

$$\begin{Vmatrix} a_{11} & \cdots\cdots\cdots & a_{1m} \\ \cdots & \cdots\cdots\cdots & \cdots \\ a_{\mu 1} & \cdots\cdots\cdots & a_{\mu m} \end{Vmatrix}$$

is μ. Here, δW is replaced by

$$(18) \qquad \delta W' = Q_1' \delta q_1 + \cdots + Q_m' \delta q_m$$

but only $m - \mu$ of the δq_r, and δt, can now be chosen arbitrarily. In what sense is δT now a "variation"? Emphatically, in *no* sense; for no definition has been laid down which reaches out to this case, and it is only from a definition that δT can derive its meaning. Nevertheless, Equations (17) and (18) continue to define the values of the terms δT, δW that appear in Equation X), and thus this equation continues to have a meaning, and to hold when a certain set of $m - \mu$ variations δq_r, and δt, are chosen arbitrarily.

Force Function. Finally, there may be a *force function*, U:

$$(19) \qquad X_i = \frac{\partial U}{\partial x_i}, \qquad Y_i = \frac{\partial U}{\partial y_i}, \qquad Z_i = \frac{\partial U}{\partial z_i},$$

where U is a function of the x_i, y_i, z_i and t. Thus the Fundamental Equation (2) becomes in this case:

$$\text{II.} \qquad \int_{t_0}^{t_1} \left(\delta T + \delta U + 2T \frac{d\,\delta t}{dt} \right) dt = 0,$$

where the (X_i, Y_i, Z_i) of (19) is the total force acting on m_i, provided U does not depend on t; otherwise we must understand by δU the *virtual variation* of U, or:

$$\delta U = \sum_{i=1}^{n} \frac{\partial U}{\partial x_i} \delta x_i + \frac{\partial U}{\partial y_i} \delta y_i + \frac{\partial U}{\partial z_i} \delta z_i.$$

Again, there may be a function $U(q_1, \cdots, q_m, t)$ such that in (15)

$$Q_r = \frac{\partial U}{\partial q_r}, \qquad r = 1, \cdots, m.$$

Then

$$\delta W = \delta U$$

and the Fundamental Equation takes on the same form, II., provided U does not depend on t; otherwise,

$$\delta U = \frac{\partial U}{\partial q_1}\delta q_1 + \cdots + \frac{\partial U}{\partial q_m}\delta q_m.$$

6. The Variational Principle. The variational principle as expressed by the Fundamental Equation I. of § 5, or even by Equation II., does not assert that the integral of some function, or physical quantity, is a minimum, or even stationary:

$$\delta \int (\text{something}) = 0 \qquad \text{or} \qquad \int \delta (\text{something}) = 0.$$

For the integrand is not a variation, in the sense of the Calculus of Variations; nor are the forces of the problem varied; they are considered only along the natural path of the system.* The Principle expresses a necessary condition for the motion of the system. In the non-holonomic case, the condition cannot be sufficient, since the first-order differential equations have not been incorporated into the formulation of the problem.

We turn now to certain further restrictions whereby Hamilton's Integral or an analogous integral does become stationary, and in fact, in a restricted region, a minimum.

7. Hamilton's Principle. If we set:

$$t(u, \epsilon) = u, \qquad t_0 \leqq u \leqq t_1,$$

then

$$\delta t \equiv 0$$

and the Fundamental Equation I. becomes:

$$\int_{t_0}^{t_1} (\delta T + W_\delta)\, dt = 0. \tag{1}$$

We can now suppress the parameter u since the time is not to be varied.

* In its leading ideas this treatment was given by Hölder, *Göttinger Nachrichten*, 1896, p. 122. Unfortunately Hölder felt impelled to defer to the primitive view of *variations* as "infinitely small quantities" in the sense of little zeros, *i.e.* infinitely small constants or functions of x_i, y_i, z_i, t. In the foot-notes on pp. 130, 131 the "neglect of infinitesimals of higher order" renders obscure — in fact, vitiates — the treatment, so far as clean-cut definitions go. The writer cannot but feel that the inner Hölder would have preferred such a treatment as that of the text, but that he did not have the courage to break with the unsound traditions of the little zeros, for fear of losing his clientèle.

Suppose that a force function U exists, which depends only on the x_i, y_i, z_i and t, or on the q_r and t:

$$(2) \qquad U = U(x_i, y_i, z_i, t) \qquad \text{or} \qquad U = U(q_r, t).$$

Since t is here the independent variable with respect to variation δ, we have

$$W_\delta = \delta U$$

in the sense of the Calculus of Variations, and (1) becomes:

$$(3) \qquad \int_{t_0}^{t_1} (\delta T + \delta U)\, dt = 0.$$

It must be remembered, however, that the variations δx_i, δy_i, δz_i or δq_r satisfy the condition of vanishing when $t = t_0$ and when $t = t^1$.

Here we meet our first example of an integral,

$$(4) \qquad \int_{t_0}^{t_1} (T + U)\, dt,$$

the variation of which vanishes:

$$(5) \qquad \delta \int_{t_0}^{t_1} (T + U)\, dt = 0.$$

This equation embodies Hamilton's Principle, which we may formulate as follows.

Hamilton's Principle. *Let T be the kinetic energy of a system of particles, and let a force function $U = U(q_r, t)$ exist. The natural path of the system is that for which Hamilton's Integral:*

$$(6) \qquad \int_{t_0}^{t_1} (T + U)\, dt,$$

is stationary:

$$(7) \qquad \delta \int_{t_0}^{t_1} (T + U)\, dt = 0.$$

Here t is the independent variable, and the variations of the dependent variables are such as vanish when $t = t_0$ and when $t = t_1$.

A necessary and sufficient condition that (7) be true is afforded by Euler's Equations, § 1, which here become:

$$\frac{d}{dt}\frac{\partial T}{\partial \dot{q}_r} - \frac{\partial T}{\partial q_r} = \frac{\partial U}{\partial q_r}, \qquad r = 1, \cdots, m.$$

But these are precisely Lagrange's Equations for the system. Incidentally we have a new proof of Lagrange's Equations, in case we make Hamilton's Principle our point of departure.

We have proved the Principle for systems of particles with m degrees of freedom, and it can be established in certain more general cases, *e.g.* for systems of rigid bodies; provided each time that a force function exists. The case is also included, in which relations of the form:

$$\varphi_\alpha(q_1, \cdots, q_m, t) = 0, \qquad \alpha = 1, \cdots, \mu,$$

exist; cf. Bolza, *Variationsrechnung*, p. 554. The most general case is that of a system having a Lagrangean Function, or kinetic potential, L. In the above cases,

$$L = T + U.$$

When it is not possible to establish it without special postulates — consider, for example, the motion of a perfect fluid or of an elastic body — it is taken as itself the postulate governing the motion of the system. The Principle consists, then, in requiring that the Lagrangean Integral:

$$\int_{t_0}^{t_1} L\,dt \tag{8}$$

be stationary; or that

$$\delta \int_{t_0}^{t_1} L\,dt = 0.$$

It will be shown in § 14 that Hamilton's Integral (8) is actually a minimum for a path lying within a suitably restricted region; but the minimum property does not necessarily hold for unrestricted paths.

8. Lagrange's Principle of Least Action. Our point of departure is the Fundamental Equation II., § 5, in which U now does not depend on t:

$$U = U(q_1, \cdots, q_m).$$

Moreover, T does not depend on t:

$$T = T(q_1, \cdots, q_m, \dot{q}_1, \cdots, \dot{q}_m).$$

Thus we have:

$$\int_{t_0}^{t_1} (\delta T + \delta U)\, dt + 2T\, d\,\delta t = 0. \tag{1}$$

Here each δ represents a variation in the sense of the Calculus of Variations, the independent functions being $q_1, \cdots, q_m, t$; but the integrand is not the variation of some function, nor is the integral the variation of some integral. Nevertheless, the equation is true when all $m + 1$ variations, $q_1, \cdots, q_m, t$, are chosen arbitrarily. Let us examine more minutely the meaning of this last statement. These variations are defined by arbitrary functions:

$$q_r(u, \epsilon), \qquad t(u, \epsilon), \tag{2}$$

such that

$$q_r(u, 0) = q_r(u), \qquad t(u, 0) = t(u).$$

Moreover:

$$q_r(0, \epsilon) = q_r(0), \qquad q_r(1, \epsilon) = q_r(1);$$

$$0 < \frac{\partial t(u, \epsilon)}{\partial u},$$

$$t(0, \epsilon) = t(0, 0) = t_0 = \text{const.}; \qquad t(1, 0) = t_1.$$

But in general $t(1, \epsilon) \neq t_1$. Thus

$$\delta t \,|_{u=0} = 0; \qquad \delta t \,|_{u=1} \neq 0.$$

In particular, then, the functions (2) may be restricted by any further conditions which are compatible merely with the general conditions of continuity. Such a condition is the one that, not merely for the natural path corresponding to $\epsilon = 0$, but also for all varied paths:

$$T = U + h, \tag{3}$$

or, more explicitly:

$$T\left[q_r(u, \epsilon), \frac{q_r'(u, \epsilon)}{t'(u, \epsilon)}\right] = U[q_r(u, \epsilon)] + h, \tag{3'}$$

where h is a constant. Since T is here a homogeneous quadratic polynomial in $\dot{q}_1, \cdots, \dot{q}_m$, it is clear that $t(u, \epsilon)$ is obtained by a quadrature when the $q_r(u, \epsilon)$, $r = 1, \cdots, m$, are chosen arbitrarily.

Let

$$F(q_r, q_r', t, t')$$

be any function. By δF we shall now mean the following:

$$\delta F = \frac{\partial F}{\partial \epsilon}\bigg|_{=0}, \tag{4}$$

where $q_r(u, \epsilon)$, $t(u, \epsilon)$ are restricted by the relation (3), *i.e.* (3′). And similarly:

$$\delta \int_0^1 F\,du = \frac{\partial}{\partial \epsilon}\int_0^1 F\,du \bigg|_{\epsilon=0}, \tag{5}$$

where the integrand on the left is formed for the arguments $q_r(u)$, etc., and the integrand on the right, for $q_r(u, \epsilon)$, etc.; Equation (3′) still holding. Thus it follows, in particular, that

$$\delta T = \delta U. \tag{6}$$

Although these definitions are in form identical with the earlier ones, where the $m + 1$ functions (2) were arbitrary, they are in substance distinct, since these $m + 1$ functions are now related by (3′).

Equation (1) now becomes, on suppressing the factor 2:

$$\int_{t_0}^{t_1} \delta T\,dt + T\,d\,\delta t = 0. \tag{7}$$

Since obviously, under our new definition of δ,

$$\delta(Tt') = \delta T \cdot t' + T\,\delta t',$$

and since $\delta t' = d\,\delta t/du$, Equation (7) takes the form:

$$\int_0^1 \delta(Tt')\,du = 0. \tag{8}$$

Hence, finally:

$$\delta \int_{t_0}^{t_1} T\,dt = 0. \tag{9}$$

We are thus led to the following Principle.

Lagrange's Principle of Least Action. *Let a system of particles have the kinetic energy T and a force function U, where U depends only on the position of the system, not on its velocity or the* *time, and where T is independent of t. Then a necessary and sufficient condition for the natural path of the system is, that*

$$\delta \int_{t_0}^{t_1} T\,dt = 0, \tag{10}$$

subject to the hypothesis that all varied paths fulfil the requirement that

$$T = U + h. \tag{11}$$

In addition, the variations of the coordinates shall vanish for $t = t_0$ and $t = t_1$.

The Principle thus formulated presents a Lagrangean problem in the Calculus of Variations with variable end points and one auxiliary condition:

$$\left\{ \begin{aligned} &\delta \int_{t_0}^{t_1} T\,dt = 0, \\ &\varphi \equiv T - U - h = 0. \end{aligned} \right. \tag{12}$$

The method of solution developed in that theory* employs Lagrange's Method of Multipliers. Briefly outlined it is as follows. Set

$$F = T + \lambda\varphi,$$

where λ is a function of t, and let $q_r(t)$, $\lambda(t)$, be determined by the $m + 1$ equations

$$\frac{\partial F}{\partial q_r} - \frac{d}{dt}\frac{\partial F}{\partial \dot{q}_r} = 0, \qquad r = 1, \cdots, m, \tag{13}$$

and the second equation (12). From Equation (13) it follows that

$$\frac{\partial T}{\partial q_r} + \lambda\frac{\partial \varphi}{\partial q_r} - \frac{d}{dt}\left[\frac{\partial T}{\partial \dot{q}_r} + \lambda\frac{\partial \varphi}{\partial \dot{q}_r}\right] = 0,$$

or

$$(1 + \lambda)\frac{\partial T}{\partial q_r} - \lambda\frac{\partial U}{\partial q_r} - \frac{d}{dt}\left[(1 + \lambda)\frac{\partial T}{\partial \dot{q}_r}\right] = 0.$$

These equations, combined with the second equation (12), give:

$$\lambda = -\tfrac{1}{2}. \tag{14}$$

* Cf. Bolza, *Variationsrechnung*, p. 586, where the case is considered that there are, in addition, relations between the coordinates, not involving the time.

Hence the $q_r(t)$ are determined from the resulting equations,

$$\frac{\partial T}{\partial q_r} + \frac{\partial U}{\partial q_r} - \frac{d}{dt}\frac{\partial T}{\partial \dot{q}_r} = 0, \qquad r = 1, \cdots, m. \tag{15}$$

The latter are Lagrange's Equations. Incidentally we have a new deduction of them, based on Lagrange's Principle of Least Action.

As in the case of Hamilton's Principle, so here we can give a direct proof of Lagrange's Principle of Least Action by means of the Calculus of Variations. For, as above pointed out, the Principle is equivalent to the Lagrangean problem represented by (12).

Recurring to the condition (3) we see that the functions $q_1(u, \epsilon)$, $\cdots$, $q_m(u, \epsilon)$ may be chosen arbitrarily, and the function $t(u, \epsilon)$ then determined by (3′). If the function $t(u, \epsilon)$ thus determined be substituted in the integral:

$$\int_0^1 Tt'\, du, \tag{16}$$

then t is completely eliminated from that integral. For

$$T = \sum_{r,\, s} A_{rs}\dot{q}_r\dot{q}_s, \qquad A_{rs} = A_{sr}, \tag{17}$$

the coefficients A_{rs}, depending only on the $q_1, \cdots, q_m$. Now,

$$\int_{t_0}^{t_1} T\, dt = \int_0^1 Tt'\, du. \tag{18}$$

Let

$$S = \sum_{r,\, s} A_{rs}q'_r q'_s \tag{19}$$

where, as usual, $q'_r = dq_r(u)/du$. Then

$$T\frac{dt^2}{du^2} = S,$$

or

$$t' = \frac{\sqrt{S}}{\sqrt{T}}. \tag{20}$$

From (20) and (3) it follows that

$$Tt' = \sqrt{U + h}\,\sqrt{S},$$

and thus t is eliminated, the integral (18) taking the form:

$$\int_0^1 \sqrt{U+h}\,\sqrt{S}\,du. \tag{21}$$

We now have before us a problem in the Calculus of Variations, of much simpler type — the simplest type of all, considered at the outset. It is the integral (21), formed for the functions $q_r(u)$, that is to be stationary, and these functions are all arbitrary. After this problem has been solved, t is determined from (20), or

$$t - t_0 = \int_0^u \frac{\sqrt{S}}{\sqrt{U+h}}\,du. \tag{22}$$

This is Jacobi's Principle of Least Action, which we will treat in the next paragraph as an independent Principle. But it is interesting to see how it can be derived from the Fundamental Equation of § 5, and proved as a particular case under Lagrange's Principle of Least Action.

EXERCISE

Show that Equation (1) under the restrictions named can be thrown into the form:

$$\int_{t_0}^{t_1} \sum_{r=1}^{m} \left(\frac{\partial T}{\partial q_r} + \frac{\partial U}{\partial q_r} - \frac{d}{dt}\frac{\partial T}{\partial \dot{q}_r} \right) \delta q_r \, dt = 0.$$

Hence deduce Lagrange's Equations.

9. Jacobi's Principle of Least Action. *Let a system of particles have the kinetic energy T and a force function U, where U depends only on the position of the system, not on its velocity or the time, and where the conditions imposed on the coordinates do not contain the time explicitly. Then a necessary and sufficient condition for the natural path of the system is, that the integral:*

$$\int_{t_0}^{t_1} \sqrt{U+h}\,\sqrt{T}\,dt \tag{1}$$

be stationary:

$$\delta \int_{t_0}^{t_1} \sqrt{U+h}\,\sqrt{T}\,dt = 0. \tag{2}$$

The time is given by the equation:

$$T = U + h, \tag{3}$$

or

$$t = \int_0^u \frac{\sqrt{S}\,du}{\sqrt{U+h}}, \tag{4}$$

where

$$\sqrt{S}\,du = \sqrt{T}\,dt. \tag{5}$$

We can give a direct proof as follows. The integral (1) has the value:

$$\int_0^1 \sqrt{U+h}\,\sqrt{S}\,du, \tag{6}$$

where $q_1(u), \cdots, q_m(u)$ are arbitrary functions. It is to be stationary:

$$\delta \int_0^1 \sqrt{U+h}\,\sqrt{S}\,du = 0. \tag{7}$$

Hence Euler's Equations must hold, or:

$$\frac{\partial}{\partial q_r}(\sqrt{U+h}\,\sqrt{S}) - \frac{d}{du}\frac{\partial}{\partial q_r'}(\sqrt{U+h}\,\sqrt{S}) = 0, \tag{8}$$

$$r = 1, \cdots, m.$$

Hence

$$\frac{\sqrt{S}}{\sqrt{U+h}}\frac{\partial U}{\partial q_r} + \frac{\sqrt{U+h}}{\sqrt{S}}\frac{\partial S}{\partial q_r} - \frac{d}{du}\left(\frac{\sqrt{U+h}}{\sqrt{S}}\frac{\partial S}{\partial q_r'}\right) = 0. \tag{9}$$

Equations (9) determine the path; t has not yet entered in the solution. Equations (3) and (5) now determine t; it is given by (4).

It follows furthermore that

$$\frac{\partial S}{\partial q_r} = \left(\frac{dt}{du}\right)^2 \frac{\partial T}{\partial q_r}, \qquad \frac{\partial S}{\partial q_r'} = \frac{dt}{du}\frac{\partial T}{\partial \dot{q}_r}. \tag{10}$$

Combining these equations with (4) and (9) we find:

$$\frac{d}{dt}\frac{\partial T}{\partial \dot{q}_r} - \frac{\partial T}{\partial q_r} = \frac{\partial U}{\partial q_r}. \tag{11}$$

Thus we arrive at Lagrange's Equations. If we assume them, then we have a proof of Jacobi's Principle. Conversely, if we assume Jacobi's Principle, we have a new proof of Lagrange's Equations.

10. Critique of the Methods. Retrospect and Prospect. The symbol δ is treacherous. It can and does mean many things, and writers on Mechanics are not careful to say what they mean by it. In d'Alembert's Principle the δx_i, δy_i, δz_i began life by being $3n$ arbitrary numbers. In their youth they were disciplined to conform to certain linear homogeneous equations. Thus still a number of them were arbitrary quantities; the rest had no choice, they were uniquely determined.

Enter, the Calculus of Variations. And now the δx_i, δy_i, δz_i, *and* δt become the *variations* of functions of a parameter, or independent variable, u. From now on these δ's must be dealt with under the sanctions of the Calculus of Variations — at least, if the findings of that branch of mathematics are to be adopted.

The Future. As the physicist fares forth over the uncharted ocean of his ever-expanding science, his compass is the Principles. He seeks an integral which in the new domain will do for him what Hamilton's Principle achieved in classical mechanics. There is mysticism about this integral. Imagination must guide him, and he will try many guesses. But he will not be helped by an undefined δ. He must make a clean-cut postulate defining the integral, and then lay down a clean-cut definition of what he means by the variation. There is no short cut. A thoroughgoing knowledge of the rudiments of the Calculus of Variations is as essential in Mechanics as perspective is in art.

11. Applications. Let a particle be acted on by a central attracting force inversely proportional to the square of the distance. Then

$$T = \frac{m}{2}\left(\dot{r}^2 + r^2\,\dot{\theta}^2\right), \qquad U = \frac{\mu}{r}, \tag{1}$$

where the pole is at the centre of force, and it is assumed that the motion takes place in a plane (cf. Exercise 4, below). Then the integral:

$$\int_0^1 \sqrt{\frac{\mu}{r} + h}\,\sqrt{r'^2 + r^2\theta'^2}\,du \tag{2}$$

must be made a minimum. Set

$$F(r, \theta, r', \theta') = \sqrt{\left(\frac{\mu}{r} + h\right)(r'^2 + r^2\theta'^2)}.$$

Then

$$\frac{d}{du}\frac{\partial F}{\partial \theta'} - \frac{\partial F}{\partial \theta} = 0.$$

Since $\partial F/\partial \theta = 0$, it follows that

$$\text{(3)} \qquad \frac{\partial F}{\partial \theta'} = \sqrt{\frac{\mu}{r} + h}\,\frac{r^2\theta'}{\sqrt{r'^2 + r^2\theta'^2}} = c.$$

If $c = 0$, then

$$\theta = \text{const.}$$

and the motion takes place in a right line. But if $c \neq 0$, θ may be taken as the variable of integration:* $u = \theta$, and (3) becomes:

$$r^2\sqrt{\frac{\mu}{r} + h} = c\sqrt{r'^2 + r^2}.$$

Hence

$$c^2\frac{dr^2}{d\theta^2} = hr^4 + \mu r^3 - c^2r^2,$$

$$\theta = \pm \int \frac{c\,dr}{r\sqrt{hr^2 + \mu r - c^2}}.$$

Change the variable of integration:

$$u = \frac{1}{r}.$$

Then

$$\theta = \mp \int \frac{c\,du}{\sqrt{h + \mu u - c^2u^2}}.$$

Performing the integration, we find:

$$u = \frac{1 - e\cos(\theta - \gamma)}{K}, \qquad r = \frac{K}{1 - e\cos(\theta - \gamma)}.$$

EXERCISES

1. Discuss in detail the case $c = 0$.

2. In the general case, determine the constants e, K, γ in terms of the initial conditions.

* It is true that the interval for u was $(0, 1)$; but it might equally well have been an arbitrary interval: $a \leqq u \leqq b$.

3. Obtain the time.

4. Allowing the particle free motion in space, show that its path is a plane orbit.

Suggestion:— Use Cartesian coordinates.

5. Discuss the motion of a particle *in vacuo* under the force of gravity. Assume the path to lie in a plane.

6. In Question 5, prove that the path must lie in a plane.

7. Explain the case of motion in a circle under the solution given in the text.

12. Hamilton's Integral a Minimum in a Restricted Region.*

THEOREM. *The integral*

$$\int_{t_0}^{t_1} L\,dt \tag{1}$$

is a minimum for the natural path, provided t_0 and t_1 are not too far apart.

The Lagrangean function:

$$L(q_1, \cdots, q_m, \dot{q}_1, \cdots, \dot{q}_m, t),$$

has the properties:

$$\frac{\partial(L_{\dot{q}_1}, \cdots, L_{\dot{q}_m})}{\partial(\dot{q}_1, \cdots, \dot{q}_m)} \neq 0; \tag{2}$$

$$\sum_{k,l} L_{\dot{q}_k \dot{q}_l} \xi_k \xi_l \tag{3}$$

is a positive definite quadratic form. Moreover,

$$H + L = \sum_r p_r \dot{q}_r, \tag{4}$$

where

$$p_r = \frac{\partial L}{\partial \dot{q}_r} \qquad r = 1, \cdots, m, \tag{5}$$

and the Hamiltonian function

$$H(q_1, \cdots, q_m, p_1, \cdots, p_m, t)$$

has the properties:

$$\frac{\partial(H_{p_1}, \cdots, H_{p_m})}{\partial(p_1, \cdots, p_m)} \neq 0; \tag{6}$$

* Carathéodory has given a proof of this theorem: Riemann-Weber, *Partielle Differentialgleichungen der mathematischen Physik*, 8. ed. 1930, Vol. I, Chap. V.

(7) $$\dot{q}_r = \frac{\partial H}{\partial p_r}.$$

A necessary condition that the integral (1) be a minimum is, that

$$\delta \int_{t_0}^{t_1} L\, dt = 0.$$

The extremals are given by Euler's equations:

(8) $$\frac{d}{dt}\frac{\partial L}{\partial \dot{q}_r} - \frac{\partial L}{\partial q_r} = 0, \qquad r = 1, \cdots, m,$$

which are precisely Lagrange's equations.

By the transformation (5), the inverse of which is given by (7), Lagrange's equations (8) go over into Hamilton's canonical equations, Chap. XI:

(9) $$\frac{dq_r}{dt} = \frac{\partial H}{\partial p_r}, \qquad \frac{dp_r}{dt} = -\frac{\partial H}{\partial q_r}, \qquad r = 1, \cdots, m.$$

The latter can be solved by means of Jacobi's equation; cf. Chap. XV and Appendix C:

(10) $$\frac{\partial V}{\partial t} + H\left(q_1, \cdots, q_m, \frac{\partial V}{\partial q_1}, \cdots, \frac{\partial V}{\partial q_m}, t\right) = 0,$$

as follows. Let (q_r^0, p_r^0, t_0) be a point in the neighborhood of which Equations (9) are to be solved. A solution of (10):

(11) $$V = S(q_1, \cdots, q_m, \alpha_1, \cdots, \alpha_m, t),$$

can be found* such that

(12) $$\frac{\partial (S_{q_1}, \cdots, S_{q_m})}{\partial (\alpha_1, \cdots, \alpha_m)} \neq 0$$

* The existence theorem in question follows at once from the theory of characteristics as applied to Equation (10). That theory tells us that there exists a solution of (10):

$$V = S(q_1, \cdots, q_m, t),$$

such that, when $t = t_0$, S reduces to a given function $\varphi(q_1, \cdots, q_m)$:

$$S(q_1, \cdots, q_m, t_0) = \varphi(q_1, \cdots, q_m).$$

Here, $\varphi(q_1, \cdots, q_m)$ is any function which, together with its first derivatives, is continuous in the neighborhood of the point $(q_1^0, \cdots, q_m^0)$. Such a function is:

$$\varphi(q_1, \cdots, q_m) = \sum_r \alpha_r q_r,$$

where the $\alpha_1, \cdots, \alpha_m$ are m arbitrary constants, or parameters. The function S thus resulting is the function required in the text.

If, as we may assume, the function $H(q_r, p_r, t)$ is analytic in the point (q_r^0, p_r^0, t_0), and if, as is here the case, $\varphi(q_r)$ is analytic in the point (q_r^0), then the fundamental existence theorem of the classical Cauchy Problem, formulated for the simplest case, applies at once, and the theory of characteristics is not needed.

in the point (q_r^0, α_r^0, t_0) and furthermore the equations:

$$(13) \qquad p_r = \frac{\partial S}{\partial q_r}, \qquad \beta_r = -\frac{\partial S}{\partial \alpha_r}, \qquad r = 1, \cdots, m$$

are satisfied — the first set, when (q_r^0, p_r^0, t_0) are given, by the values $\alpha_r = \alpha_r^0$; and then the second set determines β_r^0.

By means of this function S Equations (9) are solved. The solution is contained in (13) and is obtained explicitly by solving (13) for q_r, p_r:

$$(14) \qquad \begin{cases} q_r = f_r(\alpha_1, \cdots, \alpha_m, \beta_1, \cdots, \beta_m, t) \\ p_r = g_r(\alpha_1, \cdots, \alpha_m, \beta_1, \cdots, \beta_m, t) \end{cases}$$

New Properties of the Extremals. If $q_r = q_r(t)$ represents an extremal, and if $\dot{q}_r = dq_r/dt$, then by (5) and (13):

$$(15) \qquad L_{\dot{q}_r} = S_{q_r}, \qquad r = 1, \cdots, n.$$

Moreover:

$$(16) \qquad \sum_r S_{q_r}\dot{q}_r + S_t = L.$$

For, since S is a solution of (10), it follows, by the aid of (13), that

$$(17) \qquad S_t + H(q_1, \cdots, q_m, p_1, \cdots, p_m, t) = 0.$$

On substituting this value of H in (4), and replacing p_r in the resulting equation by its value from (13), Equation (16) results.

The Function $E(q_r, q'_r, \dot{q}_r, t)$. Consider the function

$$L' = L(q_r, q'_r, t),$$

where (q_r, q'_r, t) are $2n + 1$ independent variables. Let $(q_r, \dot{q}_r, t)$ be an arbitrary point, and develop L' about this point by Taylor's Theorem with a Remainder. We have:

$$(18) \qquad L' = L + \sum_r L_{\dot{q}_r}(q'_r - \dot{q}_r) + E(q_r, q'_r, \dot{q}_r, t),$$

where $L, L_{\dot{q}_r}$ are formed for the arguments $(q_r, \dot{q}_r, t)$, and

$$(19) \qquad E(q_r, q'_r, \dot{q}_r, t) = \tfrac{1}{2}\sum_{r,s} \tilde{L}_{\dot{q}_r\dot{q}_s}(q'_r - \dot{q}_r)(q'_s - \dot{q}_s)$$

the coefficient $\tilde{L}_{\dot{q}_r\dot{q}_s}$ being the value of $L_{\dot{q}_r\dot{q}_s}$ for a mean value of the arguments $\dot{q}_r$, namely, $\dot{q}_r + \theta(q'_r - \dot{q}_r)$, where $0 < \theta < 1$. The quadratic form (3) is positive definite. Hence

(20) $$0 < E(q_r, q_r', \dot{q}_r, t)$$

if $(q_1', \cdots, q_m')$ is distinct from $(\dot{q}_1, \cdots, \dot{q}_m)$.

Proof of the Minimum Property. Consider an arbitrary extremal $\mathscr{E}_0$ through the point P_0: (q_r^0, t_0), represented by (14):

$$\mathscr{E}_0: \qquad q_r = q_r(t), \qquad r = 1, \cdots, m.$$

Let P_1: (q_r^1, t_1) be a second point on $\mathscr{E}_0$ near by. Connect P_0 and P_1 by an arbitrary curve

$$C: \qquad q_r = \bar{q}_r(t), \qquad r = 1, \cdots, m,$$

and let $\bar{q}_r'(t) = d\bar{q}_r/dt$. The curve C shall, however, be a weak variation,

$$|\bar{q}_r(t) - q_r(t)| < \eta, \qquad |\bar{q}_r'(t) - q_r'(t)| < \eta.$$

Let

$$\bar{L} = L(\bar{q}_r, \bar{q}_r', t).$$

Let $(\bar{q}_r, t)$ be an arbitrary point on C. Through this point there passes an m-parameter family of extremals, (13) or (14). We select one of them as follows. Let $\alpha_1, \cdots, \alpha_m$ retain the values they have for $\mathscr{E}_0$; but let $\beta_1, \cdots, \beta_m$ have new values, namely, those given by the second of the equations (13), when $q_r = \bar{q}_r$, $t = t$. The corresponding value of $\dot{q}_r$ will be given by (7). It is the value found, for the α_r, β_r in question, by differentiating the first of the equations (14) with respect to t. These values of α_r, $\bar{q}_r$, $\dot{q}_r$, t satisfy Equations (15) and (16); the β_r do not enter explicitly in these equations, and so the fact that they depend on t does not complicate the equations.

We now apply Equation (18), setting $q_r = \bar{q}_r$, giving to $\dot{q}_r$ the value just found, and letting q_r' refer to C, $q_r' = \bar{q}_r'$. Thus

(21) $$\bar{L} = L(\bar{q}_r, \dot{q}_r, t) + \sum_r L_{\dot{q}_r}(\bar{q}_r, \dot{q}_r, t)(\bar{q}_r' - \dot{q}_r) + E(\bar{q}_r, \bar{q}_r', \dot{q}_r, t).$$

The first two terms on the right of (21) can be modified as follows. First,

(22) $$\frac{dS(\bar{q}_r, t)}{dt} = \sum_r S_{q_r}(\bar{q}_r, t)\,\bar{q}_r' + S_t(\bar{q}_r, t).$$

Next, from (16):

(23) $$\sum_r S_{q_r}(\bar{q}_r, t)\,\dot{q}_r + S_t(\bar{q}_r, t) - L(\bar{q}_r, \dot{q}_r, t) = 0.$$

Subtracting (23) from (22) we have:

$$\frac{dS(\bar{q}_r, t)}{dt} = \sum_r S_{q_r}(\bar{q}_r, t)(\bar{q}'_r - \dot{q}_r) + L(\bar{q}_r, \dot{q}_r, t).$$

Finally, since from (15)

$$L_{\dot{q}_r}(\bar{q}_r, \dot{q}_r, t) = S_{q_r}(\bar{q}_r, t),$$

the first two terms on the right of (21) have the value $dS(\bar{q}_r, t)/dt$, and (21) can be written:

$$\bar{L} = \frac{dS(\bar{q}_r, t)}{dt} + E(\bar{q}_r, \bar{q}'_r, \dot{q}_r, t). \tag{24}$$

We now proceed to integrate this equation from t_0 to t_1. Observe that

$$\int_{t_0}^{t_1} \frac{dS(\bar{q}_r, t)}{dt}\, dt = S(\bar{q}_r, t)\Big|_{t_0}^{t_1}$$

has precisely the value of the integral:

$$\int_{t_0}^{t_1} L\, dt,$$

taken along the natural path of the system. For, along $\mathscr{E}_0$ Equation (16) says that

$$L = \frac{dS(q_r, t)}{dt}$$

and so

$$\int_{t_0}^{t_1} L\, dt = S(q_r, t)\Big|_{t_0}^{t_1}.$$

But in the end points, $q_r(t) = \bar{q}_r(t)$. We thus arrive at the final result:

$$\int_{t_0}^{t_1} \bar{L}\, dt = \int_{t_0}^{t_1} L\, dt + \int_{t_0}^{t_1} E(\bar{q}_r, \bar{q}'_r, \dot{q}_r, t)\, dt. \tag{25}$$

If, then, C differs from $\mathscr{E}_0$, there will be points of C at which $E > 0$, and so the integral of L over C (*i.e.* the integral on the left) will be greater than the integral of L over $\mathscr{E}_0$ (*i.e.* the first integral on the right) and our theorem is proved.

13. Jacobi's Integral a Minimum in a Restricted Region. In Jacobi's Integral:

$$(1)\qquad \int_{t_0}^{t_1} \sqrt{U+h}\,\sqrt{T}\,dt,$$

the functions T and U do not contain t explicitly, and T is homogeneous in the $\dot{q}_r$:

$$(2)\qquad T = \sum_{r,s} A_{rs}\,\dot{q}_r\dot{q}_s.$$

The varied functions, $q_r(u, \epsilon)$, are arbitrary, subject merely to the condition that $\delta q_r = 0$ in each end-point, $t = t_0, t_1$. It is obvious that the integral (1) has the same value as the integral:

$$(3)\qquad \int_{t_0}^{t_1} T\,dt,$$

subject to the restriction:

$$(4)\qquad T = U + h.$$

This condition shall hold for the varied paths, too. Thus $q_r(u, \epsilon)$ is still arbitrary; but $t(u, \epsilon)$ is determined by (4). To prove, then, that the integral (1) is a minimum for the natural path, it is sufficient to show that the integral (3) has this property, if (4) holds for the varied paths.

In the present case,

$$(5)\qquad L = T + U.$$

$$(6)\qquad H = T - U.$$

From (4) and (5),

$$(7)\qquad L = 2T - h.$$

Let $\mathscr{E}_0$ be the path defined in § 12, and let C':

$$(8)\qquad \bar{q}_r = q_r(u, \epsilon), \qquad \bar{t} = t(u, \epsilon),$$

be a varied path. Consider the varied integral. From (7)

$$(9)\qquad \int_{t_0}^{\bar{t}_1} 2\bar{T}\,d\bar{t} = \int_{t_0}^{\bar{t}_1} \bar{L}\,d\bar{t} + h(\bar{t}_1 - t_0),$$

where

$$\int_{\bar{t}_0}^{\bar{t}_1} \bar{L}\, d\bar{t} = \int_{u_0}^{u_1} L\left(\bar{q}_r, \frac{\bar{q}_r'}{\bar{t}'}\right) \bar{t}'\, du.$$

The right-hand side of Equation (9) can be computed as follows. From the analysis used in § 12, Equation (24), we see that

$$\bar{L}\bar{t}' = \frac{dS(\bar{q}_r, \bar{t})}{du} + E\bar{t}'.$$

Hence

$$\int_{\bar{t}_0}^{\bar{t}_1} \bar{L}\, d\bar{t} = S(\bar{q}_r, \bar{t})\Big|_{u_0}^{u_1} + \int_{\bar{t}_0}^{\bar{t}_1} E\, d\bar{t}.$$

Since $\bar{q}_r = q_r$ for $u = u_0, u_1$, the first term on the right has the value:

$$S(q_r^1, \bar{t}_1) - S(q_r^0, t_0).$$

Hence

$$(10) \qquad \int_{\bar{t}_0}^{\bar{t}_1} 2\bar{T}\, d\bar{t} = S(q_r^1, \bar{t}_1) - S(q_r^0, t_0) + h(\bar{t}_1 - t_0) + \int_{\bar{t}_0}^{\bar{t}_1} E\, d\bar{t}.$$

Since H is independent of t:

$$H = H(q_r, p_r),$$

it follows as in Chap. XIV, § 4, that a function S of the form:

$$S = -ht + W(q_1, \cdots, q_m, h, \alpha_2, \cdots, \alpha_m)$$

can be found, where h is to be identified with α_1. Using this function S in (10), we have:

$$(11) \qquad \int_{\bar{t}_0}^{\bar{t}_1} 2\bar{T}\, d\bar{t} = W(q_r^1) - W(q_r^0) + \int_{\bar{t}_0}^{\bar{t}_1} E\, d\bar{t}.$$

If we allow C' to coincide with $\mathscr{E}_0$, then $E \equiv 0$, and

$$\int_{t_0}^{t_1} 2T\, dt = W(q_r^1) - W(q_r^0).$$

Thus (11) becomes:

$$(12) \qquad \int_{t_0}^{\bar{t}_1} 2\bar{T}\, d\bar{t} = \int_{t_0}^{t_1} 2T\, dt + \int_{t_0}^{\bar{t}_1} E\, d\bar{t}.$$

This proves the theorem. For, if C' is distinct from $\mathscr{E}_0$, then E, which is never negative, will be positive for some parts of the interval of integration, and hence the integral (3), extended over C', will exceed in value the same integral extended over the natural path, as was to be proved.

The case U = const. leads to the geodesics on a manifold for which the differential of arc is given by the equation:

$$ds^2 = \sum_{r,s} A_{rs}\, dq_r\, dq_s.$$

Thus we have a proof that a geodesic on a manifold obtained as above is the shortest line connecting two points which are not too far apart.

CHAPTER XIV

CONTACT TRANSFORMATIONS

1. Purpose of the Chapter.* The final problem before us is the integration of Hamilton's Canonical Equations:

$$\text{A)} \qquad \frac{dq_r}{dt} = \frac{\partial H}{\partial p_r}, \qquad \frac{dp_r}{dt} = -\frac{\partial H}{\partial q_r}, \qquad r = 1, \cdots, m.$$

The method consists in finding a large and important class of transformations of the variables (q_r, p_r, t) into new variables (q'_r, p'_r, t'), such that Equations A) are carried over into a new system of like form:

$$\text{A}') \qquad \frac{dq'_r}{dt'} = \frac{\partial H'}{\partial p'_r}, \qquad \frac{dp'_r}{dt'} = -\frac{\partial H'}{\partial q'_r}, \qquad r = 1, \cdots, m,$$

or, as we say, transformations with respect to which Hamilton's Equations remain *invariant*.

The most general class of such transformations we shall consider, are the so-called *Canonical Transformations*. A one-to-one transformation:

$$\text{I.} \qquad \begin{cases} q'_r = q'_r(q_1, \cdots, q_m, p_1, \cdots, p_m, t) \\ p'_r = p'_r(q_1, \cdots, q_m, p_1, \cdots, p_m, t) \\ t' = t'(q_1, \cdots, q_m, p_1, \cdots, p_m, t) \end{cases}$$

$$\frac{\partial(q'_1, \cdots, q'_m, p'_1, \cdots, p'_m, t')}{\partial(q_1, \cdots, q_m, p_1, \cdots, p_m, t)} \neq 0,$$

is said to be *canonical* if there exist two functions,

$H(q_1, \cdots, q_m, p_1, \cdots, p_m, t)$ and $H'(q'_1, \cdots, q'_m, p'_1, \cdots, p'_m, t')$

(not in general equal to each other) such that

* This introductory paragraph is designed to give an outline of the treatment contained in the following chapter. The student should read it carefully, not, however, expecting to comprehend its full meaning, but rather regarding it as a guide, to which, in his study of the detailed developments, he will turn back time and again for purposes of orientation.

$$(1) \qquad \int_{\Gamma'} \Big(\sum_r p'_r\, dq'_r - H'\, dt'\Big) = \int_{\Gamma} \Big(\sum_r p_r\, dq_r - H\, dt\Big),$$

where Γ is an arbitrary closed curve of the $(2m + 1)$-dimensional (q_r, p_r, t)-space, and Γ' is its image in the transformed (q'_r, p'_r, t')-space, these spaces being thought of as simply connected.

To a canonical transformation there corresponds a function $\Psi(q_1, \cdots, q_m, p_1, \cdots, p_m, t)$ such that

$$(2) \qquad \sum_r p'_r\, dq'_r - H'\, dt' = \sum_r p_r\, dq_r - H\, dt + d\Psi.$$

And conversely, when three functions H', H, Ψ exist, for which the latter relation is true, the transformation is canonical.

Contact Transformations. An important sub-set of these canonical transformations consists in those for which the last Equation I. is

$$(3) \qquad t' = t.$$

On equating the coefficients of dt on the two sides of Equation (2) we find:

$$(4) \qquad H' = H - \frac{\partial \Psi}{\partial t} + \sum_r p'_r \frac{\partial q'_r}{\partial t}.$$

Since $t' = t$, we may say that the variable t is not transformed, and treat it as a parameter. Equations I. thus take on the form:

$$\text{II.} \qquad \begin{cases} q'_r = q'_r(q_1, \cdots, q_m, p_1, \cdots, p_m, t) \\ p'_r = p'_r(q_1, \cdots, q_m, p_1, \cdots, p_m, t) \end{cases}$$

with

$$\frac{\partial(q'_1, \cdots, q'_m, p'_1, \cdots, p'_m)}{\partial(q_1, \cdots, q_m, p_1, \cdots, p_m)} \neq 0.$$

And now comes an important modification of Equation (2). Since we now are regarding the (q_r, p_r), and not the (q_r, p_r, t), as the independent variables, (2) can be written by the aid of (4) in the form:

$$(5) \qquad \sum_r (p'_r\, dq'_r - p_r\, dq_r) = d\Psi.$$

Of course, $d\Psi$ has different meanings in (2) and (5). In (2),

$$(6) \qquad d\Psi = \sum_r \Big(\frac{\partial \Psi}{\partial q_r} dq_r + \frac{\partial \Psi}{\partial p_r} dp_r\Big) + \frac{\partial \Psi}{\partial t} dt,$$

since here the independent variables are q_r, p_r, t, whereas in (5),

$$d\Psi = \sum_r \left(\frac{\partial \Psi}{\partial q_r} dq_r + \frac{\partial \Psi}{\partial p_r} dp_r\right), \tag{7}$$

since here the independent variables are q_r, p_r; a similar remark applying to the other differentials, dq'_r. This is not an exception, or contradiction, in principle, but only in practice, since the differential of any function, $\Psi(x_1, \cdots, x_n)$, depends on the *independent variables*:

$$d\Psi = \sum_i \frac{\partial \Psi}{\partial x_i} dx_i,$$

and it is not until we have said what these shall be — *i.e. defined our function* — that we can speak of its differential.

A transformation — we will henceforth change the notation from m to n:

$$\begin{cases} q'_r = q'_r(q_1, \cdots, q_n, p_1, \cdots, p_n) \\ p'_r = p'_r(q_1, \cdots, q_n, p_1, \cdots, p_n) \end{cases} \tag{8}$$

$$\frac{\partial(q'_1, \cdots, q'_n, p'_1, \cdots, p'_n)}{\partial(q_1, \cdots, q_n, p_1, \cdots, p_n)} \neq 0$$

such that

$$\int_{\Gamma'} p'_r\, dq'_r = \int_{\Gamma} p_r\, dq_r, \tag{9}$$

where Γ is an arbitrary closed curve of the (q_r, p_r)-space, thought of as simply connected, and Γ' is the curve into which it is transformed, shall be called a *contact transformation*. There corresponds to such a transformation a function $\Psi(q_1, \cdots, q_n, p_1, \cdots, p_n)$ for which

$$\sum_r (p'_r\, dq'_r - p_r\, dq_r) = d\Psi. \tag{10}$$

And conversely, a transformation (8) for which (10) is true satisfies (9) and so is a contact transformation.

A contact transformation may, of course, depend on certain parameters, p'_r, q'_r and Ψ thus becoming functions of these parameters as well. The transformation II. above is a case in point.

Finally, the canonical transformations form a *group*; *i.e.* the result of applying first one and then a second such transformation may itself be expressed as a canonical transformation.

The contact transformations also form a group. The group of contact transformations II. is a subgroup of the group of canonical transformations I.

The approach to the contact transformations is through the *Integral Invariants* of Poincaré.

The contact transformations, as defined generally by (8) and (9), are of especial importance in Mechanics because any such transformation carries an arbitrary system A) of Hamiltonian Equations over into a second such system, A′); cf. infra, § 4. We shall treat the application of these transformations to the integration of Hamilton's Equations at length in Chapter XV. If the student is willing to take this one property of contact transformations for granted, he can turn at once to Chapter XV, and he will find no other assumptions needed in the study of that chapter.

2. Integral Invariants. Consider the action integral:

$$\int_{t_0}^{t_1} L(q_r, q_r', t)\, dt, \tag{1}$$

and the extremals, which are the path curves, given by Lagrange's Equations:

$$\frac{d}{dt}\frac{\partial L}{\partial \dot{q}_r} - \frac{\partial L}{\partial q_r} = 0, \qquad r = 1, \cdots, n, \tag{2}$$

where L is the Lagrangean function, or the kinetic potential. The general solution can be written in the form:

$$q_r = q_r(t;\, q_1^0, \cdots, q_n^0, \dot{q}_1^0, \cdots, \dot{q}_n^0), \qquad r = 1, \cdots, n, \tag{3}$$

where q_r^0, $\dot{q}_r^0$ are the initial values of q_r, $\dot{q}_r$, i.e. their values when $t = t_0$. In the $(2n + 1)$-dimensional space of the variables $(q_1, \cdots, q_n, \dot{q}_1, \cdots, \dot{q}_n, t)$ these equations, together with the n further equations:

$$\dot{q}_r = \dot{q}_r(t;\, q_1^0, \cdots, q_n^0, \dot{q}_1^0, \cdots, \dot{q}_n^0), \tag{4}$$

represent a curve C, — or more properly, a $2n$-parameter family of curves C. Let a closed curve, Γ_0:

$$q_r^0 = q_r^0(\lambda), \qquad \dot{q}_r^0 = \dot{q}_r^0(\lambda), \qquad r = 1, \cdots, n, \qquad \lambda_0 \leqq \lambda \leqq \lambda_1, \tag{5}$$

be drawn in the plane $t = t_0$. The curves C which pass through the points of Γ_0 form a *tube* of solutions, which we will denote by S.

Let the action integral, (1), be extended along the curves C which form S. Its value is a function of λ:

(6) $$J(\lambda) = \int_{t_0}^{t_1} L(q_r, \dot{q}_r, t)\, dt,$$

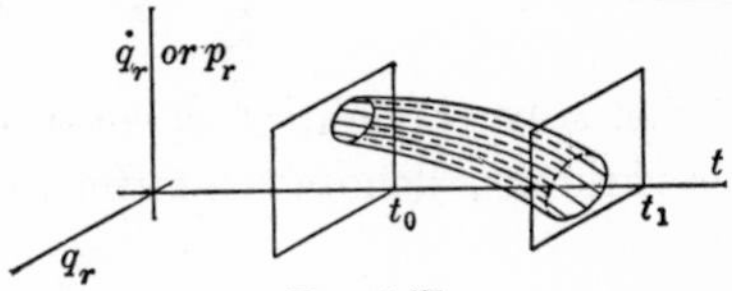

where q_r, $\dot{q}_r$ are given by (3) and (4) and q_r^0, $\dot{q}_r^0$ by (5). Differentiate $J(\lambda)$:

FIG. 147

$$J'(\lambda) = \int_{t_0}^{t_1} \sum_r \left(\frac{\partial L}{\partial q_r}\frac{\partial q_r}{\partial \lambda} + \frac{\partial L}{\partial \dot{q}_r}\frac{\partial \dot{q}_r}{\partial \lambda} \right) dt.$$

On integrating by parts, observing that

$$\frac{\partial \dot{q}_r}{\partial \lambda} = \frac{d}{dt}\frac{\partial q_r}{\partial \lambda},$$

we have:

$$\int \frac{\partial L}{\partial \dot{q}_r}\frac{\partial \dot{q}_r}{\partial \lambda}\, dt = \frac{\partial L}{\partial \dot{q}_r}\frac{\partial q_r}{\partial \lambda} - \int \frac{d}{dt}\frac{\partial L}{\partial \dot{q}_r}\frac{\partial q_r}{\partial \lambda}\, dt.$$

Hence

$$J'(\lambda) = \int_{t_0}^{t_1} \sum_r \left(\frac{\partial L}{\partial q_r} - \frac{d}{dt}\frac{\partial L}{\partial \dot{q}_r} \right) \frac{\partial q_r}{\partial \lambda}\, dt + \sum_r \frac{\partial L}{\partial \dot{q}_r}\frac{\partial q_r}{\partial \lambda}\Bigg|_{t_0}^{t_1}.$$

The integral vanishes, because q_r is by hypothesis a solution of (2).

We now make the transformation, Chapter XI, § 3:

(7) $$p_r = \frac{\partial L}{\partial \dot{q}_r}.$$

Thus

(8) $$J'(\lambda) = \sum_r p_r \frac{\partial q_r}{\partial \lambda}\Bigg|_{t_0}^{t_1}.$$

Since Γ_0 is a closed curve, $q_r^0(\lambda_1) = q_r^0(\lambda_0)$, $\dot{q}_r^0(\lambda_1) = \dot{q}_r^0(\lambda_0)$, and (6) gives:

(9) $$J(\lambda_1) = J(\lambda_0).$$

Hence

$$\int_{\lambda_0}^{\lambda_1} J'(\lambda)\, d\lambda = J(\lambda_1) - J(\lambda_0) = 0,$$

and so from (8):

$$(10)\qquad \int_{\lambda_0}^{\lambda_1} \sum_r p_r \frac{\partial q_r}{\partial \lambda}\bigg|_{t_0}^{t_1} d\lambda = 0.$$

Let t_0 be thought of as constant, but t_1, which is also arbitrary, as variable; denote the latter by t. Then

$$(11)\qquad \int_{\lambda_0}^{\lambda_1} \sum_r p_r \frac{\partial q_r}{\partial \lambda} d\lambda = \int_{\lambda_0}^{\lambda_1} \Big(\sum_r p_r \frac{\partial q_r}{\partial \lambda}\Big)_0 d\lambda.$$

This equation represents the theorem in which the whole investigation of this paragraph culminates. In substance it can be stated as follows. We may regard Equation (7), along with the $n + 1$ further identical equations, $q_r = q_r$, $t = t$, as representing a *transformation of the* $(q_r, \dot{q}_r, t)$*-space on the* (q_r, p_r, t)*-space.* Observe that the Jacobian

$$(12)\qquad \frac{\partial (p_1, \cdots, p_n)}{\partial (\dot{q}_1, \cdots, \dot{q}_n)} \neq 0,$$

Chapter XI, § 3. Thus the curves C of the first space go over into curves C' of the second space, and Γ_0 goes over into a curve Γ_0', S being transformed into a tube S'.

Let us now make the second space — the space of the variables (q_r, p_r, t) — our point of departure and, dropping the primes, consider a closed curve in the plane $t = t_0$ of that space:

$$\Gamma_0: \qquad q_r^{\,0} = q_r^{\,0}(\lambda), \qquad p_r^{\,0} = p_r^{\,0}(\lambda).$$

Consider furthermore curves C through its points, which are obtained by transforming the curves C of the earlier space. The integrals (11) now become line integrals in the present space. If we change the notation, setting

$$(13)\qquad q_r^0 = \alpha_r, \qquad p_r^0 = \beta_r,$$

then (11) assumes the form:

$$(14)\qquad \int_{\Gamma} \sum_r p_r\, dq_r = \int_{\Gamma_0} \sum_r \beta_r\, d\alpha_r,$$

where Γ is the curve of intersection of the arbitrary plane $t = t$ with the tube S determined by Γ_0, an arbitrary closed curve of the plane $t = t_0$. But this is precisely the definition of a contact transformation, t being thought of as a parameter:*

$$(15) \qquad \begin{cases} q_r = q_r(\alpha_1, \cdots, \alpha_n, \beta_1, \cdots, \beta_n, t) \\ p_r = p_r(\alpha_1, \cdots, \alpha_n, \beta_1, \cdots, \beta_n, t) \end{cases}$$

3. Consequences of the Theorem. *a) Hamilton's Canonical Equations.* Lagrange's Equations (2), § 2 form a system of n simultaneous total differential equations of the second order. By means of the transformation (7) these are carried over into a simultaneous system of $2n$ total differential equations of the first order in the (q_r, p_r, t)-space. Let these be written in the form:

$$(16) \qquad \frac{dq_r}{dt} = Q_r(q, p, t), \qquad \frac{dp_r}{dt} = P_r(q, p, t),$$

$$r = 1, \cdots, n.$$

Since the right-hand side of (14) is independent of t, the derivative of the left-hand side with respect to t must vanish. Hence

$$\frac{d}{dt}\int_{\lambda_0}^{\lambda_1} \sum_r p_r \frac{\partial q_r}{\partial \lambda} d\lambda = 0,$$

or

$$\int_{\lambda_0}^{\lambda_1} \left(\sum_r \frac{\partial p_r}{\partial t}\frac{\partial q_r}{\partial \lambda} + p_r \frac{\partial^2 q_r}{\partial t\, \partial \lambda}\right) d\lambda = 0.$$

Integrate by parts:

$$\frac{\partial}{\partial \lambda}\left(p_r \frac{\partial q_r}{\partial t}\right) = p_r \frac{\partial^2 q_r}{\partial \lambda\, \partial t} + \frac{\partial p_r}{\partial \lambda}\frac{\partial q_r}{\partial t}.$$

Since Γ_0 is a closed curve,

$$p_r \frac{\partial q_r}{\partial t}\bigg|_{\lambda_0}^{\lambda_1} = 0,$$

* The geometric picture is here slightly different from the earlier one, since the variables (α_r, β_r) and (q_r, p_r) are interpreted in different planes. But of course one may think of a cylinder on Γ_0 as directrix, with its elements parallel to the t-axis. On cutting this cylinder with the plane $t = t$, we have a curve Γ_0 lying in the same plane with Γ. Or, to look at the situation from another angle, t is only a parameter, and it is the spaces of (α_r, β_r) and (q_r, p_r) which concern us.

and we have:

$$(17) \qquad \int_{\lambda_0}^{\lambda_1} \sum_r \left(\frac{\partial p_r}{\partial t} \frac{\partial q_r}{\partial \lambda} - \frac{\partial q_r}{\partial t} \frac{\partial p_r}{\partial \lambda} \right) d\lambda = 0.$$

Here,

$$\frac{\partial q_r}{\partial t} = \frac{dq_r}{dt} = Q_r, \qquad \frac{\partial p_r}{\partial t} = \frac{dp_r}{dt} = P_r.$$

Thus Equation (17) may be written in the form:

$$(18) \qquad \int_\Gamma \sum_r (P_r\, dq_r - Q_r\, dp_r) = 0.$$

But Γ may be *any* closed curve of the plane $t = t$, since to any such curve in that plane corresponds a Γ_0 in the plane $t = t_0$.

It follows, then, that we can define a function H by means of the integral:

$$(19) \qquad H = \int_{(a,\, b)}^{(q,\, p)} \sum_r -P_r\, dq_r + Q_r\, dp_r,$$

where the fixed point $(a_1, \cdots, a_n, b_1, \cdots, b_n, t)$ of the plane $t = t$ is connected with the variable point $(q_1, \cdots, q_n, p_1, \cdots, p_n, t)$ of this same plane by a curve lying in the plane. Because of (18) the value of the integral does not depend on the path, and thus H is defined as a function of (q_r, p_r) for the particular value of t.

Let the point (a, b, t), for definiteness, lie on the extremal through the point (α', β', t_0). Then H becomes a function of (q_r, p_r, t). If (α', β', t_0) is replaced by a different point (α'', β'', t_0), the new H will differ from the old H by an additive term which is a function of t, but not of (q_r, p_r).

More generally, let H be defined by the equation:

$$(20) \qquad H = \bar{H} + f(t),$$

where $\bar{H}$ is a specific one of the functions H just defined, and $f(t)$ is an arbitrary function of t alone.

From (19) it follows that

$$(21) \qquad Q_r = \frac{\partial H}{\partial p_r}, \qquad P_r = -\frac{\partial H}{\partial q_r}.$$

Thus the system of equations (16) is seen to have the form:

$$\frac{dq_r}{dt} = \frac{\partial H}{\partial p_r}, \qquad \frac{dp_r}{dt} = -\frac{\partial H}{\partial q_r}, \qquad r = 1, \cdots, n. \tag{22}$$

First fruits of our theorem. The Hamiltonian Function H grows naturally out of Equation (14); for (18) is but another form of (14), and (18) at once suggests the definition of H by (19) and (20). Thus if we had never heard of H through the transformations of Chapter XI, we should still be led to it by the theorem of this paragraph.

The Function V and Its Relation to H. Equation (14) can be written in the form:

$$\int\limits_{\Gamma_0} \sum_r (p_r\,dq_r - \beta_r\,d\alpha_r) = 0,$$

where q_r, p_r are given by (15), the curve Γ_0 being as before any closed curve in the plane $t = t_0$. It follows, then, that the integral:

$$\int\limits_{(\alpha', \beta')}^{(\alpha, \beta)} \sum_r (p_r\,dq_r - \beta_r\,d\alpha_r), \tag{23}$$

extended over an arbitrary path in the plane $t = t_0$ joining the points (α', β'), (α, β), is independent of the path and thus defines a function of (α, β), t entering as a parameter:

$$\int\limits_{(\alpha', \beta')}^{(\alpha, \beta)} \sum_r (p_r\,dq_r - \beta_r\,d\alpha_r) = V(\alpha, \beta, t). \tag{24}$$

Differentiate this equation with respect to t:

$$\int\limits_{\lambda'}^{\lambda} \sum_r \left(\frac{dp_r}{dt}\frac{\partial q_r}{\partial \lambda} + p_r \frac{\partial}{\partial \lambda}\frac{dq_r}{dt}\right) d\lambda = \frac{dV}{dt},$$

where the italic d means differentiation along a curve (15). Transforming through integration by parts we have:

$$\int\limits_{(\alpha', \beta')}^{(\alpha, \beta)} \sum_r \left(\frac{dp_r}{dt}\,dq_r - \frac{dq_r}{dt}\,dp_r\right) + \sum_r p_r \dot{q}_r \Bigg|_{(\alpha', \beta')}^{(\alpha, \beta)} = \frac{dV}{dt}. \tag{25}$$

The integral on the left is precisely the negative of the integral (19), or $-\overline{H}(q_r, p_r, t)$. Hence

$$H = \sum_r p_r \dot{q}_r - \frac{dV}{dt}, \tag{26}$$

where H is given by (20), and

$$f(t) = \sum_r p_r \dot{q}_r \Big|_{(\alpha', \beta', t,)} = \sum_r p_r \frac{\partial H}{\partial p_r}\Big|_{(\alpha', \beta', t)}.$$

On the other hand, the Lagrangean Function $L(q_r, \dot{q}_r, t)$ is connected with $H(q_r, p_r, t)$ by the relation (cf. Chapter XI, § 3):

$$L + H = \sum_r p_r \dot{q}_r. \tag{27}$$

Hence it appears that

$$\frac{dV}{dt} = L. \tag{28}$$

Just as H was defined only save as to an additive function of t, so V can be modified by adding any function t, and the same is true of L. But it is convenient to restrict these additive functions so that (26) and (27) will hold.

From the foregoing reasoning we can draw a more general conclusion, and then supplement it with a converse.

THEOREM I. *Let*

i)
$$\frac{dq_r}{dt} = Q_r(q, p, t), \qquad \frac{dp_r}{dt} = P_r(q, p, t),$$
$$r = 1, \cdots, n,$$

be an arbitrary system of simultaneous differential equations, and let

ii)
$$\begin{cases} q_r = \varphi_r(t;\ \alpha_1, \cdots, \alpha_n, \beta_1, \cdots, \beta_n) \\ p_r = \psi_r(t;\ \alpha_1, \cdots, \alpha_n, \beta_1, \cdots, \beta_n) \end{cases}$$

be the solution, where α_r, β_r mean the initial values of q_r, p_r corresponding to $t = t_0$. Let Γ_0 be an arbitrary closed curve lying in the plane $t = t_0$ of the $(2n + 1)$-dimensional space of the variables (q_r, p_r, t). Let S be a tube consisting of the curves ii) which pass through points of Γ_0; and let Γ be the section of S by the plane $t = t$. If

iii)
$$\int_\Gamma \sum_r p_r \, dq_r$$

is an integral invariant of Equations i); i.e. if

iv) $$\int_{\Gamma} \sum_r p_r\, dq_r = \int_{\Gamma_0} \sum_r \beta_r\, d\alpha_r,$$

then Equations i) form a Hamiltonian System:

v) $$\frac{dq_r}{dt} = \frac{\partial H}{\partial p_r}, \qquad \frac{dp_r}{dt} = -\frac{\partial H}{\partial q_r}, \qquad r = 1, \cdots, n.$$

Conversely, if Equations i) form a Hamiltonian System v), then iii) will be an integral invariant, or iv) will be satisfied.

Observe, however, that Theorem I. is more general than its origin from the action integral (1) and the transformation (7) would indicate. It applies to any functions Q_r, P_r for which *iii*) is an integral invariant; or, in the converse, to any function H, provided that the determinant

$$\sum \pm H_{11} \cdots H_{nn} \neq 0, \qquad H_{ij} = \frac{\partial^2 H}{\partial p_i\, \partial p_j}.$$

But a system of Equations *v*) may conceivably not lead to a mechanical problem — why should it?

b) *Contact Transformations.* The content of Theorem I. can be restated in terms of contact transformations.

THEOREM II. *Let*

a) $$\begin{cases} q_r = g_r(\alpha_1, \cdots, \alpha_n, \beta_1, \cdots, \beta_n, t) \\ p_r = h_r(\alpha_1, \cdots, \alpha_n, \beta_1, \cdots, \beta_n, t) \end{cases}$$

where $r = 1, \cdots, n$,

$$\frac{\partial(q_1, \cdots, q_n, p_1, \cdots, p_n)}{\partial(\alpha_1, \cdots, \alpha_n, \beta_1, \cdots, \beta_n)} \neq 0,$$

and

$$\begin{cases} \alpha_r = g_r(\alpha_1, \cdots, \alpha_n, \beta_1, \cdots, \beta_n, t_0) \\ \beta_r = h_r(\alpha_1, \cdots, \alpha_n, \beta_1, \cdots, \beta_n, t_0) \end{cases}$$

be a transformation of the $2n$*-dimensional* (α_r, β_r)*-space on the* (q_r, p_r)*-space; and let*

b) $$\frac{dq_r}{dt} = Q_r(q, p, t), \qquad \frac{dp_r}{dt} = P_r(q, p, t),$$

$$r = 1, \cdots, n,$$

be the system of differential equations corresponding to a); *i.e. defined by* a). *If* a) *is a contact transformation; i.e. if*

$$\int_{\Gamma} \sum_r p_r\, dq_r = \int_{\Gamma_0} \sum_r \beta_r\, d\alpha_r,$$

or

$$\sum_r (p_r\, dq_r - \beta_r\, d\alpha_r) = dV(\alpha, \beta, t),$$

then b) *is a Hamiltonian System:*

$$\text{c)} \qquad \frac{dq_r}{dt} = \frac{\partial H}{\partial p_r}, \qquad \frac{dp_r}{dt} = -\frac{\partial H}{\partial q_r}, \qquad r = 1, \cdots, n,$$

and conversely.

4. Transformation of Hamilton's Equations by Contact Transformations. If we start out with a given system of Hamiltonian Equations:

$$(1) \qquad \frac{dq_r}{dt} = \frac{\partial H}{\partial p_r}, \qquad \frac{dp_r}{dt} = -\frac{\partial H}{\partial q_r}, \qquad r = 1, \cdots, n,$$

and make an arbitrary transformation:

$$(2) \qquad \begin{cases} q'_r = f_r(q_1, \cdots, q_n, p_1, \cdots, p_n, t) \\ p'_r = \varphi_r(q_1, \cdots, q_n, p_1, \cdots, p_n, t) \end{cases}$$

$$\frac{\partial(q'_1, \cdots, q'_n, p'_1, \cdots, p'_n)}{\partial(q_1, \cdots, q_n, p'_1, \cdots, p_n)} \neq 0,$$

the transformed equations:

$$(3) \qquad \frac{dq'_r}{dt} = F(q'_r, p'_r, t), \qquad \frac{dp'_r}{dt} = G(q'_r, p'_r, t),$$

$r = 1, \cdots, n$, will not in general be of the form (1); *i.e.* they will not have the form:

$$(4) \qquad \frac{dq'_r}{dt} = \frac{\partial H'}{\partial p'_r}, \qquad \frac{dp'_r}{dt} = -\frac{\partial H'}{\partial q'_r}, \qquad r = 1, \cdots, n,$$

where $H' = H'(q'_r, p'_r\, t)$ is some function of the arguments $q'_r, p'_r\, t$. A *sufficient* condition that (3) be Hamiltonian, *i.e.* of the form (4), is that (2) be a *contact transformation.*

The proof is based on Theorem II., § 3 and the fact that the contact transformations form a group. Let (2), then, be a contact transformation. Denote it by T. Let (α'_r, β'_r) be the initial values of (q_r', p_r') for $t = t_0$. They arise from (α_r, β_r) by T, formed for $t = t_0$; T_0, let us write it. Thus, symbolically,

$$(\alpha_r', \beta_r') = T_0(\alpha_r, \beta_r), \qquad \text{or} \qquad (\alpha_r, \beta_r) = T_0^{-1}(\alpha_r', \beta_r').$$

Again, we may write symbolically:

$$(q_r', p_r') = T\,(q_r, p_r).$$

Finally, consider the solution of (1), whereby the space of the (α_r, β_r) is carried over into the space of the (q_r, p_r). This transformation is the Transformation a) of Theorem II., § 3, and so because of (1) is a contact transformation. Denote it by D:

$$D\,(\alpha_r, \beta_r) = (q_r, p_r).$$

On the other hand, the effect of the transformation defined by the differential equations (3) is to carry the space of the (α_r', β_r') over into the space of the (q_r', p_r'). Denote it by Δ:

$$\Delta\,(\alpha_r', \beta_r') = (q_r', p_r').$$

And now we see that this result — this transformation Δ — can be obtained as follows: Perform first the contact transformation* T_0^{-1} on the (α_r', β_r')-space, thus obtaining the (α_r, β_r)-space:

$$(\alpha_r, \beta_r) = T_0^{-1}\,(\alpha_r', \beta_r').$$

Next, perform the contact transformation D on the (α_r, β_r)-space, thus obtaining the (q_r, p_r)-space:

$$(q_r, p_r) = D\,(\alpha_r, \beta_r).$$

Finally, perform the contact transformation T on the latter space, thus obtaining the (q_r', p_r')-space:

$$(q_r', p_r') = T\,(q_r, p_r)$$

We have in this way obtained Δ as the result of three contact transformations:

$$\Delta = TDT_0^{-1}.$$

Hence Δ is itself a contact transformation, and so the system (3) is Hamiltonian, by Theorem II., § 3.

This is the result on which the developments of Chapter XV depend. It may be stated as follows.

* The inverse of a contact transformation is obviously itself a contact transformation.

THEOREM. *If a system of Hamiltonian Equations* (1) *be transformed by a contact transformation* (2), *the result is a Hamiltonian system* (4). *The condition is sufficient, but not necessary.*

Computation of H'. The original system of Hamiltonian Equations (1) leads to the contact transformation D, for which the relation:

$$\sum_r p_r\, dq_r - \sum_r \beta_r\, d\alpha_r = dV(\alpha_r, \beta_r, t), \tag{5}$$

is characteristic, where

$$H = \sum_r p_r \dot{q}_r - \frac{dV}{dt}. \tag{6}$$

The transformed Hamiltonian Equations (4) lead likewise to a contact transformation $D' = \Delta$, for which the relation

$$\sum_r p'_r\, dq'_r - \sum_r \beta'_r\, d\alpha'_r = dV'(\alpha', \beta', t) \tag{7}$$

is characteristic, where

$$H' = \sum_r p'_r \dot{q}'_r - \frac{dV'}{dt}. \tag{8}$$

Let

$$\sum_r p'_r\, dq'_r - \sum_r p_r\, dq_r = dW(q_r, p_r, t) \tag{9}$$

be the characteristic relation of the contact transformation T. Then

$$\sum_r \beta'_r\, d\alpha'_r - \sum_r \beta_r\, d\alpha_r = dW(\alpha_r, \beta_r, t_0) \tag{10}$$

will be the characteristic relation corresponding to T_0.

Each of the differentials on the right is taken on the supposition that t is a parameter, and so a constant. Moreover, (q_r, p_r) are given in terms of (α_r, β_r) by equations of the type a), § 3.

From (5), (9), (10) we infer that

$$\sum_r p'_r\, dq'_r - \sum_r \beta'_r\, d\alpha'_r = \tag{11}$$

$$d[- W(\alpha_r, \beta_r, t_0) + V(\alpha_r, \beta_r, t) + W(q_r, p_r, t)]$$

$$= dV'(\alpha', \beta', t).$$

Hence

$$\frac{dV'}{dt} = \frac{dV}{dt} + \frac{dW}{dt}, \tag{12}$$

provided $V'(\alpha'_r, \beta'_r, t)$ and $W(q_r, p_r, t)$, which are determined only save as to additive functions of t, are chosen properly. From (6) and (8) we now infer, by means of (12), that

$$H' = H - \frac{dW}{dt} + \sum_r (p'_r \dot{q}'_r - p_r \dot{q}_r). \tag{13}$$

Each of the functions H', H, dW/dt was originally defined only save as to an additive function of t, and it is only when these additive functions are suitably restricted, that (13) holds.

5. Particular Contact Transformations. In applying the theory we have developed it will be convenient to denote the transformed variables by Q_r, P_r instead of by q'_r, p'_r. Thus a transformation:

$$\begin{cases} Q_r = f_r(q_1, \cdots, q_n, p_1, \cdots, p_n, t) \\ P_r = g_r(q_1, \cdots, q_n, p_1, \cdots, p_n, t) \end{cases} \tag{1}$$

where

$$\frac{\partial(Q_1, \cdots, Q_n, P_1, \cdots, P_n)}{\partial(q_1, \cdots, q_n, p_1, \cdots, p_n)} \neq 0, \tag{2}$$

is a *contact transformation* if

$$\sum_r (P_r\, dQ_r - p_r\, dq_r) = d\,W(q_r, p_r, t), \tag{3}$$

where t is regarded as a parameter and the differentials are taken with respect to (q_r, p_r) as the independent variables.

If such a transformation be applied to the Hamiltonian system:

$$\frac{dq_r}{dt} = \frac{\partial H}{\partial p_r}, \qquad \frac{dp_r}{dt} = -\frac{\partial H}{\partial q_r}, \qquad r = 1, \cdots, n, \tag{4}$$

these equations go over into a new Hamiltonian system:

$$\frac{dQ_r}{dt} = \frac{\partial H'}{\partial P_r}, \qquad \frac{dP_r}{dt} = -\frac{\partial H'}{\partial Q_r}, \qquad r = 1, \cdots, n, \tag{5}$$

where $H'(Q_r, P_r, t)$ is connected with $H(q_r, p_r, t)$ by Equation (13), § 4, or:

$$H' = H - \frac{dW}{dt} + \sum_r (P_r \dot{Q}_r - p_r \dot{q}_r). \tag{6}$$

The (q_r, Q_r, t) *as Independent Variables.* Equations (1) represent $2n$ relations between the $4n$ variables (q_r, p_r, Q_r, P_r), and when (q_r, p_r) are chosen as the independent variables, (2) and (3) hold. It may be possible to choose the $2n$ variables (q_r, Q_r) as the independent variables, t always being regarded as a parameter in (3). Write

$$W(q_r, p_r, t) = W'(q_r, Q_r, t). \tag{7}$$

Thus (3) becomes:

$$\sum_r (P_r\, dQ_r - p_r\, dq_r) = dW'(q_r, Q_r, t). \tag{8}$$

On equating the coefficients of dQ_r, dq_r in (8) we find:

$$\begin{cases} P_r = \dfrac{\partial W'}{\partial Q_r} \\[2ex] -p_r = \dfrac{\partial W'}{\partial q_r}. \end{cases} \tag{9}$$

Equation (6) can now be transformed as follows: Since

$$\frac{dW}{dt} = \frac{dW'}{dt} = \sum_r \frac{\partial W'}{\partial Q_r}\frac{dQ_r}{dt} + \sum_r \frac{\partial W'}{\partial q_r}\frac{dq_r}{dt} + \frac{\partial W'}{\partial t},$$

we have:

$$\frac{dW}{dt} = \sum_r (P_r Q_r - p_r \dot{q}_r) + \frac{\partial W'}{\partial t}.$$

Hence (6) becomes:

$$H' = H - \frac{\partial W'}{\partial t}. \tag{10}$$

The Transformation: $p_r = \dfrac{\partial S}{\partial q_r}$, $P_r = -\dfrac{\partial S}{\partial Q_r}$. We can write down a particular contact transformation, in which (q_r, Q_r) can be taken as the independent variables. Let

$$S(q_1, \cdots, q_n, \alpha_1, \cdots, \alpha_n, t)$$

be a function of the $2n + 1$ arguments such that

$$\frac{\partial(S_{q_1}, \cdots, S_{q_n})}{\partial(\alpha_1, \cdots, \alpha_n)} \neq 0. \tag{11}$$

Set $\alpha_r = Q_r$ and make the transformation:

$$(12) \qquad p_r = \frac{\partial S}{\partial q_r}, \qquad P_r = -\frac{\partial S}{\partial Q_r}, \qquad r = 1, \cdots, n.$$

The first n of these equations can be solved for the Q_r in terms of the (q_r, p_r) because of (11), and then the P_r are given by the last n equations. Thus a transformation (1) results, the Jacobian (2) not vanishing.*

The transformation will be a contact transformation, for

$$\sum_r (P_r\,dQ_r - p_r\,dq_r) = -\sum_r \left(\frac{\partial S}{\partial Q_r}\,dQ_r + \frac{\partial S}{\partial q_r}\,dq_r\right) = -\,dS,$$

and we may set $W' = -\,S$, since W and hence W' is determined only save as to an additive function of t. Equation (10) now becomes:

$$(13) \qquad H' = H + \frac{\partial S}{\partial t}.$$

How such a function S can be found, which will enable us to solve Hamilton's equations explicitly, will be shown in Chapter XV.

Conversely, the most general contact transformation (1) which can be written in the form:

$$p_r = F_r(q_r, Q_r, t), \qquad P_r = \Phi_r(q_r, Q_r, t),$$

is given by (12). For, Equations (9) must be true, and it remains only to set $S = -\,W'$. It is seen at once that the W' of (9) must satisfy (11), since otherwise there would be a relation between the P_r.

* The proof is as follows. If

$$y_r = f_r(x_1, \cdots, x_n), \qquad r = 1, \cdots, n,$$

be a transformation having an inverse

$$x_r = \varphi_r(y_1, \cdots, y_n), \qquad r = 1, \cdots, n,$$

where f_r, φ_r are all functions having continuous first derivatives, then

$$\frac{\partial(y_1, \cdots, y_n)}{\partial(x_1, \cdots, x_n)} \cdot \frac{\partial(x_1, \cdots, x_n)}{\partial(y_1, \cdots, y_n)} = 1.$$

Consequently neither Jacobian can vanish.

In the present case, the q_r, p_r can be expressed in terms of the Q_r, P_r, since the value of the determinant (11) is unchanged if the q_r, α_r are interchanged.

EXERCISES

1. *The Transformation:* $p_r = \frac{\partial S}{\partial q_r}, \quad Q_r = \frac{\partial S}{\partial P_r}.$ Study the analogous case, in which (q_r, P_r) can be taken as the independent variables, t being, as usual, a parameter. Show that, if $S(q_1, \cdots, q_n, \alpha_1, \cdots, \alpha_n, t)$ be chosen as before, and if we set $P_r = \alpha_r$, then

$$p_r = \frac{\partial S}{\partial q_r}, \qquad Q_r = \frac{\partial S}{\partial P_r}, \qquad r = 1, \cdots, n, \tag{14}$$

will give a contact transformation. Observe that (3) can be transformed by means of the identity

$$d(P_r Q_r) = P_r \, dQ_r + Q_r \, dP_r,$$

so that it takes on the equivalent form:

$$\sum_r (Q_r \, dP_r + p_r \, dq_r) = d(-W'' + \sum_r P_r Q_r).$$

Choose $W(q_r, p_r, t) = W''(q_r, P_r, t)$, therefore, so that

$$S = -W'' + \sum_r P_r Q_r.$$

Compute dW''/dt and show by the aid of (14) that (6) yields:

$$H' = H + \frac{\partial S}{\partial t}. \tag{15}$$

State also, and prove, the converse.

2. *Computation of H' in the General Case.* Let $\pi_1, \cdots, \pi_{2n}$ be any set of $2n$ variables, chosen from the $4n$ variables (q_r, p_r, Q_r, P_r), in terms of which the remaining $2n$ variables can be expressed. Show that

$$H' = H - \frac{\partial W}{\partial t} + \sum_r \left(P_r \frac{\partial Q_r}{\partial t} - p_r \frac{\partial q_r}{\partial t} \right), \tag{16}$$

where q_r, Q_r, and W are expressed as functions of (π_k, t).

3. *The Transformation:* $q_r = \frac{\partial S}{\partial p_r}, \quad P_r = \frac{\partial S}{\partial Q_r}.$ If (p_r, Q_r) can be taken as the independent variables, and if we set

$$S = W + \sum_r p_r q_r,$$

where q_r, W, and S are now functions of (p_r, Q_r, t), then the transformation takes the form:

$$(17) \qquad q_r = \frac{\partial S}{\partial p_r}, \qquad P_r = \frac{\partial S}{\partial Q_r}, \qquad r = 1, \cdots, n,$$

and (6) yields:

$$(18) \qquad H' = H - \frac{\partial S}{\partial t}.$$

Conversely, if $S(q_1, \cdots, q_n, \alpha_1, \cdots, \alpha_n)$ be chosen as before, and if we set $Q_r = \alpha_r$, then (17) will define a contact transformation.

6. The Ω-Relations. There is one case of importance still to be considered, namely, that in which W is a function of (q_r, Q_r, t), but the (q_r, Q_r, t) cannot be chosen as the independent variables. The extreme case would be that in which

$$Q_r = \omega_r(q_1, \cdots, q_n), \qquad r = 1, \cdots, n,$$

$$\frac{\partial(\omega_1, \cdots, \omega_n)}{\partial(q_1, \cdots, q_n)} \neq 0.$$

The general case is that in which $0 < m \leqq n$ independent relations between the (q_r, Q_r, t) exist, and no more:

$$(1) \qquad \Omega_1(q_r, Q_r, t) = 0, \quad \cdots, \quad \Omega_m(q_r, Q_r, t) = 0,$$

where the rank of the matrix:

$$(2) \qquad \left\| \begin{matrix} \dfrac{\partial \Omega_1}{\partial Q_1} & \cdots & \dfrac{\partial \Omega_1}{\partial Q_n} \\ \cdots & \cdots & \cdots \\ \dfrac{\partial \Omega_m}{\partial Q_1} & \cdots & \dfrac{\partial \Omega_m}{\partial Q_n} \end{matrix} \right\|,$$

is m. Thus m of the Q_k's can be expressed as functions of the remaining $\mu = n - m$ Q_l's and $q_1, \cdots, q_n, t$. As a matter of notation let the above m Q_k's be $Q_1, \cdots, Q_m$:

$$(3) \qquad Q_\nu = \omega_\nu(Q_{m+1}, \cdots, Q_n, q_1, \cdots, q_n, t), \qquad \nu = 1, \cdots, m.$$

Then the determinant whose matrix consists of the first m columns of (2) will not vanish. Among the $2n$ (p_r, P_r) it shall be possible to choose m variables, $\pi_1, \cdots, \pi_m$ such that $(\pi_1, \cdots, \pi_m, Q_{m+1}, \cdots, Q_n, q_1, \cdots, q_n, t)$ can serve as the $2n + 1$ independent variables. But the function $W(q_r, p_r, t)$, when expressed in terms of the new variables, does not depend on $\pi_1, \cdots, \pi_m$:

$$W(q_r, p_r, t) = W^*(q_r, Q_r, t). \tag{4}$$

It is not, of course, unique, because of the Ω-relations, (1).

Equation (3), § 5 now takes on the form:

$$\sum_r \left(P_r\, dQ_r - p_r\, dq_r\right) = dW^*(q_r, Q_r, t). \tag{5}$$

We will rewrite it in the form:

$$\sum_r \left(P_r - \frac{\partial W^*}{\partial Q_r}\right) dQ_r - \sum_r \left(p_r + \frac{\partial W^*}{\partial q_r}\right) dq_r = 0. \tag{6}$$

It is not, however, in general true that the coefficients of the differentials vanish.

By means of the m equations (1) the first m differentials $dQ_1, \cdots, dQ_m$ can be eliminated, the resulting equation being of the form:

$$X_{m+1}dQ_{m+1} + \cdots + X_n dQ_n + Y_1 dq_1 + \cdots + Y_n dq_n = 0. \tag{7}$$

The differentials in (7) are independent variables, and so we can infer that

$$X_{m+1} = 0, \cdots, X_n = 0, \qquad Y_1 = 0, \cdots, Y_n = 0.$$

The actual elimination can be conveniently performed by means of Lagrange's multipliers. From Equations (1) we infer that

$$\left\{\begin{array}{l} \dfrac{\partial \Omega_1}{\partial Q_1} dQ_1 + \cdots + \dfrac{\partial \Omega_1}{\partial Q_n} dQ_n + \dfrac{\partial \Omega_1}{\partial q_1} dq_1 + \cdots + \dfrac{\partial \Omega_1}{\partial q_n} dq_n = 0 \\ \cdots\cdots\cdots\cdots\cdots\cdots\cdots\cdots\cdots\cdots \\ \dfrac{\partial \Omega_m}{\partial Q_1} dQ_1 + \cdots + \dfrac{\partial \Omega_m}{\partial Q_n} dQ_n + \dfrac{\partial \Omega_m}{\partial q_1} dq_1 + \cdots + \dfrac{\partial \Omega_m}{\partial q_n} dq_n = 0 \end{array}\right. \tag{8}$$

Multiply the k-th of these equations by λ_k and add to (6). Then determine the λ_k's so that the coefficients of $dQ_1, \cdots, dQ_m$ vanish. The resulting equation is of the form (7), and so its coefficients vanish automatically. We thus arrive at the $2n$ equations:

$$\left\{\begin{array}{l} P_r = \dfrac{\partial W^*}{\partial Q_r} + \lambda_1 \dfrac{\partial \Omega_1}{\partial Q_r} + \cdots + \lambda_m \dfrac{\partial \Omega_m}{\partial Q_r}, \\ -p_r = \dfrac{\partial W^*}{\partial q_r} + \lambda_1 \dfrac{\partial \Omega_1}{\partial q_r} + \cdots + \lambda_m \dfrac{\partial \Omega_m}{\partial q_r}. \end{array}\right. \tag{9}$$

The first m of these equations determine the λ_k's. The remainder are satisfied as shown above. The result is symmetric and holds, no matter what set of m Q_k's is determined by (1); *i.e.* no matter what m-rowed determinant out of the matrix (2) is different from 0.

It is now easy to determine H' by means of (13), § 4:

$$H' = H - \frac{dW^*}{dt} + \sum_r \left(P_r \frac{dQ_r}{dt} - p_r \frac{dq_r}{dt}\right).$$

On replacing P_r, p_r here by their values from (9) and observing that

$$\frac{dW^*}{dt} = \sum_r \frac{\partial W^*}{\partial Q_r}\frac{dQ_r}{dt} + \sum_r \frac{\partial W^*}{\partial q_r}\frac{dq_r}{dt} + \frac{\partial W^*}{\partial t},$$

$$\frac{d\Omega_r}{dt} = \sum_r \frac{\partial \Omega_r}{\partial Q_r}\frac{dQ_r}{dt} + \sum_r \frac{\partial \Omega_r}{\partial q_r}\frac{dq_r}{dt} + \frac{\partial \Omega_r}{\partial t} = 0,$$

we find the following result:

$$H' = H - \frac{\partial W^*}{\partial t} - \lambda_1 \frac{\partial \Omega_1}{\partial t} - \cdots - \lambda_m \frac{\partial \Omega_m}{\partial t}. \tag{10}$$

If, in particular, the Ω's do not contain t explicitly, this equation reduces to

$$H' = H - \frac{\partial W^*}{\partial t}. \tag{11}$$

CHAPTER XV

SOLUTION OF HAMILTON'S EQUATIONS

1. The Problem and Its Treatment. We have considered a great variety of problems in mechanics, the solution of which depends, or can be made to depend, on Hamilton's Canonical Equations:

$$\frac{dq_r}{dt} = \frac{\partial H}{\partial p_r}, \qquad \frac{dp_r}{dt} = -\frac{\partial H}{\partial q_r}, \qquad r = 1, \cdots, n, \tag{1}$$

where H is a function of (q_r, p_r, t). The object of this chapter is to solve these equations explicitly in the important cases which arise in practice.

The method is that of *transformation.* By means of a suitably chosen transformation:

$$\left\{ \begin{aligned} Q_r &= F_r(q_1, \cdots, q_n, p_1, \cdots, p_n, t) \\ P_r &= G_r(q_1, \cdots, q_n, p_1, \cdots, p_n, t) \end{aligned} \right. \tag{2}$$

Equations (1) are carried over into equations of the same type:

$$\frac{dQ_r}{dt} = \frac{\partial H'}{\partial P_r}, \qquad \frac{dP_r}{dt} = -\frac{\partial H'}{\partial Q_r}, \qquad r = 1, \cdots, n, \tag{3}$$

but more easily solved. Here, H' is a function of (Q_r, P_r, t), not in general equal to H.

The determination of a convenient transformation (2) depends on a partial differential equation of the first order, due to Jacobi *:

$$\frac{\partial V}{\partial t} + H\left(q_1, \cdots, q_n, \frac{\partial V}{\partial q_1}, \cdots, \frac{\partial V}{\partial q_n}, t\right) = 0. \tag{4}$$

It is not the theory of this equation, however, but the practice, that concerns us, for all we need is a single explicit solution,

$$V = V(q_1, \cdots, q_n, \alpha_1, \cdots, \alpha_n, t), \tag{5}$$

depending in a suitable manner on n arbitrary constants, or parameters, $\alpha_1, \cdots, \alpha_n$. Such a solution is found in practice by means of simple devices, notably that of *separating the variables.*

* Hamilton came upon this equation; but its use as here set forth is due to Jacobi.

The function (5) once found, the further work consists merely in differentiation and the solution of equations defining the q_r, p_r implicitly. Two cases are especially important, namely:

a) *Reduction to the Equilibrium Problem.* Here, a solution (5) of (4) enables us so to choose (2) that the transformed H vanishes identically, $H' \equiv 0$. Equations (3) can now be integrated at sight:

$$Q_r = \alpha_r, \qquad P_r = \beta_r, \qquad r = 1, \cdots, n, \tag{6}$$

where α_r, β_r are arbitrary constants. On substituting these values in (2), the inverse transformation,

$$\begin{cases} q_r = f_r(Q_1, \cdots, Q_n, P_1, \cdots, P_n, t) \\ p_r = g_r(Q_1, \cdots, Q_n, P_1, \cdots, P_n, t) \end{cases} \tag{7}$$

yields the desired solution:

$$\begin{cases} q_r = f_r(\alpha_1, \cdots, \alpha_n, \beta_1, \cdots, \beta_n, t) \\ p_r = g_r(\alpha_1, \cdots, \alpha_n, \beta_1, \cdots, \beta_n, t) \end{cases} \tag{8}$$

The transformation (2) in this case, as will be shown in § 2, is given by the equations:

$$p_r = \frac{\partial V}{\partial q_r}, \qquad P_r = -\frac{\partial V}{\partial Q_r}, \qquad r = 1, \cdots, n, \tag{9}$$

where V is written for the arguments q_r, Q_r:

$$V = V(q_1, \cdots, q_n, Q_1, \cdots, Q_n, t).$$

Thus the solution (8) is obtained by solving the equations:

$$p_r = \frac{\partial V}{\partial q_r}, \qquad \beta_r = -\frac{\partial V}{\partial \alpha_r}, \qquad r = 1, \cdots, n,$$

where the present V has the form (5).

b) *Constant Energy*, $H(q_r, p_r) = h$. The second case is that in which H does not contain the time explicitly:

$$H = H(q_1, \cdots, q_n, p_1, \cdots, p_n).$$

It is here possible to find a transformation (2) in which F_r, G_r do not depend on t,

$$\begin{cases} Q_r = F_r(q_1, \cdots, q_n, p_1, \cdots, p_n) \\ P_r = G_r(q_1, \cdots, q_n, p_1, \cdots, p_n) \end{cases} \tag{10}$$

such that the new H will depend only on the P_r, but not on Q_r, t. In particular,

$$H' = P_1.$$

Equations (3) now take on the form:

$$\frac{dQ_1}{dt} = 1, \qquad \frac{dQ_s}{dt} = 0, \qquad s = 2, \cdots, n;$$

$$\frac{dP_r}{dt} = 0, \qquad r = 1, \cdots, n.$$

Thus*

$$(11) \qquad \begin{cases} Q_1 = t + \beta_1, \qquad Q_s = \beta_s, \qquad s = 2, \cdots, n; \\ P_r = \alpha_r, \qquad r = 1, \cdots, n. \end{cases}$$

Let the inverse of (10) be written:

$$(12) \qquad \begin{cases} q_r = f_r(Q_1, \cdots, Q_n, P_1, \cdots, P_n) \\ p_r = g_r(Q_1, \cdots, Q_n, P_1, \cdots, P_n) \end{cases}$$

Then the solution of (1) is given by the formula:

$$(13) \qquad \begin{cases} q_r = f_r(t + \beta_1, \beta_2, \cdots, \beta_n, \alpha_1, \cdots, \alpha_n) \\ p_r = g_r(t + \beta_1, \beta_2, \cdots, \beta_n, \alpha_1, \cdots, \alpha_n) \end{cases}$$

The transformation (2) in this case, as will be shown in § 4, is given by the equations:

$$(14) \qquad p_r = \frac{\partial W}{\partial q_r}, \qquad Q_r = \frac{\partial W}{\partial P_r}, \qquad r = 1, \cdots, n,$$

where W is a solution of the equation:

$$H\left(q_1, \cdots, q_n, \frac{\partial W}{\partial q_1}, \cdots, \frac{\partial W}{\partial q_n}\right) = h,$$

or:

$$W = W(q_1, \cdots, q_n, h, \alpha_2, \cdots, \alpha_n).$$

Here W depends on the arbitrary constant h, and also, in a suitable manner, on $n - 1$ further constants, or parameters, $\alpha_2, \cdots, \alpha_n$. These are set equal respectively to the P_r:

$$P_1 = h; \qquad P_s = \alpha_s, \qquad s = 2, \cdots, n.$$

* The change of notation whereby the α_r's and the β_r's are interchanged is made for the purpose of conforming to usage in the literature.

Equations (14), combined with (11), thus yield:

$$(15)\qquad \begin{cases} t + \beta_1 = \dfrac{\partial W}{\partial h} \\ \beta_s = \dfrac{\partial W}{\partial \alpha_s}, \qquad s = 2, \cdots, n. \end{cases}$$

The last $n - 1$ of these equations can be solved for $q_2, \cdots, q_n$ in terms of q_1, as will be shown in § 4, thus giving the *form* of the path; and then q_1 can be found from the first equation (15) in terms of t.

We have characterized this case by the caption: "Constant Energy," but this is not a physical hypothesis. Our hypothesis is, that H does not depend explicitly on t, and this is all we need for the mathematical development. That H then turns out to be constant along the curves of the natural path, is an important consequence; but our treatment does not depend on this hypothesis.

Contact Transformations. The transformations used in a) and b), namely, (9) and (14), are examples of contact transformations. A transformation (2) with non-vanishing Jacobian was defined in Chapter XIV, § 1, to be a *contact transformation* if

$$(16)\qquad \sum_r (P_r\,dQ_r - p_r\,dq_r) = dW\,(q_r, p_r, t),$$

where the differentials are taken with respect to the (q_r, p_r) as the independent variables, t being regarded as a parameter. Such a transformation always carries a Hamiltonian System (1) into a Hamiltonian System (3). That the transformations (9) and (14) satisfy the condition (16) is seen at once by substituting in (16), observing in the case of (14) that

$$d(P_r\,Q_r) = P_r\,dQ_r + Q_r\,dP_r.$$

This is all the theory the student need know from Chapter XIV, to enter on the study of the present chapter, and this amount of theory was all developed in §§ 1–4 of that chapter.

2. Reduction to the Equilibrium Problem. We have seen in Chapter XIV, § 5, that a transformation:

$$(1)\qquad p_r = \frac{\partial S}{\partial q_r}, \qquad P_r = -\frac{\partial S}{\partial Q_r}, \qquad r = 1, \cdots, n,$$

where

$$S = S(q_1, \cdots, q_n, \alpha_1, \cdots, \alpha_n)$$

is any function such that

$$\frac{\partial(S_{q_1}, \cdots, S_{q_n})}{\partial(\alpha_1, \cdots, \alpha_n)} \neq 0,$$

and where Q_r is set $= \alpha_r$, will carry the Hamiltonian System (1) of the last paragraph over into a Hamiltonian System (3), where

$$H' = H + \frac{\partial S}{\partial t}. \tag{2}$$

The transformed function H' can be made to vanish identically if we can find a solution V of the partial differential equation:

$$\frac{\partial V}{\partial t} + H\left(q_1, \cdots, q_n, \frac{\partial V}{\partial q_1}, \cdots, \frac{\partial V}{\partial q_n}, t\right) = 0, \tag{3}$$

which depends on n arbitrary constants, $\alpha_1, \cdots, \alpha_n$:

$$V = V(q_1, \cdots, q_n, \alpha_1, \cdots, \alpha_n, t),$$

and is such that

$$\frac{\partial(V_{q_1}, \cdots, V_{q_n})}{\partial(\alpha_1, \cdots, \alpha_n)} \neq 0. \tag{4}$$

On setting S equal to this function V, and making the transformation (1), H' as now determined vanishes identically. Thus the transformation:

$$p_r = \frac{\partial V}{\partial q_r}, \qquad P_r = -\frac{\partial V}{\partial Q_r}, \qquad r = 1, \cdots, n, \tag{5}$$

where α_r is replaced by Q_r in V, transforms the Hamiltonian System (1) to the *Equilibrium Problem:*

$$\frac{dQ_r}{dt} = 0, \qquad \frac{dP_r}{dt} = 0, \qquad r = 1, \cdots, n. \tag{6}$$

The solution of these equations is the system of equations (6), § 1. These are the values of Q_r, P_r to be substituted in the transformation (1); *i.e.* in the present case, in (5):

$$p_r = \frac{\partial V}{\partial q_r}, \qquad \beta_r = -\frac{\partial V}{\partial \alpha_r}, \qquad r = 1, \cdots, n. \tag{7}$$

The last n of these equations can be solved for the q_r's because of (4), and then the first n equations give the p_r.

There is, of course, a further requirement in the large, namely, that the α_r, β_r can be so determined as to correspond to the initial conditions: $t = t_0$, $q_r = q_r^0$, $p_r = p_r^0$. Thus the equations:

$$p_r^0 = V_{q_r}(q_1^0, \cdots, q_n^0, \alpha_1, \cdots, \alpha_n, t_0), \qquad r = 1, \cdots, n,$$

must admit a solution, $\alpha_r = \alpha_r^0$, and $V(q_1, \cdots, q_n, \alpha_1, \cdots, \alpha_n, t)$ must satisfy all the conditions of continuity, notably (4), in the neighborhood of the point $(q_r, \alpha_r) = (q_r^0, \alpha_r^0)$.

EXERCISE

Pass to the Equilibrium Problem by means of the transformation studied in Chapter XIV, § 5, Exercise 1:

$$p_r = \frac{\partial S}{\partial q_r}, \qquad Q_r = \frac{\partial S}{\partial P_r}, \qquad r = 1, \cdots, n.$$

Here,

$$H' = H + \frac{\partial S}{\partial t}.$$

Let $V = V(q_1, \cdots, q_n, \alpha_1, \cdots, \alpha_n, t)$ be the same function as that of the text — a solution of Equation (3). If, then, we replace α_r by P_r and set $S = V$, the transformed H' will vanish: $H' = 0$, and Hamilton's Equations will take on the form of the Equilibrium Problem:

$$\frac{dQ_r}{dt} = 0, \qquad \frac{dP_r}{dt} = 0, \qquad r = 1, \cdots, n.$$

If we write their solution in the form:

$$Q_r = -\beta_r, \qquad P_r = \alpha_r, \qquad r = 1, \cdots, n,$$

we are led to the same solution of the original Hamiltonian Equations as before — namely, that given by (7).

3. Example. Simple Harmonic Motion. Here the kinetic energy T and the work function U are expressible respectively in the form:

$$T = \frac{m}{2}\dot{q}^2, \qquad U = -\frac{\lambda}{2}q^2, \qquad 0 < \lambda. \tag{1}$$

Thus

$$L = T + U = \frac{m}{2}\dot{q}^2 - \frac{\lambda}{2}q^2 \tag{2}$$

$$p = \frac{\partial L}{\partial \dot{q}} = m\dot{q} \tag{3}$$

$$H = p\dot{q} - L = \frac{1}{2m}p^2 + \frac{\lambda}{2}q^2. \tag{4}$$

Hamilton's Equations assume the form:

a) $$\frac{dq}{dt} = \frac{p}{m}, \qquad \frac{dp}{dt} = -\lambda q.$$

We propose to solve them by the method of § 2. The equation for determining V, § 2, (3), here becomes:

$$\frac{\partial V}{\partial t} + \frac{1}{2m}\left(\frac{\partial V}{\partial q}\right)^2 + \frac{\lambda}{2}q^2 = 0. \tag{5}$$

We wish to find a function:

$$V = V(q, \alpha), \tag{6}$$

$$\frac{\partial^2 V}{\partial \alpha \, \partial q} \neq 0, \tag{7}$$

which satisfies this equation.* One such function is enough. Let us see if we cannot find one in the form:

$$V = \Omega + W, \tag{8}$$

where $\Omega = \Omega(t)$ is a function of t alone, and $W = W(q)$ is a function of q alone. If this be possible, we shall have:

$$\frac{d\Omega}{dt} + \frac{1}{2m}\left(\frac{dW}{dq}\right)^2 + \frac{\lambda}{2}q^2 = 0.$$

This equation can be written in the form:

$$\frac{1}{2m}\left(\frac{dW}{dq}\right)^2 + \frac{\lambda}{2}q^2 = -\frac{d\Omega}{dt}. \tag{9}$$

The left-hand side of (9) depends on q alone, the right-hand side, on t alone. Hence each is a constant — denote it by α; it is obvious that $\alpha \geqq 0$:

* Let the student disembarass himself of any fears due to his ignorance of the theory of partial differential equations. No such theory is needed in the kind of application in Physics which we are about to consider; it would not even be helpful in practice. The single function $V(q, \alpha)$ is obtained by a simple device fully explained in the text.

There is, of course, a most intimate relation between the theory of Hamilton's Equations and the theory of this partial differential equation, as is indicated, for example, by the "theory of characteristics"; cf. Appendix C. The point is, that this theory is not employed in such applications as those illustrated here. For the latter purpose, a single solution $V(q_1, \cdots, q_n, \alpha_1, \cdots, \alpha_n, t)$ is all that is required, and such a solution is obtained by ingenious devices of a homely kind, as set forth in this Chapter.

$$-\frac{d\Omega}{dt}=\alpha,$$

$$\frac{1}{2m}\left(\frac{dW}{dq}\right)^2+\frac{\lambda}{2}q^2=\alpha.$$

The first equation gives:

$$\Omega=-\alpha t,$$

no constant of integration being added because we need only a particular integral, and so choose the simplest. From the second equation,

$$\left(\frac{dW}{dq}\right)^2=2m\alpha-m\lambda q^2.$$

One solution of this equation is:

$$W=\int_0^q\sqrt{2m\alpha-m\lambda q^2}\,dq.$$

Thus

$$V=-\alpha t+\int_0^q\sqrt{2m\alpha-m\lambda q^2}\,dq. \tag{10}$$

Equations (7), § 2 here become:

$$\begin{cases} p=\dfrac{\partial V}{\partial q}=\sqrt{2m\alpha-m\lambda q^2}, \\ \beta=-\dfrac{\partial V}{\partial\alpha}=t-\sqrt{m}\displaystyle\int_0^q\frac{dq}{\sqrt{2\alpha-\lambda q^2}}. \end{cases} \tag{11}$$

This last equation gives:

$$\beta=t-\sqrt{\frac{m}{\lambda}}\sin^{-1}q\sqrt{\frac{\lambda}{2\alpha}},$$

and thus

$$q=\sqrt{\frac{2\alpha}{\lambda}}\sin\sqrt{\frac{\lambda}{m}}(t-\beta). \tag{12}$$

From the first Equation (11),

$$p=\sqrt{2m\alpha}\cos\sqrt{\frac{\lambda}{m}}(t-\beta). \tag{13}$$

Equations (12) and (13) constitute a solution of Hamilton's Equations, which, however, is at present restricted; for we have not paid heed to Condition (7) on the one hand or, on the other, considered that the second equation (11) is restricted. Here then is a difficulty.* Either we must follow the theory as hitherto developed, using *single-valued* functions W and V; then t is confined between certain fixed values. Or else we must introduce multiple-valued functions V, and then we must go back and revise and supplement the general theory.

There is, however, a third choice — a way out, whereby we can remain within the restrictions of the present theory. According to that theory the solution given by (12), (13) is valid so long as

$$-\frac{\pi}{2} \leqq \sqrt{\frac{\lambda}{m}}(t - \beta) \leqq \frac{\pi}{2}.$$

Now, from the general theory of differential equations, Equations a) admit a solution single-valued and analytic for the whole range of values $-\infty < t < +\infty$. Equations (12), (13) yield a solution for a part of this interval. Therefore, by analytic continuation, the solution (12), (13) must hold for the whole interval.

EXERCISES

1. Obtain the solution of Equations a) in the form:

$$(14) \qquad \begin{cases} q = \sqrt{\dfrac{2\alpha}{\lambda}} \cos \sqrt{\dfrac{\lambda}{m}}(t - \beta) \\[2ex] p = -\sqrt{2m\alpha} \sin \sqrt{\dfrac{\lambda}{m}}(t - \beta) \end{cases}$$

by choosing as W the function:

$$W = -\int_0^q \sqrt{2m\alpha - m\lambda q^2}\, dq + C(\alpha),$$

and suitably determining the constant of integration $C(\alpha)$.

2. Solve Equations a) directly, eliminating p and thus obtaining the equation

$$m\frac{d^2q}{dt^2} + \lambda q = 0,$$

*There is also a further difficulty, since the first equation (11) may not admit a solution (suppose $p_0 < 0$), but this difficulty can be met by choosing the negative radical,

$$p = -\sqrt{2m\alpha - m\lambda q^2}.$$

the general solution of which can be written in the form:

$$q = A \cos \sqrt{\frac{\lambda}{m}}\,(t - \beta), \qquad 0 \leqq A.$$

3. *The Simple Pendulum.* Let q be the angle of displacement from the downward vertical. Then

$$T = \frac{ml^2}{2}\dot{q}^2, \qquad U = mgl \cos q;$$

$$H = \frac{p^2}{2ml^2} - mgl \cos q.$$

Obtain the equation for motion near the point of stable equilibrium:

$$\int_0^q \frac{dq}{\sqrt{\dfrac{2\alpha}{mgl} + 2 \cos q}} = \sqrt{\frac{g}{l}}\,(t - \beta),$$

where t is restricted. Hence discuss the two cases: — a) oscillatory motion (libration); b) quasi-periodic motion, when the pendulum describes continually complete circles (limitation).

Observe that, when t passes beyond the restricted interval, the sign of the radical changes, and q changes from increasing to decreasing, or *vice versa*.

4. *Freely Falling Body,* or vertical motion under gravity. Here, q shall be measured downward from the initial position.

$$T = \frac{m}{2}\dot{q}^2, \qquad U = mgq;$$

$$p = m\dot{q},$$

$$H = \frac{1}{2m}p^2 - mgq.$$

$$\frac{\partial V}{\partial t} + \frac{1}{2m}\left(\frac{\partial V}{\partial q}\right)^2 - mgq = 0.$$

$$V = \Omega + W;$$

$$-\frac{d\Omega}{dt} = \alpha, \qquad \Omega = -\alpha t;$$

$$\left(\frac{dW}{dq}\right)^2 - 2m^2 gq = 2m\alpha,$$

$$\frac{dW}{dq} = \pm\sqrt{2m\alpha + 2m^2 gq}.$$

Since

$$p = \frac{\partial V}{\partial q} = \frac{dW}{dq},$$

we have no option as to which radical shall be taken. If the body is projected upward, $\dot{q}$ will be negative for a while, and so we *must* choose the negative radical for this stage of the motion. At the turning point, (7) is not fulfilled, since $\partial^2 V/\partial q \partial \alpha$ does not exist.

We have now a new problem, as the body descends. The choice of W *must* be made on the basis of the positive radical. Nevertheless, both stages of the motion are covered by the solution for the first stage:

$$\begin{cases} q = \dfrac{g}{2}(t-\beta)^2 - \sqrt{\dfrac{2\alpha}{m}}(t-\beta), \\ p = mg(t-\beta) - \sqrt{2\alpha m}. \end{cases}$$

Why?

4. *H*, Independent of *t*. Reduction to the Form, $H' = P_1$. We have seen in Chap. XIV, § 5, Ex. 1, that if S be an arbitrary function of the q_r, of n arbitrary constants, or parameters, the α_r, and of t:

$$S = S(q_1, \cdots, q_n, \alpha_1, \cdots, \alpha_n, t),$$

where

$$\frac{\partial(S_{q_1}, \cdots, S_{q_n})}{\partial(\alpha_1, \cdots, \alpha_n)} \neq 0, \tag{1}$$

and if we set $\alpha_r = P_r$, then the equations:

$$p_r = \frac{\partial S}{\partial q_r}, \qquad Q_r = \frac{\partial S}{\partial P_r}, \qquad r = 1, \cdots, n, \tag{2}$$

define a contact transformation whereby Hamilton's Equations (4), § 5, go over into (5), § 5, and

$$H' = H + \frac{\partial S}{\partial t}. \tag{3}$$

If S does not depend on t, this equation reduces to the following:

$$H' = H. \tag{4}$$

Suppose, furthermore, that H is also independent of t:

$$H = H(q_1, \cdots, q_n, p_1, \cdots, p_n).$$

Then

$$H' = H'(Q_1, \cdots, Q_n, P_1, \cdots, P_n).$$

We propose the problem of determining S so that H' will depend only on the P_r:

$$H' = H'(P_1, \cdots, P_n),$$

and, in fact, that H' will be an arbitrarily preassigned function of the P_r. Begin with the case:

$$H'(P_1, \cdots, P_n) = P_1. \tag{5}$$

To find such a function $S(q_1, \cdots, q_n, \alpha_1, \cdots, \alpha_n)$, consider the equation:

$$H\left(q_1, \cdots, q_n, \frac{\partial W}{\partial q_1}, \cdots, \frac{\partial W}{\partial q_n}\right) = h. \tag{6}$$

Suppose it is possible to find a solution:

$$W = W(q_1, \cdots, q_n, h, \alpha_2, \cdots, \alpha_n)$$

depending on $n - 1$ arbitrary constants $\alpha_2, \cdots, \alpha_n$ — and of course on h, which is also arbitrary — such that*

$$\frac{\partial(W_{q_2}, \cdots, W_{q_n})}{\partial(\alpha_2, \cdots, \alpha_n)} \neq 0. \tag{7}$$

It then follows, as we will show later, that

$$\frac{\partial(W_{q_1}, W_{q_2}, \cdots, W_{q_n})}{\partial(h, \alpha_2, \cdots, \alpha_n)} \neq 0. \tag{8}$$

This is the function which we will choose as S:

$$S(q_1, \cdots, q_n, \alpha_1, \cdots, \alpha_n) = W(q_1, \cdots, q_n, h, \alpha_2, \cdots, \alpha_n), \tag{9}$$

where $\alpha_1 = h$. If now we set:

$$P_1 = \alpha_1 = h; \qquad P_s = \alpha_s, \qquad s = 2, \cdots, n, \tag{10}$$

then (6) becomes, because of (2), (9), and (10):

$$H(q_1, \cdots, q_n, p_1, \cdots, p_n) = P_1, \tag{11}$$

and hence (4) gives:

$$H' = P_1,$$

as was desired.

* In practice this is done by writing down an explicit function of the nature desired, obtained by such artifices as the separation of variables.

Thus the transformed Hamiltonian Equations become:

$$(12)\qquad \begin{cases} \dfrac{dQ_1}{dt} = 1, \qquad \dfrac{dQ_s}{dt} = 0, \qquad s = 2, \cdots, n; \\[2ex] \dfrac{dP_r}{dt} = 0, \qquad r = 1, \cdots, n. \end{cases}$$

The solution of this system is obviously:

$$(13)\qquad \begin{cases} Q_1 = t + \beta_1, \qquad Q_s = \beta_s, \qquad s = 2, \cdots, n; \\ P_r = \alpha_r, \qquad r = 1, \cdots, n. \end{cases}$$

Returning, then, to the original transformation (2), which now takes on the form:

$$(14)\qquad p_r = \frac{\partial W}{\partial q_r}, \qquad Q_r = \frac{\partial W}{\partial P_r}, \qquad r = 1, \cdots, n,$$

we have:

$$(15)\qquad \begin{cases} t + \beta_1 = \dfrac{\partial W}{\partial h} \\[2ex] \beta_s = \dfrac{\partial W}{\partial \alpha_s}, \qquad s = 2, \cdots, n. \end{cases}$$

The last $n - 1$ of these equations can be solved for $q_2, \cdots, q_n$ as functions of q_1 because of (7), thus determining the *form* of the curves of the natural path of the system. And then the first equation can be solved for q_1 in terms of t. This last statement is conveniently substantiated indirectly. All n equations (15) can be solved for $q_1, \cdots, q_n$ in terms of t because of (8). These functions $q_r(t)$ satisfy the last $n - 1$ equations (15), and so the earlier solution of these equations for $q_2, \cdots, q_n$ in terms of q_1 become identities in t when q_r is replaced by $q_r(t)$ given by using all n equations.

Proof of Relation (8). Observe that Relation (6) is an identity in the h, α_s as well as in the q_r. Hence on differentiating successively with respect to $h, \alpha_2, \cdots, \alpha_n$, we find:

$$(16)\qquad \begin{cases} H_{p_1} W_{hq_1} + \cdots + H_{p_n} W_{hq_n} = 1, \\ H_{p_1} W_{\alpha_2 q_1} + \cdots + H_{p_n} W_{\alpha_2 q_n} = 0, \\ \cdots\cdots\cdots\cdots\cdots\cdots \\ H_{p_1} W_{\alpha_n q_1} + \cdots + H_{p_n} W_{\alpha_n q_n} = 0. \end{cases}$$

The determinant of these equations is the Jacobian that appears in (8). If it were 0, it would be possible to determine n multipliers $\lambda_1, \cdots, \lambda_n$, not all 0, such that the n equations:

$$(17) \qquad \begin{cases} \lambda_1 W_{hq_1} + \lambda_2 W_{\alpha_2 q_1} + \cdots + \lambda_n W_{\alpha_n q_1} = 0 \\ \cdots\cdots\cdots\cdots\cdots\cdots\cdots\cdots \\ \lambda_1 W_{hq_n} + \lambda_2 W_{\alpha_2 q_n} + \cdots + \lambda_n W_{\alpha_n q_n} = 0 \end{cases}$$

are true, and since (7) holds by hypothesis, λ_1 may be chosen at pleasure. Now multiply the k-th equation (16) by λ_k and add. The coefficient of each H_{p_j} vanishes, and so the whole left-hand side reduces to 0. But the right-hand side is λ_1, which is arbitrary. This contradiction arises from supposing that (8) is not true, and the proof is complete.

The Equation of Energy. When the kinetic energy T and the work function U are both independent of t, H is also independent of t, and H represents the total energy (sum of the kinetic energy T and the potential energy $-U$). Hence* H is constant and we may write:

$$(18) \qquad h = H(q_1, \cdots, q_n, p_1, \cdots, p_n).$$

Thus this equation appears to be derived from the physics of the problem. It is. But this derivation is not helpful in the present theory. For we are dealing with contact transformations which reduce Hamilton's equations to a desired form, and Equation (6) takes its systematic place in that theory. It expresses a condition for the function W that will make the desired transformation possible. Nevertheless, the physics of the situation throws a side light on the situation, which it is well to note.

The Symmetric Form. We have set, unsymmetrically, $h = P_1$ in Equations (10). We might equally well replace (10) by the equations:

$$(10') \qquad \Phi(P_1, \cdots, P_n) = h, \qquad P_s = \alpha_s, \qquad s = 2, \cdots, n,$$

where $\Phi(\alpha_1, \cdots, \alpha_n)$ is any function such that $\partial\Phi/\partial\alpha_1 \neq 0$. The above reasoning, with an obvious modification in detail, shows that the determinant:

*That H is here constant along a natural path follows from Chap. XI, § 3:

$$\frac{dH}{dt} = \frac{\partial H}{\partial t} = 0.$$

$$(8') \qquad \frac{\partial(W_{q_1}, \cdots, W_{q_n})}{\partial(\alpha_1, \alpha_2, \cdots, \alpha_n)} \neq 0,$$

where $W = W(q_1, \cdots, q_n, h, \alpha_1, \cdots, \alpha_n)$ is determined as before from (6), and $h = \Phi(\alpha_1, \alpha_2, \cdots, \alpha_n)$. Thus the transformation (14) is justified and Equations (12) become:

$$(12') \qquad \begin{cases} \dfrac{dQ_r}{dt} = \dfrac{\partial \Phi}{\partial P_r}, \\ \\ \dfrac{dP_r}{dt} = 0, \end{cases} \qquad r = 1, \cdots, n.$$

The solution of these equations is obvious, and symmetric. First,

$$P_r = \alpha_r, \qquad r = 1, \cdots, n,$$

where the α_r are n arbitrary constants. Next,

$$Q_r = \omega_r t + \beta_r, \qquad r = 1, \cdots, n,$$

where

$$\omega_r = \Phi_r(\alpha_1, \cdots, \alpha_n), \qquad r = 1, \cdots, n,$$

and the β_r are n arbitrary constants. Thus we have, finally:

$$(19) \qquad \begin{cases} q_r = f_r(\omega_1 t + \beta_1, \cdots, \omega_n t + \beta_n, \alpha_1, \cdots, \alpha_n) \\ p_r = g_r(\omega_1 t + \beta_1, \cdots, \omega_n t + \beta_n, \alpha_1, \cdots, \alpha_n) \end{cases}$$

a wholly symmetric solution of Hamilton's Equations.

If we should wish to use a function $\Phi(\alpha_1, \cdots, \alpha_n)$, for which some other derivative, as $\partial\Phi/\partial\alpha_2$, is $\neq 0$, then we should need a solution $W(q_1, \cdots, q_n, \alpha_1, \cdots, \alpha_n)$ such that

$$\frac{\partial(W_{q_1}, W_{q_3}, \cdots, W_{q_n})}{\partial(\alpha_1, \alpha_3, \cdots, \alpha_n)} \neq 0.$$

5. Examples. Projectile in vacuo. Let a particle of mass m be acted on solely by gravity, and let it be launched so that it will rise for a time. Let q_1, q_2, q_3 be its Cartesian coordinates, with q_1 vertical and positive downward. Then

$$T = \frac{m}{2}(\dot{q}_1^2 + \dot{q}_2^2 + \dot{q}_3^2), \qquad U = mgq_1;$$

$$p_r = \frac{\partial T}{\partial \dot{q}_r} = m\dot{q}_r,$$

$$H = \frac{1}{2m}[p_1^2 + p_2^2 + p_3^2] - mgq_1.$$

The equation for W becomes:

$$\frac{1}{2m}\left[\left(\frac{\partial W}{\partial q_1}\right)^2 + \left(\frac{\partial W}{\partial q_2}\right)^2 + \left(\frac{\partial W}{\partial q_3}\right)^2\right] - mgq_1 = h.$$

Let us try to find the desired function,

$$W = W(q_1, q_2, q_3, h, \alpha_2, \alpha_3),$$

$$\frac{\partial(W_{q_2}, W_{q_3})}{\partial(\alpha_2, \alpha_3)} \neq 0,$$

by setting

$$W = W_1 + W_2 + W_3,$$

where $W_r = W_r(q_r)$ is a function of q_r only. Thus

$$\left(\frac{dW_1}{dq_1}\right)^2 + \left(\frac{dW_2}{dq_2}\right)^2 + \left(\frac{dW_3}{dq_3}\right)^2 - 2m^2gq_1 - 2mh = 0.$$

Since it is only a *particular* function W that is needed, satisfying the Jacobian Relation of Inequality, § 4, (8), it will suffice to set

$$\left(\frac{dW_2}{dq_2}\right)^2 = 2m\alpha_2^2, \qquad \left(\frac{dW_3}{dq_3}\right)^2 = 2m\alpha_3^2,$$

$$\left(\frac{dW_1}{dq_1}\right)^2 = 2m(h - \alpha_2^2 - \alpha_3^2) + 2m^2gq_1.$$

Here, h is determined by the initial conditions from the equation $H = h$, and α_2, α_3 are any two parameters such that initially

$$2m(h - \alpha_2^2 - \alpha_3^2) + 2m^2gq_1 > 0.$$

We now may choose:

$$\frac{dW_s}{dq_s} = \sqrt{2m}\,\alpha_s, \qquad W_s = \sqrt{2m}\,\alpha_s q_s, \qquad s = 2, 3,$$

where α_s is positive, negative, or zero, subject merely to the relation of inequality. But, in the choice of W_1, it is the *negative* root,

$$\frac{dW_1}{dq_1} = -\sqrt{2m(h - \alpha_2^2 - \alpha_3^2) + 2m^2gq_1},$$

that *must* be chosen, since

$$p_1 = \frac{\partial W}{\partial q_1} = \frac{dW_1}{dq_1},$$

and $p_1 < 0$ in the stage we are considering. We may take

$$W_1 = -\int_{c_1}^{q_1} \sqrt{2m(h - \alpha_2^2 - \alpha_3^2) + 2m^2 g q_1}\, dq_1,$$

where c_1 is the initial value of q_1. Thus, finally,

$$W = -\int_{c_1}^{q_1} \sqrt{2m(h - \alpha_2^2 - \alpha_3^2) + 2m^2 g q_1}\, dq_1 + \sqrt{2m}\,\alpha_2 q_2 + \sqrt{2m}\,\alpha_3 q_3.$$

The condition (7), § 4, is satisfied.

We are now in a position to write down the solution of the problem. It is given by Equations (15), § 4:

$$t + \beta_1 = \frac{\partial W}{\partial h} = -m \int_{c_1}^{q_1} \frac{dq_1}{\sqrt{2m(h - \alpha_2^2 - \alpha_3^2) + 2m^2 g q_1}},$$

$$\beta_s = \frac{\partial W}{\partial \alpha_s} = 2m\alpha_s \int_{c_1}^{q_1} \frac{dq_1}{\sqrt{2m(h - \alpha_2^2 - \alpha_3^2) + 2m^2 g q_1}} + \sqrt{2m}\, q_s, \qquad s = 2, 3;$$

along with the equations:

$$p_r = \frac{\partial W}{\partial q_r}, \qquad r = 1, 2, 3.$$

The first of the equations in each of these sets of three is in substance identical with the one which governs the vertical motion of a falling body, § 3, Exercise 4, where now

$$\alpha = h - \alpha_2^2 - \alpha_3^2, \qquad \beta = -\beta_1;$$

and hence:

$$\begin{cases} q_1 = \dfrac{g}{2}(t + \beta_1)^2 - \sqrt{\dfrac{2(h - \alpha_2^2 - \alpha_3^2 + mg\,c_1)}{m}}\,(t + \beta_1) + c_1, \\ p_1 = mg(t + \beta_1) - \sqrt{2m(h - \alpha_2^2 - \alpha_3^2) + 2m^2 g c_1}. \end{cases}$$

In the earlier case, $q = 0$ initially, and so c_1 must be set $= 0$.

The last two equations in the first set give:

$$\beta_s = -2\alpha_s(t + \beta_1) + \sqrt{2m}\, q_s;$$

and so, finally:

$$\begin{cases} q_s = \dfrac{2\alpha_s(t+\beta_1)+\beta_s}{\sqrt{2m}}, & s = 2, 3. \\[2ex] p_s = \sqrt{2m}\,\alpha_s. & \end{cases}$$

The method we have employed gives the solution of the problem so long as the body is rising — no longer; for when it is descending, p_1 becomes positive, and $dW_1/dq_1 = \partial W/\partial q_1$ cannot be expressed by the negative radical. This second stage of the motion, in which the body is falling, could be dealt with by applying the method afresh with suitable modifications — in particular, by taking the positive radical for dW_1/dq_1. But this step can be eliminated if we observe that the equations we are integrating, Hamilton's Equations, here become:

$$\begin{cases} \dfrac{dq_r}{dt} = \dfrac{1}{m}p_r, \qquad r = 1, 2, 3; \\[2ex] \dfrac{dp_1}{dt} = -\,mg; \qquad \dfrac{dp_s}{dt} = 0, \qquad s = 2, 3. \end{cases}$$

The solution of these equations is unique, and is expressed by functions of t which are analytic for all values of t. Hence the analytic continuation of the restricted solution found above gives the general solution, and the formulas found for q_r, p_r are true generally.

EXERCISES

1. *Central Force,* two dimensions, attracting according to the law of nature. Let $q_1 = r,\quad q_2 = \varphi$. Then:

$$T = \frac{m}{2}\left[\left(\frac{dr}{dt}\right)^2 + r^2\left(\frac{d\varphi}{dt}\right)^2\right], \qquad U = \frac{\lambda}{r};$$

$$H = \frac{1}{2m}\left(p_1{}^2 + \frac{p_2{}^2}{q_1{}^2}\right) - \frac{\lambda}{q_1},$$

$$W = R + \Phi,$$

$$\frac{1}{2m}\left[\left(\frac{dR}{dr}\right)^2 + \frac{1}{r^2}\left(\frac{d\Phi}{d\varphi}\right)^2\right] - \frac{\lambda}{r} = h;$$

$$\Phi = \alpha\varphi,$$

$$\left(\frac{dR}{dr}\right)^2 = 2mh + \frac{2m\lambda}{r} - \frac{\alpha^2}{r^2}.$$

Thus

$$W = \pm \int_{r_0}^{r} \sqrt{2mh + \frac{2m\lambda}{r} - \frac{\alpha^2}{r^2}}\, dr + \alpha\varphi,$$

where either the plus sign or the minus sign holds throughout the first stage. Hence

$$t + \beta_1 = \pm\, m \int_{r_0}^{r} \frac{dr}{\sqrt{2mh + \frac{2m\lambda}{r} - \frac{\alpha^2}{r^2}}},$$

$$\beta_2 = \mp\alpha \int_{r_0}^{r} \frac{dr}{r^2\sqrt{2mh + \frac{2m\lambda}{r} - \frac{\alpha^2}{r^2}}} + \varphi.$$

Discuss the case that the radicand vanishes for two distinct positive values of r, expressing r as a periodic function of φ, and evaluate the integral that expresses t; cf. § 9.

2. The same problem in space. Let $q_1 = r, \quad q_2 = \theta, \quad q_3 = \varphi$;

$$x = r\cos\theta\cos\varphi, \qquad y = r\cos\theta\sin\varphi, \qquad z = r\sin\theta;$$

$$T = \frac{m}{2}\left[\left(\frac{dr}{dt}\right)^2 + r^2\left(\frac{d\theta}{dt}\right)^2 + r^2\cos^2\theta\left(\frac{d\varphi}{dt}\right)^2\right], \qquad U = \frac{\lambda}{r};$$

$$H = \frac{1}{2m}\left(p_1^{\,2} + \frac{p_2^{\,2}}{r^2} + \frac{p_3^{\,2}}{r^2\cos^2\theta}\right) - \frac{\lambda}{r};$$

$$W = R + \Theta + \Phi;$$

$$r^2\left[2mh + \frac{2m\lambda}{r} - \left(\frac{dR}{dr}\right)^2\right] = \left(\frac{d\Theta}{d\theta}\right)^2 + \frac{1}{\cos^2\theta}\left(\frac{d\Phi}{d\varphi}\right)^2 = G^2,$$

$$\left(\frac{d\Phi}{d\varphi}\right)^2 = \cos^2\theta\left[G^2 - \left(\frac{d\Theta}{d\theta}\right)^2\right] = L^2;$$

$$W = \int_{a}^{r} \sqrt{2mh + \frac{2m\lambda}{r} - \frac{G^2}{r^2}}\, dr + \int_{\theta_0}^{\theta} \sqrt{G^2 - L^2\sec^2\theta}\, d\theta + L\varphi.$$

Complete the solution and discuss the cases that the radicands have distinct roots.

3. Discuss the problem of § 4 when $n = 1$. Show that W is given by solving the equation:

$$H\left(q, \frac{\partial W}{\partial q}\right) = h,$$

and integrating:

$$\frac{\partial W}{\partial q} = f(q, h), \qquad \frac{\partial f}{\partial h} \neq 0;$$

$$W = \int_{c_0}^{q} f(q, h)\, dq.$$

Then

$$Q = \frac{\partial W}{\partial P} = \int_{c}^{q} \frac{\partial f}{\partial h}\, dq.$$

Thus

$$q = g(Q, P);$$

$$Q = t + \beta, \qquad P = h;$$

$$\begin{cases} q = g(t + \beta, h) \\ p = \dfrac{\partial W}{\partial q} = f(q, h). \end{cases}$$

6. Comparison of the Two Methods. We have studied two methods of solving Hamilton's Equations, a) Reduction to the Equilibrium Problem; b), when H does not depend on t, Reduction to the Form, $H' = P_1$.

The first method, being general, must apply to the second case. It does. Let us treat this case by the first method, as set forth in the Exercise of § 2. We will choose as V the function:

$$V = -ht + W, \tag{1}$$

where $W = W(q_1, \cdots, q_n, h, \alpha_2, \cdots, \alpha_n)$ is the function of § 4, and h, α_s have been replaced by P_1, P_s. The transformation of that Exercise,

$$p_r = \frac{\partial V}{\partial q_r}, \qquad Q_r = \frac{\partial V}{\partial P_r}, \qquad r = 1, \cdots, n, \tag{2}$$

yields an H' that vanishes identically. The transformed Hamiltonian Equations thus take the form:

$$\frac{dQ_r}{dt} = 0, \qquad \frac{dP_r}{dt} = 0, \qquad r = 1, \cdots, n. \tag{3}$$

Departing from the notation of the Exercise, write their integrals in the form:

$$(4)\qquad \begin{cases} Q_r = \beta_r, & r = 1, \cdots, n; \\ P_1 = h, \quad P_s = \alpha_s, & s = 2, \cdots, n. \end{cases}$$

The solution of the original Hamiltonian Equations is now given by substituting these values in (2):

$$(5)\qquad \begin{cases} p_r = \dfrac{\partial V}{\partial q_r}, & r = 1, \cdots, n; \\[2ex] \beta_1 = \dfrac{\partial V}{\partial h}, \quad \beta_s = \dfrac{\partial V}{\partial \alpha_s}, & s = 2, \cdots, n. \end{cases}$$

But

$$(6)\qquad \frac{\partial V}{\partial q_r} = \frac{\partial W}{\partial q_r}, \qquad \frac{\partial V}{\partial h} = -\,t + \frac{\partial W}{\partial h}, \qquad \frac{\partial V}{\partial \alpha_s} = \frac{\partial W}{\partial \alpha_s}.$$

Hence Equations (5) agree not only in substance, but even in form, save for one exception, with Equations (15), § 4. The equation arising from differentiation with respect to h in the earlier case read:

$$t + \beta_1 = \frac{\partial W}{\partial h}.$$

Here it is:

$$\beta_1 = -\,t + \frac{\partial W}{\partial h}.$$

7. Cyclic Coordinates. It frequently happens that H, besides being independent of t, contains fewer than n q's. Begin with the case of one q,

$$(1)\qquad H = H(q_1, p_1, \cdots, p_n).$$

From Hamilton's Equations,

$$(2)\qquad \frac{dp_s}{dt} = -\frac{\partial H}{\partial q_s} = 0, \qquad s = 2, \cdots, n,$$

and hence

$$(3)\qquad p_s = \alpha_s, \qquad s = 2, \cdots, n.$$

It is not difficult to complete the solution by means of Hamilton's Equations and the integral of energy,

$$(4)\qquad h = H(q_1, p_1, \cdots, p_n);$$

but this is not the form of solution in which we are interested. We desire a discussion by the methods of § 4; in particular, by the transformation:

$$p_r = \frac{\partial W}{\partial q_r}, \qquad Q_r = \frac{\partial W}{\partial P_r}, \qquad r = 1, \cdots, n, \tag{5}$$

where

$$W = W(q_1, \cdots, q_n, h, \alpha_2, \cdots, \alpha_n) \tag{6}$$

is a solution of the equation:

$$h = H\left(q_1, \frac{\partial W}{\partial q_1}, \cdots, \frac{\partial W}{\partial q_n}\right), \tag{7}$$

$$\frac{\partial(W_{q_1}, W_{q_2}, \cdots, W_{q_n})}{\partial(h, \alpha_2, \cdots, \alpha_n)} \neq 0, \tag{8}$$

and $P_1 = h$, $P_s = \alpha_s$, $s = 2, \cdots, n$.

To find such a solution we turn to the Method of Separation of Variables, which has rendered such good service in the past. Let

$$W = W_1 + \cdots + W_n, \tag{9}$$

where $W_r = W_r(q_r)$ is a function of q_r alone — and of the n parameters, $h, \alpha_2, \cdots, \alpha_n$. From (5) and (3) we see that

$$p_s = \frac{dW_s}{dq_s} = \alpha_s, \qquad s = 2, \cdots, n,$$

and so we try:

$$W_s = \alpha_s q_s, \qquad s = 2, \cdots, n.$$

Let W_1 be denoted more simply by v:

$$W_1 = W_1(q_1, h, \alpha_2, \cdots, \alpha_n) = v. \tag{10}$$

Then (7) becomes:

$$H\left(q_1, \frac{\partial v}{\partial q_1}, \alpha_2, \cdots, \alpha_n\right) = h. \tag{11}$$

If we assume that

$$\frac{\partial H}{\partial p_1} = H_{p_1}(q_1, p_1, \alpha_2, \cdots, \alpha_n) \neq 0, \tag{12}$$

and solve the equation:

$$H(q_1, p_1, \alpha_2, \cdots, \alpha_n) = h \tag{13}$$

for p_1:

$$p_1 = \Psi(q_1, h, \alpha_2, \cdots, \alpha_n), \tag{14}$$

we have:

$$\frac{\partial v}{\partial q_1} = \Psi(q_1, h, \alpha_2, \cdots, \alpha_n). \tag{15}$$

Now choose as v:

$$v = \int_c^{q_1} \Psi(q_1, h, \alpha_2, \cdots, \alpha_n)\, dq_1, \tag{16}$$

where c is a numerical constant.

We are thus led to a function

$$W = v + \alpha_2 q_2 + \cdots + \alpha_n q_n \tag{17}$$

of the desired kind, provided the Jacobian relation (8) is satisfied. The Jacobian here reduces to

$$\frac{\partial^2 v}{\partial q_1 \partial h} = \frac{\partial \Psi}{\partial h}, \quad \text{or} \quad \frac{\partial p_1}{\partial h},$$

where p_1 is determined by (13). On differentiating (13) we find:

$$\frac{\partial H}{\partial p_1}\frac{\partial p_1}{\partial h} = 1,$$

and so the Jacobian does not vanish.

Solution of Hamilton's Equations. We can now apply the general theory of § 4. The transformation (5) of the present paragraph carries Hamilton's Equations over into the form:

$$\frac{dQ_1}{dt} = 1, \qquad \frac{dQ_s}{dt} = 0, \qquad s = 2, \cdots, n;$$

$$\frac{dP_r}{dt} = 0, \qquad r = 1, \cdots, n,$$

the solution of which is:

$$Q_1 = t + \beta_1, \qquad Q_s = \beta_s, \qquad s = 2, \cdots, n;$$

$$P_1 = h, \qquad P_s = \alpha_s, \qquad s = 2, \cdots, n.$$

These values for Q_r, P_r are to be substituted in (5), and the resulting equations solved for q_r, p_r:

$$
(18)\quad
\begin{cases}
t + \beta_1 = \dfrac{\partial W}{\partial h} = \dfrac{\partial v}{\partial h} = \displaystyle\int_c^{q_1} \dfrac{\partial \Psi}{\partial h}\, dq_1\,; \\[2ex]
\beta_s = \dfrac{\partial W}{\partial \alpha_s} = \displaystyle\int_c^{q_1} \dfrac{\partial \Psi}{\partial \alpha_s}\, dq_1 + q_s, \qquad s = 2, \cdots, n. \\[2ex]
p_1 = \dfrac{\partial W}{\partial q_1} = \Psi(q_1, h, \alpha_2, \cdots, \alpha_n), \\[2ex]
p_s = \dfrac{\partial W}{\partial q_s} = \alpha_s, \qquad s = 2, \cdots, n.
\end{cases}
$$

The equations of the second line determine q_s as a function of q_1:

$$
(19)\qquad q_s = \beta_s - \int_c^{q_1} \frac{\partial \Psi}{\partial \alpha_s}\, dq_1, \qquad s = 2, \cdots, n.
$$

The first equation gives q_1 as a function of t.

EXERCISE

Obtain the final result (18) directly from Hamilton's Equations.

8. Continuation. The General Case. Let H depend on $1 < \nu < n$ arguments q_k:

$$
(1)\qquad H = H(q_1, \cdots, q_\nu, p_1, \cdots, p_n).
$$

The method of treatment is similar, though the solution cannot in general be obtained by quadratures. Equations (3) of § 7 now become:

$$
(2)\qquad p_s = \alpha_s, \qquad s = \nu + 1, \cdots, n.
$$

By analogy we now seek to determine W in the form:

$$
(3)\qquad W = v + \alpha_{\nu+1} q_{\nu+1} + \cdots + \alpha_n q_n,
$$

$$
(4)\qquad v = \Phi(q_1, \cdots, q_\nu, h, \alpha_2, \cdots, \alpha_n),
$$

$$
(5)\quad
\begin{cases}
\dfrac{\partial(v_{q_1}, v_{q_2}, \cdots, v_{q_n})}{\partial(h, \alpha_2, \cdots, \alpha_n)} = \dfrac{\partial(v_{q_1}, v_{q_2}, \cdots, v_{q_\nu})}{\partial(h, \alpha_2, \cdots, \alpha_\nu)} \neq 0; \\[2ex]
\dfrac{\partial(v_{q_2}, \cdots, v_{q_n})}{\partial(\alpha_2, \cdots, \alpha_n)} = \dfrac{\partial(v_{q_2}, \cdots, v_{q_\nu})}{\partial(\alpha_2, \cdots, \alpha_\nu)} \neq 0.
\end{cases}
$$

Equation (7), § 7, for W now becomes:

$$h = H\left(q_1, \cdots, q_\nu, \frac{\partial v}{\partial q_1}, \cdots, \frac{\partial v}{\partial q_\nu}, \alpha_{\nu+1}, \cdots, \alpha_n\right). \tag{6}$$

This is an equation of the same type as (6), § 4, but with $\nu < n$ variables q_r. As in the earlier case, only a particular solution is sought, and such a solution may be found by special devices, notably the method of separation of variables.

A function v once found, the solution proceeds as before.

$$\begin{cases} t + \beta_1 = \dfrac{\partial v}{\partial h}; \\ \beta_k = \dfrac{\partial v}{\partial \alpha_k}, \qquad k = 2, \cdots, \nu; \\ \beta_l = \dfrac{\partial v}{\partial \alpha_l} + q_l, \qquad l = \nu + 1, \cdots, n. \end{cases} \tag{7}$$

From the equations of the second line q_k can be found in terms of q_1, $k = 2, \cdots, \nu$. From the first equation q_1 is now found in terms of t. Finally, q_l is given by the last line, $l = \nu + 1, \cdots, n$.

9. Examples. The Two-Body Problem. Consider the motion of two bodies (particles) that attract each other according to the law of nature and are acted on by no other forces. Their centre of gravity travels in a right line with constant velocity, or else remains permanently at rest. We will assume the latter case. Then each of the bodies moves as if attracted by a force at O, the centre of gravity, which is inversely proportional to the square of the distance of the body from O.

We will first discuss the motion in a plane — later, in space.

Let the particle be referred to polar coordinates, $q_1 = r$, $q_2 = \varphi$. Then

$$T = \frac{m}{2}\left(\frac{dr^2}{dt^2} + r^2\frac{d\varphi^2}{dt^2}\right), \qquad U = \frac{\lambda}{r}.$$

Hence

$$H = \frac{1}{2m}\left(p_1{}^2 + \frac{p_2{}^2}{q_1{}^2}\right) - \frac{\lambda}{q_1}.$$

Let

$$W = v + \alpha_2 q_2.$$

Then v is given by the equation:

$$\frac{1}{2m}\left[\left(\frac{\partial v}{\partial q_1}\right)^2 + \frac{\alpha_2^2}{q_1^2}\right] - \frac{\lambda}{q_1} = h,$$

or

$$v = \int_c^r \pm\sqrt{2mh + \frac{2m\lambda}{r} - \frac{\alpha_2^2}{r^2}}\, dr,$$

where a definite one of the two signs holds for the first stage of the motion.

Equations (18) of § 7 now give the solution of Hamilton's Equations in the form:

$$t + \beta_1 = \frac{\partial v}{\partial h}, \qquad \beta_2 = \frac{\partial v}{\partial \alpha_2} + q_2;$$

or

$$(1) \qquad \begin{cases} t + \beta_1 = m\displaystyle\int_c^r \frac{dr}{\pm\sqrt{2mh + \frac{2m\lambda}{r} - \frac{\alpha_2^2}{r^2}}} \\ \beta_2 = \alpha_2\displaystyle\int_c^r \frac{dr}{\mp r^2\sqrt{2mh + \frac{2m\lambda}{r} - \frac{\alpha_2^2}{r^2}}} + \varphi. \end{cases}$$

The directness of the result is particularly noteworthy. It has not been necessary to make use of skillful devices or to effect complicated eliminations. From the evaluation of the second integral r can be expressed as a trigonometric function of φ. But the discussion of r in terms of t is more complicated; cf. below the reference to Charlier.

The Orbit in Space. To treat the motion in three dimensions let

$$x = r\cos\theta\cos\varphi, \qquad y = r\cos\theta\sin\varphi, \qquad z = r\sin\theta.$$

Then

$$T = \frac{m}{2}\left[\left(\frac{dr}{dt}\right)^2 + r^2\left(\frac{d\theta}{dt}\right)^2 + r^2\cos^2\theta\left(\frac{d\varphi}{dt}\right)^2\right], \qquad U = \frac{\lambda}{r}.$$

Let $q_1 = r,\ q_2 = \theta,\ q_3 = \varphi$. Then

$$p_1 = m\dot{q}_1, \qquad p_2 = mq_1^2\dot{q}_2, \qquad p_3 = mq_1^2\dot{q}_3\cos^2 q_2.$$

$$H = \frac{1}{2m}\left(p_1{}^2 + \frac{p_2{}^2}{q_1{}^2} + \frac{p_3{}^2}{q_1{}^2 \cos^2 q_2}\right) - \frac{\lambda}{q_1}.$$

Since $H = H(q_1, q_2, p_1, p_2, p_3)$, we see that $p_3 = \alpha_3$ (const.). Thus

$$W = v + \alpha_3 q_3,$$

where v is given by the equation:

$$\frac{1}{2m}\left[\left(\frac{\partial v}{\partial q_1}\right)^2 + \frac{1}{q_1{}^2}\left(\frac{\partial v}{\partial q_2}\right)^2 + \frac{\alpha_3{}^2}{q_1{}^2 \cos^2 q_2}\right] - \frac{\lambda}{q_1} = h.$$

Here, there are only two independent variables, $q_1 = r$ and $q_2 = \theta$. The equation can be written in the form:

$$r^2\left(\frac{\partial v}{\partial r}\right)^2 + \left(\frac{\partial v}{\partial \theta}\right)^2 + \frac{\alpha_3{}^2}{\cos^2 \theta} - 2mhr^2 - 2m\lambda r = 0.$$

On setting

$$v = R + \Theta,$$

the variables can be separated:

$$-r^2\left(\frac{dR}{dr}\right)^2 + 2mhr^2 + 2m\lambda r = \left(\frac{d\Theta}{d\theta}\right)^2 + \alpha_3{}^2 \sec^2 \theta = \alpha_2{}^2.$$

Hence

$$R = \int_c^r \pm\sqrt{2mh + \frac{2m\lambda}{r} - \frac{\alpha_2{}^2}{r^2}}\, dr,$$

$$\Theta = \int_\gamma^\theta \pm\sqrt{\alpha_2{}^2 - \alpha_3{}^2 \sec^2 \theta}\, d\theta,$$

where the signs are determined for a particular stage of the motion, and c, γ are arbitrary numerical constants. Adding the further term $\alpha_3 q_3$, we have:

$$W = v + \alpha_3 q_3, \qquad v = R + \Theta.$$

We are thus led to the solution of the problem in the form given by (7), § 8:

$$(2)\quad \begin{cases} t + \beta_1 = m \displaystyle\int_c^r \frac{dr}{\pm\sqrt{2mh + \frac{2m\lambda}{r} - \frac{\alpha_2^2}{r^2}}}; \\ \beta_2 = -\alpha_2 \displaystyle\int_c^r \frac{dr}{\pm r^2\sqrt{2mh + \frac{2m\lambda}{r} - \frac{\alpha_2^2}{r^2}}} \\ \qquad + \alpha_2 \displaystyle\int_\gamma^\theta \frac{d\theta}{\pm\sqrt{\alpha_2^2 - \alpha_3^2 \sec^2\theta}}; \\ \beta_3 = -\alpha_3 \displaystyle\int_\gamma^\theta \frac{\sec^2\theta\, d\theta}{\pm\sqrt{\alpha_2^2 - \alpha_3^2\sec^2\theta}} + \varphi. \end{cases}$$

The discussion of this solution on the hand of the explicit evaluation of the integrals and the inverse functions thus arising presents practical difficulties. The problem is of so great importance in Astronomy that it has been treated at length by Charlier, *Mechanik des Himmels*, Vol. I, Chap. 4, p. 167. On p. 171, Equations (7) are identical with our solution, save as to notation.

Failure of the Method. There are cases in which the method breaks down. Consider, for example, motion in a plane. Suppose the body is projected from a point A, distant a from the centre of force, O, at right angles to the line OA and with a velocity v_0 such that

$$\frac{mv_0^2}{a} = \frac{\lambda}{a^2}.$$

It will then describe a circle, $r = a$. But the Equations (1) or (2) can obviously never yield this solution. Why?

The function v was determined from the equation:

$$\left(\frac{\partial v}{\partial q_1}\right)^2 = 2mh + \frac{2m\lambda}{r} - \frac{\alpha_2^2}{r^2}.$$

In the present case,

$$h = -\frac{\lambda}{2a}, \qquad \alpha_2 = mav_0,$$

and hence

$$\frac{\partial v}{\partial q_1} \equiv 0.$$

Thus the condition

$$\frac{\partial^2 v}{\partial h\,\partial q_1} \neq 0$$

is not fulfilled, and so, of course, there is no reason why the method should apply, since the hypotheses on which it depends do not hold.

10. Continuation. The Top. We take over from Chapter VI, § 18, the expression for the kinetic energy,

$$T = \tfrac{1}{2}[A(p^2 + q^2) + Cr^2].$$

By Euler's Geometrical Equations, this becomes:

$$T = \tfrac{1}{2}[A(\dot\theta^2 + \dot\psi^2 \sin^2\theta) + C(\dot\varphi + \dot\psi\cos\theta)^2].$$

Let

$$q_1 = \theta, \qquad q_2 = \varphi, \qquad q_3 = \psi.$$

Then, since

$$p_r = \frac{\partial T}{\partial \dot q_r},$$

we have:

$$p_1 = A\dot\theta,$$

$$p_2 = C(\dot\varphi + \dot\psi\cos\theta),$$

$$p_3 = C\dot\varphi\cos\theta + (A\sin^2\theta + C\cos^2\theta)\dot\psi.$$

Thus T, expressed in terms of the p's and q's, becomes:

$$T = \frac{1}{2}\left[\frac{1}{A}p_1^2 + \frac{1}{C}p_2^2 + \frac{1}{A}\left(\frac{p_2\cos q_1 - p_3}{\sin q_1}\right)^2\right].$$

Furthermore,*

$$U = -Mgb\cos\theta.$$

Thus

$$H = \frac{1}{2}\left[\frac{1}{A}p_1^2 + \frac{1}{C}p_2^2 + \frac{1}{A}\left(\frac{p_2\cos q_1 - p_3}{\sin q_1}\right)^2\right] + Mgb\cos q_1.$$

Hence it appears that the problem comes under the case of cyclic coordinates treated in § 7. First, then,

$$p_2 = \alpha_2, \qquad p_3 = \alpha_3.$$

* It is necessary to change from the earlier notation h for the distance from the peg to the centre of gravity, since h plays so important a rôle in the present theory. Let the distance be denoted by b.

To determine v we have:

$$\frac{1}{2}\left[\frac{1}{A}\frac{\partial v^2}{\partial q_1{}^2} + \frac{1}{C}\alpha_2{}^2 + \frac{1}{A}\left(\frac{\alpha_2 \cos q_1 - \alpha_3}{\sin q_1}\right)^2\right] + Mgb \cos q_1 = h,$$

$$\sin^2 q_1 \frac{\partial v^2}{\partial q_1{}^2} = (2Ah - L\alpha_2{}^2 - N \cos q_1) \sin^2 q_1 - (\alpha_2 \cos q_1 - \alpha_3)^2,$$

$$v = \int_{\epsilon}^{q_1} \frac{\pm\sqrt{(2Ah - L\alpha_2{}^2 - N \cos q_1) \sin^2 q_1 - (\alpha_2 \cos q_1 - \alpha_3)^2}}{\sin q_1}\, dq_1,$$

where ϵ is an arbitrary numerical constant, not a parameter, and the sign of the radical must be chosen with respect to the special stage of the motion under consideration. Moreover, for brevity,

$$L = \frac{A}{C}, \qquad N = 2AMgb.$$

The solution of the problem, as given in § 7, now takes on the form:

$$t + \beta_1 = \frac{\partial v}{\partial h},$$

$$\beta_2 = \frac{\partial v}{\partial \alpha_2} + q_2, \qquad \beta_3 = \frac{\partial v}{\partial \alpha_3} + q_3.$$

Thus

$$t + \beta_1 = \int_{\epsilon}^{q_1} \frac{A \sin q_1\, dq_1}{\pm\sqrt{(2Ah - L\alpha_2{}^2 - N \cos q_1) \sin^2 q_1 - (\alpha_2 \cos q_1 - \alpha_3)^2}}.$$

Let

$$u = \cos q_1.$$

Then this equation becomes:

$$t + \beta_1 = \int_{c}^{u} \frac{du}{\mp\sqrt{F(u)}}, \qquad c = \cos \epsilon,$$

where

$$F(u) = \left(\frac{2h}{A} - \frac{\alpha_2{}^2}{AC} - \frac{2Mgbu}{A}\right)(1 - u^2) - \frac{(\alpha_2 u - \alpha_3)^2}{A^2}.$$

This is the same result obtained by elementary methods, Chap. VI, § 18. But compare the technique. With only Euler's

Dynamical and Geometrical Equations to work with,* eliminations had to be made by ingenious devices, whereas the present advanced methods free the treatment from all artifice. The fundamental equation in desired form is evolved naturally, directly, from the general theory, not untangled from a snarl of equations. Instead of having to solve three equations for $\dot{\theta}$, $\dot{\varphi}$, $\dot{\psi}$ by more or less ingenious methods of elimination, the functions T, U, and hence H are obtained without the use of any artifice whatever, and the method of § 7 yields $q_1 = \theta$ at once as a function of t, the further equations giving $q_3 = \varphi$ and $q_3 = \psi$ immediately.

EXERCISE

Study the motion of a top with hemispherical peg, spinning and sliding on a smooth table. Show that

$$t + \beta_1 = \int_c^u \frac{\sqrt{A + Mb^2(1 - u^2)}\,du}{\pm\sqrt{F(u)}},$$

where

$$F(u) = \left(2h - \frac{\alpha_2^{\,2}}{C} - 2Mgbu\right)(1 - u^2) - (\alpha_2 u - \alpha_3)^2.$$

11. Perturbations. Variation of Constants. In the problem of perturbations the motion which the system would execute if only the major forces acted is regarded as fundamental, and then the variation from this motion due to the disturbing forces, thought of as slight, is studied.

This analysis of the physical problem is mirrored mathematically by writing down Hamilton's Equations for the actual motion:

$$(1) \qquad \frac{dq_r}{dt} = \frac{\partial H}{\partial p_r}, \quad \frac{dp_r}{dt} = -\frac{\partial H}{\partial q_r}, \qquad r = 1, \cdots, n,$$

and then setting the characteristic function H of the actual problem equal to the H_0 of the problem due to the major forces, plus a remainder, H_1:

$$(2) \qquad H = H_0 + H_1.$$

* It is true that in the earlier treatment we had two integrals of the differential equations of motion to work with at the outset, namely; the equation of energy, $T = U + h$, and the equation arising from the fact that the vector moment of momentum σ is always horizontal. But even so there were three equations in $\dot{\theta}$, $\dot{\psi}$, $\dot{\varphi}$ to integrate.

Transformation of the Major Problem to the Equilibrium Problem. First, the major problem, represented by Hamilton's Equations in the form:

$$\frac{dq_r}{dt} = \frac{\partial H_0}{\partial p_r}, \qquad \frac{dp_r}{dt} = -\frac{\partial H_0}{\partial q_r}, \qquad r = 1, \cdots, n, \tag{3}$$

is solved by reducing it, through a contact transformation, to the Equilibrium Problem. The contact transformation is given by the equations:

$$p_r = \frac{\partial V^0}{\partial q_r}, \qquad Q_r = \frac{\partial H_0}{\partial P_r}, \qquad r = 1, \cdots, n, \tag{4}$$

where

$$V^0 = V^0(q_1, \cdots, q_n, P_1, \cdots, P_n, t) \tag{5}$$

is obtained as follows. Write down Jacobi's Equation, corresponding to Hamilton's Equations (3):

$$\frac{\partial V^0}{\partial t} + H_0\left(q_1, \cdots, q_n, \frac{\partial V^0}{\partial q_1}, \cdots, \frac{\partial V^0}{\partial q_n}, t\right) = 0. \tag{6}$$

Let

$$V^0 = V^0(q_1, \cdots, q_n, \alpha_1, \cdots, \alpha_n, t)$$

be a solution of this equation such that the Jacobian

$$\frac{\partial(V^0_{q_1}, \cdots, V^0_{q_n})}{\partial(\alpha_1, \cdots, \alpha_n)} \neq 0.$$

In this function, replace α_r by P_r. The resulting function is the function (5). [In practise, the function $V^0(q_1, \cdots, q_n, \alpha_1, \cdots, \alpha_n, t)$ is obtained, not from an elaborate theory of partial differential equations, but by means of simple devices, *ad hoc*.]

Let the transformation (4) be written in the explicit form:

$$\begin{cases} Q_r = F_r(p_1, \cdots, p_n, q_1, \cdots, q_n, t) \\ P_r = G_r(p_1, \cdots, p_n, q_1, \cdots, q_n, t) \end{cases} \tag{4'}$$

or

$$\begin{cases} q_r = f_r(P_1, \cdots, P_n, Q_1, \cdots, Q_n, t) \\ p_r = g_r(P_1, \cdots, P_n, Q_1, \cdots, Q_n, t) \end{cases} \tag{4''}$$

To say that the major problem is thereby transformed to the Equilibrium Problem means that, when the variables q_r, p_r that form the solution of Equations (3) are subjected to the transformation (4), the resulting Hamiltonian Equations become:

$$\frac{dQ_r}{dt} = 0, \qquad \frac{dP_r}{dt} = 0, \qquad r = 1, \cdots, n. \tag{7}$$

The solution of these equations can be written in the form:

$$Q_r = \beta_r, \qquad P_r = \alpha_r, \qquad r = 1, \cdots, n, \tag{8}$$

where α_r, β_r are constants. Now transform the variables Q_r, P_r that are the solution of Equations (7), namely, the functions given by (8), back by means of the transformation (4″), and we have the solution of Equations (3) in the form:

$$\begin{cases} q_r = f_r(\alpha_1, \cdots, \alpha_n, \beta_1, \cdots, \beta_n, t) \\ p_r = g_r(\alpha_1, \cdots, \alpha_n, \beta_1, \cdots, \beta_n, t) \end{cases} \tag{9}$$

Thus the transformations (4′) or (4″), and (9), identical except in notation, represent two distinct things:

a) In the form (4″) these equations represent the Contact Transformation (4).

b) In the form (9) they represent the Solution of the Hamiltonian Equations of the Major Problem, or (3).

Transformation of the Actual Problem by the Same Contact Transformation. We now proceed to apply the contact transformation (4), not to the variables (q_r, p_r) which satisfy Equations (3), but to the variables (q_r, p_r) of the original problem, which satisfy Equations (1). Since this is a contact transformation, we know that Equations (1) will go over into new equations of the same form:

$$\frac{dQ_r}{dt} = \frac{\partial H'}{\partial P_r}, \qquad \frac{dP_r}{dt} = -\frac{\partial H'}{\partial Q_r}, \qquad r = 1, \cdots, n. \tag{10}$$

Here $H' = H'(Q_r, P_r, t)$ has the value, cf. Chap. XIV, § 5, Ex. 1, (15):

$$H' = H + \frac{\partial V^0}{\partial t}. \tag{11}$$

But from (6):

$$0 = H_0 + \frac{\partial V^0}{\partial t}.$$

Hence

$$H' = H - H_0.$$

Finally, from (2) it follows that

$$H' = H_1. \tag{12}$$

Thus Equations (10) take the form:

$$\frac{dQ_r}{dt} = \frac{\partial H_1}{\partial P_r}, \quad \frac{dP_r}{dt} = -\frac{\partial H_1}{\partial Q_r}, \qquad r = 1, \cdots, n. \tag{13}$$

The result may be stated as follows. *When the variables q_r, p_r which form the solution of the actual problem represented by Equations* (1) *are transformed by the contact transformation* (4) *or* (4′), *the transformed equations take the form* (13), *where H_1 is the given, or known, function of Equation* (2), *now expressed through* (4) *or* (4″) *in terms of Q_r, P_r, t.*

The Final Solution. It is now but a step to the solution of Equations (1), which represent the actual problem. Solve Equations (13), thus determining Q_r, P_r as functions of t. Then transform these functions, the solution of (13), back by means of (4) or (4″) to the variables q_r, p_r. The latter satisfy Equations (1).

The result can be expressed in the form:

$$\begin{cases} q_r = f_r(P_1, \cdots, P_n, Q_1, \cdots, Q_n, t) \\ p_r = g_r(P_1, \cdots, P_n, Q_1, \cdots, Q_n, t) \end{cases} \tag{14}$$

where Q_r, P_r are determined by Equations (13).

Variation of Constants. The method above set forth has been called the "variation of constants." This expression is a mathematical pun. It is a pun on the letters α_r, β_r. These, in Equations (9), are constants — the equations there representing the solution of the major problem, (3). On the other hand, they can be identified with the variables P_r, Q_r of (14), these variables being determined by (13), and then Equations (14) represent the solution of the actual problem, (1).

We can attain complete confusion of ideas, as is done in the literature, by changing the notation in (13) and (14) from Q_r, P_r to β_r, α_r. Thus (14) goes over into the form of (9), and (13) is replaced by the equations:

$$\frac{d\alpha_r}{dt} = \frac{\partial(-H_1)}{\partial \beta_r}, \quad \frac{d\beta_r}{dt} = -\frac{\partial(-H_1)}{\partial \alpha_r}, \quad r = 1, \cdots, n,$$

where $H_1 = H_1(Q_r, P_r, t)$ is now written as $H_1(\beta_1, \alpha_1, t)$, the Hamiltonian function now being $-H_1$ instead of H_1.

Thus the pun is explained — but it is a poor pun that has to be explained.

12. Continuation. A Second Method. It is possible to treat the problem of perturbations in still another manner. Let $\varphi(\alpha_1, \cdots, \alpha_n)$ be any given function whose first partial derivatives are not all 0. Let the Hamiltonian Equations for the undisturbed motion, namely, (3), be transformed by a new contact transformation:

$$(15) \qquad p_r = \frac{\partial S}{\partial q_r}, \qquad P_r = -\frac{\partial S}{\partial Q_r}, \qquad r = 1, \cdots, n,$$

where S is defined as follows. Consider the equation:

$$(16) \quad \varphi(\alpha_1, \cdots, \alpha_n) = H_0\left(q_1, \cdots, q_n, \frac{\partial S}{\partial q_1}, \cdots, \frac{\partial S}{\partial q_n}, t\right) + \frac{\partial S}{\partial t}.$$

Let

$$S = S(q_1, \cdots, q_n, \alpha_1, \cdots, \alpha_n, t)$$

be a solution such that *

$$\frac{\partial(S_{q_1}, \cdots, S_{q_n})}{\partial(\alpha_1, \cdots, \alpha_n)} \neq 0.$$

Now, make the contact transformation:

$$(17) \qquad p_r = \frac{\partial S}{\partial q_r}, \qquad P_r = -\frac{\partial S}{\partial Q_r}, \qquad r = 1, \cdots, n,$$

where

$$S = S(q_1, \cdots, q_n, Q_1, \cdots, Q_n, t).$$

* In order to find such a solution, begin with the equation:

$$h = H\left(q_1, \cdots, q_n, \frac{\partial S}{\partial q_1}, \cdots, \frac{\partial S}{\partial q_n}, t\right) + \frac{\partial S}{\partial t},$$

where h is an arbitrary constant, and seek a solution:

$$S = S(q_1, \cdots, q_n, h, \alpha_2, \cdots, \alpha_n, t),$$

such that

$$\frac{\partial(S_{q_1}, S_{q_2}, \cdots, S_{q_n})}{\partial(h, \alpha_2, \cdots, \alpha_n)} \neq 0,$$

where $\alpha_2, \cdots, \alpha_n$ are arbitrary. Substitute

$$h = \varphi(\alpha_1, \cdots, \alpha_n)$$

in S. If $\partial\varphi/\partial\alpha_1 \neq 0$, this will be the function desired.

This transformation, applied to Equations (1), carries these over into equations of the same type:

$$\frac{dQ_r}{dt} = \frac{\partial H'}{\partial P_r}, \quad \frac{dP_r}{dt} = -\frac{\partial H'}{\partial Q_r}, \qquad r = 1, \cdots, n, \tag{18}$$

where, by Chap. XIV, § 5:

$$H' = H + \frac{\partial S}{\partial t}. \tag{19}$$

But, by (16) and (17):

$$\varphi = H_0 + \frac{\partial S}{\partial t}.$$

Hence, with the aid of (2):

$$H' = H_1 + \varphi,$$

the arguments now being the Q_r, P_r into which q_r, p_r have been transformed by (17). Thus Equations (18) take the form:

$$\frac{dQ_r}{dt} = \frac{\partial H_1}{\partial P_r}, \quad \frac{dP_r}{dt} = -\frac{\partial H_1}{\partial Q_r} - \frac{\partial \varphi}{\partial Q_r}, \quad r = 1, \cdots, n. \tag{20}$$

Solve these equations and substitute the functions of t thus obtained, namely, the Q_r, P_r, in (17). The functions q_r, p_r of t obtained from these equations are the solution of the actual problem, or Equations (1).

Carathéodory* treats Equations (20) as follows. He writes λH_1 instead of H_1:

$$\frac{dQ_r}{dt} = \lambda \frac{\partial H_1}{\partial P_r}, \quad \frac{dP_r}{dt} = -\lambda \frac{\partial H_1}{\partial Q_r} - \varphi_{Q_r}. \tag{21}$$

He then develops the solution into a power series in λ:

$$\begin{cases} Q_r = \alpha_r + \lambda C_{1r} + \lambda^2 C_{2r} + \cdots, \\ P_r = \beta_r - \varphi_{\alpha_r} t + \lambda D_{1r} + \cdots, \end{cases} \tag{22}$$

where C_{kr}, D_{kr} are functions of t, vanishing when $t = 0$ (for simplicity we have set $t_0 = 0$). On substituting these values for Q_r, P_r in (21) and equating coefficients of like powers of λ, the coefficients C_{kr}, D_{kr} can then be obtained by quadratures.

* Cf. reference above, p. 381. The page in R.-W. is 211.

APPENDIX A

VECTOR ANALYSIS

In Rational Mechanics only a slight knowledge of Vector Analysis is needed. It is important that this knowledge be based on a postulational treatment of vectors. The system of vectors is a set of elements, forming a logical class. Certain functions of these elements are defined, whereby two elements are transformed into a third element. These functions are called *addition, multiplication by a real number* (here, only one element enters as the independent variable), the *inner product* (scalar multiplication), and the *outer product* (vector multiplication). The functions obey certain functional, or formal, laws, which happen to be a subset of the formal laws of algebra:

$$A + B = B + A$$

$$A + (B + C) = (A + B) + C$$

$$AB = BA$$

$$A(BC) = (AB)C$$

$$A(B + C) = AB + AC$$

$$(B + C)A = BA + CA$$

A brief, systematic treatment such as is here required is given in the Author's *Advanced Calculus,* Chap. XIII. For a first approach to the subject the Hamiltonian notation of S and V for the scalar and vector products has the great advantage of clearness in emphasizing the functional idea — the concept: *transformation.* On the other hand the notation pretty generally adopted at the present day is the designation of vectors by Clarendon or boldface, the scalar product being written as $\mathbf{a} \cdot \mathbf{b}$ or $\mathbf{ab}$ (read: $\mathbf{a}$ dot $\mathbf{b}$), and the vector product as $\mathbf{a} \times \mathbf{b}$ (read: $\mathbf{a}$ cross $\mathbf{b}$). It is useful, therefore, to have a syllabus of definitions and essential formulas in this notation.

1. Vectors and Their Addition. By a *vector* is meant a directed line segment, situated anywhere in space. Vectors will usually be denoted by boldface letters **a**, **A**, or by parentheses; thus a vector angular velocity may be written (ω).

Two vectors, **A** and **B**, are defined as *equal* if they are parallel and have the same sense, and moreover are of equal length:

$$\mathbf{A} = \mathbf{B}.$$

By the *absolute value* of a vector **A** is meant its length; it is denoted by $|\mathbf{A}|$, or by A.

Addition. By the *sum* of two vectors, **A** and **B**, is meant their geometric sum, or the vector **C** obtained by the parallelogram law:

$$\mathbf{A} + \mathbf{B} = \mathbf{C}.$$

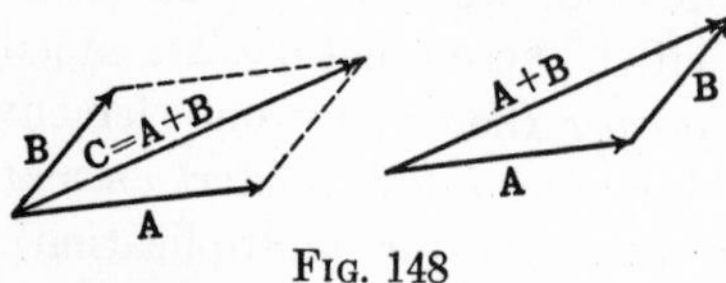

FIG. 148

In order that this definition may apply in all cases, it is necessary to enlarge the system of vectors above defined by a *nul vector*, represented by the symbol 0.

If **B** is parallel to **A** and of the same length, but opposite in sense, then

$$\mathbf{A} + \mathbf{B} = 0, \qquad \text{or} \qquad \mathbf{B} = -\mathbf{A}.$$

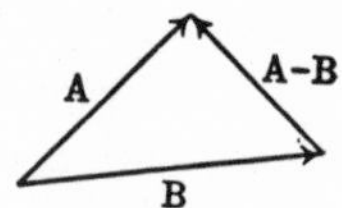

FIG. 149

Moreover, we understand by $m\mathbf{A}$, where m is any real number, a vector parallel to **A** and m times as long; its sense being the same as that of **A**, or opposite, according as m is positive or negative. If $m = 0$, then $m\mathbf{A}$ is a nul vector: $0\mathbf{A} = 0$. The notation $\mathbf{A}m$ means $m\mathbf{A}$, and also

$$\frac{a\mathbf{A} + b\mathbf{B}}{a + b} \qquad \text{means} \qquad \frac{a}{a+b}\mathbf{A} + \frac{b}{a+b}\mathbf{B}.$$

Vector addition obeys the *commutative* and the *associative law* of ordinary algebra:

$$\mathbf{A} + \mathbf{B} = \mathbf{B} + \mathbf{A}$$

$$\mathbf{A} + (\mathbf{B} + \mathbf{C}) = (\mathbf{A} + \mathbf{B}) + \mathbf{C}$$

Subtraction. By $\mathbf{A} - \mathbf{B}$ is meant that vector, **X**, which added to **B** will give **A**:

$$\mathbf{B} + \mathbf{X} = \mathbf{A}, \qquad \mathbf{X} = \mathbf{A} - \mathbf{B}.$$

To obtain **X** geometrically, construct **A** and **B** with the same initial point; then **A** − **B** is the vector whose initial point is the terminal point of **B**, and whose terminal point is the terminal point of **A**; Fig. 149.

Cartesian Representation of a Vector. Let a system of Cartesian axes be chosen, and let **i**, **j**, **k** be three unit vectors lying along these axes. Let **A** be an arbitrary vector, whose components along the axes are A_1, A_2, A_3. Then evidently

$$\mathbf{A} = A_1\,\mathbf{i} + A_2\,\mathbf{j} + A_3\,\mathbf{k}.$$

If

$$\mathbf{B} = B_1\,\mathbf{i} + B_2\,\mathbf{j} + B_3\,\mathbf{k},$$

then

$$\mathbf{A} + \mathbf{B} = (A_1 + B_1)\,\mathbf{i} + (A_2 + B_2)\,\mathbf{j} + (A_3 + B_3)\,\mathbf{k}.$$

Also:

$$|\,\mathbf{A}\,| = \sqrt{A_1^{\,2} + A_2^{\,2} + A_3^{\,2}}\,.$$

Resultant. If n forces, $\mathbf{F}_1$, $\mathbf{F}_2$, $\cdots$, $\mathbf{F}_n$, act at a point, their resultant, **F**, is equal to their vector sum:

$$\mathbf{F} = \mathbf{F}_1 + \mathbf{F}_2 + \cdots + \mathbf{F}_n.$$

If n couples, $\mathbf{M}_1$, $\mathbf{M}_2$, $\cdots$, $\mathbf{M}_n$, act on a body, the resultant couple, **M**, is equal to their vector sum:

$$\mathbf{M} = \mathbf{M}_1 + \mathbf{M}_2 + \cdots + \mathbf{M}_n.$$

Two or more vectors are said to be *collinear* if there is a line in space to which they are all parallel. In particular, a nul vector is said to be collinear with any vector. Three or more vectors are said to be *complanar* if there is a plane in space to which they are all parallel. In particular, a nul vector is said to be parallel to any plane. If three vectors, **A**, **B**, and **C**, are non-complanar, then no one of them can vanish (*i.e.* be a nul vector) and any vector, **X**, can be expressed in the form:

$$\mathbf{X} = l\mathbf{A} + m\mathbf{B} + n\mathbf{C},$$

where l, m, n are uniquely determined.

Differentiation. Velocity. Acceleration. Osculating Plane. A variable vector can be expressed in the form:

$$\mathbf{A} = \mathbf{i}\,f(t) + \mathbf{j}\,\varphi(t) + \mathbf{k}\,\psi(t),$$

where **i**, **j**, **k** are three fixed vectors mutually perpendicular. If $f(t)$, $\varphi(t)$, $\psi(t)$ have derivatives, the vector **A** will have a derivative defined as

$$\lim_{\Delta t = 0} \frac{\Delta \mathbf{A}}{\Delta t} = D_t \mathbf{A}.$$

Its value is:

$$D_t \mathbf{A} = \mathbf{i} f'(t) + \mathbf{j}\varphi'(t) + \mathbf{k}\psi'(t).$$

Moreover,

$$d\mathbf{A} = D_t \mathbf{A}\, dt.$$

If m is a function of x and **A** is a vector depending on x, and if each has a derivative, then $m\mathbf{A}$ will have a derivative, and

$$\frac{d(m\mathbf{A})}{dx} = \frac{dm}{dx}\mathbf{A} + m\frac{d\mathbf{A}}{dx}.$$

If a point P move in any manner in space, its coordinates being given by the equations:

$$x = f(t), \qquad y = \varphi(t), \qquad z = \psi(t),$$

where f, φ, ψ are continuous functions of the time, having continuous derivatives, and if

$$\mathbf{r} = x\mathbf{i} + y\mathbf{j} + z\mathbf{k},$$

the vector velocity of P is represented by

$$\dot{\mathbf{r}} = \frac{d\mathbf{r}}{dt}.$$

If f, φ, ψ have continuous second derivatives, the vector acceleration of P is given by

$$\ddot{\mathbf{r}} = \frac{d^2\mathbf{r}}{dt^2}.$$

The plane determined by the vectors $\ddot{\mathbf{r}}$ and $\dot{\mathbf{r}}$ drawn from P (on the assumption that neither is a nul vector) is the *osculating plane*. Thus the vector acceleration always lies in the osculating plane.

2. The Scalar or Inner Product. The *scalar* or *inner product* of two vectors, **A** and **B**, is defined as the product of their absolute values by the cosine of the angle between them. It is denoted by $\mathbf{A} \cdot \mathbf{B}$ or $\mathbf{AB}$ and is read: "**A** dot **B**." Thus

$$\mathbf{A} \cdot \mathbf{B} = \mathbf{AB} = |\mathbf{A}|\,|\mathbf{B}| \cos \epsilon.$$

If one of the factors is a nul vector, the scalar product is defined as 0.

The *commutative* and the *distributive* laws hold:

$$\mathbf{AB} = \mathbf{BA}$$

$$\mathbf{A}(\mathbf{B} + \mathbf{C}) = \mathbf{AB} + \mathbf{AC}.$$

The *associative* law has no meaning.

The scalar product vanishes when either factor is a nul vector; otherwise when and only when the vectors are perpendicular to each other. Furthermore:

$$\mathbf{A}^2 = |\,\mathbf{A}\,|^2,$$

often called the *norm* of the vector.

$$\mathbf{i}^2 = 1, \qquad \mathbf{j}^2 = 1, \qquad \mathbf{k}^2 = 1,$$

$$\mathbf{jk} = 0, \qquad \mathbf{ki} = 0, \qquad \mathbf{ij} = 0.$$

Cartesian Form of the Scalar Product:

$$\mathbf{AB} = A_1B_1 + A_2B_2 + A_3B_3.$$

$$\cos\epsilon = \frac{\mathbf{AB}}{|\,\mathbf{A}\,|\,|\,\mathbf{B}\,|}.$$

Differentiation:

$$\frac{d}{dx}\mathbf{AB} = \mathbf{A}\frac{d\mathbf{B}}{dx} + \mathbf{B}\frac{d\mathbf{A}}{dx}.$$

If $\mathbf{a}$ is a unit vector, *i.e.* if $|\,\mathbf{a}\,| = 1$, then

$$\mathbf{a}^2 = 1, \qquad \text{and} \qquad \mathbf{aa}' = 0.$$

3. The Vector or Outer Product. Let two vectors, **A** and **B**, be drawn from the same initial point. Then they determine a plane, M, and a parallelogram in that plane. The *vector* or *outer product* is defined as a vector perpendicular to M and of length equal to the area of the parallelogram. Its sense is arbitrary. It is defined with reference to the particular system of Cartesian axes to be used later. It is denoted by

$$\mathbf{A} \times \mathbf{B}$$

and is read: "**A** cross **B**."

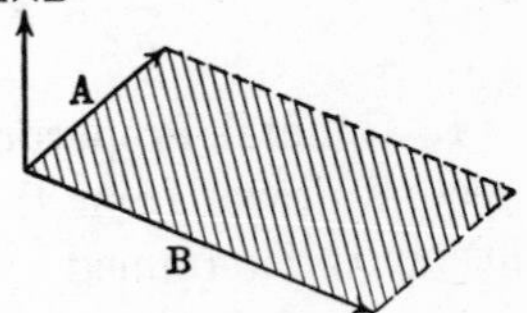

FIG. 150

If one of these vectors is 0, or if the vectors are collinear, neither being 0, the vector product is defined as 0, and these are the only cases in which it is 0. Otherwise, let ϵ be the angle between the vectors. Then

$$|\mathbf{A} \times \mathbf{B}| = |\mathbf{A}|\,|\mathbf{B}| \sin \epsilon.$$

The *commutative* law does not hold in general, for

$$\mathbf{A} \times \mathbf{B} = -\,\mathbf{B} \times \mathbf{A}.$$

The *associative* law does not hold; *e.g.* $(\mathbf{i} \times \mathbf{j}) \times \mathbf{j} \neq \mathbf{i} \times (\mathbf{j} \times \mathbf{j})$. But the *distributive* law is true:

$$\mathbf{A} \times (\mathbf{B} + \mathbf{C}) = \mathbf{A} \times \mathbf{B} + \mathbf{A} \times \mathbf{C}$$

and

$$(\mathbf{B} + \mathbf{C}) \times \mathbf{A} = \mathbf{B} \times \mathbf{A} + \mathbf{C} \times \mathbf{A},$$

as can be proved geometrically, or still more simply, analytically, by means of the Cartesian form; cf. infra.

It is convenient to choose the sense of the vector product so that

$$\mathbf{i} \times \mathbf{j} = \mathbf{k}, \qquad \mathbf{j} \times \mathbf{k} = \mathbf{i}, \qquad \mathbf{k} \times \mathbf{i} = \mathbf{j}.$$

In any case

$$\mathbf{A} \times \mathbf{A} = 0,$$

and so, in particular,

$$\mathbf{i} \times \mathbf{i} = 0, \qquad \mathbf{j} \times \mathbf{j} = 0, \qquad \mathbf{k} \times \mathbf{k} = 0.$$

Cartesian Form of the Vector Product:

$$\mathbf{A} \times \mathbf{B} =$$

$$(A_2B_3 - A_3B_2)\,\mathbf{i} + (A_3B_1 - A_1B_3)\,\mathbf{j} + (A_1B_2 - A_2B_1)\,\mathbf{k}$$

$$= \begin{vmatrix} \mathbf{i} & \mathbf{j} & \mathbf{k} \\ A_1 & A_2 & A_3 \\ B_1 & B_2 & B_3 \end{vmatrix}$$

Differentiation:

$$\frac{d}{dx}\mathbf{A} \times \mathbf{B} = \mathbf{A} \times \frac{d\mathbf{B}}{dx} - \mathbf{B} \times \frac{d\mathbf{A}}{dx}.$$

4. General Properties. Let **A**, **B**, **C** be three non-complanar vectors drawn from the same point. The volume of the parallelepiped determined by these vectors is numerically

$$\mathbf{A} \cdot (\mathbf{B} \times \mathbf{C}).$$

A necessary and sufficient condition that three vectors **A**, **B**, **C** be complanar is:

$$\mathbf{A} \cdot (\mathbf{B} \times \mathbf{C}) = 0.$$

Linear Velocity in Terms of Angular Velocity. Let space be rotating about an axis I with vector angular velocity (ω). Then the velocity $\mathbf{v}$ of an arbitrary point P will be:

$$\mathbf{v} = (\omega) \times \mathbf{r},$$

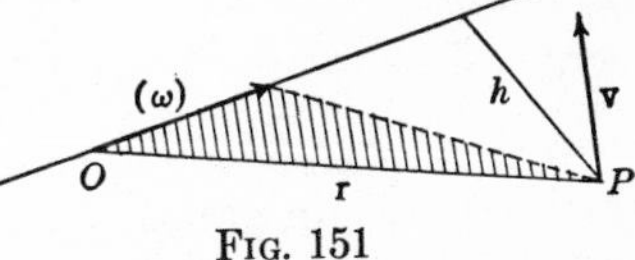

FIG. 151

where $\mathbf{r}$ is the vector drawn from any point O of the axis to the point P. If the axis passes through the origin, then

$$\mathbf{v} = \begin{vmatrix} \mathbf{i} & \mathbf{j} & \mathbf{k} \\ \omega_x & \omega_y & \omega_z \\ x & y & z \end{vmatrix}$$

and

$$\begin{cases} v_x = z\omega_y - y\omega_z \\ v_y = x\omega_z - z\omega_x \\ v_z = y\omega_x - x\omega_y \end{cases}$$

If it passes through the point (a, b, c), then

$$\begin{cases} v_x = (z - c)\,\omega_y - (y - b)\,\omega_z \\ v_y = (x - a)\,\omega_z - (z - c)\,\omega_x \\ v_z = (y - b)\,\omega_x - (x - a)\,\omega_y \end{cases}$$

In the general case of motion of a rigid body (*i.e.* motion of rigid space), let $O' : (x_0, y_0, z_0)$ be a point fixed in the body, and let (ξ, η, ζ) be the coordinates of any point P fixed in the body, the origin being at O'; but otherwise the (ξ, η, ζ)-axes may move in any manner. Then

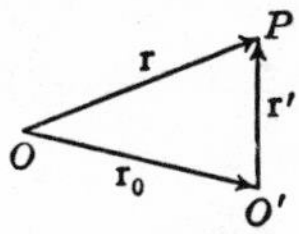

$$\mathbf{v} = \mathbf{v}_0 + \mathbf{v}',$$

where

$$\mathbf{v}_0 = \dot{x}_0\mathbf{i} + \dot{y}_0\mathbf{j} + \dot{z}_0\mathbf{k},$$

FIG. 152

$$\mathbf{v}' = \begin{vmatrix} \alpha & \beta & \gamma \\ \omega_\xi & \omega_\eta & \omega_\zeta \\ \xi & \eta & \zeta \end{vmatrix}, \qquad \begin{cases} v'_\xi = \zeta\omega_\eta - \eta\omega_\zeta \\ v'_\eta = \xi\omega_\zeta - \zeta\omega_\xi \\ v'_\zeta = \eta\omega_\xi - \xi\omega_\eta \end{cases}$$

Localized Vectors. It is sometimes convenient to prescribe the initial point of a vector, or the line in which the vector shall lie, as in the case of a force acting on a particle, or a force acting on a rigid body. It is with reference to such vectors that the following definitions are framed.

By the *moment of a vector* **F** *with respect to a point* O is meant the vector

$$\mathbf{M} = \mathbf{r} \times \mathbf{F},$$

where **r** is the vector drawn from O to any point of the line in which **F** lies. In practice, **F** may be a force acting on a rigid body, or **F** may be the vector momentum, $m\mathbf{v}$, of a particle.

The *moment of a couple* can be expressed as

$$\mathbf{r}_1 \times \mathbf{F}_1 + \mathbf{r}_2 \times \mathbf{F}_2,$$

where $\mathbf{F}_1$, $\mathbf{F}_2$ are the forces of the couple and $\mathbf{r}_1$, $\mathbf{r}_2$ are vectors drawn from any point O of space to any points P_1, P_2 of the lines of action of $\mathbf{F}_1$, $\mathbf{F}_2$, respectively.

By the *moment of a vector* **F** *about a directed line* L is meant the vector

$$\mathbf{M} = M\mathbf{a}, \qquad M = \mathbf{a} \cdot (\mathbf{r} \times \mathbf{F}),$$

where **a** is a unit vector having the direction and sense of L, and **r** is the vector drawn from any point of L to any point of the line in which **F** lies. Thus if **F** is a force acting on a rigid body, let its point of application be transferred to the point P nearest to L, and let O be the point of L nearest to **F**; *i.e.* OP is the common perpendicular of L and the line of action of **F**. Decompose **F** at P into a force parallel to L and one perpendicular to L. The vector moment of the latter component at P with respect to O is $M\mathbf{a}$.

5. Rotation of the Axes. Direction Cosines. A transformation from one set of Cartesian axes to a second having the same origin (both systems being right-handed, or both left-handed) is characterized by the scheme of direction cosines:

	ξ	η	ζ
x	l_1	l_2	l_3
y	m_1	m_2	m_3
z	n_1	n_2	n_3

	α	β	γ
i	l_1	l_2	l_3
j	m_1	m_2	m_3
k	n_1	n_2	n_3

Between the nine direction cosines there exist the following relations:

$$\begin{cases} l_1^2 + l_2^2 + l_3^2 = 1 \\ m_1^2 + m_2^2 + m_3^2 = 1 \\ n_1^2 + n_2^2 + n_3^2 = 1 \end{cases} \qquad \begin{cases} l_1^2 + m_1^2 + n_1^2 = 1 \\ l_2^2 + m_2^2 + n_2^2 = 1 \\ l_3^2 + m_3^2 + n_3^2 = 1 \end{cases}$$

$$\begin{cases} l_2 l_3 + m_2 m_3 + n_2 n_3 = 0 \\ l_3 l_1 + m_3 m_1 + n_3 n_1 = 0 \\ l_1 l_2 + m_1 m_2 + n_1 n_2 = 0 \end{cases} \qquad \begin{cases} m_1 n_1 + m_2 n_2 + m_3 n_3 = 0 \\ n_1 l_1 + n_2 l_2 + n_3 l_3 = 0 \\ l_1 m_1 + l_2 m_2 + l_3 m_3 = 0 \end{cases}$$

$$\begin{cases} l_1 = m_2 n_3 - m_3 n_2 \\ l_2 = m_3 n_1 - m_1 n_3 \\ l_3 = m_1 n_2 - m_2 n_1 \end{cases} \quad \begin{cases} m_1 = n_2 l_3 - n_3 l_2 \\ m_2 = n_3 l_1 - n_1 l_3 \\ m_3 = n_1 l_2 - n_2 l_1 \end{cases} \quad \begin{cases} n_1 = l_2 m_3 - l_3 m_2 \\ n_2 = l_3 m_1 - l_1 m_3 \\ n_3 = l_1 m_2 - l_2 m_1 \end{cases}$$

$$\begin{cases} l_1 = m_2 n_3 - m_3 n_2 \\ m_1 = n_2 l_3 - n_3 l_2 \\ n_1 = l_2 m_3 - l_3 m_2 \end{cases} \quad \begin{cases} l_2 = m_3 n_1 - m_1 n_3 \\ m_2 = n_3 l_1 - n_1 l_3 \\ n_2 = l_3 m_1 - l_1 m_3 \end{cases} \quad \begin{cases} l_3 = m_1 n_2 - m_2 n_1 \\ m_3 = n_1 l_2 - n_2 l_1 \\ n_3 = l_1 m_2 - l_2 m_1 \end{cases}$$

$$\begin{vmatrix} l_1 & l_2 & l_3 \\ m_1 & m_2 & m_3 \\ n_1 & n_2 & n_3 \end{vmatrix} = 1.$$

APPENDIX B

THE DIFFERENTIAL EQUATION: $\left(\frac{du}{dt}\right)^2 = f(u)$

Differential equations of the form:

$$\left(\frac{du}{dt}\right)^2 = f(u), \tag{1}$$

where

I. $$f(u) = (u - a)(b - u)\,\psi(u)$$

or

II. $$f(u) = (u - a)(b - u)^2\psi(u)$$

and $\psi(u)$ is continuous and positive in the interval

$$a \leqq u \leqq b,$$

play an important rôle in Mechanics. Let us study their integrals.

Case I. A particular integral of (1) is found by extracting the square root:

$$\frac{du}{dt} = \sqrt{f(u)},$$

and separating the variables:

$$dt = \frac{du}{\sqrt{f(u)}},$$

$$t = \int\limits_a^u \frac{du}{\sqrt{f(u)}}, \qquad a \leqq u \leqq b. \tag{2}$$

Geometrically, the function on the right of (2) can be interpreted as the area under the curve,

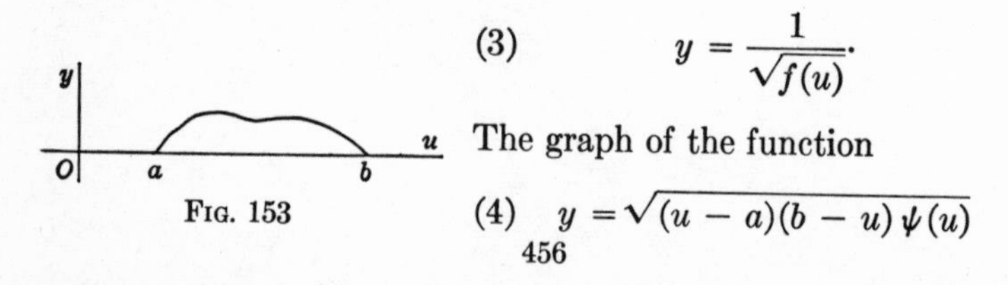

Fig. 153

$$y = \frac{1}{\sqrt{f(u)}}. \tag{3}$$

The graph of the function

$$y = \sqrt{(u - a)(b - u)\,\psi(u)} \tag{4}$$

is represented by Fig. 153. The reciprocal of an ordinate of this curve gives the corresponding ordinate of the graph of the function (3), Fig. 154:

$$(5) \qquad y = \frac{1}{\sqrt{(u-a)(b-u)\psi(u)}}.$$

The area under the curve (5), shaded in the figure, represents the integral (2), or:

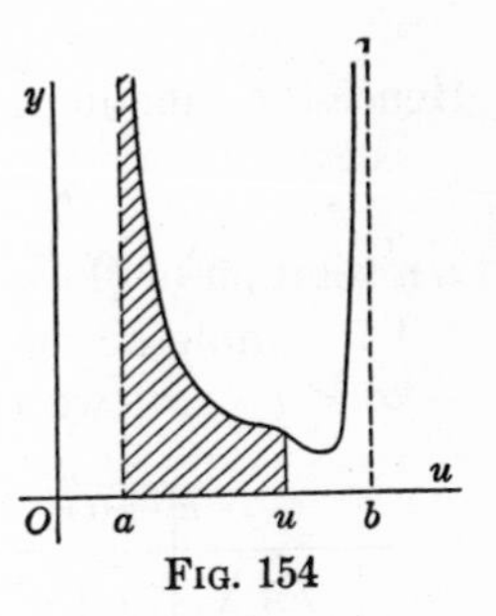

FIG. 154

$$(6) \qquad t = \int_a^u \frac{du}{\sqrt{(u-a)(b-u)\psi(u)}}.$$

Thus this area expresses t and brings out the fact that t increases as u increases. Conversely, u increases as t increases. Let A be defined by the equation:

$$(7) \qquad A = \int_a^b \frac{du}{\sqrt{(u-a)(b-u)\psi(u)}}.$$

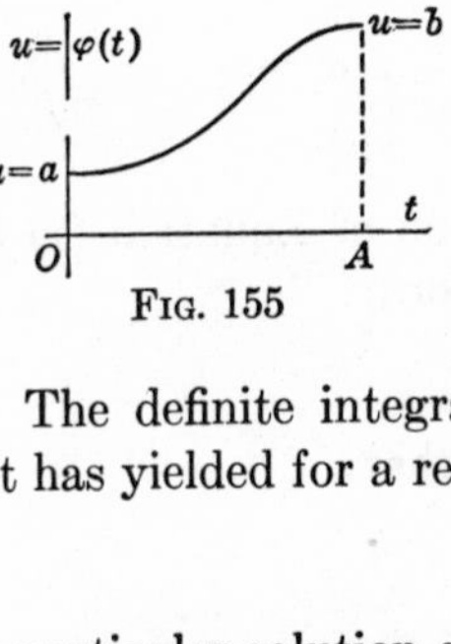

FIG. 155

Then the graph of u, regarded as a function of t,

$$(8) \qquad u = \varphi(t), \qquad 0 \leqq t \leqq A,$$

is as shown in Fig. 155. Its slope is 0 at each extremity and positive in between.

The definite integral, (2) or (6), has now served its purpose. It has yielded for a restricted interval,

$$0 \leqq t \leqq A,$$

a particular solution of (1).

Continuation by Reflection. Reflect the graph of the function (8), Fig. 155, in the axis of ordinates, and let the curve thus obtained define a continuation of the function $\varphi(t)$ throughout the interval $-A \leqq t \leqq 0$. Analytically the reflection is represented by the transformation:

$$t' = -t, \qquad u' = u.$$

Thus

$$\varphi(t') = \varphi(-t), \qquad \begin{cases} 0 \leqq t \leqq A \\ -A \leqq t' \leqq 0 \end{cases}$$

The extended function:

$$u' = \varphi(t'),$$

is seen to satisfy the differential equation

$$\left(\frac{du'}{dt'}\right)^2 = f(u').$$

Hence the function $\varphi(t)$ thus defined in the interval $(-A, A)$, or

$$u = \varphi(t), \qquad -A \leqq t \leqq A,$$

is a solution of (1).

To complete the definition of $\varphi(t)$ for all values of t, *i.e.* $-\infty < t < \infty$, we could repeat the process of reflection, using next the lines $t = A$ and $t = -A$; and so on. But it is simpler to introduce the idea of periodicity.

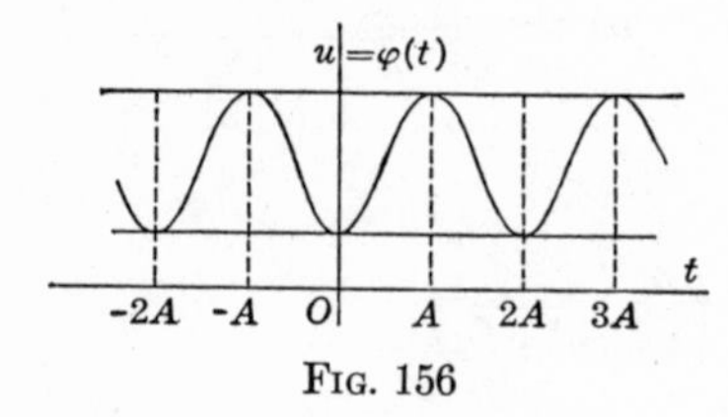

FIG. 156

Periodicity. Let the function $\varphi(t)$ now be extended to all values of t by the requirement of periodicity:

$$\varphi(t + 2A) = \varphi(t), \qquad -\infty < t < \infty. \tag{9}$$

Then we have *one* solution of Equation (1).

The General Solution. The general solution of Equation (1) in the present case can now be written in the form:

$$u = \varphi(t + \gamma), \tag{10}$$

where γ is an arbitrary constant. Observe that

$$\varphi(-t) = \varphi(t). \tag{11}$$

Hence

$$\varphi'(-t) = -\varphi(t). \tag{12}$$

To an arbitrary value u_0 of u such that $a < u_0 < b$ there correspond two and only two values of t in the interval $(-A, A)$, for which

$$u_0 = \varphi(t), \tag{13}$$

namely

$$0 < t_0 < A, \qquad -A < t_0' < 0, \qquad t_0' = -t_0.$$

If $u_0 = a$, there is only one value, namely, $t = 0$; and if $u_0 = b$, then $t = A, -A$. But only one should be counted, since the fundamental interval of periodicity should be taken as an open interval,

$$c < t \leqq c + 2A \qquad \text{or} \qquad c \leqq t < c + 2A,$$

where c is arbitrary. Moreover, du/dt has opposite signs in t_0 and t_0', because of (12).

We can now prove that there is a solution of the given differential equation, which corresponds to arbitrary initial conditions: $u = u_1$, $t = t_1$, provided merely that

$$a \leqq u_1 \leqq b.$$

Suppose that it is known from the physics of the problem that du/dt is negative initially. Now, set $u_0 = u_1$ and determine t_0 as above so that

$$u_0 = \varphi(t_0), \qquad \varphi'(t_0) < 0.$$

Finally, define γ by the equation:

$$t_1 + \gamma = t_0, \qquad \gamma_1 = t_0 - t_1.$$

Thus $\gamma = \gamma_1$ is uniquely determined and the function

$$u = \varphi(t + \gamma_1) \tag{14}$$

is the solution we set out to obtain.

But is this solution unique, or are there still other solutions which satisfy the same initial conditions? If $a < u_1 < b$, the answer is affirmative for values of t near t_1; but for remote values, the question of *singular solutions* arises, to which we now turn.

Singular Solutions. The given differential equation admits, furthermore, singular solutions. The functions

$$u = a, \qquad u = b$$

are obviously solutions of the differential equation:

$$\left(\frac{du}{dt}\right)^2 = (u - a)(b - u)\psi(u), \tag{15}$$

each being considered in any interval for t, finite or infinite. Such a solution, moreover, may be combined with a solution (10) at any point. The solution now may follow (10) indefinitely; or it may switch off on a singular solution again.

These solutions do not, however, have any validity in the problems of mechanics, for which the above study has been made. The mechanical problems depend each time on differential equations of the *second* order, and these have unique solutions, depending on the initial or boundary conditions. Equation (14) represents an integral of these equations. But the converse is not true, namely, that every integral of (15) is an integral of the second order equations — why should it be? We see, then, that we may be on dangerous ground when we replace the latter equations, in part, by the integral of energy, for example; since the modified system may have solutions other than that of the given mechanical problem. Cf. the Author's *Advanced Calculus*, p. 349.

Does this remark not call into question the validity of the treatment in Chap. XV, since the equation:

$$H(q_1, \cdots, q_n, p_1, \cdots, p_n) = h,$$

is essentially the integral of energy? Not if we apply that method as set forth in the text. For *in a suitably restricted region* there is *only one solution* yielded by those methods, and we were careful to point out that it is the *analytical continuation* of this solution that yields the solution of the mechanical problem beyond this region. Thus the singular solutions are automatically eliminated.

CASE II. This case:

$$\left(\frac{du}{dt}\right)^2 = (u-a)(b-u)^2\psi(u), \tag{16}$$

is more easily dealt with. A particular integral of (16) is given by the formula

$$t = \int_a^u \frac{du}{(b-u)\sqrt{(u-a)\,\psi(u)}}, \qquad a \leqq u < b \tag{17}$$

The inverse function,

$$u = \varphi(t), \qquad 0 \leqq t < \infty, \tag{18}$$

represents an integral of (16) in the interval indicated. And now this solution can be completed by the definition:

$$\varphi(-t) = \varphi(t). \tag{19}$$

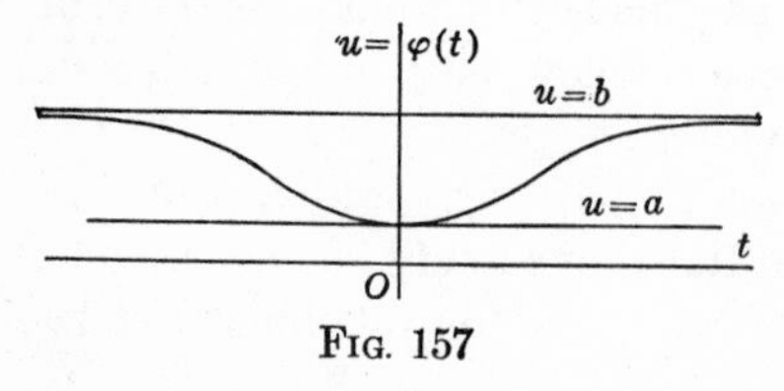

FIG. 157

Thus we have *one* solution:

$$u = \varphi(t), \qquad -\infty < t < \infty. \tag{20}$$

It is now shown as before that the general solution is

$$u = \varphi(t + \gamma). \tag{21}$$

A further case, namely:

$$\left(\frac{du}{dt}\right)^2 = (u - a)^2(b - u)\,\psi(u), \tag{22}$$

can be treated in a similar manner; or, more simply, be thrown back on the case just considered by a linear transformation.

Finally, the case (not mentioned above):

$$\left(\frac{du}{dt}\right)^2 = (u - a)^2(b - u)^2\,\psi(u), \tag{23}$$

breaks up into the two distinct equations:

a) $$\frac{du}{dt} = +\,(u - a)(b - u)\sqrt{\psi(u)};$$

b) $$\frac{du}{dt} = -\,(u - a)(b - u)\sqrt{\psi(u)}.$$

Each of these is solved at once by a quadrature.

Further Study of Case I. There is another treatment of Case I which brings out the important fact that the function $\varphi(t)$ is essentially a sine or cosine function:

$$u = C\cos\theta + C', \tag{24}$$

where θ, in the simplest case, is proportional to the time:

$$\theta = \frac{\pi}{A}t,$$

and in the general case is of the form:

$$\theta = \frac{\pi}{A}t + h(t),$$

where $h(t)$ is periodic with the period $2A$:

$$h(t + 2A) = h(t).$$

This method, moreover, may simplify the computation in case it is desired to tabulate the function $\varphi(t)$.

The given differential equation:

$$\left(\frac{du}{dt}\right)^2 = (u - a)(b - u)\,\psi(u), \tag{25}$$

can be reduced by a linear transformation:

$$u' = \frac{2u - a - b}{b - a},$$

to the form, after dropping the accent:

$$\left(\frac{du}{dt}\right)^2 = (1 - u^2)\,\psi(u), \qquad -1 \leqq u \leqq 1. \tag{26}$$

Make the substitution:

$$u = \cos\theta, \qquad 0 < \theta < \pi. \tag{27}$$

Equation (26) becomes, on suppressing the factor* $\sin^2\theta$:

$$\left(\frac{d\theta}{dt}\right)^2 = \psi(\cos\theta). \tag{28}$$

This equation is equivalent to the two equations:

$$\frac{d\theta}{dt} = \sqrt{\psi(\cos\theta)}\,; \tag{29}$$

$$\frac{d\theta}{dt} = -\sqrt{\psi(\cos\theta)}. \tag{30}$$

The solution of (30) is obtained from the solution of (29) by changing the sign of t.

A particular solution of (29) is given by the quadrature:

$$t = \int_0^\theta \frac{d\theta}{\sqrt{\psi(\cos\theta)}}, \qquad -\infty < \theta < \infty. \tag{31}$$

Write

$$\int_0^\theta \frac{d\theta}{\sqrt{\psi(\cos\theta)}} = \frac{A\theta}{\pi} + g(\theta), \tag{32}$$

where

$$\int_{-\pi}^{\pi} \frac{d\theta}{\sqrt{\psi(\cos\theta)}} = 2A. \tag{33}$$

Then $g(\theta)$ is periodic with the period 2π. For,

$$\int_\theta^{\theta+2\pi} \frac{d\theta}{\sqrt{\psi(\cos\theta)}} = \left[\frac{A(\theta + 2\pi)}{\pi} + g(\theta + 2\pi)\right] - \left[\frac{A\theta}{\pi} + g(\theta)\right] \tag{34}$$

$$= 2A + g(\theta + 2\pi) - g(\theta).$$

* In so doing we suppress the singular solutions of (26).

But the value of the integral, because of the periodicity of the integrand, is $2A$ for all values of θ. Hence

$$g(\theta + 2\pi) = g(\theta). \tag{35}$$

Equation (31), or its equivalent,

$$t = \frac{A\theta}{\pi} + g(\theta) \tag{36}$$

defines θ as a single-valued function of t, since the integral (31) represents a monotonic function of θ. Let θ be written in the form:

$$\theta = \frac{\pi t}{A} + h(t). \tag{37}$$

Then $h(t)$ has the period $2A$:

$$h(t + 2A) = h(t). \tag{38}$$

For, let θ have an arbitrary value θ_0 in (36) and let the corresponding value of t be t_0:

$$t_0 = \frac{A\theta_0}{\pi} + g(\theta_0).$$

Let $\theta = \theta_0 + 2\pi$, and let t' be the new value of t:

$$t' = \frac{A(\theta_0 + 2\pi)}{\pi} + g(\theta_0 + 2\pi).$$

By virtue of (35),

$$t' = \frac{A\theta_0}{\pi} + g(\theta_0) + 2A,$$

or

$$t' = t_0 + 2A.$$

From (37) we now infer:

$$\theta_0 = \frac{\pi t_0}{A} + h(t_0);$$

$$\theta_0 + 2\pi = \frac{\pi(t_0 + 2A)}{A} + h(t_0 + 2A).$$

Hence

$$h(t_0 + 2A) = h(t_0),$$

and the proof is complete.

If we multiply (36) by π and (37) by A and add, we find:

$$\pi g(\theta) + A\,h(t) = 0. \tag{39}$$

We are now ready to express u in terms of t. In Equation (27) θ was restricted. Now, set generally:

$$u = \cos\left[\frac{\pi t}{A} + h(t)\right]. \tag{40}$$

This function is seen by direct substitution to be a solution of (26). If we denote it by $\varphi(t)$, the general solution of (26) will be:

$$u = \varphi(t + \gamma). \tag{41}$$

The other equation, (30), leads to the same result.

If it is a question actually of computing $h(t)$, then the integral (31) can be tabulated for values of θ from 0 to π, the reckoning being performed by the ordinary methods for evaluating definite integrals — Simpson's Rule, etc.

Integral of a Periodic Function. Let $f(x)$ be a continuous periodic function:

$$f(x + A) = f(x), \qquad -\infty < x < \infty,$$

where A is a primitive period, corresponding to $2A$ above. Let

$$C = \int_0^A f(x)\,dx.$$

Then

$$\int_x^{x+A} f(x)\,dx = C,$$

where x is arbitrary. For,

$$\frac{d}{dx}\int_x^{x+A} f(x)\,dx = f(x + A) - f(x) = 0.$$

Let

$$\varphi(x) = \int_c^x f(x)\,dx - \frac{C}{A}x.$$

Then $\varphi(x)$ is periodic:

$$\varphi(x + A) = \varphi(x).$$

For,

$$\varphi(x + A) - \varphi(x) = \int_x^{x+A} f(x)\,dx - \frac{C}{A}(x + A) + \frac{C}{A}x = 0.$$

Hence

$$\int_c^x f(x)\,dx = \varphi(x) + \lambda x,$$

where $\lambda = C/A$. The result may be stated as follows.

THEOREM. *The integral of a periodic function is the sum of a periodic function and a linear function:*

$$\int_c^x f(x)\,dx = \varphi(x) + \lambda x,$$

where

$$f(x + A) = f(x), \qquad \varphi(x + A) = \varphi(x),$$

and

$$C = \int_0^A f(x)\,dx, \qquad \lambda = \frac{C}{A}.$$

Instead of the linear function λx we may write

$$\lambda x + \gamma \qquad \text{or} \qquad \lambda(x - x_0),$$

the function $\varphi(x)$ being changed by an additive constant. In particular, $\lambda = 0$ if and only if

$$\int_0^A f(x)\,dx = 0.$$

APPENDIX C

CHARACTERISTICS OF JACOBI'S EQUATION

Although Jacobi's partial differential equation of the first order:

$$\text{A)}\qquad \frac{\partial V}{\partial t} + H\left(q_1, \cdots, q_m, \frac{\partial V}{\partial q_1}, \cdots, \frac{\partial V}{\partial q_m}, t\right) = 0,$$

has played an important rôle in the solution of Hamilton's Equations:

$$\text{B)}\qquad \frac{dq_r}{dt} = \frac{\partial H}{\partial p_r}, \qquad \frac{dp_r}{dt} = -\frac{\partial H}{\partial q_r}, \qquad r = 1, \cdots, m,$$

where $H = H(q_1, \cdots, q_m, p_1, \cdots, p_m, t)$, we have not found it necessary to refer to the theory of characteristics, partly because we have sought certain explicit solutions by means of ingenious devices (separation of variables, for example); partly because, when we have needed an existence theorem, it was supplied at once by reference to the Cauchy Problem. Nevertheless it is of interest for completeness to connect the equation with its characteristics.

1. The Analytic Theorem. Consider the general partial differential equation of the first order:

$$\text{I.}\qquad F\left(x_1, \cdots, x_n, z, \frac{\partial z}{\partial x_1}, \cdots, \frac{\partial z}{\partial x_n}\right) = 0.$$

Let $F(x_1, \cdots, x_n, z, y_1, \cdots, y_n)$, together with its partial derivatives of the first two orders, be continuous for those values of the arguments for which $(x_1, \cdots, x_n, z)$ is an interior point of an $(n + 1)$-dimensional region* R of the space of the variables $(x_1, \cdots, x_n, z)$, and the y_k are wholly unrestricted. Use the notation:

$$(1)\qquad X_k = \frac{\partial F}{\partial x_k}, \qquad Z = \frac{\partial F}{\partial z}, \qquad Y_k = \frac{\partial F}{\partial y_k}.$$

At a given point $A: (a_1, \cdots, a_n, c, b_1, \cdots, b_n) = (a, c, b)$ of R let the Y_k not all vanish; in particular, let $Y_n \neq 0$.

* R shall not include any of its boundary points.

The characteristic strips are defined by the system of $2n$ ordinary differential equations:

$$\text{II.} \qquad \frac{dx_k}{Y_k} = \frac{dz}{\Sigma\, y_k Y_k} = \frac{-\,dy_k}{X_k + y_k Z}, \qquad k = 1, \cdots, n.$$

The solution of II. shall go through the point (x^0, z^0, y^0), which shall lie in the neighborhood of A and moreover on the manifold $F = 0$, or

$$F(x_1^0, \cdots, x_n^0, z^0, y_1^0, \cdots, y_n^0) = 0.$$

Although there are $2n + 1$ initial values — the (x^0, z^0, y^0) — there is only a $2n$-parameter family of solutions of II., for we may without loss of generality set $x_n^0 = a_n$ once for all. The solution of II. can now be written in the form:

$$(2) \quad \begin{cases} x_i = \quad f_i(x_n;\ x_1^0, \cdots, x_{n-1}^0, z^0, y_1^0, \cdots, y_n^0), \\ \qquad\qquad\qquad\qquad\qquad\qquad i = 1, \cdots, n-1; \\ z = \quad f(x_n;\ x_1^0, \cdots, x_{n-1}^0, z^0, y_1^0, \cdots, y_n^0); \\ y_k = f_{n+k}(x_n;\ x_1^0, \cdots, x_{n-1}^0, z^0, y_1^0, \cdots, y_n^0), \\ \qquad\qquad\qquad\qquad\qquad\qquad k = 1, \cdots\cdots, n. \end{cases}$$

Along any curve (2) the function $F(x_1, \cdots, x_n, z, y_1, \cdots, y_n)$ is constant, since

$$dF = \sum_k X_k\, dx_k + Z\, dz + \sum_k Y_k\, dy_k.$$

On subjecting dx_k, dz, dy_k to the conditions imposed by II. it appears that $dF = 0$. Hence

$$(3) \qquad F(x_1, \cdots, x_n, z, y_1, \cdots, y_n) = C$$

is an integral of the system of differential equations II.

Characteristic strips are curves (2) for which

$$(4) \qquad F(x_1^0, \cdots, x_{n-1}^0, a_n, z^0, y_1^0, \cdots, y_n^0) = 0,$$

i.e. $C = 0$ in (3). This equation can be solved for y_n^0 since $Y_n \neq 0$:

$$(4') \qquad y_n^0 = \chi(x_1^0, \cdots, x_{n-1}^0, z^0, y_1^0, \cdots, y_{n-1}^0).$$

Thus there is a $(2n - 1)$-parameter family of characteristic strips.

Consider now the $(n + 1)$-dimensional space of the variables $(x_1, \cdots, x_n, z)$, in which a solution:

$$(5) \qquad z = \Psi(x_1, \cdots, x_n),$$

of the partial differential equation I. will lie. In the (hyper-) plane $x_n = a_n$ of this space let a manifold be defined by the equation:

$$(6) \qquad z^0 = \omega(x_1^0, \cdots, x_{n-1}^0),$$

where $\omega(x_1, \cdots, x_{n-1})$, together with its first partial derivatives, is continuous in the neighborhood of the point $(a_1, \cdots, a_{n-1})$, and

$$c = \omega(a_1, \cdots, a_{n-1}).$$

Furthermore, let

$$(7) \qquad y_i^0 = \frac{\partial \omega}{\partial x_i^0}, \qquad i = 1, \cdots, n-1;$$

and let y_n^0 be given by (4) or (4′). If, now, we regard the

$$x_1^0 = u_1, \cdots, x_{n-1}^0 = u_{n-1}$$

as $n - 1$ independent parameters and for symmetry in notation set

$$x_n = u_n,$$

the first n equations (2), combined with this last equation, will represent a (hyper-) surface parametrically, the equation of which can be thrown into the form (5) by eliminating the $(u_1, \cdots, u_n)$, and this function (5) is a solution of the given partial differential equation I. Moreover it is the most general solution; *i.e.* any solution (5), such that Ψ satisfies the above requirements of continuity, can be obtained in this manner.*

This is the general theorem of the solution of I. by means of characteristics. We proceed to apply the result to Jacobi's Equation A).

2. Jacobi's Equation. Let

$$(8) \qquad \begin{cases} x_r = q_r, & y_r = p_r, \qquad r = 1, \cdots, m = n-1; \\ z = V, & x_n = t. \end{cases}$$

As regards y_n, we see from I. and A) that it is given by the equation:

$$(9) \qquad F \equiv H(x_1, \cdots, x_m, *, y_1, \cdots, y_m, x_n) + y_n = 0.$$

Equations II. now take the form:

* For the proof cf. Goursat-Hedrick, *Mathematical Analysis*, or the *Advanced Calculus*, Chap. XIV, p. 366.

$$\text{(10)} \qquad \frac{dq_r}{\frac{\partial H}{\partial p_r}} = \frac{dt}{1} = \frac{dV}{\sum_r p_r \frac{\partial H}{\partial p_r} + y_n} = \frac{-\,dp_r}{\frac{\partial H}{\partial q_r}} = \frac{-\,dy_n}{\frac{\partial H}{\partial t}}.$$

The initial values are:

$$\text{(11)} \qquad \begin{cases} x_r^0 = q_r^0, & y_r^0 = p_r^0, \qquad r = 1, \cdots, m; \\ x_n^0 = t_0, & y_n^0 = -\,H^0; \qquad V_0. \end{cases}$$

From (10) follow first Hamilton's Equations:

$$\text{(12)} \qquad \frac{dq_r}{dt} = \frac{\partial H}{\partial p_r}, \qquad \frac{dp_r}{dt} = -\frac{\partial H}{\partial q_r}, \qquad r = 1, \cdots, m.$$

Furthermore, by the aid of (9),

$$\text{(13)} \qquad \frac{dH}{dt} = \frac{\partial H}{\partial t};$$

and finally, since from (12)

$$\frac{\partial H}{\partial p_r} = \dot{q}_r,$$

$$\text{(14)} \qquad \frac{dV}{dt} = \sum_r p_r \dot{q}_r - H.$$

Observe in passing that the right-hand side of (14) is the Lagrangean Function, L:

$$H + L = \sum_r p_r \dot{q}_r,$$

and so

$$\text{(15)} \qquad \frac{dV}{dt} = L.$$

But we have anticipated the results of the general theory and although obtaining the facts of the case in Equations (12), (13), and (14), we have not brought out the direct testimony of the general theory in the present case. Let us turn back, then, to Equations (2) and Condition (4) or (4′). It appears that the solution of Equations (10) takes the form:

$$\text{(16)} \qquad \begin{cases} q_r = f_r(t;\, q_1^0, \cdots, q_m^0, V_0, p_1^0, \cdots, p_m^0, y_n^0) \\ p_r = f_{n+r}(t;\, q_1^0, \cdots, q_m^0, V_0, p_1^0, \cdots, p_m^0, y_n^0) \\ V = f(t;\, q_1^0, \cdots, q_m^0, V_0, p_1^0, \cdots, p_m^0, y_n^0) \\ y_n = f_{2n}(t;\, q_1^0, \cdots, q_m^0, V_0, p_1^0, \cdots, p_m^0, y_n^0) \end{cases}$$

where $r = 1, \cdots, m = n - 1$ and

$$y_n^0 = -H(q_1^0, \cdots, q_m^0, p_1^0, \cdots, p_m^0, t_0) \tag{17}$$

In the case before us the functions f_r, f_{n+r}, $r = 1, \cdots, m$, arising as they do from the solution of Equations (12), do not depend on V_0, and y_n^0 is given by (17). Thus we have

$$\begin{cases} q_r = \varphi_r(t;\, q_1^0, \cdots, q_m^0, p_1^0, \cdots, p_m^0) \\ p_r = \varphi_{n+r}(t;\, q_1^0, \cdots, q_m^0, p_1^0, \cdots, p_m^0) \end{cases} \tag{18}$$

and also the further integral of (10), given by (9):

$$y_n = -H(q_1, \cdots, q_m, p_1, \cdots, p_m, t). \tag{19}$$

But V in (16) does depend on V_0. It is given by (14):

$$V = \int_{t_0}^{t} \left(\sum_r p_r \dot{q}_r - H \right) dt + V_0, \tag{20}$$

or

$$V = \int_{t_0}^{t} L\, dt + V_0.$$

So much, then, for the discussion of the solution of Equations II., *i.e.* (10). As regards now the solution of Equation I., *i.e.* A):

$$\frac{\partial V}{\partial t} + H\left(q_1, \cdots, q_m, \frac{\partial V}{\partial q_1}, \cdots, \frac{\partial V}{\partial q_m}, t\right) = 0, \tag{21}$$

we choose ω subject to the conditions under (6):

$$V_0 = \omega(q_1^0, \cdots, q_m^0) \tag{22}$$

and set

$$p_r^0 = \frac{\partial \omega}{\partial q_r^0}, \qquad r = 1, \cdots, m.$$

3. Application. We have seen in Chap. XV, § 2, that Hamilton's Equations can be solved by a contact transformation:

$$p_r = \frac{\partial S}{\partial q_r}, \qquad P_r = -\frac{\partial S}{\partial Q_r}, \qquad r = 1, \cdots, m, \tag{23}$$

which transforms the given dynamical problem into the Equilibrium Problem, the solution of which is

$$Q_r = \alpha_r, \qquad P_r = \beta_r, \qquad r = 1, \cdots, m,$$

where α_r, β_r are arbitrary constants, wholly unrestricted so far as the transformed Hamiltonian Equations:

$$\frac{dQ_r}{dt} = 0, \qquad \frac{dP_r}{dt} = 0, \qquad r = 1, \cdots, m,$$

are concerned.

The demands that the function $S(q_1, \cdots, q_m, Q_1, \cdots, Q_m, t)$ fulfil are the following. First, it must be possible to solve the equations:

$$b_r = S_{q_r}(a_1, \cdots, a_m, \alpha_1, \cdots, \alpha_m, t_0) \tag{24}$$

for the α_r:

$$\alpha_1 = \alpha_1{}^0, \cdots, \alpha_m = \alpha_m{}^0, \tag{25}$$

where $a_r = q_r{}^0$, $b_r = p_r{}^0$ are an arbitrary set of initial values of q_r, p_r.

Furthermore, $S(q_1, \cdots, q_m, \alpha_1, \cdots, \alpha_m, t)$ shall be continuous, together with its derivatives:

$$\frac{\partial S}{\partial q_r}, \qquad \frac{\partial S}{\partial \alpha_r}, \qquad \frac{\partial^2 S}{\partial q_r \partial \alpha_s},$$

in the neighborhood of the point $(a_r, \alpha_r{}^0, t_0)$, and

$$\frac{\partial(S_{q_1}, \cdots, S_{q_m})}{\partial(\alpha_1, \cdots, \alpha_m)} \equiv \frac{\partial(S_{\alpha_1}, \cdots, S_{\alpha_m})}{\partial(q_1, \cdots, q_m)} \neq 0. \tag{26}$$

Finally, the function $V = S(q_r, \alpha_r, t)$ shall satisfy Jacobi's Equation A).

The proof of the existence of such a function S is given by the theorem of § 2 by setting

$$\omega(q_1{}^0, \cdots, q_m{}^0) = \alpha_1 q_1{}^0 + \cdots + \alpha_m q_m{}^0. \tag{27}$$

For now the corresponding solution of Jacobi's Equation A):

$$V = S(q_1, \cdots, q_m, \alpha_1, \cdots, \alpha_m, t), \tag{28}$$

has the property that

$$\left(\frac{\partial S}{\partial q_r}\right)_0 = \alpha_r.$$

Moreover, the Jacobian determinant (26) is seen to have the value 1, and we are through.

We have obtained this existence theorem for the function S by means of the theorem of § 2, the proof of which is based on characteristics. But it might equally well have been derived directly from the existence theorem which is usually referred to as Cauchy's Problem, § 5 below, provided we are willing to assume that $H(q_r, p_r, t)$ is analytic in the point (q_r^0, p_r^0, t_0).

4. Jacobi's Equation: *H*, Independent of *t*. Consider the case that H does not depend on t:

(1) $$H = H(q_1, \cdots, q_m, p_1, \cdots, p_m).$$

Jacobi's Equation now takes the form:

A′) $$\frac{\partial V}{\partial t} + H\left(q_1, \cdots, q_m, \frac{\partial V}{\partial q_1}, \cdots, \frac{\partial V}{\partial q_m}\right) = 0.$$

We seek the special solution

(2) $$V = S(q_1, \cdots, q_m, \alpha_1, \cdots, \alpha_m, t)$$

demanded in § 3.

It is possible to obtain S as follows. A solution of A′) can be found by setting

(3) $$V = -ht + W,$$

where

$$W = W(q_1, \cdots, q_m)$$

does not depend on t. Then W will satisfy the partial differential equation:

C) $$H\left(q_1, \cdots, q_m, \frac{\partial W}{\partial q_1}, \cdots, \frac{\partial W}{\partial q_m}\right) = h.$$

The derivatives of $H(q_1, \cdots, q_m, p_1, \cdots, p_m)$ with respect to the p_r are not all 0.* Let

(4) $$\frac{\partial H}{\partial p_1} \neq 0.$$

Then the equation

(5) $$H(q_1, \cdots, q_m, p_1, \cdots, p_m) = h$$

can be solved for p_1:

(6) $$p_1 = \chi(q_1, \cdots, q_m, h, p_2, \cdots, p_m),$$

and C) is equivalent to the equation:

C′) $$\frac{\partial W}{\partial q_1} = \chi\left(q_1, \cdots, q_m, h, \frac{\partial W}{\partial q_2}, \cdots, \frac{\partial W}{\partial q_m}\right).$$

* Either because of the hypotheses of Chap. XI, § 3 or because H is a positive definite quadratic function of the p_r's.

Let

$$W = W(q_1, \cdots, q_m, h, \alpha_2, \cdots, \alpha_m) \tag{7}$$

be that solution of C) which reduces to

$$W_0 = \omega(q_2^0, \cdots, q_m^0) = \alpha_2 q_2^0 + \cdots + \alpha_m q_m^0 \tag{8}$$

when $q_1 = q_1^0$. If, now, we set $h = \alpha_1$, the desired function S is given by the equation:

$$S(q_1, \cdots, q_m, \alpha_1, \cdots, \alpha_m, t) = \\ -\alpha_1 t + W(q_1, \cdots, q_m, \alpha_1, \cdots, \alpha_m). \tag{9}$$

For,

$$\left(\frac{\partial S}{\partial q_r}\right)_0 = \left(\frac{\partial W}{\partial q_r}\right)_0 = \alpha_r, \qquad r = 2, \cdots, m.$$

Hence

$$p_r^0 = \alpha_r, \qquad r = 2, \cdots, m,$$

are a system of equations which can be solved for the α_r, $r = 2, \cdots, m$, and α_1 is given by (5).

It remains to examine the Jacobian,

$$\frac{\partial(S_{q_1}, \cdots, S_{q_m})}{\partial(\alpha_1, \cdots, \alpha_m)}. \tag{10}$$

Since

$$\left(\frac{\partial^2 S}{\partial \alpha_r \partial q_s}\right)_0 = \begin{cases} 0, & r \neq s \\ 1, & r = s \end{cases} \qquad r, s = 2, \cdots, m,$$

we have only to show that

$$\frac{\partial^2 S}{\partial \alpha_1 \partial q_1} = \frac{\partial}{\partial h}\frac{\partial W}{\partial q_1} \neq 0. \tag{11}$$

Now,

$$\left(\frac{\partial W}{\partial q_1}\right)_0 = p_1^0 \tag{12}$$

is given by the equation (5):

$$H(q_1^0, \cdots, q_m^0, p_1^0, \alpha_2, \cdots, \alpha_m) = h.$$

Hence with the aid of (4)

$$\left(\frac{\partial}{\partial h}\frac{\partial W}{\partial q_1}\right)_0 = \left(\frac{\partial H}{\partial p_1}\right)_0^{-1} \neq 0, \tag{13}$$

and the proof is complete.

Summary of Results. To sum up, then: — the solution of Hamilton's Equations B) is given by the equations:

$$p_r = \frac{\partial S}{\partial q_r}, \qquad \beta_r = -\frac{\partial S}{\partial \alpha_r}, \qquad r = 1, \cdots, m; \tag{14}$$

or

$$\begin{cases} p_r = \dfrac{\partial W}{\partial q_r}, & r = 1, \cdots, m; \\ \beta_r = -\dfrac{\partial W}{\partial \alpha_r}, & r = 2, \cdots, m; \\ \beta_1 = t - \dfrac{\partial W}{\partial h}. \end{cases} \tag{15}$$

The α_r are determined in terms of the (q^0, p^0) by the equations:

$$\begin{cases} \alpha_r = p_r^0, \qquad r = 2, \cdots, m; \\ \alpha_1 = h = H(q_1^0, \cdots, q_m^0, p_1^0, \cdots, p_m^0). \end{cases} \tag{16}$$

The β_r are now given by (15) on setting $q_r = q_r^0$ and substituting for α_r the value given by (16).

The Function W. The total differential equations which determine the characteristics of C) are:

$$\frac{dq_r}{\dfrac{\partial H}{\partial p_r}} = \frac{dW}{\Sigma\, p_r \dfrac{\partial H}{\partial p_r}} = \frac{-\,dp_r}{\dfrac{\partial H}{\partial q_r}}, \qquad r = 1, \cdots, m. \tag{17}$$

Since

$$\frac{\partial H}{\partial p_r} = \dot{q}_r,$$

we have

$$\frac{dW}{dt} = \sum_r p_r \dot{q}_r \tag{18}$$

or

$$W = \int_{t_0}^{t} \sum_r p_r \dot{q}_r \, dt + W_0. \tag{19}$$

If H is a homogeneous quadratic function of $p_1, \cdots, p_m$, then

$$W = 2\int_{t_0}^{t} H\, dt + W_0. \tag{20}$$

5. The Cauchy Problem. Let $F(x_1, \cdots, x_n, z, y_1, \cdots, y_n)$ be analytic in the point $(a, c, b) = (a_1, \cdots, a_n, c, b_1, \cdots, b_n)$ and let $\partial F/\partial x_1 \neq 0$ there. Consider the partial differential equation:

$$F\left(x_1, \cdots, x_n, z, \frac{\partial z}{\partial x_1}, \cdots, \frac{\partial z}{\partial x_n}\right) = 0.$$

Let

$$z_0 = \psi(x_2, \cdots, x_n)$$

be analytic in the point $(a_2, \cdots, a_n)$ and let

$$\psi(a_2, \cdots, a_n) = c,$$

$$\psi_k(a_2, \cdots, a_n) = b_k, \qquad k = 2, \cdots, n.$$

Then there exists one and only one function,

$$z = \varphi(x_1, \cdots, x_n),$$

which is analytic in the point $(a_1, \cdots, a_n)$, has

$$\varphi(a_1, \cdots, a_n) = c,$$

$$\varphi_j(a_1, \cdots, a_n) = b_j, \qquad j = 1, \cdots, n,$$

and satisfies the given differential equation in the neighborhood of the point $(a_1, \cdots, a_n)$.

This is the existence theorem known as the Cauchy Problem. Cf. Goursat-Hedrick, *Mathematical Analysis*, Vol. II, § 446.

APPENDIX D

THE GENERAL PROBLEM OF RATIONAL MECHANICS

I

PATHS

Consider* a system of n particles $m_i : (x_i, y_i, z_i)$ acted on by forces (X_i, Y_i, Z_i). Their motion is governed by Newton's Law:

$$\text{A)} \qquad m_i \ddot{x}_i = X_i, \qquad m_i \ddot{y}_i = Y_i, \qquad m_i \ddot{z}_i = Z_i.$$

Here are $6n$ dependent variables, the $x_i, y_i, z_i, X_i, Y_i, Z_i$, connected by $3n$ equations. The *problem of motion* is to find $3n$ supplementary conditions whereby these $6n$ variables will be determined as functions of the time, t, and suitable initial conditions, and to solve for these functions. Each member of the family which forms the solution, namely the curve:

$$(1) \qquad \begin{cases} x_i = x_i(t), & y_i = y_i(t), & z_i = z_i(t) \\ X_i = X_i(t), & Y_i = Y_i(t), & Z_i = Z_i(t) \end{cases}$$

determines a curve:

$$(2) \qquad \begin{cases} x_i = x_i(t), & y_i = y_i(t), & z_i = z_i(t) \\ \dot{x}_i = \dfrac{dx_i}{dt}, & \dot{y}_i = \dfrac{dy_i}{dt}, & \dot{z}_i = \dfrac{dz_i}{dt} \end{cases}$$

in the $(6n + 1)$-dimensional space of the $(x_i, y_i, z_i, \dot{x}_i, \dot{y}_i, \dot{z}_i, t)$, and such a curve is called a *path*. Obviously the paths (2) stand in a one-to-one relation to the curves (1).

The problem of motion as so formulated transcends the domain of Rational Mechanics. In order to restrict our attention to the latter field, we now lay down the further postulate — which, be it noted, is not satisfied by certain systems which occur in nature, viz., certain systems in which electro-magnetic phenomena are present.

* The following treatment is the result of a joint study of the problem by Professor Bernard Osgood Koopman and myself.

POSTULATE I. DYNAMICAL DETERMINATENESS. *In a given dynamical system, when* $6n + 1$ *constants* $(x_i^0, y_i^0, z_i^0, \dot{x}_i^0, \dot{y}_i^0, \dot{z}_i^0, t_0)$ *are arbitrarily assigned, not more than one path* (2) *exists which passes through this point:*

$$(3) \qquad \begin{cases} x_i(t_0) = x_i^0, & y_i(t_0) = y_i^0, & z_i(t_0) = z_i^0, \\ \dot{x}_i(t_0) = \dot{x}_i^0, & \dot{y}_i(t_0) = \dot{y}_i^0, & \dot{z}_i(t_0) = \dot{z}_i^0. \end{cases}$$

THE DOMAIN D. Those points $(x_i, y_i, z_i, \dot{x}_i, \dot{y}_i, \dot{z}_i, t)$ of the $(6n + 1)$-dimensional space, through which paths pass, constitute the *domain* D. This domain may consist of the entire space, or of a region of it; but in general neither of these things will be the case. It is a point set, concerning the constitution of which we need make no hypothesis at the present moment. It will be restricted by later postulates.

THEOREM I. *The variables* X_i, Y_i, Z_i *are uniquely determined in the points of* D*:*

$$(4) \qquad \begin{cases} X_i = X_i(x_j, y_j, z_j, \dot{x}_j, \dot{y}_j, \dot{z}_j, t) \\ Y_i = Y_i(x_j, y_j, z_j, \dot{x}_j, \dot{y}_j, \dot{z}_j, t) \\ Z_i = Z_i(x_j, y_j, z_j, \dot{x}_j, \dot{y}_j, \dot{z}_j, t) \end{cases}$$

where $(x_j, y_j, z_j, \dot{x}_j, \dot{y}_j, \dot{z}_j, t)$ *is any point of* D.

For, through each point of D passes a path, unique in virtue of Postulate I. Along a given path X_i, Y_i, Z_i are uniquely determined as functions of t by A). Hence X_i, Y_i, Z_i are uniquely determined at the point of D in question, but not in general in points not lying on D.

THE DOMAIN R. In the $(3n + 1)$-dimensional space of the variables (x_i, y_i, z_i, t) those points which participate in paths form a point set R, which may be described as the orthogonal projection on this space of the domain D. In particular R may consist of the whole space, or of a $(3n + 1)$-dimensional region in it. But in general neither of these things will be the case.

Let P be a point of R. To P there corresponds at least one path given by (2). The points (x_i, y_i, z_i, t) represented by the first line of (2), namely:

$$(5) \qquad x_i = x_i(t), \qquad y_i = y_i(t), \qquad z_i = z_i(t),$$

all belong to R. Hence the curve (5) lies wholly in R.

Consider an arbitrary line through P, but not perpendicular to the axis of t. Let its direction components be $\alpha_i, \beta_i, \gamma_i, \kappa$, where $\kappa \neq 0$. There may be a path corresponding to P, such that at this point

$$\dot{x}_i : \dot{y}_i : \dot{z}_i = \alpha_i : \beta_i : \gamma_i.$$

When this is not the case, not all lines through P correspond to paths, and so certain relations between the direction components $(\alpha_i, \beta_i, \gamma_i, \kappa)$ must exist. Thus we are led to a second postulate.

POSTULATE II. *The direction components at points of R, to which paths correspond, are given by the equations:*

$$(6) \qquad \sum_{i=1}^{n} (A_{si}\alpha_i + B_{si}\beta_i + C_{si}\gamma_i) + D_s\kappa = 0, \qquad s = 1, \cdots, \sigma,$$

where $A_{si}, B_{si}, C_{si}, D_s$ are functions of (x_i, y_i, z_i, t) such that the rank of the matrix:*

$$(7) \qquad \begin{Vmatrix} A_{11} \cdots A_{1n} & B_{11} \cdots B_{1n} & C_{11} \cdots C_{1n} \\ \cdots\cdots\cdots & \cdots\cdots\cdots & \cdots\cdots\cdots \\ A_{\sigma 1} \cdots A_{\sigma n} & B_{\sigma 1} \cdots B_{\sigma n} & C_{\sigma 1} \cdots C_{\sigma n} \end{Vmatrix}$$

is σ.

Since along a curve (5)

$$\frac{dx_i}{\alpha_i} = \frac{dy_i}{\beta_i} = \frac{dz_i}{\gamma_i} = \frac{dt}{\kappa}$$

at the point P, it follows that

$$\text{B)} \qquad \sum_{i=1}^{n} (A_{si}\dot{x}_i + B_{si}\dot{y}_i + C_{si}\dot{z}_i) + D_s = 0, \qquad s = 1, \cdots, \sigma.$$

These equations form a necessary and sufficient condition for $(\dot{x}_i, \dot{y}_i, \dot{z}_i)$ if $(x_i, y_i, z_i, \dot{x}_i, \dot{y}_i, \dot{z}_i, t)$ is to be a point of D.

It may happen that the system of Equations B) (a system of Pfaffians) admits certain integrals:

$$\text{C)} \qquad F_k(x_i, y_i, z_i, t) = C_k, \qquad k = 1, \cdots, l,$$

where the rank of the matrix:

*Throughout the whole treatment, the continuity of the functions which enter, and the existence and continuity of such derivatives as it may be convenient to use, are assumed.

$$(8) \qquad \left\| \begin{matrix} \frac{\partial F_1}{\partial x_1} & \cdots\cdots & \frac{\partial F_l}{\partial x_1} \\ \cdots & \cdots & \cdots \\ \frac{\partial F_1}{\partial z_n} & \cdots\cdots & \frac{\partial F_l}{\partial z_n} \end{matrix} \right\|$$

is l. Since the system B) may obviously be replaced by any non-specialized linear combination of these equations, it is clear that Equations B) may be so chosen that the last l of them are:

$$(9) \qquad \frac{dF_k}{dt} = 0, \qquad k = 1, \cdots, l.$$

The constants C_k come to us as constants of integration in the system of integrals C) of the Pfaffians B). They contribute toward determining the particular dynamical system we are defining, different choices of the C_k leading to separate dynamical systems. They are not to be confused with constants of integration that are determined by the initial conditions within a particular dynamical system.

Holonomic and Non-Holonomic Systems. If, in particular, $l = \sigma$, Equations B) can be replaced by Equations C) and thus become completely integrable. The dynamical system we are in process of defining is then said to be *holonomic.* But if there remain $\sigma - l = \mu > 0$ Equations B), which then are non-integrable, the system is said to be *non-holonomic.* Equations B) shall now be replaced by the first μ of them, and Equations C):

$$\text{B}') \qquad \sum_{i=1}^{n} (A_{\alpha i}\, \dot{x}_i + B_{\alpha i}\, \dot{y}_i + C_{\alpha i}\, \dot{z}_i) + D_\alpha = 0, \qquad \alpha = 1, \cdots, \mu;$$

$$\text{C}) \qquad F_k(x_i, y_i, z_i, t) = C_k, \qquad k = 1, \cdots, l;$$

$$\sigma = \mu + l.$$

II

THE FORCES. D'ALEMBERT'S PRINCIPLE

The force X_i, Y_i, Z_i which acts on m_i is made up in general of a force X'_i, Y'_i, Z'_i which is known in terms of $x_i, y_i, z_i, \dot{x}_i, \dot{y}_i, \dot{z}_i, t$, and of further forces X'_{ij}, Y'_{ij}, Z'_{ij}, where $j = 1, 2, \cdots, p_i$, the components of these latter forces being wholly or in part unknown. Denote the unknown components by $S_1, \cdots, S_\kappa$. Then our postulates must provide for enough known equations between the S's and the $x_i, y_i, z_i, \dot{x}_i, \dot{y}_i, \dot{z}_i, t$ to make possible the elimination of the S's between these equations and Equations A), with the result that the equations thus obtained, combined with Equations B′) and C), will just suffice to determine x_i, y_i, z_i as functions of t and the initial conditions. We proceed to the details.

D'Alembert's Principle

In practice the equations which the $S_1, \cdots, S_\kappa$ satisfy are usually linear. Our problem shall be restricted to systems which obey the following postulate.

Postulate III. *The force X_i, Y_i, Z_i is the sum of two forces:*

$$(10)\quad X_i = X_i' + X_i^*, \qquad Y_i = Y_i' + Y_i^*, \qquad Z_i = Z_i' + Z_i^*,$$

where X_i', Y_i', Z_i' are known in terms of the coordinates $x_i, y_i, z_i, \dot{x}_i, \dot{y}_i, \dot{z}_i, t$ of an arbitrary point of D, and where

$$(11)\qquad \sum_{i=1}^{n} X_i^* \xi_i + Y_i^* \eta_i + Z_i^* \zeta_i = 0$$

for all ξ_i, η_i, ζ_i such that

$$(12)\qquad \sum_{i=1}^{n} A'_{\beta i} \xi_i + B'_{\beta i} \eta_i + C'_{\beta i} \zeta_i = 0, \qquad \beta = 1, \cdots, \nu.$$

Here, $A'_{\alpha i}, B'_{\alpha i}, C'_{\alpha i}$ are known functions of the above $x_i, y_i, z_i, \dot{x}_i, \dot{y}_i, \dot{z}_i, t$, and the rank of the matrix:

$$(13)\qquad \left\| \begin{matrix} A'_{11} \cdots A'_{1n} & B'_{11} \cdots B'_{1n} & C'_{11} \cdots C'_{1n} \\ \cdots\cdots & \cdots\cdots & \cdots\cdots \\ A'_{\nu 1} \cdots A'_{\nu n} & B'_{\nu 1} \cdots B'_{\nu n} & C'_{\nu 1} \cdots C'_{\nu n} \end{matrix} \right\|$$

is ν. Conversely, when Equations (12) *are satisfied, Equation* (11) *shall hold.*

Turning now to Equations A), we have what is known as the General Equation of Mechanics:

$$(14)\quad \sum_{i=1}^{n} (m_i \ddot{x}_i - X_i)\, \xi_i + (m_i \ddot{y}_i - Y_i)\, \eta_i + (m_i \ddot{z}_i - Z_i)\, \zeta_i = 0,$$

where ξ_i, η_i, ζ_i are $3n$ arbitrary quantities. Under the sanction of Postulate III. this equation can be replaced by the following:

$$(15) \quad \sum_{i=1}^{n} (m_i \ddot{x}_i - X'_i)\, \xi_i + (m_i \ddot{y}_i - Y'_i)\, \eta_i + (m_i \ddot{z}_i - Z'_i)\, \zeta_i = 0,$$

where ξ_i, η_i, ζ_i are any $3n$ quantities which satisfy the condition (12).

Multiply the β-th equation (12) by λ_β and subtract the resulting equation from (15):

$$(16) \quad \sum_{i=1}^{n} (m_i \ddot{x}_i - X'_i - \sum_{\beta=1}^{\nu} A'_{\beta i} \lambda_\beta)\, \xi_i + (m_i \ddot{y}_i - Y'_i - \sum_{\beta=1}^{\nu} B'_{\beta i} \lambda_\beta)\, \eta_i$$
$$+ (m_i \ddot{z}_i - Z'_i - \sum_{\beta=1}^{\nu} C'_{\beta i} \lambda_\beta)\, \zeta_i = 0.$$

Suppose for definiteness that the determinant whose matrix consists of the first ν columns of the matrix (13) is $\neq 0$. Then the λ's can be so determined that the coefficients of the first ν of the quantities $\xi_1, \cdots, \xi_n, \eta_1, \cdots, \eta_n, \zeta_1, \cdots, \zeta_n$ in (16) will vanish. Substitute these values of $\lambda_1, \cdots, \lambda_\nu$ in the remaining coefficients of (16). Thus a new linear equation in the ξ_i, η_i, ζ_i arises, in which only the last $3n - \nu$ of these quantities appear. But the latter are arbitrary. Hence each coefficient must vanish.

The $3n - \nu$ equations thus obtained express the result of eliminating the unknown $S_1 \cdots, S_\kappa$, *i.e.* the X^*_i, Y^*_i, Z^*_i, from the problem. They contain only $x_i, y_i, z_i, \dot{x}_i, \dot{y}_i, \dot{z}_i, \ddot{x}_i, \ddot{y}_i, \ddot{z}_i, t$, and can be written in the form:

$$\text{E)} \quad \sum_{i=1}^{n} (E_{\gamma i}\, \ddot{x}_i + F_{\gamma i}\, \ddot{y}_i + G_{\gamma i}\, \ddot{z}_i) + H_\gamma = 0, \quad \gamma = 1, \cdots, 3n - \nu,$$

where the coefficients $E_{\gamma i}, F_{\gamma i}, G_{\gamma i}, H_\gamma$ are known functions of $x_i, y_i, z_i, \dot{x}_i, \dot{y}_i, \dot{z}_i, t$ at each point of D.

Equations E) and B) form a necessary condition for the functions $x_i(t)$, $y_i(t)$, $z_i(t)$ which define a path (2). Hence if P_0: $(x_i^0, y_i^0, z_i^0, \dot{x}_i^0, \dot{y}_i^0, \dot{z}_i^0, t_0)$ is an arbitrary point of D, Equations E) and B) admit a solution having as its initial values the coordinates of P_0. Furthermore, by virtue of Postulate I., this solution is unique. We have now arrived at a complete analytical formulation of the problem, for we can retrace our steps. Let

$$\alpha) \qquad x_i = f_i(t), \qquad y_i = \varphi_i(t), \qquad z_i = \psi_i(t)$$

be a curve lying on C) and satisfying B′). Then α) gives rise to a curve Γ which lies on D. Consequently all the coefficients in B′), (12), and E) are determined in the points of α). Let α) also satisfy E).

Since E) holds, it follows that $\lambda_1, \cdots, \lambda_\nu$ can be determined so as to make each parenthesis in (16) vanish. Next, determine X_i^*, Y_i^*, Z_i^* from these λ's by the equations:

$$X_i^* = \sum_{\beta=1}^{\nu} A'_{\beta i}\,\lambda_\beta, \qquad Y_i^* = \sum_{\beta=1}^{\nu} B'_{\beta i}\,\lambda_\beta, \qquad Z_i^* = \sum_{\beta=1}^{\nu} C'_{\beta i}\,\lambda_\beta.$$

These quantities satisfy (11) and (12). On substituting them in (10), values of X_i, Y_i, Z_i are obtained for which A) is true, because each parenthesis in (16) vanishes, and so Γ is a path. But there is only one path through an arbitrary point P_0 of D. Hence α) is unique.

Retrospect. These Postulates complete the formulation of the class of problems in Rational Mechanics which we set out to isolate. The rôle which d'Alembert's Principle * plays is twofold. First, it requires that the relations between the unknown $S_1, \cdots, S_\kappa$ shall be *linear*. Secondly, it performs the elimination by a technique such that the multipliers ξ_i, η_i, ζ_i can always be interpreted as *virtual displacements* of the system of particles m_i : (x_i, y_i, z_i) by setting

$$(17) \qquad \delta x_i = \xi_i \qquad \delta y_i = \eta_i, \qquad \delta z_i = \zeta_i.$$

Remark. In general there is no relation between the coefficients A_{si}, B_{si}, C_{si} of Equations B) and the $A'_{\beta i}, B'_{\beta i}, C'_{\beta i}$ of Equations (12). Hence the virtual displacements $\delta x_i, \delta y_i, \delta z_i$ of (17) will not coincide save as to infinitesimals of higher order with any possible displacement $\Delta x_i, \Delta y_i, \Delta z_i$ due to an actual motion of the system in time Δt.

In a sub-class of cases it happens, however, that the A_{si}, B_{si}, C_{si} in B) and the $A'_{\beta i}, B'_{\beta i}, C'_{\beta i}$ in (12) are respectively equal to each

* Historically d'Alembert's Principle took its start in the assumption of a condition, necessary and sufficient, that a system of forces, acting on a system of particles, be in equilibrium, namely, that the virtual work corresponding to a virtual velocity be nil. When a system of forces not in equilibrium acts on a system of particles, the former can be replaced by a system of forces in equilibrium through the introduction of "counter effective forces" or "forces of inertia" (sic), and thus d'Alembert arrived at the General Equation of Dynamics.

other. But even so, if the D_s are not all 0, the virtual displacement will not tally save as to infinitesimals of higher order with any possible actual displacement.

Finally it can happen that, in addition, the D_s are all 0. Then the virtual displacement corresponds to a possible displacement. But this is a very special, though highly important, case.

III

LAGRANGE'S EQUATIONS

Let

$$(18)\qquad \begin{cases} x_i = f_i(q_1, \cdots, q_m, t) \\ y_i = \varphi_i(q_1, \cdots, q_m, t) \\ z_i = \psi_i(q_1, \cdots, q_m, t) \end{cases}$$

where the rank of the matrix:

$$(19)\qquad \left\| \begin{matrix} \dfrac{\partial x_1}{\partial q_1} & \cdots\cdots & \dfrac{\partial x_1}{\partial q_m} \\ \cdots & \cdots & \cdots \\ \dfrac{\partial z_n}{\partial q_1} & \cdots\cdots & \dfrac{\partial z_n}{\partial q_m} \end{matrix} \right\|$$

is m, and where, moreover, the region of the (x_i, y_i, z_i, t)-space which corresponds to the points $(q_1, \cdots, q_m, t)$ in which f_i, φ_i, ψ_i are defined, at least includes the points of R.

Let T denote the kinetic energy:

$$T = \sum_{i=1}^{n} \frac{m_i}{2} (\dot{x}_i^2 + \dot{y}_i^2 + \dot{z}_i^2).$$

Then, for an arbitrary choice of the $\dot{q}_r$, since

$$\dot{x}_i = \sum_{r=1}^{m} \frac{\partial f_i}{\partial q_r} \dot{q}_r + \frac{\partial f_i}{\partial t}, \qquad \dot{y}_i = \text{etc.},$$

T goes over into a function of $q_r, \dot{q}_r, t$:

$$T = T(q_r, \dot{q}_r, t).$$

Conversely, if $\dot{x}_i, \dot{y}_i, \dot{z}_i$, are any set of numbers for which these equations are true, the $\dot{q}_r$ are uniquely determined.

Consider a path (2). Since the points $x_i = x_i(t)$, $y_i = y_i(t)$, $z_i = z_i(t)$ all lie in R, a curve of the (q_r, t)-space is thus defined:

$$q_r = q_r(t), \qquad r = 1, \cdots, m. \tag{20}$$

For the path in question we have:

$$\frac{d}{dt}\frac{\partial T}{\partial \dot{q}_r} - \frac{\partial T}{\partial q_r} = Q_r, \qquad r = 1, \cdots, m, \tag{21}$$

where

$$Q_r = \sum_{i=1}^{n}\left[X_i\frac{\partial x_i}{\partial q_r} + Y_i\frac{\partial y_i}{\partial q_r} + Z_i\frac{\partial z_i}{\partial q_r}\right].$$

Equations (21) are always true under the foregoing restrictions. They will be sufficient to determine the motion if the system is holonomic and if

$$\sum_{i=1}^{n}\left[X_i^*\frac{\partial x_i}{\partial q_r} + Y_i^*\frac{\partial y_i}{\partial q_r} + Z_i^*\frac{\partial z_i}{\partial q_r}\right] = 0, \tag{22}$$

$r = 1, \cdots, m$. For then

$$Q_r = \sum_{i=1}^{n}\left[X_i'\frac{\partial x_i}{\partial q_r} + Y_i'\frac{\partial y_i}{\partial q_r} + Z_i'\frac{\partial z_i}{\partial q_r}\right], \tag{23}$$

and thus Q_r is known — first, in terms of $x_i, y_i, z_i, \dot{x}_i, \dot{y}_i, \dot{z}_i, t$, and so finally in terms of $q_r, \dot{q}_r, t$. That Equations (21) can be solved for $\ddot{q}_1, \cdots, \ddot{q}_m$ follows from the fact that T is a positive definite quadratic form in the $\dot{q}_1, \cdots, \dot{q}_m$.

This means in terms of the foregoing treatment that a suitable choice of the multipliers ξ_i, η_i, ζ_i in Equation (12) is:

$$\xi_i = \sum_{r=1}^{m}\frac{\partial x_i}{\partial q_r}\delta q_r, \qquad \eta_i = \sum_{r=1}^{m}\frac{\partial y_i}{\partial q_r}\delta q_r, \qquad \zeta_i = \sum_{r=1}^{m}\frac{\partial z_i}{\partial q_r}\delta q_r, \tag{24}$$

where the δq_r are arbitrary. Equations C), if present, are all satisfied identically when the x_i, y_i, z_i are expressed in terms of the q_r and t by Equation (18). Equations B′) are not present in the problem. The system is, to be sure, holonomic, but it is not the only case in which this is so.

The General Case. We assumed in Postulate III. that X_i', Y_i', Z_i' are free from the S's, and that the S's coincide with the X_i^*, Y_i^*, Z_i^*. We now divide the S's into two categories:

i) a sub-set, denoted anew by X_i^*, Y_i^*, Z_i^*, which fulfil the former requirements (11), (12), (13);

ii) a second sub-set, $R_1, \cdots, R_\tau$, on which the X'_i, Y'_i, Z'_i shall now depend linearly.

Thus Equation (21) holds, where Q_r is given by (23). Let

$$Q_r = Q'_r + Q^*_r, \qquad r = 1, \cdots, m, \tag{25}$$

where Q'_r is known in terms of such values of $q_r, \dot{q}_r, t$ as correspond to points of D.

In the particular case before us, namely, Equations (18), it can happen that the equation:

$$Q^*_1 \pi_1 + \cdots + Q^*_m \pi_m = 0 \tag{26}$$

is true for all values of the multipliers π_r for which the following equations hold:

$$a'_{\beta 1} \pi_1 + \cdots + a'_{\beta m} \pi_m = 0, \qquad \beta = 1, \cdots, \nu_1, \tag{27}$$

where $a'_{\beta r}$ depends on values of $q_r, \dot{q}_r, t$, which correspond to points of D, and the rank of the matrix:

$$\begin{Vmatrix} a'_{11} & \cdots\cdots & a'_{1m} \\ \cdots & \cdots\cdots & \cdots \\ a'_{\nu_1 1} & \cdots\cdots & a'_{\nu_1 m} \end{Vmatrix} \tag{28}$$

is ν_1; and conversely, when Equations (27) hold, then (26) is true.

Equations B′), C) go over in the present case into:

$$B'_q) \qquad a_{\alpha 1}\dot{q}_1 + \cdots + a_{\alpha m}\dot{q}_m + a_\alpha = 0, \qquad \alpha = 1, \cdots, \mu_1;$$

$$C_q) \qquad \Phi_k(q_r, t) = \Gamma_k, \qquad k = 1, \cdots, l_1,$$

where $\mu_1 \leqq \mu$; $l_1 \leqq l$, and where the rank of the matrix:

$$\begin{Vmatrix} a_{11} & \cdots\cdots & a_{1m} \\ \cdots & \cdots\cdots & \cdots \\ a_{\mu_1 1} & \cdots\cdots & a_{\mu_1 m} \end{Vmatrix} \tag{29}$$

is μ_1; the rank of the matrix:

$$\begin{Vmatrix} \dfrac{\partial \Phi_1}{\partial q_1} & \cdots\cdots & \dfrac{\partial \Phi_1}{\partial q_m} \\ \cdots & \cdots\cdots & \cdots \\ \dfrac{\partial \Phi_{l_1}}{\partial q_1} & \cdots\cdots & \dfrac{\partial \Phi_{l_1}}{\partial q_m} \end{Vmatrix} \tag{30}$$

being l_1.

Finally it can happen that the $R_1, \cdots, R_\tau$ can be eliminated between these equations, thus leaving a system of equations between the $q_r, \dot{q}_r, \ddot{q}_r$. Such a system yields a unique solution, corresponding to each path of the dynamical system with which we set out. Thus the dynamical problem is completely formulated by means of Lagrange's Equations.

All of the foregoing assumptions are in tentative form — "It may happen · · ·" At the one extreme, the choice of the functions f_i, φ_i, ψ_i can always be made so that all these things do happen; for the q_r can, in particular, be identified with the x_i, y_i, z_i:

$$q_{3i} = x_i, \qquad q_{3i+1} = y_i, \qquad q_{3i+2} = z_i.$$

At the other extreme, m may be chosen so small that Equations (21), though true, will contain unknown functions which cannot be eliminated — namely, the $R_1, \cdots, R_\tau$. This means that, for such a choice of the functions (18), the ξ_i, η_i, ζ_i as given by (24) are too restricted. The ξ_i, η_i, ζ_i of Equations (11) are quantities which must be able to take on *every* set of values which satisfy (12). The ξ_i, η_i, ζ_i which here figure, given by (24), are not free under the condition (24), but are unwarrantably restricted by (24).

In a given problem the desideratum usually is, to choose m as small as possible, subject to the requirement that the same degree of elimination of the S's through (24) shall have been attained, as if Equations (11) and (12) had been used.

IV

NOTES

Consider the dynamical system that consists of a bead sliding on a fixed circular wire and acted on by no other forces than the reaction of the wire. Equations A) take the form:

$$m\ddot{x} = X, \qquad m\ddot{y} = Y, \qquad m\ddot{z} = Z.$$

Let the wire be a circle whose axis is the axis of z. Then Equations B) become:

$$\text{B)} \qquad \begin{cases} x\dot{x} + y\dot{y} = 0 \\ \dot{z} = 0 \end{cases}$$

This system of Pfaffians is completely integrable:

$$\text{C)} \qquad \begin{cases} x^2 + y^2 = a^2 \\ z = c \end{cases}$$

Different values of the constants of integration, a and c, give different systems of paths, (2); but a path of one such system has no point in common with a path of a second system.

Proceeding to the forces we see that $Z = 0$, since $\dot{z} = 0$, and so we have a two-dimensional problem.

The Smooth Wire. Assume first that the wire is smooth. Then the reaction is along the inner normal.

$$X = X^*, \qquad Y = Y^*,$$

and

$$X^* \xi + Y^* \eta = 0$$

provided

$$x \xi + y \eta = 0.$$

Turning to Lagrange's Equations we set $m = 1$ and take

$$x = a \cos q, \qquad y = a \sin q.$$

Then

$$T = \frac{m}{2}(\dot{x}^2 + \dot{y}^2) = \frac{ma^2}{2}\dot{q}^2:$$

$$Q = X^* \frac{\partial x}{\partial q} + Y^* \frac{\partial y}{\partial q}$$

$$= X^*(-a \sin q) + Y^*(a \cos q)$$

$$= X^*(-y) + Y^* x = 0.$$

Hence, finally:

$$ma^2 \frac{d^2 q}{dt^2} = 0,$$

and it remains merely to integrate this differential equation.

The Rough Wire. Suppose, however, the wire is rough. Let $\dot{q} > 0$. Then

$$X^* = -R \cos q + \mu R \sin q$$

$$Y^* = -R \sin q - \mu R \cos q.$$

Lagrange's Equation:

$$\frac{d}{dt}\frac{\partial T}{\partial \dot{q}} - \frac{\partial T}{\partial q} = Q,$$

is still true. But

$$Q = X^* \frac{\partial x}{\partial q} + Y^* \frac{\partial y}{\partial q} = -\mu a R$$

(a result at once obvious) and Lagrange's Equation becomes:

$$ma^2 \frac{d^2 q}{dt^2} = - \mu a R.$$

We have not enough equations to solve the problem. This is the case in which Lagrange's Equations are said to "fail" or be "inapplicable." The failure lies, not in Lagrange's Equations, but in a misuse of them. We should take $m = 2$. Let us first treat the problem, however, by the methods of Parts I., II., before Lagrange's Equations were introduced in Part III. Here, then,

$$m \frac{d^2 x}{dt^2} = X^* = - \quad R \cos \theta + \mu R \sin \theta$$

$$m \frac{d^2 y}{dt^2} = Y^* = - \mu R \cos \theta - \quad R \sin \theta.$$

Equation (11) now takes the form:

$$X^* \xi + Y^* \eta = 0,$$

or

$$(- R \cos \theta + \mu R \sin \theta)\, \xi + (- \mu R \cos \theta - R \sin \theta)\, \eta = 0,$$

or, finally,

$$(- x + \mu y)\, \xi + (- \mu x - y)\, \eta = 0,$$

and this is the form of Equation (12). Hence we may take

$$\xi = \mu x + y, \qquad \eta = - x + \mu y.$$

On substituting these values in the General Equation of Dynamics we have:

$$(\mu x + y) \frac{d^2 x}{dt^2} + (- x + \mu y) \frac{d^2 y}{dt^2} = 0.$$

This equation and Equation C), namely:

$$x^2 + y^2 = a^2,$$

provide us with two equations for determining x and y as functions of t, and thus the problem is reduced to a purely mathematical problem in differential equations. Observe, however, that the virtual displacement used in this solution:

$$\delta x = \epsilon \xi = (\mu x + y)\, \epsilon, \qquad \delta y = \epsilon \eta = (- x + \mu y)\, \epsilon,$$

is not one which is compatible with the constraints, *i.e.* the circular wire — even save as to infinitesimals of higher order than ϵ. It corresponds to a displacement along a line at right angles to the resultant of R and μR.

Turning now to Lagrange's Equations let us choose q_1 and q_2 as the polar coordinates of the mass m. Then Lagrange's Equations (21) become:

$$(31)\qquad \left\{\begin{aligned} &m\left(\frac{d^2r}{dt^2} - r\,\frac{d\theta^2}{dt^2}\right) = -R \\ &\frac{m}{r}\frac{d}{dt}\left(r^2\,\frac{d\theta}{dt}\right) = -\mu R. \end{aligned}\right.$$

Now, Equations C_q) here become:

C_q) $$r = a.$$

On the other hand Equation (26):

$$Q_1^*\pi_1 + Q_1^*\pi_2 = 0,$$

here becomes:

$$(-R)\,\pi_1 + (-\mu R)\,\pi_2 = 0,$$

and thus Equation (27) takes the form:

$$\pi_1 + \mu\pi_2 = 0.$$

If, then, we set:

$$\pi_1 = -\mu, \qquad \pi_2 = 1,$$

Equations (31) and C_q) yield:

$$-ma\frac{d\theta^2}{dt^2}\pi_1 + ma\frac{d^2\theta}{dt^2}\pi_2 = 0$$

or

$$\frac{d^2\theta}{dt^2} + \mu\frac{d\theta^2}{dt^2} = 0,$$

and it remains merely to integrate this equation.

As a further illustration of the use and abuse of Lagrange's Equations may be mentioned the Ladder Problems of pages 322 and 323.

INDEX

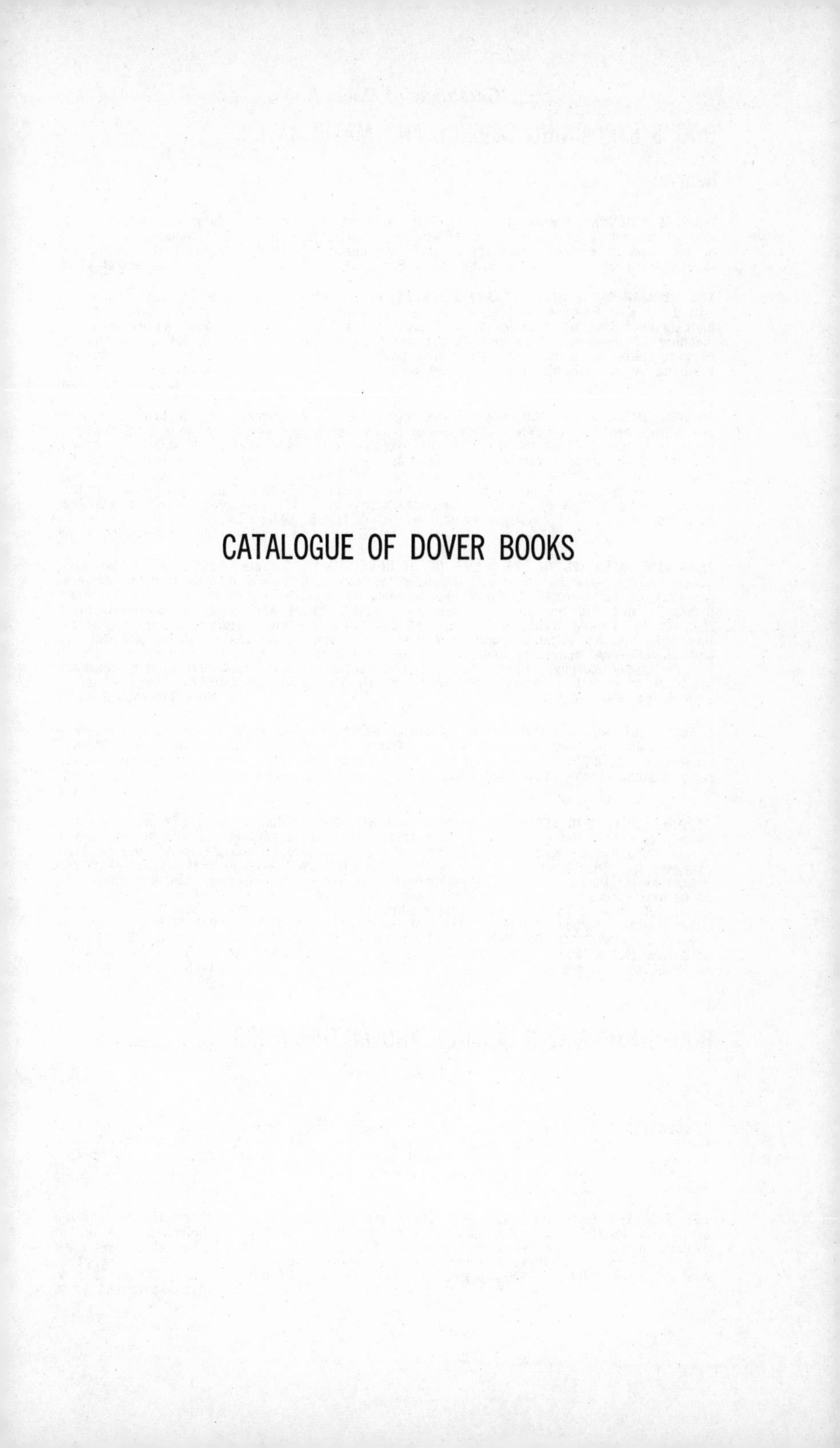

CATALOGUE OF DOVER BOOKS

BOOKS EXPLAINING SCIENCE AND MATHEMATICS

General

WHAT IS SCIENCE?, Norman Campbell. This excellent introduction explains scientific method, role of mathematics, types of scientific laws. Contents: 2 aspects of science, science & nature, laws of science, discovery of laws, explanation of laws, measurement & numerical laws, applications of science. 192pp. 5⅜ x 8. S43 Paperbound **$1.25**

THE COMMON SENSE OF THE EXACT SCIENCES, W. K. Clifford. Introduction by James Newman, edited by Karl Pearson. For 70 years this has been a guide to classical scientific and mathematical thought. Explains with unusual clarity basic concepts, such as extension of meaning of symbols, characteristics of surface boundaries, properties of plane figures, vectors, Cartesian method of determining position, etc. Long preface by Bertrand Russell. Bibliography of Clifford. Corrected, 130 diagrams redrawn. 249pp. 5⅜ x 8. T61 Paperbound **$1.60**

SCIENCE THEORY AND MAN, Erwin Schrödinger. This is a complete and unabridged reissue of SCIENCE AND THE HUMAN TEMPERAMENT plus an additional essay: "What is an Elementary Particle?" Nobel laureate Schrödinger discusses such topics as nature of scientific method, the nature of science, chance and determinism, science and society, conceptual models for physical entities, elementary particles and wave mechanics. Presentation is popular and may be followed by most people with little or no scientific training. "Fine practical preparation for a time when laws of nature, human institutions . . . are undergoing a critical examination without parallel," Waldemar Kaempffert, N. Y. TIMES. 192pp. 5⅜ x 8. T428 Paperbound **$1.35**

FADS AND FALLACIES IN THE NAME OF SCIENCE, Martin Gardner. Examines various cults, quack systems, frauds, delusions which at various times have masqueraded as science. Accounts of hollow-earth fanatics like Symmes; Velikovsky and wandering planets; Hoerbiger; Bellamy and the theory of multiple moons; Charles Fort; dowsing, pseudoscientific methods for finding water, ores, oil. Sections on naturopathy, iridiagnosis, zone therapy, food fads, etc. Analytical accounts of Wilhelm Reich and orgone sex energy; L. Ron Hubbard and Dianetics; A. Korzybski and General Semantics; many others. Brought up to date to include Bridey Murphy, others. Not just a collection of anecdotes, but a fair, reasoned appraisal of eccentric theory. Formerly titled IN THE NAME OF SCIENCE. Preface. Index. x + 384pp. 5⅜ x 8. T394 Paperbound **$1.50**

A DOVER SCIENCE SAMPLER, edited by George Barkin. 64-page book, sturdily bound, containing excerpts from over 20 Dover books, explaining science. Edwin Hubble, George Sarton, Ernst Mach, A. d'Abro, Galileo, Newton, others, discussing island universes, scientific truth, biological phenomena, stability in bridges, etc. Copies limited; no more than 1 to a customer, FREE

POPULAR SCIENTIFIC LECTURES, Hermann von Helmholtz. Helmholtz was a superb expositor as well as a scientist of genius in many areas. The seven essays in this volume are models of clarity, and even today they rank among the best general descriptions of their subjects ever written. "The Physiological Causes of Harmony in Music" was the first significant physiological explanation of musical consonance and dissonance. Two essays, "On the Interaction of Natural Forces" and "On the Conservation of Force," were of great importance in the history of science, for they firmly established the principle of the conservation of energy. Other lectures include "On the Relation of Optics to Painting," "On Recent Progress in the Theory of Vision," "On Goethe's Scientific Researches," and "On the Origin and Significance of Geometrical Axioms." Selected and edited with an introduction by Professor Morris Kline. xii + 286pp. 5⅜ x 8½. T799 Paperbound **$1.45**

BOOKS EXPLAINING SCIENCE AND MATHEMATICS

Physics

CONCERNING THE NATURE OF THINGS, Sir William Bragg. Christmas lectures delivered at the Royal Society by Nobel laureate. Why a spinning ball travels in a curved track; how uranium is transmuted to lead, etc. Partial contents: atoms, gases, liquids, crystals, metals, etc. No scientific background needed; wonderful for intelligent child. 32pp. of photos, 57 figures. xii + 232pp. 5⅜ x 8. T31 Paperbound **$1.35**

THE RESTLESS UNIVERSE, Max Born. New enlarged version of this remarkably readable account by a Nobel laureate. Moving from sub-atomic particles to universe, the author explains in very simple terms the latest theories of wave mechanics. Partial contents: air and its relatives, electrons & ions, waves & particles, electronic structure of the atom, nuclear physics. Nearly 1000 illustrations, including 7 animated sequences. 325pp. 6 x 9. T412 Paperbound **$2.00**

FROM EUCLID TO EDDINGTON: A STUDY OF THE CONCEPTIONS OF THE EXTERNAL WORLD, Sir Edmund Whittaker. A foremost British scientist traces the development of theories of natural philosophy from the western rediscovery of Euclid to Eddington, Einstein, Dirac, etc. The inadequacy of classical physics is contrasted with present day attempts to understand the physical world through relativity, non-Euclidean geometry, space curvature, wave mechanics, etc. 5 major divisions of examination: Space; Time and Movement; the Concepts of Classical Physics; the Concepts of Quantum Mechanics; the Eddington Universe. 212pp. 5⅜ x 8. T491 Paperbound **$1.35**

PHYSICS, THE PIONEER SCIENCE, L. W. Taylor. First thorough text to place all important physical phenomena in cultural-historical framework; remains best work of its kind. Exposition of physical laws, theories developed chronologically, with great historical, illustrative experiments diagrammed, described, worked out mathematically. Excellent physics text for self-study as well as class work. Vol. 1: Heat, Sound: motion, acceleration, gravitation, conservation of energy, heat engines, rotation, heat, mechanical energy, etc. 211 illus. 407pp. 5⅜ x 8. Vol. 2: Light, Electricity: images, lenses, prisms, magnetism, Ohm's law, dynamos, telegraph, quantum theory, decline of mechanical view of nature, etc. Bibliography. 13 table appendix. Index. 551 illus. 2 color plates. 508pp. 5⅜ x 8.
Vol. 1 S565 Paperbound **$2.00**
Vol. 2 S566 Paperbound **$2.00**
The set **$4.00**

A SURVEY OF PHYSICAL THEORY, Max Planck. One of the greatest scientists of all time, creator of the quantum revolution in physics, writes in non-technical terms of his own discoveries and those of other outstanding creators of modern physics. Planck wrote this book when science had just crossed the threshold of the new physics, and he communicates the excitement felt then as he discusses electromagnetic theories, statistical methods, evolution of the concept of light, a step-by-step description of how he developed his own momentous theory, and many more of the basic ideas behind modern physics. Formerly "A Survey of Physics." Bibliography. Index. 128pp. 5⅜ x 8. S650 Paperbound **$1.15**

THE ATOMIC NUCLEUS, M. Korsunsky. The only non-technical comprehensive account of the atomic nucleus in English. For college physics students, etc. Chapters cover: Radioactivity, the Nuclear Model of the Atom, the Mass of Atomic Nuclei, the Disintegration of Atomic Nuclei, the Discovery of the Positron, the Artificial Transformation of Atomic Nuclei, Artificial Radioactivity, Mesons, the Neutrino, the Structure of Atomic Nuclei and Forces Acting Between Nuclear Particles, Nuclear Fission, Chain Reaction, Peaceful Uses, Thermoculear Reactions. Slightly abridged edition. Translated by G. Yankovsky. 65 figures. Appendix includes 45 photographic illustrations. 413 pp. 5⅜ x 8. S1052 Paperbound **$2.00**

PRINCIPLES OF MECHANICS SIMPLY EXPLAINED, Morton Mott-Smith. Excellent, highly readable introduction to the theories and discoveries of classical physics. Ideal for the layman who desires a foundation which will enable him to understand and appreciate contemporary developments in the physical sciences. Discusses: Density, The Law of Gravitation, Mass and Weight, Action and Reaction, Kinetic and Potential Energy, The Law of Inertia, Effects of Acceleration, The Independence of Motions, Galileo and the New Science of Dynamics, Newton and the New Cosmos, The Conservation of Momentum, and other topics. Revised edition of "This Mechanical World." Illustrated by E. Kosa, Jr. Bibliography and Chronology. Index. xiv + 171pp. 5⅜ x 8½. T1067 Paperbound **$1.00**

THE CONCEPT OF ENERGY SIMPLY EXPLAINED, Morton Mott-Smith. Elementary, non-technical exposition which traces the story of man's conquest of energy, with particular emphasis on the developments during the nineteenth century and the first three decades of our own century. Discusses man's earlier efforts to harness energy, more recent experiments and discoveries relating to the steam engine, the engine indicator, the motive power of heat, the principle of excluded perpetual motion, the bases of the conservation of energy, the concept of entropy, the internal combustion engine, mechanical refrigeration, and many other related topics. Also much biographical material. Index. Bibliography. 33 illustrations. ix + 215pp. 5⅜ x 8½. T1071 Paperbound **$1.25**

HEAT AND ITS WORKINGS, Morton Mott-Smith. One of the best elementary introductions to the theory and attributes of heat, covering such matters as the laws governing the effect of heat on solids, liquids and gases, the methods by which heat is measured, the conversion of a substance from one form to another through heating and cooling, evaporation, the effects of pressure on boiling and freezing points, and the three ways in which heat is transmitted (conduction, convection, radiation). Also brief notes on major experiments and discoveries. Concise, but complete, it presents all the essential facts about the subject in readable style. Will give the layman and beginning student a first-rate background in this major topic in physics. Index. Bibliography. 50 illustrations. x + 165pp. 5⅜ x 8½. T978 Paperbound **$1.00**

THE STORY OF ATOMIC THEORY AND ATOMIC ENERGY, J. G. Feinberg. Wider range of facts on physical theory, cultural implications, than any other similar source. Completely non-technical. Begins with first atomic theory, 600 B.C., goes through A-bomb, developments to 1959. Avogadro, Rutherford, Bohr, Einstein, radioactive decay, binding energy, radiation danger, future benefits of nuclear power, dozens of other topics, told in lively, related, informal manner. Particular stress on European atomic research. "Deserves special mention . . . authoritative," Saturday Review. Formerly "The Atom Story." New chapter to 1959. Index. 34 illustrations. 251pp. 5⅜ x 8. T625 Paperbound **$1.45**

THE STRANGE STORY OF THE QUANTUM, AN ACCOUNT FOR THE GENERAL READER OF THE GROWTH OF IDEAS UNDERLYING OUR PRESENT ATOMIC KNOWLEDGE, B. Hoffmann. Presents lucidly and expertly, with barest amount of mathematics, the problems and theories which led to modern quantum physics. Dr. Hoffmann begins with the closing years of the 19th century, when certain trifling discrepancies were noticed, and with illuminating analogies and examples takes you through the brilliant concepts of Planck, Einstein, Pauli, de Broglie, Bohr, Schroedinger, Heisenberg, Dirac, Sommerfeld, Feynman, etc. This edition includes a new, long postscript carrying the story through 1958. "Of the books attempting an account of the history and contents of our modern atomic physics which have come to my attention, this is the best," H. Margenau, Yale University, in "American Journal of Physics." 32 tables and line illustrations. Index. 275pp. 5⅜ x 8. T518 Paperbound **$1.50**

THE EVOLUTION OF SCIENTIFIC THOUGHT FROM NEWTON TO EINSTEIN, A. d'Abro. Einstein's special and general theories of relativity, with their historical implications, are analyzed in non-technical terms. Excellent accounts of the contributions of Newton, Riemann, Weyl, Planck, Eddington, Maxwell, Lorentz and others are treated in terms of space and time, equations of electromagnetics, finiteness of the universe, methodology of science. 21 diagrams. 482pp. 5⅜ x 8. T2 Paperound **$2.00**

THE RISE OF THE NEW PHYSICS, A. d'Abro. A half-million word exposition, formerly titled THE DECLINE OF MECHANISM, for readers not versed in higher mathematics. The only thorough explanation, in everyday language, of the central core of modern mathematical physical theory, treating both classical and modern theoretical physics, and presenting in terms almost anyone can understand the equivalent of 5 years of study of mathematical physics. Scientifically impeccable coverage of mathematical-physical thought from the Newtonian system up through the electronic theories of Dirac and Heisenberg and Fermi's statistics. Combines both history and exposition; provides a broad yet unified and detailed view, with constant comparison of classical and modern views on phenomena and theories. "A must for anyone doing serious study in the physical sciences," JOURNAL OF THE FRANKLIN INSTITUTE. "Extraordinary faculty . . . to explain ideas and theories of theoretical physics in the language of daily life," ISIS. First part of set covers philosophy of science, drawing upon the practice of Newton, Maxwell, Poincaré, Einstein, others, discussing modes of thought, experiment, interpretations of causality, etc. In the second part, 100 pages explain grammar and vocabulary of mathematics, with discussions of functions, groups, series, Fourier series, etc. The remainder is devoted to concrete, detailed coverage of both classical and quantum physics, explaining such topics as analytic mechanics, Hamilton's principle, wave theory of light, electromagnetic waves, groups of transformations, thermodynamics, phase rule, Brownian movement, kinetics, special relativity, Planck's original quantum theory, Bohr's atom, Zeeman effect, Broglie's wave mechanics, Heisenberg's uncertainty, Eigen-values, matrices, scores of other important topics. Discoveries and theories are covered for such men as Alembert, Born, Cantor, Debye, Euler, Foucault, Galois, Gauss, Hadamard, Kelvin, Kepler, Laplace, Maxwell, Pauli, Rayleigh, Volterra, Weyl, Young, more than 180 others. Indexed. 97 illustrations. ix + 982pp. 5⅜ x 8. T3 Volume 1, Paperbound **$2.00**
T4 Volume 2, Paperbound **$2.00**

SPINNING TOPS AND GYROSCOPIC MOTION, John Perry. Well-known classic of science still unsurpassed for lucid, accurate, delightful exposition. How quasi-rigidity is induced in flexible and fluid bodies by rapid motions; why gyrostat falls, top rises; nature and effect on climatic conditions of earth's precessional movement; effect of internal fluidity on rotating bodies, etc. Appendixes describe practical uses to which gyroscopes have been put in ships, compasses, monorail transportation. 62 figures. 128pp. 5⅜ x 8. T416 Paperbound **$1.00**

THE UNIVERSE OF LIGHT, Sir William Bragg. No scientific training needed to read Nobel Prize winner's expansion of his Royal Institute Christmas Lectures. Insight into nature of light, methods and philosophy of science. Explains lenses, reflection, color, resonance, polarization, x-rays, the spectrum, Newton's work with prisms, Huygens' with polarization, Crookes' with cathode ray, etc. Leads into clear statement of 2 major historical theories of light, corpuscle and wave. Dozens of experiments you can do. 199 illus., including 2 full-page color plates. 293pp. 5⅜ x 8. S538 Paperbound **$1.85**

THE STORY OF X-RAYS FROM RÖNTGEN TO ISOTOPES, A. R. Bleich. Non-technical history of x-rays, their scientific explanation, their applications in medicine, industry, research, and art, and their effect on the individual and his descendants. Includes amusing early reactions to Röntgen's discovery, cancer therapy, detections of art and stamp forgeries, potential risks to patient and operator, etc. Illustrations show x-rays of flower structure, the gall bladder, gears with hidden defects, etc. Original Dover publication. Glossary. Bibliography. Index. 55 photos and figures. xiv + 186pp. 5⅜ x 8. T662 Paperbound **$1.35**

ELECTRONS, ATOMS, METALS AND ALLOYS, Wm. Hume-Rothery. An introductory-level explanation of the application of the electronic theory to the structure and properties of metals and alloys, taking into account the new theoretical work done by mathematical physicists. Material presented in dialogue-form between an "Old Metallurgist" and a "Young Scientist." Their discussion falls into 4 main parts: the nature of an atom, the nature of a metal, the nature of an alloy, and the structure of the nucleus. They cover such topics as the hydrogen atom, electron waves, wave mechanics, Brillouin zones, co-valent bonds, radioactivity and natural disintegration, fundamental particles, structure and fission of the nucleus,etc. Revised, enlarged edition. 177 illustrations. Subject and name indexes. 407pp. 5⅜ x 8½. S1046 Paperbound **$2.25**

OUT OF THE SKY, H. H. Nininger. A non-technical but comprehensive introduction to "meteoritics", the young science concerned with all aspects of the arrival of matter from outer space. Written by one of the world's experts on meteorites, this work shows how, despite difficulties of observation and sparseness of data, a considerable body of knowledge has arisen. It defines meteors and meteorites; studies fireball clusters and processions, meteorite composition, size, distribution, showers, explosions, origins, craters, and much more. A true connecting link between astronomy and geology. More than 175 photos, 22 other illustrations. References. Bibliography of author's publications on meteorites. Index. viii + 336pp. 5⅜ x 8. T519 Paperbound **$1.85**

SATELLITES AND SCIENTIFIC RESEARCH, D. King-Hele. Non-technical account of the manmade satellites and the discoveries they have yielded up to the autumn of 1961. Brings together information hitherto published only in hard-to-get scientific journals. Includes the life history of a typical satellite, methods of tracking, new information on the shape of the earth, zones of radiation, etc. Over 60 diagrams and 6 photographs. Mathematical appendix. Bibliography of over 100 items. Index. xii + 180pp. 5⅜ x 8½. T703 Paperbound **$2.00**

BOOKS EXPLAINING SCIENCE AND MATHEMATICS

Mathematics

CHANCE, LUCK AND STATISTICS: THE SCIENCE OF CHANCE, Horace C. Levinson. Theory of probability and science of statistics in simple, non-technical language. Part I deals with theory of probability, covering odd superstitions in regard to "luck," the meaning of betting odds, the law of mathematical expectation, gambling, and applications in poker, roulette, lotteries, dice, bridge, and other games of chance. Part II discusses the misuse of statistics, the concept of statistical probabilities, normal and skew frequency distributions, and statistics applied to various fields—birth rates, stock speculation, insurance rates, advertising, etc. "Presented in an easy humorous style which I consider the best kind of expository writing," Prof. A. C. Cohen, Industry Quality Control. Enlarged revised edition. Formerly titled "The Science of Chance." Preface and two new appendices by the author. Index. xiv + 365pp. 5⅜ x 8. T1007 Paperbound **$1.85**

PROBABILITIES AND LIFE, Emile Borel. Translated by M. Baudin. Non-technical, highly readable introduction to the results of probability as applied to everyday situations. Partial contents: Fallacies About Probabilities Concerning Life After Death; Negligible Probabilities and the Probabilities of Everyday Life; Events of Small Probability; Application of Probabilities to Certain Problems of Heredity; Probabilities of Deaths, Diseases, and Accidents; On Poisson's Formula. Index. 3 Appendices of statistical studies and tables. vi + 87pp. 5⅜ x 8½. T121 Paperbound **$1.00**

GREAT IDEAS OF MODERN MATHEMATICS: THEIR NATURE AND USE, Jagjit Singh. Reader with only high school math will understand main mathematical ideas of modern physics, astronomy, genetics, psychology, evolution, etc., better than many who use them as tools, but comprehend little of their basic structure. Author uses his wide knowledge of non-mathematical fields in brilliant exposition of differential equations, matrices, group theory, logic, statistics, problems of mathematical foundations, imaginary numbers, vectors, etc. Original publication. 2 appendices. 2 indexes. 65 illustr. 322pp. 5⅜ x 8. S587 Paperbound **$1.75**

MATHEMATICS IN ACTION, O. G. Sutton. Everyone with a command of high school algebra will find this book one of the finest possible introductions to the application of mathematics to physical theory. Ballistics, numerical analysis, waves and wavelike phenomena, Fourier series, group concepts, fluid flow and aerodynamics, statistical measures, and meteorology are discussed with unusual clarity. Some calculus and differential equations theory is developed by the author for the reader's help in the more difficult sections. 88 figures. Index. viii + 236pp. 5⅜ x 8. T440 Clothbound **$3.50**

THE FOURTH DIMENSION SIMPLY EXPLAINED, edited by H. P. Manning. 22 essays, originally Scientific American contest entries, that use a minimum of mathematics to explain aspects of 4-dimensional geometry: analogues to 3-dimensional space, 4-dimensional absurdities and curiosities (such as removing the contents of an egg without puncturing its shell), possible measurements and forms, etc. Introduction by the editor. Only book of its sort on a truly elementary level, excellent introduction to advanced works. 82 figures. 251pp. 5⅜ x 8. T711 Paperbound **$1.35**

MATHEMATICS—INTERMEDIATE TO ADVANCED

General

INTRODUCTION TO APPLIED MATHEMATICS, Francis D. Murnaghan. A practical and thoroughly sound introduction to a number of advanced branches of higher mathematics. Among the selected topics covered in detail are: vector and matrix analysis, partial and differential equations, integral equations, calculus of variations, Laplace transform theory, the vector triple product, linear vector functions, quadratic and bilinear forms, Fourier series, spherical harmonics, Bessel functions, the Heaviside expansion formula, and many others. Extremely useful book for graduate students in physics, engineering, chemistry, and mathematics. Index. 111 study exercises with answers. 41 illustrations. ix + 389pp. 5⅜ x 8½.
S1042 Paperbound **$2.00**

OPERATIONAL METHODS IN APPLIED MATHEMATICS, H. S. Carslaw and J. C. Jaeger. Explanation of the application of the Laplace Transformation to differential equations, a simple and effective substitute for more difficult and obscure operational methods. Of great practical value to engineers and to all workers in applied mathematics. Chapters on: Ordinary Linear Differential Equations with Constant Coefficients;; Electric Circuit Theory; Dynamical Applications; The Inversion Theorem for the Laplace Transformation; Conduction of Heat; Vibrations of Continuous Mechanical Systems; Hydrodynamics; Impulsive Functions; Chains of Differential Equations; and other related matters. 3 appendices. 153 problems, many with answers. 22 figures. xvi + 359pp. 5⅜ x 8½.
S1011 Paperbound **$2.25**

APPLIED MATHEMATICS FOR RADIO AND COMMUNICATIONS ENGINEERS, C. E. Smith. No extraneous material here!—only the theories, equations, and operations essential and immediately useful for radio work. Can be used as refresher, as handbook of applications and tables, or as full home-study course. Ranges from simplest arithmetic through calculus, series, and wave forms, hyperbolic trigonometry, simultaneous equations in mesh circuits, etc. Supplies applications right along with each math topic discussed. 22 useful tables of functions, formulas, logs, etc. Index. 166 exercises, 140 examples, all with answers. 95 diagrams. Bibliography. x + 336pp. 5⅜ x 8.
S141 Paperbound **$1.75**

Algebra, group theory, determinants, sets, matrix theory

ALGEBRAS AND THEIR ARITHMETICS, L. E. Dickson. Provides the foundation and background necessary to any advanced undergraduate or graduate student studying abstract algebra. Begins with elementary introduction to linear transformations, matrices, field of complex numbers; proceeds to order, basal units, modulus, quaternions, etc.; develops calculus of linears sets, describes various examples of algebras including invariant, difference, nilpotent, semi-simple. "Makes the reader marvel at his genius for clear and profound analysis," Amer. Mathematical Monthly. Index. xii + 241pp. 5⅜ x 8.
S616 Paperbound **$1.50**

THE THEORY OF EQUATIONS WITH AN INTRODUCTION TO THE THEORY OF BINARY ALGEBRAIC FORMS, W. S. Burnside and A. W. Panton. Extremely thorough and concrete discussion of the theory of equations, with extensive detailed treatment of many topics curtailed in later texts. Covers theory of algebraic equations, properties of polynomials, symmetric functions, derived functions, Horner's process, complex numbers and the complex variable, determinants and methods of elimination, invariant theory (nearly 100 pages), transformations, introduction to Galois theory, Abelian equations, and much more. Invaluable supplementary work for modern students and teachers. 759 examples and exercises. Index in each volume. Two volume set. Total of xxiv + 604pp. 5⅜ x 8.
S714 Vol I Paperbound **$1.85**
S715 Vol II Paperbound **$1.85**
The set **$3.70**

COMPUTATIONAL METHODS OF LINEAR ALGEBRA, V. N. Faddeeva, translated by **C. D. Benster.** First English translation of a unique and valuable work, the only work in English presenting a systematic exposition of the most important methods of linear algebra—classical and contemporary. Shows in detail how to derive numerical solutions of problems in mathematical physics which are frequently connected with those of linear algebra. Theory as well as individual practice. Part I surveys the mathematical background that is indispensable to what follows. Parts II and III, the conclusion, set forth the most important methods of solution, for both exact and iterative groups. One of the most outstanding and valuable features of this work is the 23 tables, double and triple checked for accuracy. These tables will not be found elsewhere. Author's preface. Translator's note. New bibliography and index. x + 252pp. 5⅜ x 8.
S424 Paperbound **$1.95**

ALGEBRAIC EQUATIONS, E. Dehn. Careful and complete presentation of Galois' theory of algebraic equations; theories of Lagrange and Galois developed in logical rather than historical form, with a more thorough exposition than in most modern books. Many concrete applications and fully-worked-out examples. Discusses basic theory (very clear exposition of the symmetric group); isomorphic, transitive, and Abelian groups; applications of Lagrange's and Galois' theories; and much more. Newly revised by the author. Index. List of Theorems. xi + 208pp. 5⅜ x 8.
S697 Paperbound **$1.45**

Differential equations, ordinary and partial; integral equations

INTRODUCTION TO THE DIFFERENTIAL EQUATIONS OF PHYSICS, L. Hopf. Especially valuable to the engineer with no math beyond elementary calculus. Emphasizing intuitive rather than formal aspects of concepts, the author covers an extensive territory. Partial contents: Law of causality, energy theorem, damped oscillations, coupling by friction, cylindrical and spherical coordinates, heat source, etc. Index. 48 figures. 160pp. 5⅜ x 8.
S120 Paperbound **$1.25**

INTRODUCTION TO THE THEORY OF LINEAR DIFFERENTIAL EQUATIONS, E. G. Poole. Authoritative discussions of important topics, with methods of solution more detailed than usual, for students with background of elementary course in differential equations. Studies existence theorems, linearly independent solutions; equations with constant coefficients; with uniform analytic coefficients; regular singularities; the hypergeometric equation; conformal representation; etc. Exercises. Index. 210pp. 5⅜ x 8. S629 Paperbound **$1.65**

DIFFERENTIAL EQUATIONS FOR ENGINEERS, P. Franklin. Outgrowth of a course given 10 years at M. I. T. Makes most useful branch of pure math accessible for practical work. Theoretical basis of D.E.'s; solution of ordinary D.E.'s and partial derivatives arising from heat flow, steady-state temperature of a plate, wave equations; analytic functions; convergence of Fourier Series. 400 problems on electricity, vibratory systems, other topics. Formerly "Differential Equations for Electrical Engineers." Index 41 illus. 307pp. 5⅜ x 8.
S601 Paperbound **$1.65**

DIFFERENTIAL EQUATIONS, F. R. Moulton. A detailed, rigorous exposition of all the non-elementary processes of solving ordinary differential equations. Several chapters devoted to the treatment of practical problems, especially those of a physical nature, which are far more advanced than problems usually given as illustrations. Includes analytic differential equations; variations of a parameter; integrals of differential equations; analytic implicit functions; problems of elliptic motion; sine-amplitude functions; deviation of formal bodies; Cauchy-Lipschitz process; linear differential equations with periodic coefficients; differential equations in infinitely many variations; much more. Historical notes. 10 figures. 222 problems. Index. xv + 395pp. 5⅜ x 8. S451 Paperbound **$2.00**

DIFFERENTIAL AND INTEGRAL EQUATIONS OF MECHANICS AND PHYSICS (DIE DIFFERENTIAL- UND INTEGRALGLEICHUNGEN DER MECHANIK UND PHYSIK), edited by P. Frank and R. von Mises. Most comprehensive and authoritative work on the mathematics of mathematical physics available today in the United States: the standard, definitive reference for teachers, physicists, engineers, and mathematicians—now published (in the original German) at a relatively inexpensive price for the first time! Every chapter in this 2,000-page set is by an expert in his field: Carathéodory, Courant, Frank, Mises, and a dozen others. Vol I, on mathematics, gives concise but complete coverages of advanced calculus, differential equations, integral equations, and potential, and partial differential equations. Index. xxiii + 916pp. Vol. II (physics): classical mechanics, optics, continuous mechanics, heat conduction and diffusion, the stationary and quasi-stationary electromagnetic field, electromagnetic oscillations, and wave mechanics. Index. xxiv + 1106pp. Two volume set. Each volume available separately. 5⅝ x 8⅜.
S787 Vol I Clothbound **$7.50**
S788 Vol II Clothbound **$7.50**
The set **$15.00**

LECTURES ON CAUCHY'S PROBLEM, J. Hadamard. Based on lectures given at Columbia, Rome, this discusses work of Riemann, Kirchhoff, Volterra, and the author's own research on the hyperbolic case in linear partial differential equations. It extends spherical and cylindrical waves to apply to all (normal) hyperbolic equations. Partial contents: Cauchy's problem, fundamental formula, equations with odd number, with even number of independent variables; method of descent. 32 figures. Index. iii + 316pp. 5⅜ x 8. S105 Paperbound **$1.75**

THEORY OF DIFFERENTIAL EQUATIONS, A. R. Forsyth. Out of print for over a decade, the complete 6 volumes (now bound as 3) of this monumental work represent the most comprehensive treatment of differential equations ever written. Historical presentation includes in 2500 pages every substantial development. Vol. 1, 2: EXACT EQUATIONS, PFAFF'S PROBLEM; ORDINARY EQUATIONS, NOT LINEAR: methods of Grassmann, Clebsch, Lie, Darboux; Cauchy's theorem; branch points; etc. Vol. 3, 4: ORDINARY EQUATIONS, NOT LINEAR; ORDINARY LINEAR EQUATIONS: Zeta Fuchsian functions, general theorems on algebraic integrals, Brun's theorem, equations with uniform periodic coffiecients, etc. Vol. 4, 5: PARTIAL DIFFERENTIAL EQUATIONS: 2 existence-theorems, equations of theoretical dynamics, Laplace transformations, general transformation of equations of the 2nd order, much more. Indexes. Total of 2766pp. 5⅜ x 8. S576-7-8 Clothbound: the set **$15.00**

PARTIAL DIFFERENTIAL EQUATIONS OF MATHEMATICAL PHYSICS, A. G. Webster. A keystone work in the library of every mature physicist, engineer, researcher. Valuable sections on elasticity, compression theory, potential theory, theory of sound, heat conduction, wave propagation, vibration theory. Contents include: deduction of differential equations, vibrations, normal functions, Fourier's series, Cauchy's method, boundary problems, method of Riemann-Volterra. Spherical, cylindrical, ellipsoidal harmonics, applications, etc. 97 figures. vii + 440pp. 5⅜ x 8. S263 Paperbound **$2.00**

ELEMENTARY CONCEPTS OF TOPOLOGY, P. Alexandroff. First English translation of the famous brief introduction to topology for the beginner or for the mathematician not undertaking extensive study. This unusually useful intuitive approach deals primarily with the concepts of complex, cycle, and homology, and is wholly consistent with current investigations. Ranges from basic concepts of set-theoretic topology to the concept of Betti groups. "Glowing example of harmony between intuition and thought," David Hilbert. Translated by A. E. Farley. Introduction by D. Hilbert. Index. 25 figures. 73pp. 5⅜ x 8. S747 Paperbound **$1.00**

Number theory

INTRODUCTION TO THE THEORY OF NUMBERS, L. E. Dickson. Thorough, comprehensive approach with adequate coverage of classical literature, an introductory volume beginners can follow. Chapters on divisibility, congruences, quadratic residues & reciprocity, Diophantine equations, etc. Full treatment of binary quadratic forms without usual restriction to integral coefficients. Covers infinitude of primes, least residues, Fermat's theorem, Euler's phi function, Legendre's symbol, Gauss's lemma, automorphs, reduced forms, recent theorems of Thue & Siegel, many more. Much material not readily available elsewhere. 239 problems. Index. I figure. viii + 183pp. 5⅜ x 8. S342 Paperbound **$1.65**

ELEMENTS OF NUMBER THEORY, I. M. Vinogradov. Detailed 1st course for persons without advanced mathematics; 95% of this book can be understood by readers who have gone no farther than high school algebra. Partial contents: divisibility theory, important number theoretical functions, congruences, primitive roots and indices, etc. Solutions to both problems and exercises. Tables of primes, indices, etc. Covers almost every essential formula in elementary number theory! Translated from Russian. 233 problems, 104 exercises. viii + 227pp. 5⅜ x 8. S259 Paperbound **$1.60**

THEORY OF NUMBERS and DIOPHANTINE ANALYSIS, R. D. Carmichael. These two complete works in one volume form one of the most lucid introductions to number theory, requiring only a firm foundation in high school mathematics. "Theory of Numbers," partial contents: Eratosthenes' sieve, Euclid's fundamental theorem, G.C.F. and L.C.M. of two or more integers, linear congruences, etc "Diophantine Analysis": rational triangles, Pythagorean triangles, equations of third, fourth, higher degrees, method of functional equations, much more. "Theory of Numbers": 76 problems. Index. 94pp. "Diophantine Analysis": 222 problems. Index. 118pp. 5⅜ x 8. S529 Paperbound **$1.35**

Numerical analysis, tables

MATHEMATICAL TABLES AND FORMULAS, Compiled by Robert D. Carmichael and Edwin R. Smith. Valuable collection for students, etc. Contains all tables necessary in college algebra and trigonometry, such as five-place common logarithms, logarithmic sines and tangents of small angles, logarithmic trigonometric functions, natural trigonometric functions, four-place antilogarithms, tables for changing from sexagesimal to circular and from circular to sexagesimal measure of angles, etc. Also many tables and formulas not ordinarily accessible, including powers, roots, and reciprocals, exponential and hyperbolic functions, ten-place logarithms of prime numbers, and formulas and theorems from analytical and elementary geometry and from calculus. Explanatory introduction. viii + 269pp. 5⅜ x 8½. S111 Paperbound **$1.00**

MATHEMATICAL TABLES, H. B. Dwight. Unique for its coverage in one volume of almost every function of importance in applied mathematics, engineering, and the physical sciences. Three extremely fine tables of the three trig functions and their inverse functions to thousandths of radians; natural and common logarithms; squares, cubes; hyperbolic functions and the inverse hyperbolic functions; $(a^2 + b^2)$ exp. ½a; complete elliptic integrals of the 1st and 2nd kind; sine and cosine integrals; exponential integrals Ei(x) and Ei(—x); binomial coefficients; factorials to 250; surface zonal harmonics and first derivatives; Bernoulli and Euler numbers and their logs to base of 10; Gamma function; normal probability integral; over 60 pages of Bessel functions; the Riemann Zeta function. Each table with formulae generally used, sources of more extensive tables, interpolation data, etc. Over half have columns of differences, to facilitate interpolation. Introduction. Index. viii + 231pp. 5⅜ x 8. S445 Paperbound **$1.75**

TABLES OF FUNCTIONS WITH FORMULAE AND CURVES, E. Jahnke & F. Emde. The world's most comprehensive 1-volume English-text collection of tables, formulae, curves of transcendent functions. 4th corrected edition, new 76-page section giving tables, formulae for elementary functions—not in other English editions. Partial contents: sine, cosine, logarithmic integral; factorial function; error integral; theta functions; elliptic integrals, functions; Legendre, Bessel, Riemann, Mathieu, hypergeometric functions, etc. Supplementary books. Bibliography. Indexed. "Out of the way functions for which we know no other source," SCIENTIFIC COMPUTING SERVICE, Ltd. 212 figures. 400pp. 5⅜ x 8. S133 Paperbound **$2.00**

CHEMISTRY AND PHYSICAL CHEMISTRY

ORGANIC CHEMISTRY, F. C. Whitmore. The entire subject of organic chemistry for the practicing chemist and the advanced student. Storehouse of facts, theories, processes found elsewhere only in specialized journals. Covers aliphatic compounds (500 pages on the properties and synthetic preparation of hydrocarbons, halides, proteins, ketones, etc.), alicyclic compounds, aromatic compounds, heterocyclic compounds, organophosphorus and organometallic compounds. Methods of synthetic preparation analyzed critically throughout. Includes much of biochemical interest. "The scope of this volume is astonishing," INDUSTRIAL AND ENGINEERING CHEMISTRY. 12,000-reference index. 2387-item bibliography. Total of x + 1005pp. 5⅜ x 8. Two volume set. S700 Vol I Paperbound **$2.00**
S701 Vol II Paperbound **$2.00**
The set **$4.00**

THE MODERN THEORY OF MOLECULAR STRUCTURE, Bernard Pullman. A reasonably popular account of recent developments in atomic and molecular theory. Contents: The Wave Function and Wave Equations (history and bases of present theories of molecular structure); The Electronic Structure of Atoms (Description and classification of atomic wave functions, etc.); Diatomic Molecules; Non-Conjugated Polyatomic Molecules; Conjugated Polyatomic Molecules; The Structure of Complexes. Minimum of mathematical background needed. New translation by David Antin of "La Structure Moleculaire." Index. Bibliography. vii + 87pp. 5⅜ x 8½. S987 Paperbound **$1.00**

CATALYSIS AND CATALYSTS, Marcel Prettre, Director, Research Institute on Catalysis. This brief book, translated into English for the first time, is the finest summary of the principal modern concepts, methods, and results of catalysis. Ideal introduction for beginning chemistry and physics students. Chapters: Basic Definitions of Catalysis (true catalysis and generalization of the concept of catalysis); The Scientific Bases of Catalysis (Catalysis and chemical thermodynamics, catalysis and chemical kinetics); Homogeneous Catalysis (acid-base catalysis, etc.); Chain Reactions; Contact Masses; Heterogeneous Catalysis (Mechanisms of contact catalyses, etc.); and Industrial Applications (acids and fertilizers, petroleum and petroleum chemistry, rubber, plastics, synthetic resins, and fibers). Translated by David Antin. Index. vi + 88pp. 5⅜ x 8½. S998 Paperbound **$1.00**

POLAR MOLECULES, Pieter Debye. This work by Nobel laureate Debye offers a complete guide to fundamental electrostatic field relations, polarizability, molecular structure. Partial contents: electric intensity, displacement and force, polarization by orientation, molar polarization and molar refraction, halogen-hydrides, polar liquids, ionic saturation, dielectric constant, etc. Special chapter considers quantum theory. Indexed. 172pp. 5⅜ x 8.
S64 Paperbound **$1.50**

THE ELECTRONIC THEORY OF ACIDS AND BASES, W. F. Luder and Saverio Zuffanti. The first full systematic presentation of the electronic theory of acids and bases—treating the theory and its ramifications in an uncomplicated manner. Chapters: Historical Background; Atomic Orbitals and Valence; The Electronic Theory of Acids and Bases; Electrophilic and Electrodotic Reagents; Acidic and Basic Radicals; Neutralization; Titrations with Indicators; Displacement; Catalysis; Acid Catalysis; Base Catalysis; Alkoxides and Catalysts; Conclusion. Required reading for all chemists. Second revised (1961) eidtion, with additional examples and references. 3 figures. 9 tables. Index. Bibliography xii + 165pp. 5⅜ x 8.
S201 Paperbound **$1.50**

KINETIC THEORY OF LIQUIDS, J. Frenkel. Regarding the kinetic theory of liquids as a generalization and extension of the theory of solid bodies, this volume covers all types of arrangements of solids, thermal displacements of atoms, interstitial atoms and ions, orientational and rotational motion of molecules, and transition between states of matter. Mathematical theory is developed close to the physical subject matter. 216 bibliographical footnotes. 55 figures. xi + 485pp. 5⅜ x 8. S95 Paperbound **$2.55**

THE PRINCIPLES OF ELECTROCHEMISTRY, D. A. MacInnes. Basic equations for almost every subfield of electrochemistry from first principles, referring at all times to the soundest and most recent theories and results; unusually useful as text or as reference. Covers coulometers and Faraday's Law, electrolytic conductance, the Debye-Hueckel method for the theoretical calculation of activity coefficients, concentration cells, standard electrode potentials, thermodynamic ionization constants, pH, potentiometric titrations, irreversible phenomena, Planck's equation, and much more. "Excellent treatise," AMERICAN CHEMICAL SOCIETY JOURNAL. "Highly recommended," CHEMICAL AND METALLURGICAL ENGINEERING. 2 Indices. Appendix. 585-item bibliography. 137 figures. 94 tables. ii + 478pp. 5⅝ x 8⅜. S52 Paperbound **$2.45**

THE PHASE RULE AND ITS APPLICATION, Alexander Findlay. Covering chemical phenomena of 1, 2, 3, 4, and multiple component systems, this "standard work on the subject" (NATURE, London), has been completely revised and brought up to date by A. N. Campbell and N. O. Smith. Brand new material has been added on such matters as binary, tertiary liquid equilibria, solid solutions in ternary systems, quinary systems of salts and water. Completely revised to triangular coordinates in ternary systems, clarified graphic representation, solid models, etc. 9th revised edition. Author, subject indexes. 236 figures. 505 footnotes, mostly bibliographic. xii + 494pp. 5⅜ x 8. S91 Paperbound **$2.45**

PHYSICS

General physics

FOUNDATIONS OF PHYSICS, R. B. Lindsay & H. Margenau. Excellent bridge between semi-popular works & technical treatises. A discussion of methods of physical description, construction of theory; valuable for physicist with elementary calculus who is interested in ideas that give meaning to data, tools of modern physics. Contents include symbolism, mathematical equations; space & time foundations of mechanics; probability; physics & continua; electron theory; special & general relativity; quantum mechanics; causality. "Thorough and yet not overdetailed. Unreservedly recommended," NATURE (London). Unabridged, corrected edition. List of recommended readings. 35 illustrations. xi + 537pp. 5⅜ x 8.
S377 Paperbound **$2.75**

FUNDAMENTAL FORMULAS OF PHYSICS, ed. by D. H. Menzel. Highly useful, fully inexpensive reference and study text, ranging from simple to highly sophisticated operations. Mathematics integrated into text—each chapter stands as short textbook of field represented. Vol. 1: Statistics, Physical Constants, Special Theory of Relativity, Hydrodynamics, Aerodynamics, Boundary Value Problems in Math. Physics; Viscosity, Electromagnetic Theory, etc. Vol. 2: Sound, Acoustics, Geometrical Optics, Electron Optics, High-Energy Phenomena, Magnetism, Biophysics, much more. Index. Total of 800pp. 5⅜ x 8.
Vol. 1 S595 Paperbound **$2.00**
Vol. 2 S596 Paperbound **$2.00**

MATHEMATICAL PHYSICS, D. H. Menzel. Thorough one-volume treatment of the mathematical techniques vital for classic mechanics, electromagnetic theory, quantum theory, and relativity. Written by the Harvard Professor of Astrophysics for junior, senior, and graduate courses, it gives clear explanations of all those aspects of function theory, vectors, matrices, dyadics, tensors, partial differential equations, etc., necessary for the understanding of the various physical theories. Electron theory, relativity, and other topics seldom presented appear here in considerable detail. Scores of definitions, conversion factors, dimensional constants, etc. "More detailed than normal for an advanced text . . . excellent set of sections on Dyadics, Matrices, and Tensors," JOURNAL OF THE FRANKLIN INSTITUTE. Index. 193 problems, with answers. x + 412pp. 5⅜ x 8.
S56 Paperbound **$2.00**

THE SCIENTIFIC PAPERS OF J. WILLARD GIBBS. All the published papers of America's outstanding theoretical scientist (except for "Statistical Mechanics" and "Vector Analysis"). Vol I (thermodynamics) contains one of the most brilliant of all 19th-century scientific papers—the 300-page "On the Equilibrium of Heterogeneous Substances," which founded the science of physical chemistry, and clearly stated a number of highly important natural laws for the first time; 8 other papers complete the first volume. Vol II includes 2 papers on dynamics, 8 on vector analysis and multiple algebra, 5 on the electromagnetic theory of light, and 6 miscellaneous papers. Biographical sketch by H. A. Bumstead. Total of xxxvi + 718pp. 5⅝ x 8⅜.
S721 Vol I Paperbound **$2.00**
S722 Vol II Paperbound **$2.00**
The set **$4.00**

BASIC THEORIES OF PHYSICS, Peter Gabriel Bergmann. Two-volume set which presents a critical examination of important topics in the major subdivisions of classical and modern physics. The first volume is concerned with classical mechanics and electrodynamics: mechanics of mass points, analytical mechanics, matter in bulk, electrostatics and magnetostatics, electromagnetic interaction, the field waves, special relativity, and waves. The second volume (Heat and Quanta) contains discussions of the kinetic hypothesis, physics and statistics, stationary ensembles, laws of thermodynamics, early quantum theories, atomic spectra, probability waves, quantization in wave mechanics, approximation methods, and abstract quantum theory. A valuable supplement to any thorough course or text.
Heat and Quanta: Index. 8 figures. x + 300pp. 5⅜ x 8½. S968 Paperbound **$1.75**
Mechanics and Electrodynamics: Index. 14 figures. vii + 280pp. 5⅜ x 8½.
S969 Paperbound **$1.75**

THEORETICAL PHYSICS, A. S. Kompaneyets. One of the very few thorough studies of the subject in this price range. Provides advanced students with a comprehensive theoretical background. Especially strong on recent experimentation and developments in quantum theory. Contents: Mechanics (Generalized Coordinates, Lagrange's Equation, Collision of Particles, etc.), Electrodynamics (Vector Analysis, Maxwell's equations, Transmission of Signals, Theory of Relativity, etc.), Quantum Mechanics (the Inadequacy of Classical Mechanics, the Wave Equation, Motion in a Central Field, Quantum Theory of Radiation, Quantum Theories of Dispersion and Scattering, etc.), and Statistical Physics (Equilibrium Distribution of Molecules in an Ideal Gas, Boltzmann statistics, Bose and Fermi Distribution, Thermodynamic Quantities, etc.). Revised to 1961. Translated by George Yankovsky, authorized by Kompaneyets. 137 exercises. 56 figures. 529pp. 5⅜ x 8½. S972 Paperbound **$2.50**

ANALYTICAL AND CANONICAL FORMALISM IN PHYSICS, André Mercier. A survey, in one volume, of the variational principles (the key principles—in mathematical form—from which the basic laws of any one branch of physics can be derived) of the several branches of physical theory, together with an examination of the relationships among them. Contents: the Lagrangian Formalism, Lagrangian Densities, Canonical Formalism, Canonical Form of Electrodynamics, Hamiltonian Densities, Transformations, and Canonical Form with Vanishing Jacobian Determinant. Numerous examples and exercises. For advanced students, teachers, etc. 6 figures. Index. viii + 222pp. 5⅜ x 8½.
S1077 Paperbound **$1.75**

MATHEMATICAL PUZZLES AND RECREATIONS

AMUSEMENTS IN MATHEMATICS, Henry Ernest Dudeney. The foremost British originator of mathematical puzzles is always intriguing, witty, and paradoxical in this classic, one of the largest collections of mathematical amusements. More than 430 puzzles, problems, and paradoxes. Mazes and games, problems on number manipulation, unicursal and other route problems, puzzles on measuring, weighing, packing, age, kinship, chessboards, joining, crossing river, plane figure dissection, and many others. Solutions. More than 450 illustrations. vii + 258pp. 5⅜ x 8. T473 Paperbound **$1.25**

SYMBOLIC LOGIC and THE GAME OF LOGIC, Lewis Carroll. "Symbolic Logic" is not concerned with modern symbolic logic, but is instead a collection of over 380 problems posed with charm and imagination, using the syllogism, and a fascinating diagrammatic method of drawing conclusions. In "The Game of Logic," Carroll's whimsical imagination devises a logical game played with 2 diagrams and counters (included) to manipulate hundreds of tricky syllogisms. The final section, "Hit or Miss" is a lagniappe of 101 additional puzzles in the delightful Carroll manner. Until this reprint edition, both of these books were rarities costing up to $15 each. Symbolic Logic: Index, xxxi + 199pp. The Game of Logic: 96pp. Two vols. bound as one. 5⅜ x 8. T492 Paperbound **$1.50**

MAZES AND LABYRINTHS: A BOOK OF PUZZLES, W. Shepherd. Mazes, formerly associated with mystery and ritual, are still among the most intriguing of intellectual puzzles. This is a novel and different collection of 50 amusements that embody the principle of the maze: mazes in the classical tradition; 3-dimensional, ribbon, and Möbius-strip mazes; hidden messages; spatial arrangements; etc.—almost all built on amusing story situations. 84 illustrations. Essay on maze psychology. Solutions. xv + 122pp. 5⅜ x 8. T731 Paperbound **$1.00**

MATHEMATICAL RECREATIONS, M. Kraitchik. Some 250 puzzles, problems, demonstrations of recreational mathematics for beginners & advanced mathematicians. Unusual historical problems from Greek, Medieval, Arabic, Hindu sources: modern problems based on "mathematics without numbers," geometry, topology, arithmetic, etc. Pastimes derived from figurative numbers, Mersenne numbers, Fermat numbers; fairy chess, latruncles, reversi, many topics. Full solutions. Excellent for insights into special fields of math. 181 illustrations. 330pp. 5⅜ x 8. T163 Paperbound **$1.75**

MATHEMATICAL PUZZLES OF SAM LOYD, Vol. I, selected and edited by M. Gardner. Puzzles by the greatest puzzle creator and innovator. Selected from his famous "Cyclopedia of Puzzles," they retain the unique style and historical flavor of the originals. There are posers based on arithmetic, algebra, probability, game theory, route tracing, topology, counter, sliding block, operations research, geometrical dissection. Includes his famous "14-15" puzzle which was a national craze, and his "Horse of a Different Color" which sold millions of copies. 117 of his most ingenious puzzles in all, 120 line drawings and diagrams. Solutions. Selected references. xx + 167pp. 5⅜ x 8. T498 Paperbound **$1.00**

MY BEST PUZZLES IN MATHEMATICS, Hubert Phillips ("Caliban"). Caliban is generally considered the best of the modern problemists. Here are 100 of his best and wittiest puzzles, selected by the author himself from such publications as the London Daily Telegraph, and each puzzle is guaranteed to put even the sharpest puzzle detective through his paces. Perfect for the development of clear thinking and a logical mind. Complete solutions are provided for every puzzle. x + 107pp. 5⅜ x 8½. T91 Paperbound **$1.00**

MY BEST PUZZLES IN LOGIC AND REASONING, H. Phillips ("Caliban"). 100 choice, hitherto unavailable puzzles by England's best-known problemist. No special knowledge needed to solve these logical or inferential problems, just an unclouded mind, nerves of steel, and fast reflexes. Data presented are both necessary and just sufficient to allow one unambiguous answer. More than 30 different types of puzzles, all ingenious and varied, many one of a kind, that will challenge the expert, please the beginner. Original publication. 100 puzzles, full solutions. x + 107pp. 5⅜ x 8½. T119 Paperbound **$1.00**

MATHEMATICAL PUZZLES FOR BEGINNERS AND ENTHUSIASTS, G. Mott-Smith. 188 mathematical puzzles to test mental agility. Inference, interpretation, algebra, dissection of plane figures, geometry, properties of numbers, decimation, permutations, probability, all enter these delightful problems. Puzzles like the Odic Force, How to Draw an Ellipse, Spider's Cousin, more than 180 others. Detailed solutions. Appendix with square roots, triangular numbers, primes, etc. 135 illustrations. 2nd revised edition. 248pp. 5⅜ x 8. T198 Paperbound **$1.00**

MATHEMATICS, MAGIC AND MYSTERY, Martin Gardner. Card tricks, feats of mental mathematics, stage mind-reading, other "magic" explained as applications of probability, sets, theory of numbers, topology, various branches of mathematics. Creative examination of laws and their applications with scores of new tricks and insights. 115 sections discuss tricks wtih cards, dice, coins; geometrical vanishing tricks, dozens of others. No sleight of hand needed; mathematics guarantees success. 115 illustrations. xii + 174pp. 5⅜ x 8. T335 Paperbound **$1.00**

RECREATIONS IN THE THEORY OF NUMBERS: THE QUEEN OF MATHEMATICS ENTERTAINS, Albert H. Beiler. The theory of numbers is often referred to as the "Queen of Mathematics." In this book Mr. Beiler has compiled the first English volume to deal exclusively with the recreational aspects of number theory, an inherently recreational branch of mathematics. The author's clear style makes for enjoyable reading as he deals with such topics as: perfect numbers, amicable numbers, Fermat's theorem, Wilson's theorem, interesting properties of digits, methods of factoring, primitive roots, Euler's function, polygonal and figurate numbers, Mersenne numbers, congruence, repeating decimals, etc. Countless puzzle problems, with full answers and explanations. For mathematicians and mathematically-inclined laymen, etc. New publication. 28 figures. 9 illustrations. 103 tables. Bibliography at chapter ends. vi + 247pp. 5⅜ x 8½.
T1096 Paperbound **$1.85**

PAPER FOLDING FOR BEGINNERS, W. D. Murray and F. J. Rigney. A delightful introduction to the varied and entertaining Japanese art of origami (paper folding), with a full crystal-clear text that anticipates every difficulty; over 275 clearly labeled diagrams of all important stages in creation. You get results at each stage, since complex figures are logically developed from simpler ones. 43 different pieces are explained: place mats, drinking cups, bonbon boxes, sailboats, frogs, roosters, etc. 6 photographic plates. 279 diagrams. 95pp. 5⅝ x 8⅜.
T713 Paperbound **$1.00**

1800 RIDDLES, ENIGMAS AND CONUNDRUMS, Darwin A. Hindman. Entertaining collection ranging from hilarious gags to outrageous puns to sheer nonsense—a welcome respite from sophisticated humor. Children, toastmasters, and practically anyone with a funny bone will find these zany riddles tickling and eminently repeatable. Sample: "Why does Santa Claus always go down the chimney?" "Because it soots him." Some old, some new—covering a wide variety of subjects. New publication. iii + 154pp. 5⅜ x 8½. T1059 Paperbound **$1.00**

EASY-TO-DO ENTERTAINMENTS AND DIVERSIONS WITH CARDS, STRING, COINS, PAPER AND MATCHES, R. M. Abraham. Over 300 entertaining games, tricks, puzzles, and pastimes for children and adults. Invaluable to anyone in charge of groups of youngsters, for party givers, etc. Contains sections on card tricks and games, making things by paperfolding—toys, decorations, and the like; tricks with coins, matches, and pieces of string; descriptions of games; toys that can be made from common household objects; mathematical recreations; word games; and 50 miscellaneous entertainments. Formerly "Winter Nights Entertainments." Introduction by Lord Baden Powell. 329 illustrations. v + 186pp. 5⅜ x 8.
T921 Paperbound **$1.00**

DIVERSIONS AND PASTIMES WITH CARDS, STRING, PAPER AND MATCHES, R. M. Abraham. Another collection of amusements and diversion for game and puzzle fans of all ages. Many new paperfolding ideas and tricks, an extensive section on amusements with knots and splices, two chapters of easy and not-so-easy problems, coin and match tricks, and lots of other parlor pastimes from the agile mind of the late British problemist and gamester. Corrected and revised version. Illustrations. 160pp. 5⅜ x 8½. T1127 Paperbound **$1.00**

STRING FIGURES AND HOW TO MAKE THEM: A STUDY OF CAT'S-CRADLE IN MANY LANDS, Caroline Furness Jayne. In a simple and easy-to-follow manner, this book describes how to make 107 different string figures. Not only is looping and crossing string between the fingers a common youthful diversion, but it is an ancient form of amusement practiced in all parts of the globe, especially popular among primitive tribes. These games are fun for all ages and offer an excellent means for developing manual dexterity and coordination. Much insight also for the anthropological observer on games and diversions in many different cultures. Index. Bibliography. Introduction by A. C. Haddon, Cambridge University. 17 full-page plates. 950 illustrations. xxiii + 407pp. 5⅜ x 8½. T152 Paperbound **$2.00**

CRYPTANALYSIS, Helen F. Gaines. (Formerly ELEMENTARY CRYPTANALYSIS.) A standard elementary and intermediate text for serious students. It does not confine itself to old material, but contains much that is not generally known, except to experts. Concealment, Transposition, Substitution ciphers; Vigenere, Kasiski, Playfair, multafid, dozens of other techniques. Appendix with sequence charts, letter frequencies in English, 5 other languages, English word frequencies. Bibliography. 167 codes. New to this edition: solution to codes. vi + 230pp. 5⅜ x 8.
T97 Paperbound **$1.95**

MAGIC SQUARES AND CUBES, W. S. Andrews. Only book-length treatment in English, a thorough non-technical description and analysis. Here are nasik, overlapping, pandiagonal, serrated squares; magic circles, cubes, spheres, rhombuses. Try your hand at 4-dimensional magical figures! Much unusual folklore and tradition included. High school algebra is sufficient. 754 diagrams and illustrations. viii + 419pp. 5⅜ x 8. T658 Paperbound **$1.85**

CALIBAN'S PROBLEM BOOK: MATHEMATICAL, INFERENTIAL, AND CRYPTOGRAPHIC PUZZLES, H. Phillips ("Caliban"), S. T. Shovelton, G. S. Marshall. 105 ingenious problems by the greatest living creator of puzzles based on logic and inference. Rigorous, modern, piquant, and reflecting their author's unusual personality, these intermediate and advanced puzzles all involve the ability to reason clearly through complex situations; some call for mathematical knowledge, ranging from algebra to number theory. Solutions. xi + 180pp. 5⅜ x 8.
T736 Paperbound **$1.25**

FICTION

THE LAND THAT TIME FORGOT and THE MOON MAID, Edgar Rice Burroughs. In the opinion of many, Burroughs' best work. The first concerns a strange island where evolution is individual rather than phylogenetic. Speechless anthropoids develop into intelligent human beings within a single generation. The second projects the reader far into the future and describes the first voyage to the Moon (in the year 2025), the conquest of the Earth by the Moon, and years of violence and adventure as the enslaved Earthmen try to regain possession of their planet. "An imaginative tour de force that keeps the reader keyed up and expectant," NEW YORK TIMES. Complete, unabridged text of the original two novels (three parts in each). 5 illustrations by J. Allen St. John. vi + 552pp. 5⅜ x 8½.
T1020 Clothbound **$3.75**
T358 Paperbound **$2.00**

AT THE EARTH'S CORE, PELLUCIDAR, TANAR OF PELLUCIDAR: THREE SCIENCE FICTION NOVELS BY EDGAR RICE BURROUGHS. Complete, unabridged texts of the first three Pellucidar novels. Tales of derring-do by the famous master of science fiction. The locale for these three related stories is the inner surface of the hollow Earth where we discover the world of Pellucidar, complete with all types of bizarre, menacing creatures, strange peoples, and alluring maidens—guaranteed to delight all Burroughs fans and a wide circle of advenutre lovers. Illustrated by J. Allen St. John and P. F. Berdanier. vi + 433pp. 5⅜ x 8½.
T1051 Paperbound **$2.00**

THE PIRATES OF VENUS and LOST ON VENUS: TWO VENUS NOVELS BY EDGAR RICE BURROUGHS. Two related novels, complete and unabridged. Exciting adventure on the planet Venus with Earthman Carson Napier broken-field running through one dangerous episode after another. All lovers of swashbuckling science fiction will enjoy these two stories set in a world of fascinating societies, fierce beasts, 5000-ft. trees, lush vegetation, and wide seas. Illustrations by Fortunino Matania. Total of vi + 340pp. 5⅜ x 8½. T1053 Paperbound **$1.75**

A PRINCESS OF MARS and A FIGHTING MAN OF MARS: TWO MARTIAN NOVELS BY EDGAR RICE BURROUGHS. "Princess of Mars" is the very first of the great Martian novels written by Burroughs, and it is probably the best of them all; it set the pattern for all of his later fantasy novels and contains a thrilling cast of strange peoples and creatures and the formula of Olympian heroism amidst ever-fluctuating fortunes which Burroughs carries off so successfully. "Fighting Man" returns to the same scenes and cities—many years later. A mad scientist, a degenerate dictator, and an indomitable defender of the right clash—with the fate of the Red Planet at stake! Complete, unabridged reprinting of original editions. Illustrations by F. E. Schoonover and Hugh Hutton. v + 356pp. 5⅜ x 8½.
T1140 Paperbound **$1.75**

THREE MARTIAN NOVELS, Edgar Rice Burroughs. Contains: Thuvia, Maid of Mars; The Chessmen of Mars; and The Master Mind of Mars. High adventure set in an imaginative and intricate conception of the Red Planet. Mars is peopled with an intelligent, heroic human race which lives in densely populated cities and with fierce barbarians who inhabit dead sea bottoms. Other exciting creatures abound amidst an inventive framework of Martian history and geography. Complete unabridged reprintings of the first edition. 16 illustrations by J. Allen St. John. vi + 499pp. 5⅜ x 8½. T39 Paperbound **$1.85**

THREE PROPHETIC NOVELS BY H. G. WELLS, edited by E. F. Bleiler. Complete texts of "When the Sleeper Wakes" (1st book printing in 50 years), "A Story of the Days to Come," "The Time Machine" (1st complete printing in book form). Exciting adventures in the future are as enjoyable today as 50 years ago when first printed. Predict TV, movies, intercontinental airplanes, prefabricated houses, air-conditioned cities, etc. First important author to foresee problems of mind control, technological dictatorships. "Absolute best of imaginative fiction," N. Y. Times. Introduction. 335pp. 5⅜ x 8. T605 Paperbound **$1.50**

28 SCIENCE FICTION STORIES OF H. G. WELLS. Two full unabridged novels, MEN LIKE GODS and STAR BEGOTTEN, plus 26 short stories by the master science-fiction writer of all time. Stories of space, time, invention, exploration, future adventure—an indispensable part of the library of everyone interested in science and adventure. PARTIAL CONTENTS: Men Like Gods, The Country of the Blind, In the Abyss, The Crystal Egg, The Man Who Could Work Miracles, A Story of the Days to Come, The Valley of Spiders, and 21 more! 928pp. 5⅜ x 8.
T265 Clothbound **$4.50**

THE WAR IN THE AIR, IN THE DAYS OF THE COMET, THE FOOD OF THE GODS: THREE SCIENCE FICTION NOVELS BY H. G. WELLS. Three exciting Wells offerings bearing on vital social and philosophical issues of his and our own day. Here are tales of air power, strategic bombing, East vs. West, the potential miracles of science, the potential disasters from outer space, the relationship between scientific advancement and moral progress, etc. First reprinting of "War in the Air" in almost 50 years. An excellent sampling of Wells at his storytelling best. Complete, unabridged reprintings. 16 illustrations. 645pp. 5⅜ x 8½.
T1135 Paperbound **$2.00**

SEVEN SCIENCE FICTION NOVELS, H. G. Wells. Full unabridged texts of 7 science-fiction novels of the master. Ranging from biology, physics, chemistry, astronomy to sociology and other studies, Mr. Wells extrapolates whole worlds of strange and intriguing character. "One will have to go far to match this for entertainment, excitement, and sheer pleasure . . . ," NEW YORK TIMES. Contents: The Time Machine, The Island of Dr. Moreau, First Men in the Moon, The Invisible Man, The War of the Worlds, The Food of the Gods, In the Days of the Comet. 1015pp. 5⅜ x 8. T264 Clothbound **$4.50**

BEST GHOST STORIES OF J. S. LE FANU, Selected and introduced by E. F. Bleiler. LeFanu is deemed the greatest name in Victorian supernatural fiction. Here are 16 of his best horror stories, including 2 nouvelles: "Carmilla," a classic vampire tale couched in a perverse eroticism, and "The Haunted Baronet." Also: "Sir Toby's Will," "Green Tea," "Schalken the Painter," "Ultor de Lacy," "The Familiar," etc. The first American publication of about half of this material: a long-overdue opportunity to get a choice sampling of LeFanu's work. New selection (1964). 8 illustrations. 5⅜ x 8⅜. T415 Paperbound **$1.85**

THE WONDERFUL WIZARD OF OZ, L. F. Baum. Only edition in print with all the original W. W. Denslow illustrations in full color—as much a part of "The Wizard" as Tenniel's drawings are for "Alice in Wonderland." "The Wizard" is still America's best-loved fairy tale, in which, as the author expresses it, "The wonderment and joy are retained and the heartaches and nightmares left out." Now today's young readers can enjoy every word and wonderful picture of the original book. New introduction by Martin Gardner. A Baum bibliography. 23 full-page color plates. viii + 268pp. 5⅜ x 8. T691 Paperbound **$1.45**

GHOST AND HORROR STORIES OF AMBROSE BIERCE, Selected and introduced by E. F. Bleiler. 24 morbid, eerie tales—the cream of Bierce's fiction output. Contains such memorable pieces as "The Moonlit Road," "The Damned Thing," "An Inhabitant of Carcosa," "The Eyes of the Panther," "The Famous Gilson Bequest," "The Middle Toe of the Right Foot," and other chilling stories, plus the essay, "Visions of the Night" in which Bierce gives us a kind of rationale for his aesthetic of horror. New collection (1964). xxii + 199pp. 5⅜ x 8⅜. T767 Paperbound **$1.00**

HUMOR

MR. DOOLEY ON IVRYTHING AND IVRYBODY, Finley Peter Dunne. Since the time of his appearance in 1893, "Mr. Dooley," the fictitious Chicago bartender, has been recognized as America's most humorous social and political commentator. Collected in this volume are 102 of the best Dooley pieces—all written around the turn of the century, the height of his popularity. Mr. Dooley's Irish brogue is employed wittily and penetratingly on subjects which are just as fresh and relevant today as they were then: corruption and hypocrisy of politicans, war preparations and chauvinism, automation, Latin American affairs, superbombs, etc. Other articles range from Rudyard Kipling to football. Selected with an introduction by Robert Hutchinson. xii + 244pp. 5⅜ x 8½. T626 Paperbound **$1.00**

RUTHLESS RHYMES FOR HEARTLESS HOMES and MORE RUTHLESS RHYMES FOR HEARTLESS HOMES, Harry Graham ("Col. D. Streamer"). A collection of Little Willy and 48 other poetic "disasters." Graham's funniest and most disrespectful verse, accompanied by original illustrations. Nonsensical, wry humor which employs stern parents, careless nurses, uninhibited children, practical jokers, single-minded golfers, Scottish lairds, etc. in the leading roles. A precursor of the "sick joke" school of today. This volume contains, bound together for the first time, two of the most perennially popular books of humor in England and America. Index. vi + 69pp. 5⅜ x 8. T930 Paperbound **75¢**

A WHIMSEY ANTHOLOGY, Collected by Carolyn Wells. 250 of the most amusing rhymes ever written. Acrostics, anagrams, palindromes, alphabetical jingles, tongue twisters, echo verses, alliterative verses, riddles, mnemonic rhymes, interior rhymes, over 40 limericks, etc. by Lewis Carroll, Edward Lear, Joseph Addison, W. S. Gilbert, Christina Rossetti, Chas. Lamb, James Boswell, Hood, Dickens, Swinburne, Leigh Hunt, Harry Graham, Poe, Eugene Field, and many others. xiv + 221pp. 5⅜ x 8½. T195 Paperbound **$1.25**

MY PIOUS FRIENDS AND DRUNKEN COMPANIONS and MORE PIOUS FRIENDS AND DRUNKEN COMPANIONS, Songs and ballads of Conviviality Collected by Frank Shay. Magnificently illuminated by John Held, Jr. 132 ballads, blues, vaudeville numbers, drinking songs, cowboy songs, sea chanties, comedy songs, etc. of the Naughty Nineties and early 20th century. Over a third are reprinted with music. Many perennial favorites such as: The Band Played On, Frankie and Johnnie, The Old Grey Mare, The Face on the Bar-room Floor, etc. Many others unlocatable elsewhere: The Dog-Catcher's Child, The Cannibal Maiden, Don't Go in the Lion's Cage Tonight, Mother, etc. Complete verses and introductions to songs. Unabridged republication of first editions, 2 Indexes (song titles and first lines and choruses). Introduction by Frank Shay. 2 volumes bounds as 1. Total of xvi + 235pp. 5⅜ x 8½. T946 Paperbound **$1.00**

MAX AND MORITZ, Wilhelm Busch. Edited and annotated by H. Arthur Klein. Translated by H. Arthur Klein, M. C. Klein, and others. The mischievous high jinks of Max and Moritz, Peter and Paul, Ker and Plunk, etc. are delightfully captured in sketch and rhyme. (Companion volume to "Hypocritical Helena.") In addition to the title piece, it contians: Ker and Plunk; Two Dogs and Two Boys; The Egghead and the Two Cut-ups of Corinth; Deceitful Henry; The Boys and the Pipe; Cat and Mouse; and others. (Original German text with accompanying English translations.) Afterword by H. A. Klein. vi + 216pp. 5⅜ x 8½.
T181 Paperbound **$1.00**

THROUGH THE ALIMENTARY CANAL WITH GUN AND CAMERA: A FASCINATING TRIP TO THE INTERIOR, Personally Conducted by George S. Chappell. In mock-travelogue style, the amusing account of an imaginative journey down the alimentary canal. The "explorers" enter the esophagus, round the Adam's Apple, narrowly escape from a fierce Amoeba, struggle through the impenetrable Nerve Forests of the Lumbar Region, etc. Illustrated by the famous cartoonist, Otto Soglow, the book is as much a brilliant satire of academic pomposity and professional travel literature as it is a clever use of the facts of physiology for supremely comic purposes. Preface by Robert Benchley. Author's Foreword. 1 Photograph. 17 illustrations by O. Soglow. xii + 114pp. 5⅜ x 8½. T376 Paperbound **$1.00**

THE BAD CHILD'S BOOK OF BEASTS, MORE BEASTS FOR WORSE CHILDREN, and A MORAL ALPHABET, H. Belloc. Hardly an anthology of humorous verse has appeared in the last 50 years without at least a couple of these famous nonsense verses. But one must see the entire volumes—with all the delightful original illustrations by Sir Basil Blackwood—to appreciate fully Belloc's charming and witty verses that play so subacidly on the platitudes of life and morals that beset his day—and ours. A great humor classic. Three books in one. Total of 157pp. 5⅜ x 8. T749 Paperbound **$1.00**

THE DEVIL'S DICTIONARY, Ambrose Bierce. Sardonic and irreverent barbs puncturing the pomposities and absurdities of American politics, business, religion, literature, and arts, by the country's greatest satirist in the classic tradition. Epigrammatic as Shaw, piercing as Swift, American as Mark Twain, Will Rogers, and Fred Allen. Bierce will always remain the favorite of a small coterie of enthusiasts, and of writers and speakers whom he supplies with "some of the most gorgeous witticisms of the English language." (H. L. Mencken) Over 1000 entries in alphabetical order. 144pp. 5⅜ x 8. T487 Paperbound **$1.00**

THE COMPLETE NONSENSE OF EDWARD LEAR. This is the only complete edition of this master of gentle madness available at a popular price. A BOOK OF NONSENSE, NONSENSE SONGS, MORE NONSENSE SONGS AND STORIES in their entirety with all the old favorites that have delighted children and adults for years. The Dong With A Luminous Nose, The Jumblies, The Owl and the Pussycat, and hundreds of other bits of wonderful nonsense. 214 limericks, 3 sets of Nonsense Botany, 5 Nonsense Alphabets. 546 drawings by Lear himself, and much more. 320pp. 5⅜ x 8. T167 Paperbound **$1.00**

SINGULAR TRAVELS, CAMPAIGNS, AND ADVENTURES OF BARON MUNCHAUSEN, R. E. Raspe, with 90 illustrations by Gustave Doré. The first edition in over 150 years to reestablish the deeds of the Prince of Liars exactly as Raspe first recorded them in 1785—the genuine Baron Munchausen, one of the most popular personalities in English literature. Included also are the best of the many sequels, written by other hands. Introduction on Raspe by J. Carswell. Bibliography of early editions. xliv + 192pp. 5⅜ x 8. T698 Paperbound **$1.00**

HOW TO TELL THE BIRDS FROM THE FLOWERS, R. W. Wood. How not to confuse a carrot with a parrot, a grape with an ape, a puffin with nuffin. Delightful drawings, clever puns, absurd little poems point out farfetched resemblances in nature. The author was a leading physicist. Introduction by Margaret Wood White. 106 illus. 60pp. 5⅜ x 8.
T523 Paperbound **75¢**

JOE MILLER'S JESTS OR, THE WITS VADE-MECUM. The original Joe Miller jest book. Gives a keen and pungent impression of life in 18th-century England. Many are somewhat on the bawdy side and they are still capable of provoking amusement and good fun. This volume is a facsimile of the original "Joe Miller" first published in 1739. It remains the most popular and influential humor book of all time. New introduction by Robert Hutchinson. xxi + 70pp. 5⅜ x 8½. T423 Paperbound **$1.00**

Prices subject to change without notice.

Dover publishes books on art, music, philosophy, literature, languages, history, social sciences, psychology, handcrafts, orientalia, puzzles and entertainments, chess, pets and gardens, books explaining science, intermediate and higher mathematics, mathematical physics, engineering, biological sciences, earth sciences, classics of science, etc. Write to:

Dept. catrr.
Dover Publications, Inc.
180 Varick Street, N.Y. 14, N.Y.